Edexcel GCSE

Mathematics A
Linear
Foundation

Student Book

Series Director: Keith Pledger
Series Editor: Graham Cumming

Authors:
Chris Baston
Julie Bolter
Gareth Cole
Gill Dyer
Michael Flowers
Karen Hughes
Peter Jolly
Joan Knott
Jean Linsky
Graham Newman
Rob Pepper
Joe Petran
Keith Pledger
Rob Summerson
Kevin Tanner
Brian Western

A PEARSON COMPANY

Published by Pearson Education Limited, a company incorporated in England and Wales, having its registered office at Edinburgh Gate, Harlow, Essex, CM20 2JE. Registered company number: 872828

Edexcel is a registered trademark of Edexcel Limited

Text © Chris Baston, Julie Bolter, Gareth Cole, Gill Dyer, Michael Flowers, Karen Hughes, Peter Jolly, Joan Knott, Jean Linsky, Graham Newman, Rob Pepper, Joe Petran, Keith Pledger, Rob Summerson, Kevin Tanner, Brian Western and Pearson Education Limited 2010

The rights of Chris Baston, Julie Bolter, Gareth Cole, Gill Dyer, Michael Flowers, Karen Hughes, Peter Jolly, Joan Knott, Jean Linsky, Graham Newman, Rob Pepper, Joe Petran, Keith Pledger, Rob Summerson, Kevin Tanner and Brian Western to be identified as the authors of this Work have been asserted by them in accordance with the Copyright, Designs and Patent Act, 1988.

First published 2010

15 14 13
10 9 8 7 6 5

British Library Cataloguing in Publication Data
A catalogue record for this book is available from the British Library

ISBN 978 1 84690 088 4

Typeset by Tech-Set Ltd, Gateshead
Picture research by Rebecca Sodergren
Printed in China (GCC/05)

Acknowledgements
The publisher would like to thank the following for their kind permission to reproduce their photographs:
(Key: b-bottom; c-centre; l-left; r-right; t-top)

4Corners Images: Schmid Reinhard 473tr; **Alamy Images:** Aflo Foto Agency 299tr; Vito Arcomano 300-301; Bildagentur-online.com / th-foto 146; CandyBox Photography 100; ICP 301tl; H. Mark Weidman Photography 299tc; roboxford 308-309; Chris Rout 472tcr; Stephen Shepherd 299tl; Sinibomb Images 471; Mark Titterton 375; Justin Kase zsixz 301b; **Corbis:** Artiga Photo 473tc; Deborah Betz Collection 61; Alan Schein 28; Paul Seheult 390; **Getty Images:** Farjana K. Godhuly 219; Hisham Ibrahim 472bl; Ko fujiwara 123; Oscar Mattsson 472c; Mike Powell 81; Antonio M. Rosario 163; Martin Rose 445; Terry Vine 363; Digital Vision 304cl; **iStockphoto:** 299cl, 301, 307t, 331, 472cl, 472bc, 473bc; Ana Abejon 472cr; Caitlin Cahill 298; Anthony Collins 307b; Joe Gough 307c; Wayne Howard 306tr; Michael Krinke 299cr; Susan McKenzie 473br; Katherine Moffitt 472tcl; Stuart Monk 516; Dmitry Mordvintsev 309tr, 309cr; Muharrem öner 473tl; Jack Puccio 302-303; Brian Sullivan 472br; Peeter Viisimaa 1; Natalia Vasina 472tr; **Pearson Education:** Corbis 472tl; **Photolibrary.com:** Brand X 420; Angelo Cavalli 553; Foodfolio Foodfolio / Imagestate 44; Amish Patel 353; Ron Chapple Stock 567; **Press Association Images:** Anthony Devlin 264; Ezra Shaw 499; **Rex Features:** Michael Fresco 310; Alisdair Macdonald 473bl; **Science Photo Library Ltd:** 245; Adam Hart-Davis 176; Reed Timmer 534; Detlev Van Ravenswaay 190; **Shutterstock:** 298t, 298tc, 298bl, 298br, 299bc; Sandra Cunningham 306; Foodpics 309t; Klaus Kaulitzki 299br; Martin Kemp 304; Glue Stock 305; Lisa F. Young 301tc, **Thinkstock:** 302bl, 302br.
All other images © Pearson Education.

We are grateful to the following for permission to reproduce copyright material:

Tables
Table on page 45 from NHL Team names and statistics, 08–09 season, http://www.nhl.com/ice/app, NHL and NHL Team marks are the property of the NHL and its teams. Copyright © NHL 2010. All Rights Reserved. Used with permission. Table on page 55 (top) adapted from 'UK population figures 1981-2021', Crown Copyright material is reproduced with the permission of the Controller, Office of Public Sector Information (OPSI).; Table on page 55 (bottom) adapted from 'Weather in Aspatria, Cumbria Jan-Oct', Crown Copyright material is reproduced with the permission of the Controller, Office of Public Sector Information (OPSI); Information on EMA, NIC and Income tax taken from www.directgov.uk © Crown Copyright, material is reproduced with the permission of the Controller, Office of Public Sector Information (OPSI).

Every effort has been made to trace the copyright holders and we apologise in advance for any unintentional omissions. We would be pleased to insert the appropriate acknowledgement in any subsequent edition of this publication.

Contents

About this book

All set to make the grade!

Edexcel GCSE Mathematics is specially written to help you get your best grade in the exams.

Section objectives show what you'll be learning.

Loads of practice to help you feel secure before you move on.

Crystal-clear worked examples – step-by-step guides to answering questions correctly, with helpful hints and reminders.

Non-calculator indicates questions where students must not use a calculator to find the answer. It does NOT indicate that the subject area covered by the question will only appear in the Non-Calculator paper of the exam.

Graded questions – so you know what you're achieving.

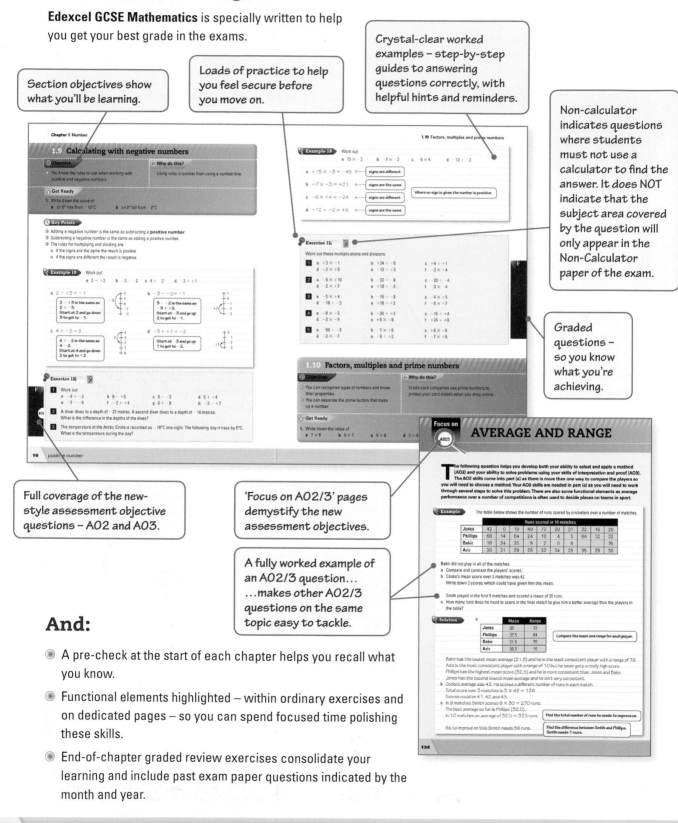

Full coverage of the new-style assessment objective questions – AO2 and AO3.

'Focus on AO2/3' pages demystify the new assessment objectives.

A fully worked example of an AO2/3 question...
...makes other AO2/3 questions on the same topic easy to tackle.

And:

- A pre-check at the start of each chapter helps you recall what you know.

- Functional elements highlighted – within ordinary exercises and on dedicated pages – so you can spend focused time polishing these skills.

- End-of-chapter graded review exercises consolidate your learning and include past exam paper questions indicated by the month and year.

About ActiveTeach

Use **ActiveTeach** to view and present the course on screen with exciting interactive content.

BBC ACTIVE — ActiveTeach is enriched with BBC Active video clips to bring maths to life.

Scribble pad adds space to enable on-screen working.

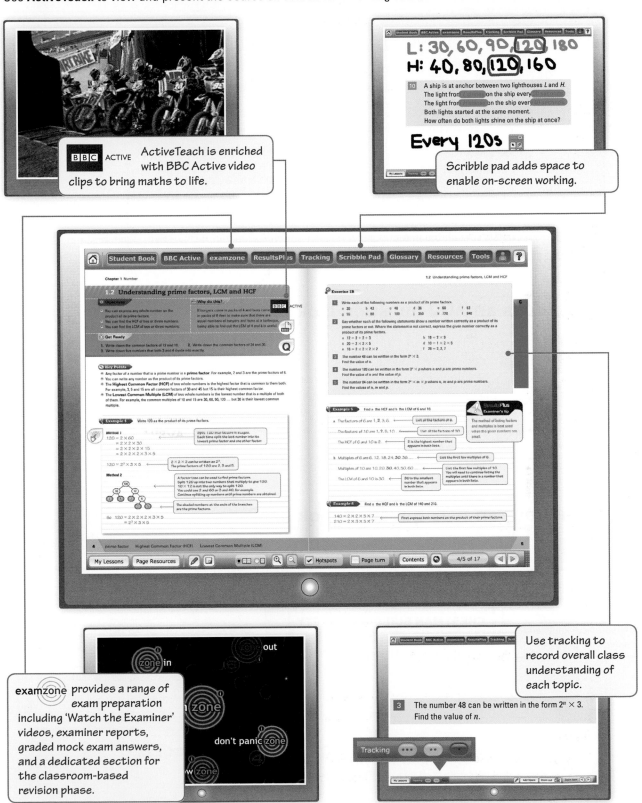

examzone provides a range of exam preparation including 'Watch the Examiner' videos, examiner reports, graded mock exam answers, and a dedicated section for the classroom-based revision phase.

Use tracking to record overall class understanding of each topic.

About Assessment Objectives

Assessment Objectives define the types of question that are set in the exam.

Assessment Objective	What it is	What this means	Range % of marks in the exam
AO1	**Recall** and use knowledge of the prescribed content.	Standard questions testing your knowledge of each topic.	45-55
AO2	**Select** and apply mathematical methods in a range of contexts.	Deciding what method you need to use to get to the correct solution to a contextualised problem.	25-35
AO3	**Interpret** and analyse problems and generate strategies to solve them.	Solving problems by deciding how and explaining why.	15-25

The proportion of marks available in the exam varies with each Assessment Objective. Don't miss out, make sure you know how to do AO2 and AO3 questions!

What does an AO2 question look like?

16 Katie wants to buy a car.
She decides to borrow £3500 from her father. She adds interest of 3.5% to the loan and this total is the amount she must repay her father. How much will Katie pay back to her father in total?

> This just needs you to
> (a) read and understand the question and
> (b) decide how to get the correct answer.

What does an AO3 question look like?

17 Rashida wishes to invest £2000 in a building society account for one year. The Internet offers two suggestions. Which of these two investments gives Rashida the greatest return?

> Here you need to read and analyse the question. Then use your mathematical knowledge to solve this problem.

CHESTMAN BUILDING SOCIETY
£3.50 per month
Plus 1% bonus at the end of the year

DUNSTAN BUILDING SOCIETY
4% per annum. Paid yearly by cheque

Focus on

We give you extra help with AO2 and AO3 on pages 284–297.

About functional elements

What does a question with functional maths look like?

Functional maths is about being able to apply maths in everyday, real-life situations.

GCSE Tier	Range % of marks in the exam
Foundation	30-40
Higher	20-30

The proportion of functional maths marks in the GCSE exam depends on which tier you are taking. Don't miss out, make sure you know how to do functional maths questions!

In the exercises...

20 The Wildlife Trust are doing a survey into the number of field mice on a farm of size 240 acres. They look at one field of size 6 acres. In this field they count 35 field mice.

a Estimate how many field mice there are on the whole farm.

b Why might this be an unreliable estimate?

You need to read and understand the question. Follow your plan.

Think what maths you need and plan the order in which you'll work.

Check your calculations and make a comment if required.

...and on our special functional maths pages: 298–309!

Quality of written communication

There will be marks in the exam for showing your working 'properly' and explaining clearly. In the exam paper, such questions will be marked with a star (*). You need to:

● use the correct mathematical notation and vocabulary, to show that you can communicate effectively

● organise the relevant information logically.

ResultsPlus

ResultsPlus features combine exam performance data with examiner insight to give you more information on how to succeed. ResultsPlus tips in the **student books** show students how to avoid errors in solutions to questions.

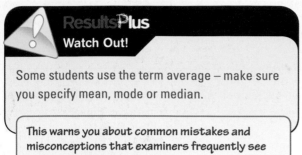

ResultsPlus
Watch Out!

Some students use the term average – make sure you specify mean, mode or median.

This warns you about common mistakes and misconceptions that examiners frequently see students make.

ResultsPlus
Exam Question Report

91% of students scored poorly on this question because they did not use the midpoint of the range to find the mean of grouped data.

This gives a breakdown of how students did on real past exam questions.

ResultsPlus
Examiner's Tip

Make sure the angles add up to 360°.

This gives exam advice, useful checks, and methods to remember key facts.

ResultsPlus in the **ActiveTeach** provides interactive practice for AO2 and AO3 questions…

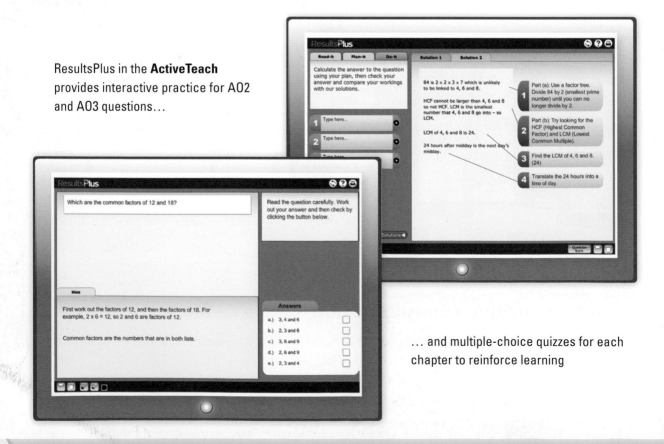

… and multiple-choice quizzes for each chapter to reinforce learning

1 NUMBER

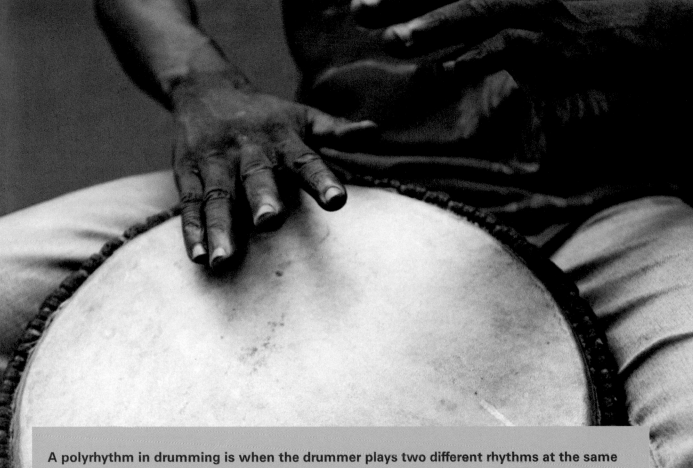

A polyrhythm in drumming is when the drummer plays two different rhythms at the same time, one with each hand. A common polyrhythm is 3 : 2. The lowest common multiple of 2 and 3 is 6, so the drummer counts from 1 to 6. On the first, third and fifth beat he plays with his left hand; on the first and fourth he uses his right. The drum is silent on the sixth beat.

left: **1** 2 **3** 4 **5** 6

right: **1** 2 3 **4** 5 6

◎ Objectives

In this chapter you will:

- ◉ see the importance of place value
- ◉ see the usefulness of a number line
- ◉ see why whole numbers can be negative as well as positive
- ◉ use the rules of addition, subtraction, multiplication and division for combining numbers
- ◉ learn some of the language of mathematics connected with numbers.

◈ Before you start

You need to know:

- ◉ the addition number bonds to $9 + 9$
- ◉ the times tables to 10×10.

1.1 Understanding digits and place value

Objective

You understand the number system.

Why do this?

To carry out an everyday task such as buying something in a shop you need to understand how the number system works.

Get Ready

1. What numbers come before and after 509?
2. Write down the answer to 7 + 6.
3. Write down the answer to 7 × 8.

Key Points

- Although you can keep on counting to very high numbers, you only use ten **digits**, often called figures.

  0 1 2 3 4 5 6 7 8 9

- Each digit has a **value** that depends on its position in the number. This is its **place value**.

Example 1   Draw a place value diagram for:

   a   a three-digit number with a 6 in the tens column

   b   a five-digit number with a 6 in the thousands column.

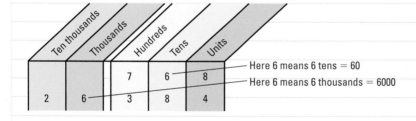

Here 6 means 6 tens = 60
Here 6 means 6 thousands = 6000

Exercise 1A

Questions in this chapter are targeted at the grades indicated.

G

1   Draw a place value diagram and write in:

   a   a four-digit number with a 4 in the thousands column

   b   a two-digit number with a 3 in the tens column

   c   a five-digit number with a 1 in the hundreds column

   d   a three-digit number with a 9 in the units column

   e   a four-digit number with a 0 in the tens column

   f   a five-digit number with a 4 in the hundreds column

   g   a four-digit number with a 7 in every column except the tens column

   h   a five-digit number with a 6 in the thousands column and the units column.

2 For each teacher, write down five different numbers that they could be thinking about.

3 Write down the value of the 6 in each of these numbers.

   a 63   **b** 3642   **c** 63 214   **d** 2546   **e** 56 345

1.2 **Reading, writing and ordering whole numbers**

⊙ Objectives

● You can read and write down whole numbers.
● You can put numbers in order of size.

❓ Why do this?

You need to be able to compare the size of numbers when you want to decide which mobile phone handset is cheaper.

⬥ Get Ready

1. Write down the value of the 4 in each of these numbers.

   a 46   **b** 2034   **c** 65 403

🔑 Key Points

◉ You can read and write numbers by thinking about the place value of each of the digits.
◉ You can use your knowledge of place value to put numbers in order of size.

🔍 Example 2

Write in figures the numbers shown in these newspaper cuttings.

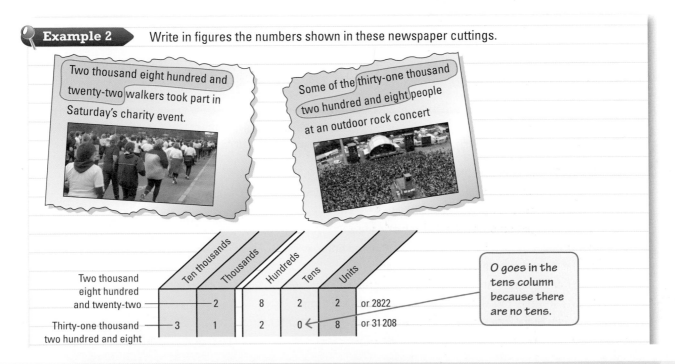

	Ten thousands	Thousands	Hundreds	Tens	Units	
Two thousand eight hundred and twenty-two		2	8	2	2	or 2822
Thirty-one thousand two hundred and eight	3	1	2	0	8	or 31 208

0 goes in the tens column because there are no tens.

Example 3 Write in words the numbers shown in the place value diagram.

In 16 412 the thin space separates thousands from hundreds and makes it easier to read the number.

Ten thousands	Thousands	Hundreds	Tens	Units	
		9	8	7	— Nine hundred and eighty-seven
1	6	4	1	2	— Sixteen thousand four hundred and twelve

Example 4 Write the numbers 8400, 6991, 15, 2410, 84 000, 2406 in order of size.

Start with the smallest.

15 is the smallest. ← It has no digits above the tens column.

84 000 is the largest. ← It is the only one with a digit in the ten thousands column.

This leaves 8400, 6991, 2410 and 2406.
2000 is smaller than 6000, which is smaller than 8000.
Lastly, 2406 is smaller than 2410.
The order is 15, 2406, 2410, 6991, 8400, 84 000.

Exercise 1B

1   Write these numbers in figures.
   a Three hundred and twenty-five
   b One thousand seven hundred and eighteen
   c Six thousand two hundred and four
   d Nineteen thousand four hundred and twenty

2   Write these numbers in words.
   a 237   b 321   c 1792   d 6502   e 1053

3   Write each set of numbers in order of size, starting with the smallest.
   a 183, 235, 190, 73, 179   b 2510, 2015, 970, 2105, 2439
   c 30 300, 3033, 3000, 3003, 2998   d 56 762, 59 342, 56 745, 56 321

4   Write the numbers that have been highlighted in blue in figures.
   a The numbers of people employed by a local police force are:
      Traffic wardens: sixty-nine
      Civilian support staff: one thousand and ten
      Police officers: two thousand three hundred and six
   b The tonnages of three cruise liners are:
      Aurora: seventy-six thousand one hundred and fifty-two
      QE2: seventy thousand three hundred and sixty-three
      Queen Mary 2: one hundred and fifty-one thousand four hundred

G

G

5   This table gives the populations of five member states of the European Union in 2009.
    Write the numbers in words.

	Country	Population
a	Czech Republic	10 467 542
b	Cyprus	793 963
c	Estonia	1 340 415
d	France	64 351 000
e	Portugal	10 627 250

6   The table gives the prices of some second-hand cars.

Car	Price
Peugeot 505	£7995
Focus	£11 495
Ka	£4835
Mini	£6549
Sharan	£13 205

    a   Write down the price of each car in words.
    b   Rewrite the list in price order, starting with the most expensive.

1.3 The number line

◉ Objectives

● You can see numbers in context.
● You can use a simple number line to increase or decrease numbers and to work out the difference between numbers.

❓ Why do this?

You need to be able to read scales you meet, such as rulers, thermometers and speedometers.

◈ Get Ready

Place these numbers in order, starting with the lowest.
1. £5490   £3645   £5250   £4190
2. 10 348 276   462 690   40 280 780   10 524 145   60 424 213
3. 76 152   70 363   150 400

◉ Key Points

● You can use a number line to help you to increase or decrease a number.
● Number lines might be read:
    ● from left to right
    ● from bottom to top
    ● clockwise.

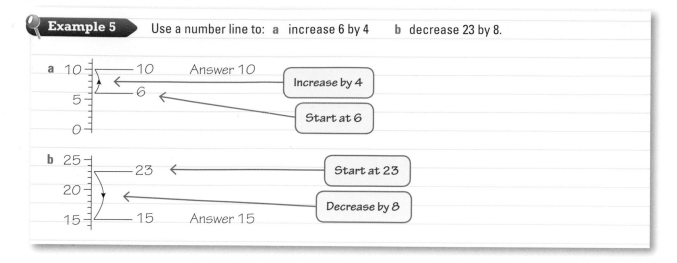

Example 5 Use a number line to: **a** increase 6 by 4 **b** decrease 23 by 8.

a Answer 10 | Increase by 4 | Start at 6

b Start at 23 | Decrease by 8 | Answer 15

Exercise 1C

G

1 Draw a number line from 0 to 30.
Mark these numbers on your number line.
  a 6   **b** 23   **c** 15   **d** 0   **e** 29

2 Use a number line from 0 to 25 to
  a increase 6 by 3   **b** decrease 15 by 7   **c** increase 11 by 7
  d decrease 19 by 13   **e** increase 17 by 8   **f** decrease 24 by 17.

3 For each of these moves, write down the difference between them, and whether it is an increase or decrease.
  a 6 to 10   **b** 15 to 21   **c** 10 to 3   **d** 29 to 24   **e** 5 to 27   **f** 19 to 2

1.4 Adding and subtracting

◉ Objective

• You can add and subtract without a calculator.

❓ Why do this?

When you are shopping you might want to know how much you have spent before you get to the checkout. You might not always have a calculator on hand to help you!

◈ Get Ready

1. Use a number line to:
  a increase 29 by 33   **b** decrease 34 by 27   **c** find the difference between 20 and 3.

🌐 Key Points

◉ Words that show you have to add numbers are: add, plus, total and sum.
◉ Words that show you have to take away are: take away, subtract, minus and find the difference.

Example 6   Work out 23 + 693 + 8.

```
    23
   693
 +   8
 ─────
   724
   1 1
```

Step 1   Put the digits in their correct columns.
Step 2   Add the units column: $3 + 3 + 8 = 14$.
          Write 4 in the units column and carry the 1 ten into the tens column.
Step 3   Repeat for the tens column: $2 + 9 = 11$ plus the 1 that was carried across $= 12$.
          Write 2 in the tens column and carry the 1 into the hundreds column.
Step 4   Add the 6 and the 1 that was carried across: 7.

Exercise 1D

1.   Find the total of 26 and 17.

2.   Work out 58 plus 22.

3.   Work out 236 + 95.

4.   In four maths tests, Anna scored 61 marks, 46 marks, 87 marks and 76 marks.
   How many marks did she score altogether?

5.   Find the sum of all the single-digit numbers.

6.   In a fishing competition, five competitors caught 16 fish, 31 fish, 8 fish, 19 fish and 22 fish.
   Find the total number of fish caught.

7.   The number of passengers on a bus was 36 downstairs and 48 upstairs.
   How many passengers were on the bus altogether?

8.   On her MP3 player, Lena had 86 pop songs, 58 rock songs and 72 dance songs.
   How many songs did she have in total?

9.   Work out 38 + 96 + 127 + 92 + 48.

10.   On six days in June, 86, 43, 75, 104, 38 and 70 people went bungee jumping over a gorge.
    How many people jumped in total?

Example 7   Work out 423 − 274.

Method 1

```
  3 11 1
  4 2 3
 − 2 7 4
 ───────
   1 4 9
```

Step 1   Put the digits in their correct columns.
Step 2   In the units column try taking 4 from 3 (not possible). So exchange 1 ten for 10 units.
          You now have $3 + 10 = 13$ units.
Step 3 $13 − 4 = 9$ in the units column.
Step 4   The tens column has 1 take away 7, which is not possible, so exchange 1 hundred for
          10 tens.
Step 5 $11 − 7 = 4$ in the tens column.
Step 6   Finally there is $3 − 2 = 1$ in the hundreds column.

Method 2

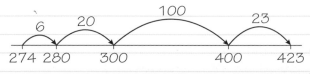

Step 1   Count on from 274 to 280 =   6
Step 2   Count on from 280 to 300 =   20
Step 3   Count on from 300 to 400 = 100
Step 4   Count on from 400 to 423 =   23
Step 5   Add   149

G

G

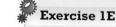

Exercise 1E

1   Work out 611 − 306.

2   How much is 7260 − 4094?

3   Take 1007 from 2010.

4   In a car boot sale Alistair sells 17 of his 29 CDs.
   How many does he have left?

A03

5   When playing darts, James scored 111 with his first three darts, Sunita scored 94 with her first three
   darts and Nadine scored 75 with her first three darts.
   a   How many more than Nadine did Sunita score?
   b   What is the difference between James's and Sunita's scores?
   c   How many less than James did Nadine score?

A03

6   The winner of the darts match is the first one to reach 501.
   James has now scored 413, Sunita 442 and Nadine 368.
   a   How many does James need to win the game?
   b   How many more does Sunita need to win the game?
   c   How many is Nadine short of winning the game?

1.5   Multiplying and dividing

◎ Objective

● You can multiply and divide without a calculator.

❓ Why do this?

If you buy several T-shirts or CDs that are the same price, it is useful if you can work out an estimate of the cost without a calculator.

◆ Get Ready

1. Work out
   a   123 + 23 + 23
   b   57 − 19 − 19 − 19

🕐 Key Points

● Words that show you have to multiply numbers are: product, times and multiply.

● Divide and share are words that show you have to divide.

● You multiply by 10, 100 and 1000 by moving digits 1, 2 or 3 places to the left.

● You divide by 10, 100 and 1000 by moving digits 1, 2 or 3 places to the right.

Example 8 ▸ Work out 23 × 10.

H	T	U
	2	3
2	3	0	= 230

The 3 moves from the units into the tens column because 3 × 10 = 30.
The 2 moves from the tens into the hundreds column because 20 × 10 = 200.

Example 9 ▸ a Work out 43 × 6. b Work out 256 × 37.

```
    43
 ×   6
   258
```

3 × 6 = 18 (1 ten and 8 units)
40 × 6 = 240
Total 258

```
    256
 ×   37
   1792
   3 4
  7680
   1
   9472
   1 1
```

Step 1 256 × 7
Step 2 256 × 30
Step 3 add

Exercise 1F

1 Multiply each of these numbers by i 10   ii 100   iii 1000
   a 27   b 8   c 301   d 60   e 5020

2 Work out
   a 35 × 20   b 26 × 200   c 122 × 30
   d 43 × 500   e 115 × 40   f 214 × 3000

3 Work out
   a 65 × 7   b 53 × 3   c 314 × 6
   d 523 × 8   e 237 × 5   f 399 × 4

4 Work out
   a 34 × 12   b 65 × 15   c 53 × 33   d 314 × 16
   e 523 × 47   f 221 × 64   g 146 × 53   h 29 × 38

5 Find the product of 71 and 13.

6 How many is 321 multiplied by 14?

7 John travels 17 miles to work and 17 miles back each day.
   In July he went to work and back 23 times. How far did he travel altogether?

8 Bocton Football Club hires 23 coaches to take supporters to an away match.
   Each coach can take 63 passengers. How many supporters can be taken to the match?

9 Each box of matches contains 43 matches.
   How many matches are there in 36 boxes?

10 Tariq buys tins of soup which are packed in cases of 48.
    He buys 8 cases. How many tins of soup does he buy?

G

A03

Example 10 Work out 3200 ÷ 100.

Th H T U
3 2 0 0
    3 2 = 32

> The 3 moves from the thousands into the tens column because 3000 ÷ 100 = 30.
>
> The 2 moves from the hundreds into the units column because 200 ÷ 100 = 2.

Example 11 Work out 256 ÷ 8.

Traditional short division

    32
  8)25¹6

> Step 3 8 into 16 goes 2 times.

> Step 1 8 into 2 does not go.

> Step 2 8 into 25 goes 3 times remainder 1. Carry remainder into units column.

Example 12 Work out 8704 ÷ 17.

Traditional long division

> Step 1
> 17 divides into 87 five times remainder 2.

> Step 2
> 17 divides into 20 one time remainder 3.

> Step 3
> 17 divides into 34 two times exactly.

```
                5                51               512
17)8704      17)8704          17)8704          17)8704
             − 85             − 85↓            − 85
               2                20               20
                              − 17             − 17↓
                                 3               34
                                                 34
                                                  0
```

So 17 divides into 8704 512 times.

Short division method

    5 1 2
17)8 7²0³4

> This is a shorter way of setting out the steps than in the long division method.

Exercise 1G

1 Work out
   a 3660 ÷ 10   b 4300 ÷ 100   c 9000 ÷ 10   d 87 000 ÷ 1000

2 Work out
   a 48 ÷ 2   b 69 ÷ 3   c 56 ÷ 4   d 96 ÷ 6
   e 640 ÷ 4   f 565 ÷ 5   g 72 ÷ 4   h 712 ÷ 8
   i 828 ÷ 9   j 637 ÷ 7   k 408 ÷ 2   l 1020 ÷ 3

3 Work out
   a 256 ÷ 16   b 660 ÷ 15   c 512 ÷ 32   d 861 ÷ 21
   e 756 ÷ 36   f 1020 ÷ 30   g 1440 ÷ 36   h 7500 ÷ 25

4 a Work out 315 ÷ 15.   b How many 50s make 750?   c Work out $\frac{680}{17}$.
   d Work out 600 divided by 30.   e Divide 8 into 112.

5 Five people shared a prize draw win of £2400 equally.
   How much did each person receive?

6 In an online computer game tournament, players are put into groups of 24, with the group winners going through to the final.
   How many finalists will there be if there are
   a 240 players   b 720 players   c 864 players?

7 An aeroplane can hold 18 parachute jumpers at a time.
   How many trips does the plane have to make for
   a 126 jumps   b 234 jumps   c 648 jumps?

8 A packing case will hold 72 economy-size boxes, 24 large-size boxes or 12 family-size boxes.
   How many packing cases would be needed to pack
   a 864 economy-size boxes   b 984 large-size boxes   c 960 family-size boxes?

1.6 Rounding

Objective
You can write numbers to a suitable accuracy.

Why do this?
When you want to estimate a shopping bill, it is useful to be able to round the numbers to make the calculation easier.

Get Ready
Work out the total cost of these.
1. Two T-shirts at £15 each and two pairs of shorts at £4 each
2. Four cookies at 75p each and two cups of coffee at £1 each

Key Points
- There are situations where an estimate of numbers may be good enough.
  - If you are working out how long a journey is going to take, then an answer to the nearest hour may be good enough.
  - If you are recording the number of customers a supermarket has in a day, then the nearest 100 may be good enough.
- **Rounding** means reducing the accuracy of a number by replacing right-hand digits with zeros.
- To round a number, look at the digit before the place value you are rounding to (so for rounding to the nearest 10 you would look at the units digit). If it is less than 5, round down. If it is 5 or more, round up.

Example 13   Round 7650 to the nearest   **a** thousand   **b** hundred.

a 8000 ⟵   7650 is nearer 8000 than 7000 on the number line. Rounded to the nearest thousand 7650 is 8000.

b 7700 ⟵   7650 is exactly halfway between 7600 and 7700. When this happens you round up.

Example 14   Round 314 to the nearest 100.

314 to the nearest 100 is 300. ⟵   Look at the tens digit. It is less than 5 so round down.

Exercise 1H

1   Round these numbers to the nearest ten.
   a 57   **b** 63   **c** 185   **d** 194   **e** 991   **f** 2407

2   Round these numbers to the nearest hundred.
   a 312   **b** 691   **c** 2406   **d** 3094   **e** 8777   **f** 29 456

3   Round these numbers to the nearest thousand.
   a 2116   **b** 36 161   **c** 28 505   **d** 321 604   **e** 717 171   **f** 2 246 810

4

	Length (ft)	Cruising speed (mph)	Takeoff weight (lb)
Airbus A310	153	557	36 095
Boeing 737	94	577	130 000
Saab 2000	89	403	50 265
Dornier 228	54	266	12 566
Lockheed L1011	177	615	496 000

For each of these aircraft, round:
   a the length to the nearest ten feet
   b the takeoff weight to the nearest hundred pounds
   c the cruising speed to the nearest ten mph.

1.7 Negative numbers

Objective

- You can use negative numbers to represent quantities that are less than zero.

Why do this?

Negative numbers are used to record temperatures below zero and distances below sea level.

Get Ready

1. Place the numbers 557, 577, 403, 266 and 615 in order. Start with the smallest.

Key Points

⦿ The further a number is from zero on the left-hand side of a number line, the smaller it is, so -25 is smaller than -3.

⦿ For **negative numbers** you count backwards from zero on the number line.

  ⦿ $-31°C$ is 31 degrees below zero.

  ⦿ -396 m is 396 m below sea level.

Example 15   Write down: **a** the highest   **b** the lowest number in this list.

$-19, 7, -10, 0, 4, -3$

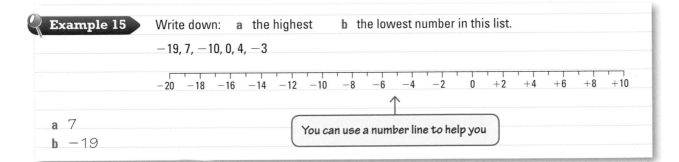

a 7

b -19

You can use a number line to help you

Exercise 1I

1. For each list of numbers write down **i** the highest and the lowest number
                              ii the numbers in order, starting with the lowest.

   a $5, -10, -3, 0, 4$ **b** $-7, -2, -9, -13, 0$

   c $-3, 6, 13, -15, -6$ **d** $-13, -2, -20, -21, -5$

2. Write down the two missing numbers in each sequence.

   a $4, 3, 2, 1, \_, \_, -2$ **b** $10, 7, 4, 1, \_, \_, -8$

   c $-13, -9, -5, -1, \_, \_, 11$ **d** $13, 8, 3, -2, \_, \_, -17$

   e $21, 12, 3, -6, \_, \_, -33$ **f** $-13, -10, -7, -4, \_, \_, 5$

3. Use the number line to find the number that is:

   a 5 more than 2   **b** 4 more than -7 **c** 7 less than 6

   d 2 less than -3 **e** 6 less than 0   **f** 10 more than -7

   g 6 more than -6 **h** 4 less than -3 **i** 10 less than 5

   j 1 less than -1.

4. What number is:

   a 30 more than -70 **b** 50 less than -20 **c** 80 greater than -50

   d 90 smaller than 60   **e** 130 smaller than -30 **f** 70 bigger than 200

   g 170 bigger than -200 **h** 100 bigger than -100 **i** 140 more than -20

   j 200 less than -200?

F

5 The table gives the highest and lowest temperatures recorded in five cities during one year.

	New York	Brussels	Tripoli	Minsk	Canberra
Highest temperature	27°C	32°C	34°C	28°C	34°C
Lowest temperature	−9°C	−6°C	8°C	−21°C	7°C

a Which city recorded the lowest temperature?
b Which city recorded the biggest difference between its highest and lowest temperatures?
c Which city recorded the smallest difference between its highest and lowest temperatures?

6 The temperature of the fridge compartment of a fridge-freezer is set at 4°C.
The freezer compartment is set at −18°C.
What is the difference between these temperature settings?

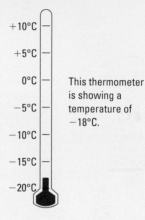

This thermometer is showing a temperature of −18°C.

7 The temperature of a shop freezer should be set at −18°C. It is set to −12°C by mistake.
What is the difference between these temperature settings?

1.8 Working with negative numbers

Objective

● You can work with negative numbers to find differences and make changes.

Why do this?

You may wish to find the difference between the temperatures at two ski resorts to see which is colder. To do this you need to work with negative numbers.

Get Ready

1. Which ski resort has the highest temperature?
2. Which resort has the lowest wind chill?

Resort	Kitzbühel	Val d'Isère	Civetta
Max (°C)	−5	−10	−3
Wind chill (°C)	−9	−20	−10

Key Point

● You can add and subtract negative numbers using a number line to help you.

Example 16 At 12 noon the temperature at the top of a mountain was 2°C.

By 6 pm it had fallen by 8°C.

What was the new temperature at 6 pm?

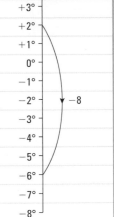

$2°C - 8°C = -6°C$

> **ResultsPlus**
> **Examiner's Tip**
>
> A number line can help you when working with negative numbers.

Example 17 a The temperature is 5°C. It falls by 8 degrees.

What is the new temperature?

b What is the difference in temperature between 4°C and −4°C?

a $-3°C$ ← From 5°C count 8 degrees down to −3°C.

b $8°C$ ← From 4°C count to −4°C. There is a difference of 8 degrees between the two temperatures.

Exercise 1J

Use this number line going from −10°C to +10°C to help you with these questions.

1 Find the number of degrees between each pair of temperatures.

a −3°C, 2°C   b −4°C, −1°C

c 2°C, 8°C   d −6°C, 4°C

e 7°C, −3°C   f 1°C, 9°C

g −3°C, −8°C   h −7°C, 6°C

2 Find the new temperature after:

a a 2° rise from −4°C   b a 7° fall from 4°C

c 8°C falls by 15°   d −4°C rises by 7°

e −5°C rises by 8°   f 4°C falls by 10°

g −3°C falls by 6°.

F

1.9 Calculating with negative numbers

◎ Objective

○ You know the rules to use when working with positive and negative numbers.

❓ Why do this?

Using rules is quicker than using a number line.

⬆ Get Ready

1. Write down the value of
   a an 8° rise from −10°C
   b an 8° fall from −2°C

🔍 Key Points

◉ Adding a negative number is the same as subtracting a **positive number**.
◉ Subtracting a negative number is the same as adding a positive number.
◉ The rules for multiplying and dividing are:
   ◉ if the signs are the same the result is positive
   ◉ if the signs are different the result is negative.

🔍 Example 18 Work out

   a $2 - +3$ **b** $-3 - -2$ **c** $4 + -2$ **d** $-3 + +1$

a $2 - +3 = -1$

> $2 - +3$ is the same as $2 + -3$.
> Start at 2 and go down 3 to get to -1.

b $-3 - -2 = -1$

> $-3 - -2$ is the same as $-3 + 2$.
> Start at -3 and go up 2 to get to -1.

c $4 + -2 = 2$

> $4 + -2$ is the same as $4 - 2$.
> Start at 4 and go down 2 to get to $+2$.

d $-3 + +1 = -2$

> Start at -3 and go up 1 to get to -2.

⚙ Exercise 1K

F

1 Work out
   a $-4 + -3$ **b** $9 - +5$ **c** $8 - -2$ **d** $5 + +4$
   e $-7 - -6$ **f** $-2 + +4$ **g** $6 + -8$ **h** $-3 - +7$

A03

2 A diver dives to a depth of −27 metres. A second diver dives to a depth of −16 metres. What is the difference in the depths of the dives?

3 The temperature at the Arctic Circle is recorded as −18°C one night. The following day it rises by 6°C. What is the temperature during the day?

🔍 **Example 19**   Work out

      a 15×-3 **b** -7×-3 **c** -6×4 **d** $-12 \div -2$

a $+15 \times -3 = -45$ ← | signs are different

b $-7 \times -3 = +21$ ← | signs are the same

> Where no sign is given the number is positive.

c $-6 \times +4 = -24$ ← | signs are different

d $-12 \div -2 = +6$ ← | signs are the same

⚙️ **Exercise 1L**

Work out these multiplications and divisions.

1   **a** $+3 \times -1$ **b** $+24 \div -8$ **c** $+4 \div +1$
     d $+2 \times +6$ **e** $-12 \div +3$ **f** $-3 \times +4$

2   **a** $-9 \times +10$ **b** $-32 \div -8$ **c** $-20 \div -4$
     d $-2 \times +7$ **e** $+10 \div -5$ **f** -3×-4

3   **a** $-5 \times +4$ **b** $-16 \div -8$ **c** $-4 \times +5$
     d $-18 \div -3$ **e** $+18 \div +2$ **f** $-6 \times +7$

4   **a** -8×-3 **b** $-30 : +2$ **c** $-16 \div +4$
     d -3×-9 **e** $+5 \times -8$ **f** $+24 \div +8$

5   **a** $-50 \div -5$ **b** $-7 \times +8$ **c** $+6 \times +6$
     d -3×-7 **e** $-9 \div +3$ **f** $-7 \times +6$

F

1.10 Factors, multiples and prime numbers

◎ **Objectives**

● You can recognise types of numbers and know their properties.
● You can separate the prime factors that make up a number.

❓ **Why do this?**

Credit card companies use prime numbers to protect your card details when you shop online.

⬦ **Get Ready**

1. Write down the value of
   a 7×6 **b** 9×7 **c** 8×8 **d** 3×6

Key Points

- Even numbers are whole numbers that divide exactly by 2.
- Any number that ends in 2, 4, 6, 8 or 0 is even.
- Odd numbers do not divide exactly by 2 and always end in 1, 3, 5, 7 or 9.
- The **factors** of a number are whole numbers that divide exactly into the number. They include 1 and the number itself.
- **Multiples** of a number are the results of multiplying the number by a positive whole number.
- A **common multiple** is a number that is a multiple of two or more numbers.
- A **prime number** is a whole number greater than 1 whose only factors are 1 and the number itself.
- A **prime factor** is a factor that is also a prime number. For example, the prime factors of 18 are 2 and 3.
- A **common factor** is a number that is a factor of two or more numbers.
- A number can be written as a product of its prime factors.

Example 20 Separate 3, 11, 14, 22, 23, 36, 39, 40, 52, 57, 60 into odd and even numbers.

14, 22, 36, 40, 52 and 60 end in an even number (2, 4, 6, 8, 0). These are the even numbers.
3, 11, 23, 39 and 57 end in an odd number (1, 3, 5, 7, 9). These are the odd numbers.

Exercise 1M

G

1 Write down all the even numbers from this list.
42, 18, 37, 955, 1110, 73 536, 500 000

2 Write down all the odd numbers from this list.
105, 537, 9216, 811, 36 225, 300 000

3 Write down the next two even numbers after:
a 12   b 28   c 196.

4 Write down the odd number that comes before:
a 5   b 31   c 200.

5 Write down all the even four-digit numbers that can be made using only the numbers on the cards below.

4   7   6   5

A03 6 Using the digits 2, 7, 3 and 8 only once each,
a write down the largest even number that can be made
b write down the smallest odd number that can be made.

A03 7 A postman has letters to deliver to house numbers
16, 9, 4, 3, 22, 17, 14, 8, 16, 3, 12, 14, 17, 1, 42, 15, 16, 22, 9, 23, 31, 15 and 12.
He is going to walk down the side with even numbers first, starting at number 2, and come back on the opposite side with odd numbers. Arrange the house numbers in order for him.

factors   multiple   common multiple   prime number   prime factor   common factor

Example 21 ▶ Find the factors of 12.

12 can be made from 1 × 12, 2 × 6, 3 × 4. The factors of 12 are 1, 2, 3, 4, 6 and 12.

Example 22 ▶ Write down the multiples of 3 that are between 20 and 29.

7 × 3 = 21 is the first multiple of 3 after 20.
Then comes 21 + 3 = 24 and 24 + 3 = 27.
As the next multiple of 3 is 30 the answer is 21, 24 and 27.

ResultsPlus
Watch Out!

Students sometimes get confused between factors and multiples – remember multiples are from multiplying.

Exercise 1N

1   Write down all the factors of the following numbers.
    a 15   b 20   c 24   d 18   e 13   f 90

2   List the first five multiples of the following numbers.
    a 2   b 5   c 10   d 7   e 13

3   Write down three multiples of 10 that are larger than 50.

4   Write down the numbers in the cloud that are:
    a factors of 24
    b multiples of 5
    c factors of 16
    d multiples of 3.

    5   13   16   20
         12   1
    9   8
       4   6

5   Find the two prime numbers that are between 30 and 40.

6   Find the next prime number after 91.

7   Find the largest number with a factor of 4 that can be made using the digits 3, 2, 4, 5.

8   A florist has 216 roses.
    She makes these into bunches with an equal number of roses.
    Each bunch has more than 10 roses.
    Find a possible number for the roses in her bunches:
    a if all the roses are used
    b if 6 roses are left over.

G

F
A03

A03

Example 23 Find the common factors of 12 and 18.

The factors of 12 are 1, 2, 3, 4, 6, 12.
The factors of 18 are 1, 2, 3, 6, 9, 18.
1, 2, 3 and 6 are all factors of both 12 and 18.
They are the common factors of 12 and 18.

Example 24 Write 36 as a product of its prime factors.

Method 1
$36 = 2 \times 18$
$\quad = 2 \times 2 \times 9$
$\quad = 2 \times 2 \times 3 \times 3$

In this method, use 2 as often as possible, then move onto 3, then 5 and work through the prime numbers in order.

Method 2

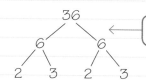

This is called a factor tree.

$36 = 2 \times 3 \times 2 \times 3$

Exercise 10

1 Find the common factors of:

a 4 and 6
b 10 and 15
c 24 and 36
d 10, 30 and 60
e 16 and 24
f 15 and 40
g 12 and 28
h 30 and 42
i 18 and 25.

2 Find all the prime factors of the following numbers.

a 30
b 25
c 42
d 39
e 105

3 Write these numbers as products of their prime factors.

a 45
b 36
c 28
d 80
e 72

1.11 Finding lowest common multiple (LCM) and highest common factor (HCF)

Objectives

- You can find the highest common factor (HCF).
- You can find the lowest common multiple (LCM).

Why do this?

If you know the number of seats per row at a cinema, you can use multiples to work out the total number of seats.

Get Ready

1. Write these numbers as products of their prime factors.

a 99
b 324
c 175

Key Points

◉ The **lowest common multiple (LCM)** is the lowest multiple that is common to two or more numbers.

◉ The **highest common factor (HCF)** is the highest factor that is common to two or more numbers.

Example 25

a Find the highest common factor (HCF) of 36 and 24.

b Find the lowest common multiple (LCM) of 6 and 8.

a $36 = 2 \times 2 \times 3 \times 3$
$24 = 2 \times 2 \times 2 \times 3$
$2 \times 2 \times 3 = 12$
12 is the HCF

> Write each number in prime factor form.
> Pick out the factors common to both numbers.

b 6: 6, 12, 18, 24, 30, 36, 42, 48, 54
8: 8, 16, 24, 32, 40, 48, 56
There are two common multiples so far,
but 24 is the lower.
The LCM of 6 and 8 is 24.

> Write a list of multiples for each number.
> The LCM is the lowest number that appears in both lists.

Exercise 1P

1 Find the highest common factor of:

a 4 and 8   b 9 and 12   c 18 and 24

d 14 and 30   e 21 and 35.

2 Find the lowest common multiple of:

a 3 and 4   b 4 and 6   c 12 and 15

d 36 and 16   e 50 and 85.

3 Find the LCM and HCF of:

a 12 and 18   b 120 and 180   c 24 and 84

d 91 and 130   e 72 and 96   f 40 and 60.

4 Two lighthouses off the Cornish coast can be recognised by the different intervals between their flashes. One flashes every 24 seconds and the other every 40 seconds. A ship's captain sees them flash at the same time. How long will it be before this happens again?

5 The light on a motorway service vehicle flashes every 20 seconds. The light on a tractor flashes every 10 seconds. As they pass each other on a road, the lights flash together. How long will it be before this happens again?

D

C

A02

A02

1.12 Finding square numbers and cube numbers

⊙ Objective

○ You can calculate the square and cube of a whole number.

⊘ Why do this?

Square and cube numbers are found in everyday life. There are 8×8 squares on a chessboard, and $3 \times 3 \times 3$ cubes on a Rubik's Cube.

⊘ Get Ready

1. Find these.
   a 11×11 b 13×13 c $3 \times 3 \times 3$

⊘ Key Points

◉ A **square number** is the result of multiplying a whole number by itself.
4×4 can be written as the square of 4, 4 squared or 4^2. (For more on indices, see Section 9.1.)
◉ You need to be able to recall the squares of all whole numbers up to 15×15.
◉ A **cube number** comes from multiplying a number by itself and then multiplying the result by the original number.
$4 \times 4 \times 4$ can be written as the cube of 4, 4 cubed or 4^3.
◉ You need to be able to recall the cubes of 2, 3, 4, 5 and 10.

⊙ Example 26   Find   a the first four square numbers,   b the first four cube numbers.

a The first four square numbers are
$1 \times 1 = 1$, $2 \times 2 = 4$, $3 \times 3 = 9$, $4 \times 4 = 16$.

b The first four cube numbers are
$1 \times 1 \times 1 = 1$, $2 \times 2 \times 2 = 8$, $3 \times 3 \times 3 = 27$, $4 \times 4 \times 4 = 64$.

ResultsPlus
Watch Out!

Students often double rather than square or multiply by 3 rather than cube.

⚙ Exercise 1Q

G

F

1 Work out the square of the following numbers.
   a 3   b 6   c 10   d 2   e 20

2 Work out the cube of the following numbers.
   a 2   b 5   c 7   d 20   e 12

3 Work out
   a 5 squared   b the cube of 6   c 9^2 d 3^3

Chapter review

◉ Although you can keep on counting, you only use ten **digits**, often called figures.

◉ Each digit has a **value** that depends on its position in the number. This is its **place value**.

◉ You can read and write numbers by thinking about the place value of each of the digits.

◉ You can use your knowledge of place value to put numbers in order of size.

- You can use a number line to help you to increase or decrease a number.
- Number lines might be read:
  - from left to right
  - from bottom to top
  - clockwise.
- Words that show you have to add numbers are: add, plus, total and sum.
- Words that show you have to take away are: take away, subtract, minus and find the difference.
- Words that show you have to multiply are: product, times and multiply.
- Divide and share are words that show you have to divide.
- You multiply by 10, 100 or 1000 by moving digits 1, 2 or 3 places to the left.
- You divide by 10, 100 or 1000 by moving digits 1, 2 or 3 places to the right.
- There are situations where an estimate of numbers could be good enough.
- **Rounding** means reducing the accuracy of a number by replacing right-hand digits with zeros.
- To round a number, look at the digit before the place value you are rounding to (so for rounding to the nearest 10 you would look at the units digit). If it is less than 5, round down. If it is 5 or more, round up.
- The further a number is from zero on the left-hand side of a number line, the smaller it is.
- For **negative numbers** you count backwards from zero on the number line.
- You can add and subtract negative numbers using a number line to help you.
- Adding a negative number is the same as subtracting a **positive number**.
- Subtracting a negative number is the same as adding a positive number.
- When multiplying or dividing numbers:
  - if the signs are the same the result is positive
  - if the signs are different the result is negative.
- Even numbers are numbers that divide by 2. They always end in 2, 4, 6, 8 or 0.
- Odd numbers are numbers that do not divide by 2. They always end in 1, 3, 5, 7 or 9.
- The **factors** of a number are whole numbers that divide exactly into the number. They include 1 and the number itself.
- Factors are pairs of numbers which, when multiplied together, give the number.
- **Multiples** of a number are the results of multiplying the number by a positive whole number.
- A **common multiple** is a number that is a multiple of two or more numbers.
- A **prime number** is a whole number greater than 1 whose only factors are 1 and the number itself.
- A **prime factor** is a factor that is also a prime number.
- A **common factor** is a number that is a factor of two or more numbers.
- A number can be written as a product of its prime factors.
- The **lowest common multiple (LCM)** is the lowest multiple that is common to two or more numbers.
- The **highest common factor (HCF)** is the highest factor that is common to two or more numbers.
- A **square number** is the result of multiplying a whole number by itself.
- You need to be able to recall the squares of all whole numbers up to 15×15.
- A **cube number** comes from multiplying a whole number by itself and then multiplying the result by the original number.
- You need to be able to recall the cubes of 2, 3, 4, 5 and 10.

Review exercise

1 Draw a place value diagram and write in a number with five digits and a 2 in the thousands column and a 3 in the units column.

2 Write down the value of the 5 in the following numbers.

   a 651   **b** 5302   **c** 253 101   **d** 10 050   **e** 175

3 Write down these numbers in words.

   a 3723   **b** 107   **c** 2007   **d** 15 071

4 Write these numbers in figures.

   a twenty-one thousand two hundred and thirty-one

   b five hundred and seven

   c seventy thousand two hundred and three

5 This table shows the number of people who were seriously injured in road accidents in a part of Britain.

Year	2001	2002	2003	2004	2005
Number	37 346	33 645	31 456	29 788	26 466

In which year were:

   a the smallest number of people seriously injured

   b more than 35 000 seriously injured

   c between 30 000 and 32 000 seriously injured

   d fewer than 28 000 seriously injured?

6 A shop's takings for March were £34 176.
The takings for April were £58 358.
Work out the total takings for the two months.

7 A school has 1321 students. 738 are boys. How many are girls?

8 Using only the numbers in the cloud, write down:

   a all the multiples of 6

   b all the square numbers

   c all the factors of 12

   d all the cube numbers.

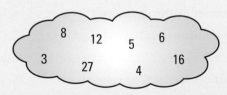

9 4, 5, 8, 9, 12, 14, 16, 20, 27, 35, 36, 37
From the numbers in the list write down:

   a the odd numbers   **b** the multiples of 3   **c** the factors of 48

   d the prime numbers   **e** the square numbers   **f** the cube numbers.

10 Using the numbers 4, 1, 6, 7 once each, make a number which is

   a a multiple of 3   **b** a multiple of 4.

11 Cans of cola are delivered in packs of 24.
Copy and complete the table.

Number of packs	Number of cans
1	24
2	
3	
4	
5	

If you need at least 75 cans of cola, how many packs should you order?

12

cube	multiple	factor	product

Use a word from the box to complete this sentence correctly.

12 is a .... of 36

13 Dan writes down the numbers from 2 to 30.
He crosses out all the multiples of 2.
He crosses out all the multiples of 3 and then all the multiples of 5.
a How many numbers have been crossed out more than once?
b How many numbers have not been crossed out at all?

14 Write down all the multiples of 15 that are less than 100.

15 Round these numbers to the nearest multiple of 10 given in the brackets.
a 27 (10)
b 349 (100)
c 2047 (100)
d 78 939 (10)
e 7 813 076 (million)
f 83.7 (10)

16 There are 376 passengers on a train.
At its first stop 27 passengers get off.
295 passengers get on.
How many passengers are now on the train?

17 Peter buys 273 stamps costing 38p each.
How much do they cost altogether?

18 There are 14 winners in a lottery.
They share the winnings of £10 332 equally.
How much does each get?

19 a The temperature during an Autumn morning went up from -3°C to 6°C.
By how many degrees did the temperature rise?
b During the afternoon the temperature fell by 11°C from 6°C.
What was the temperature at the end of the afternoon?

20 Work out
a $3 - 7$
b $-3 + 5$
c $-11 - 4$
d $4 - (-6)$
e $(-5) + (+3)$

21 Work out
a $+3 \times -7$
b -4×-5
c $16 \div -2$
d $-15 \div -3$
e $-28 \div +4$

G

A02

A03 F

22 Find three different numbers each below 10 which have a sum of 20.

23

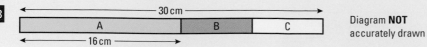

Diagram **NOT** accurately drawn

Here is a picture of a stick. The stick is in three parts, A, B and C.
The total length of the stick is 30 cm.
The length of part A is 16 cm.
The length of part B is the same as the length of part C.
Work out the length of part C.

May 2009 adapted

24

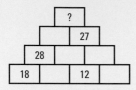

In this set of squares, each number in a square is obtained by adding the two numbers immediately underneath.
What number should go in the top square?

25 The table shows the cost of two different models of the Eiffel Tower.

Small	£2.40
Large	£4.50

Pierre buys 10 small models and 5 large models.
He pays with a £50 note.
Work out how much change he should get.

26

City	Temperature
Cardiff	−2°C
Edinburgh	−4°C
Leeds	2°C
London	−1°C
Plymouth	5°C

The table gives information about the temperatures at midnight in 5 cities.
a Write down the lowest temperature.
b Work out the difference in temperature between Cardiff and Plymouth.
c Which city has a temperature halfway between London and Plymouth?

May 2009

27 Jill says
'If you multiply any two prime numbers together, the answer will always be an odd number'.
Write down an example to show that Jill is wrong.

June 2006

28 Work out 10^3.

29 Charlie writes down the numbers from 1 to 50. Ben puts a red spot on all the even numbers. Alex puts a blue spot on all the multiples of 3.

   a What is the largest number that has both a red and a blue spot?

   b How many numbers have neither a blue nor a red spot?

   Sophie puts a green spot on all the multiples of 5.

   c How many numbers have exactly two coloured spots on them?

30 Two whole numbers are each less than 10 and greater than 0. Work out the greatest possible difference between their product and their sum.

31 Here is some information about the coaches in a coach hire company.

Type of coach	Number of passengers	Number of coaches available
Small	14	5
Medium	32	3
Large	44	2
Touring	56	3

   Jim hires two medium and two large coaches to take people to a show.

   a Work out the total number of passengers that could go to the show.

   Becky wants to take 300 people to a show.

   b Work out the smallest number of coaches she would need to hire from the company.

32 Perfume bottles are sold in boxes which measure 4 cm by 5 cm by 20 cm.

   They are supplied to shops in cartons containing 12 bottles.

   Work out two arrangements of the bottles in these cartons. (There are lots of sensible answers.)

33 A buzzer buzzes every 4 seconds and a bell rings every 6 seconds. The buzzer and the bell start at the same time. How many times in the first minute will they make a sound at the same instant?

34 **a** Express 108 as the product of its prime factors.

   b Find the highest common factor of 108 and 24.

35 **a** Express the following numbers as products of their prime factors.

     i 60   **ii** 96

   b Find the highest common factor of 60 and 96.

   c Work out the lowest common multiple of 60 and 96.

36 Doughnuts are sold in packs of 8. Cakes are sold in packets of 6.

   What is the smallest number of packs of doughnuts and the smallest number of packets of cakes that can be bought so that the number of doughnuts is equal to the number of cakes?

37 A chocolate company wishes to produce a presentation box of 36 chocolates for Valentine's Day.

   It decides that a rectangular shaped box is the most efficient, but needs to decide how to arrange the chocolates.

   How many different possible arrangements are there:

   a using one layer   **b** using two layers   **c** using three layers?

   Which one do you think would look best?

38 A car's service book states that the air filter must be replaced every 10 000 miles and the diesel fuel filter every 24 000 miles. After how many miles will both need replacing at the same time?

F A03

E A02 A03

A02 A03

D A02 A03

A03

C

A02 A03

A02 A03

A03

2 ANGLES 1

Makers of fairground rides often make them more thrilling by sending the riders through a series of sharp bends and around loops at high speed. The turns in the track are angles and they can be measured in degrees.

Objectives

In this chapter you will:
- learn how to measure angles
- name different types of angles
- use angle facts to solve problems.

Before you start

You need to know:
- how to add and subtract numbers to 360
- how to draw lines, using a ruler, accurate to the nearest 2 mm.

2.1 Fractions of a turn and degrees

◉ Objective

◉ You know that turns are measured in degrees.

⟐ Why do this?

Understanding how turns are measured is a useful skill. You use fractions of a turn when map reading or orienteering.

⟐ Get Ready

1. **a** $360 \div 4$ **b** 60×3 **c** $180 \div 4$ **d** $360 \div 2$

◉ Key Points

◉ There are 360 **degrees** in a full turn. This is written as 360°.

◉ A half turn is 180°.

 180°

◉ A quarter turn is 90°. A quarter turn is called a right angle.

90°

◉ A right angle is shown on diagrams with a small square.

⚙ Exercise 2A

Questions in this chapter are targeted at the grades indicated.

1 Write down the size of the following angles in degrees.

a

b

c

2 Write down how much a compass turns between:
   a N and E
   b E and SW
   c SE and SW.

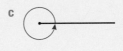

3 On a clock, how many degrees does the hour hand turn between:
   a 2 pm and 8 pm   b 7 am and 10 am   c midnight and midday?

4 Karen goes for a walk. She walks 2 km due west and then 2 km due south. She then walks back to the start by the shortest distance. What compass bearing does she use to walk back to the start?

G

F

A02 A03 C

C A02 A03

5   At 10 am Jean drives to a shop. It takes 5 minutes to drive to the shop.
Jean is at the shop for 40 minutes. She then takes 5 minutes to drive home.
On a clock, how many degrees has the minute hand turned from the time Jean drives to the shop to the time she gets home?

2.2 **What is an angle?**

◉ Objectives

○ You know what an angle is.
○ You can identify acute, obtuse and reflex angles.

❓ Why do this?

Many people use and measure angles, particularly architects, designers and artists.

⬆ Get Ready

1. Write down how many degrees a compass point makes between:
   a   S and W
   b   N and SE
   c   SW and NE.

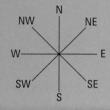

🔑 Key Points

◉ An **angle** is a measure of turn.
An angle is formed when two lines meet.
These three angles are all the same
size. The length of the line and position of
the angle do not change the size.

◉ There are different types of angle.

Acute angle
Less than a $\frac{1}{4}$ turn.

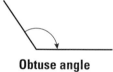

Obtuse angle
More than a $\frac{1}{4}$ turn.

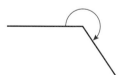

Reflex angle
More than a $\frac{1}{2}$ turn.

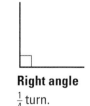

Right angle
$\frac{1}{4}$ turn.

🔍 Example 1   Name the different types of angle in this diagram.

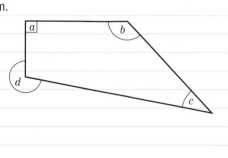

a = right angle   ←   It is a $\frac{1}{4}$ turn.

b = obtuse angle   ←   It is more than a $\frac{1}{4}$ turn.

c = acute angle   ←   It is less than a $\frac{1}{4}$ turn.

d = reflex angle   ←   It is more than a $\frac{1}{2}$ turn.

angle   acute angle   obtuse angle   reflex angle   right angle

Exercise 2B

1 Which angle is the odd one out?

a    b   c

2 Write down the special name for each of the following angles.

a   b   c

d   e   f

g   h   i

2.3 Naming sides and angles

◎ Objective

○ You can name sides and angles.

⑦ Why do this?

You can use letters to represent sides of shapes on a map or scale drawing.

⬦ Get Ready

1. Which angle is the odd one out?

   a 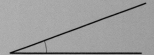   b    c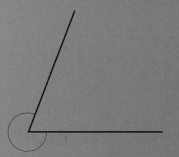

2. Draw an acute angle.
3. Write down the special name of this angle.

🔍 **Key Point**

◉ You can use letters to name the sides and angles of shapes.
  ◉ This shape is named ABCD using the letters for the corners and going round clockwise.
  ◉ Lines are named using the letters they start and finish with.
  ◉ Angles are named using the three letters of the lines that make the angle. The angle is always at the middle letter.

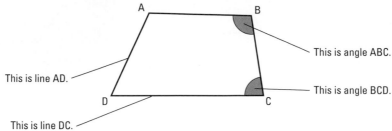

This is angle ABC.

This is line AD.

This is angle BCD.

This is line DC.

🔎 **Example 2**   **a** Use letters to name the marked angles.

   b Which is the longest side of the triangle?

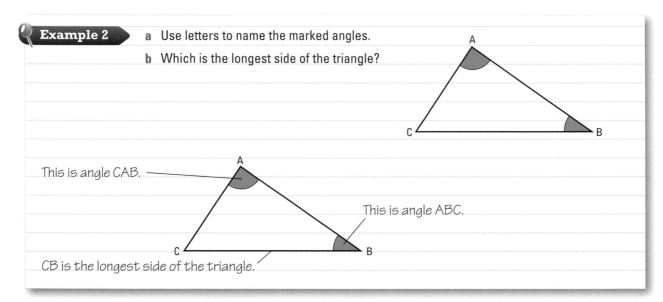

This is angle CAB.

This is angle ABC.

CB is the longest side of the triangle.

⚙ **Exercise 2C**

G

1 Write down the names of the marked angles in each of these diagrams.

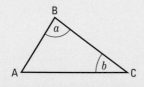

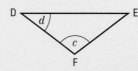

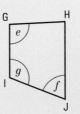

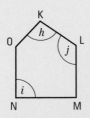

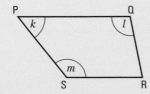

2

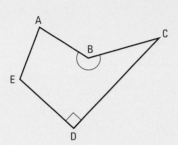

Write down the names of the

a acute angles

b right angle

c obtuse angle

d reflex angle

e longest side

f shortest side.

G

3 Draw a sketch of triangle ABC, where AB is the longest side and angle BCA is an obtuse angle.

A02 **F**

2.4 **Estimating angles**

◎ Objective

● You can estimate the size of angles.

❓ Why do this?

Sports players need to estimate angles in order to pass the ball and score goals.

⬆ Get Ready

1. Write down the size of the following angles in degrees.

a

b

c

🔑 Key Point

● You should always estimate the size of angles before measuring them. This enables you to check that your answer is sensible.

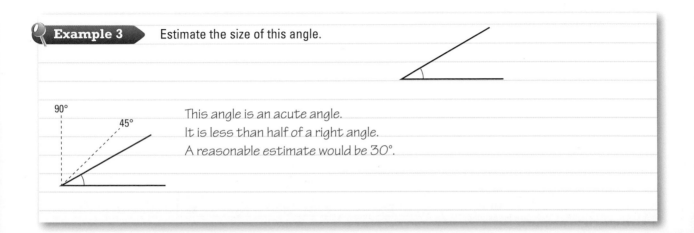

Example 3 Estimate the size of this angle.

This angle is an acute angle.
It is less than half of a right angle.
A reasonable estimate would be 30°.

90°

45°

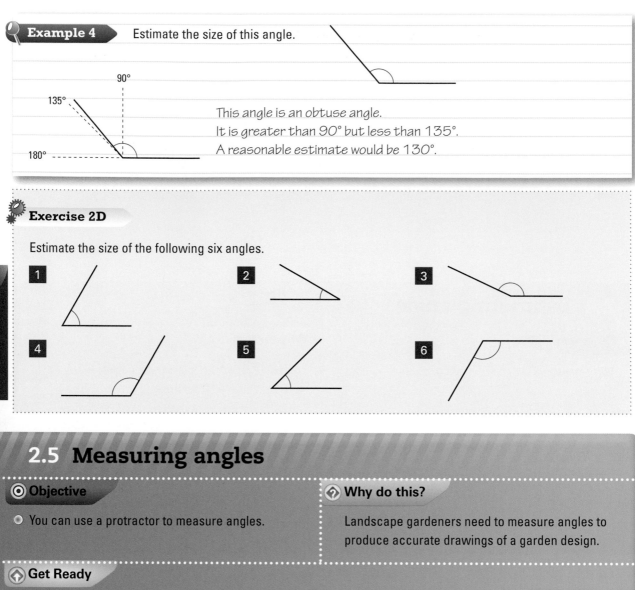

Example 4 — Estimate the size of this angle.

This angle is an obtuse angle.
It is greater than 90° but less than 135°.
A reasonable estimate would be 130°.

Exercise 2D

Estimate the size of the following six angles.

1

2

3

4

5

6

2.5 Measuring angles

Objective

You can use a protractor to measure angles.

Why do this?

Landscape gardeners need to measure angles to produce accurate drawings of a garden design.

Get Ready

1. Estimate the size of the following angles.

a

b

c

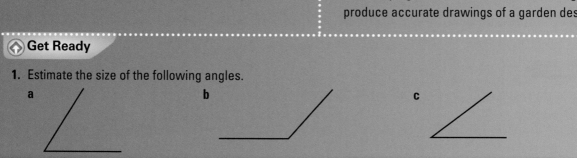

Key Point

A protractor is an instrument used to measure angles.

Use the inside scale to measure anticlockwise turns

Use the outside scale to measure clockwise turns

Place the cross at the point of the angle you are measuring.

Example 5
a Use a protractor to measure the angle CBA.
b Use a protractor to measure the angle BCD.

Here the lines of angle CBA are long enough to reach the outer edge of the protractor.

Use the inside scale to measure angle BCD. When the line is too short to reach the scale, extend it with a straight edge like this piece of paper.

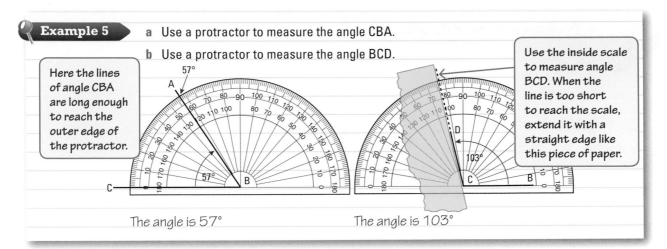

The angle is 57°

The angle is 103°

Exercise 2E

1 Measure the following angles.

a

b

2 Measure the angles:
a ABC
b CDA
c DAB.

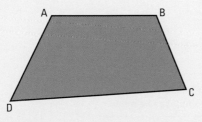

3 Measure the angles:
a STU
b UVR
c RST.

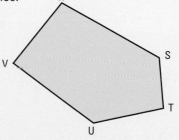

4 The table shows some information about the angles used in cutting different materials with a chisel.

Material	Aluminium	Medium steel	Mild steel	Brass	Copper	Cast iron
Cutting angle	30°	65°	55°	50°	45°	60°
Angle of inclination	22°	39.5°	34.5°	32°	29.5°	37°

The diagram shows the angle of inclination for the chisel.

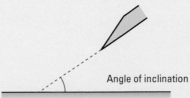

Angle of inclination

By measuring the angle shown in the diagram, write down the material being cut by the chisel.

F

E

2.6 Drawing angles

⊙ Objective

⊙ You can use a protractor to draw angles.

⊘ Why do this?

A manufacturer of toys would have to be able to draw accurate angles and shapes when designing the toys.

⟳ Get Ready

1. Draw a line 6 cm long.

2. Draw a line 2.5 cm long.

3. Draw a line 8.3 cm long.

🔍 Key Points

◉ You will be expected to draw angles accurately. Angles must be accurate to within 2°.

🔍 Example 6 Draw an angle of 125°.

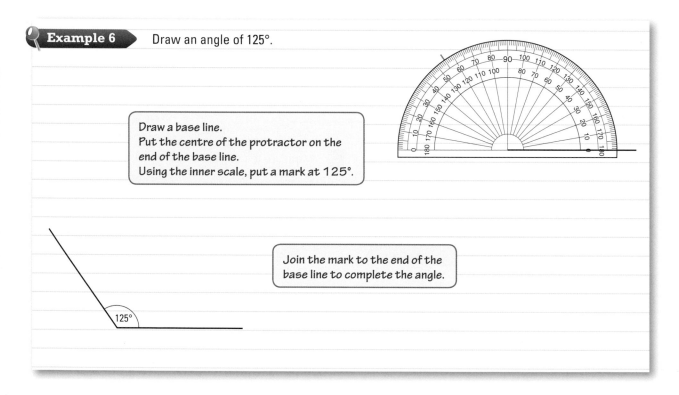

Draw a base line.
Put the centre of the protractor on the end of the base line.
Using the inner scale, put a mark at 125°.

Join the mark to the end of the base line to complete the angle.

125°

⚙ Exercise 2F

1 Use a protractor to draw the following angles.

a 50° b 160° c 55° d 115° e 43°

f 67° g 117° h 163° i 17° j 84°

2 Draw and label the following angles.

a ABC = 30° b DEF = 105° c GHK = 65°

d LMN = 48° e PQR = 162° f STU = 97°

2.7 Special triangles

- You can identify right-angled, equilateral and isosceles triangles.
- You can find missing angles in triangles.

Why do this?

You can see examples of special triangles in fashion, construction and art.

Get Ready

1. Make an accurate drawing of this triangle.
2. Measure the two unmarked angles and the unmarked side on your diagram.
3. What do you notice about the sides and angles?

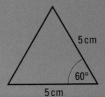

Key Points

- The **interior angles** of a triangle add up to 180°.
  You can see this if you cut out a triangle and tear the corners off as in the diagram.

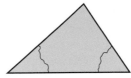

Tear these corners off.

Put all three corners together. They make a straight line which is an angle of 180°.

- You need to be able to recognise the following special types of triangles:

Isosceles

Two equal sides
Two equal angles

Equilateral

Three equal sides
All angles equal 60°

Right-angled

One angle 90°

Example 7   Work out the missing angle in this triangle.

The two given angles add up to 135°.
The missing angle is 180° − 135° = 45°.

60°   75°

Example 8   Work out the missing angles in this triangle.

Triangle ABC has two equal sides.
ABC is an isosceles triangle.
angle ACB = angle ABC
angle ABC = 50°
angle CAB = 180° − (50° + 50°)
angle CAB = 80°

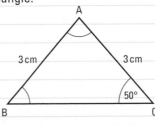

Exercise 2G

1 Work out the missing angles in the triangles below.

a

b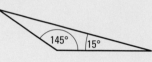

c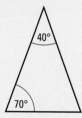

d

e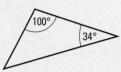

f

2 Work out the missing angles in the following triangles.

a

b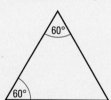

c

3 Use your answers to question 2 to write down the special name for each triangle.

4 Find angle ABC.

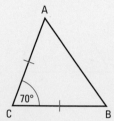

5 Find angles DEF and FDE.

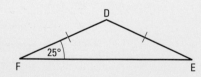

2.8 Angle facts

◉ Objective

◉ You can use angle facts to work out missing angles.

❓ Why do this?

Triangles are used in the support structures of fairground rides.

◈ Get Ready

Without using a calculator, write down the answers to the following questions.

1. **a** 180 − 40   **b** 180 − 67   **c** 180 − 132
2. **a** 360 − 47   **b** 360 − 108   **c** 360 − 247

◈ Key Points

◉ The **angles on a straight line** add up to 180°.

 $a + b = 180°$

◉ The **angles around a point** add up to 360°.

◉ Where two straight lines cross, the **opposite angles** are equal. They are called **vertically opposite angles**.

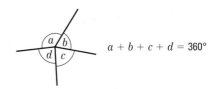 $a + b + c + d = 360°$

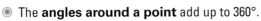

Angles a and b are the same.
Angles c and d are the same.

◈ Example 9

a What size is angle a?

The angles make a straight line so
$58° + a = 180°$
$a = 180° − 58° = 122°$

b What size is angle b?

The three angles make a straight line so
$45° + b + 67° = 180°$
$b = 180° − 45° − 67° = 68°$

◈ Example 10

a What size is angle a?

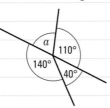

All the angles add up to 360°.
So, $140° + 40° + 110° + a = 360°$
$290° + a = 360°$
$a = 360° − 290° = 70°$

b What size is angle b?

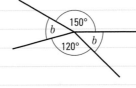

All the angles add up to 360°.
So, $120° + b + 150° + b = 360°$
$270° + 2b = 360°$
$2b = 360° − 270° = 90°$
$b = 90° ÷ 2 = 45°$

---

angles on a straight line   angles around a point   opposite angles   vertically opposite angles   **39**

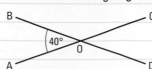

Example 11   Find all the missing angles in the diagram below. Give reasons for your answers.

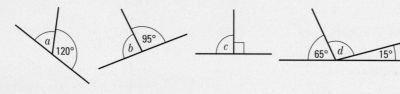

ResultsPlus
Examiner's Tip

Always remember to give reasons for your answers when you can.

Angle AOB = 40°
So, angle COD = 40° (vertically opposite angle AOB)
Angle AOD = 180° − 40° = 140° (the angles make a straight line)
So, angle BOC = 140° (vertically opposite angle AOD)

Exercise 2H

F

1   In each diagram, find the value of the letter.

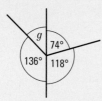

2   Find the value of the letters in each of the following diagrams.

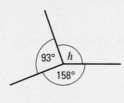  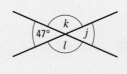 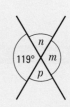

3   Find the value of the letters in the diagrams below.
       Give reasons for your answers.

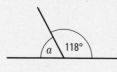 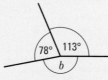

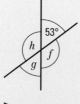  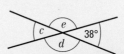

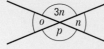 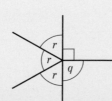 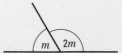

Chapter review

- There are 360° in a full turn.
- A half turn is 180°.
- A quarter turn is 90° and is called a **right angle**, shown with a small square.
- An **angle** is a measure of turn. An angle is formed when two lines meet.
- An angle that is less than a right angle is called an **acute angle**.
- An angle that is more than a quarter turn is called an **obtuse angle**.
- An angle that is more than a half turn is called a **reflex angle**.
- You can use letters to name the sides and angles of shapes.
- You should always estimate the size of angles before measuring them to check that your answer is sensible.
- A protractor is an instrument used to measure angles.
- You will be expected to draw angles. They must be accurate to within 2°.
- The **interior angles** of a triangle add up to 180°.
- An isosceles triangle has 2 equal angles and 2 equal sides.
- An equilateral triangle has 3 equal sides and 3 equal angles of 60°.
- A right-angled triangle contains an angle of 90°.
- The **angles on a straight line** add up to 180°.
- The **angles at a point** add up to 360°.
- When two lines cross, the **opposite angles** are equal. They are called **vertically opposite angles**.

Review exercise

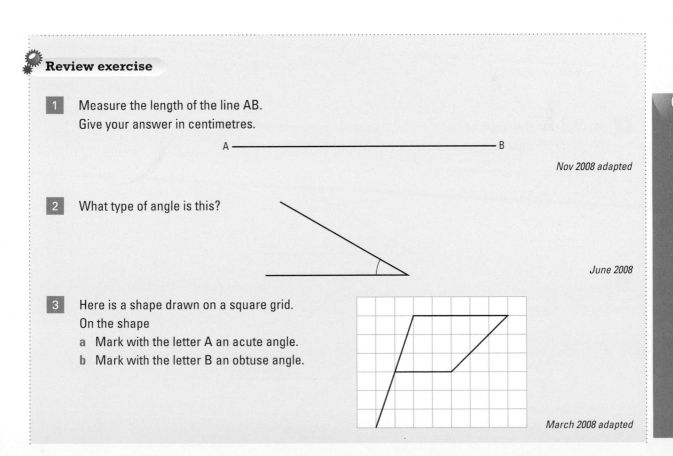

1. Measure the length of the line AB.
   Give your answer in centimetres.

   A ———————————————————————— B

   Nov 2008 adapted

2. What type of angle is this?

   June 2008

3. Here is a shape drawn on a square grid.
   On the shape
   a Mark with the letter A an acute angle.
   b Mark with the letter B an obtuse angle.

   March 2008 adapted

G

G

4 Here is a diagram drawn on a square grid.
  a Mark, with the letter A, an acute angle.
  b Mark, with the letter O, an obtuse angle.

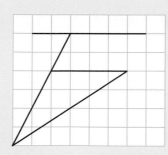

Nov 2007 adapted

5 a Draw a line 6 cm long.
    Start the line from a point labelled A.

  b Mark with a cross (×) the point on your line which is 2 cm from the point A.   *June 2007*

F

6 a i Write down the value of x.
    ii Give a reason for your answer.

Diagram **NOT** accurately drawn

A03

  This diagram is **wrong**.
  b Explain why.   *June 2008*

Diagram **NOT** accurately drawn

7 Work out the size of angle y.

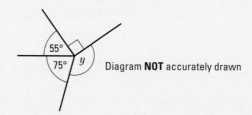

Diagram **NOT** accurately drawn

March 2008

8 a Measure the length of PQ.
    State the units with your answer.

  P ——————————————— Q

  b Measure the size of angle a.

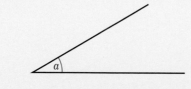

June 2007

9

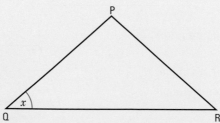

  a Measure the length, in centimetres, of QR.
  b Measure the size of angle x.   *March 2007*

10 Estimate the size of these angles.

a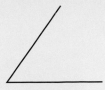

b

11 Draw and label these angles.

a ABC = 50°   b DEF = 170°   c GHI = 37°
d JKL = 90°   e MNO = 143°   f PQR = 77°

12

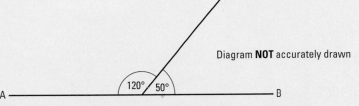

Diagram **NOT** accurately drawn

AB is a straight line.
This diagram is wrong. Explain why.

Nov 2008

13 Work out the missing angles in these triangles.

a 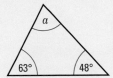   b   c

A school canteen wishes to introduce a range of healthy food options. It has a list of possible dishes that could be included on the menu. How could it find out which dishes students would prefer to eat? How could the students' choices be recorded?

◎ Objectives

In this chapter you will:
- consider the various stages of the statistical problem solving process, including how to collect, classify and interpret data
- learn how to collect and record data
- learn how samples are used and identify bias
- design and use tables of data.

◈ Before you start

You need to:
- be able to do simple arithmetic in your head
- be able to carry out simple measurements using rulers and weighing scales.

3.1 Introduction to data

Objectives

- You can understand the stages of an investigation.
- You can classify data as qualitative or quantitative.
- You can classify quantitative data as discrete or continuous.

Why do this?

Companies need to collect and use data for a variety of reasons. For example, a crisp manufacturer might want to find out which flavours people prefer.

Get Ready

1. How can you find the following information?
   a The average amount of dinner money for classmates.
   b What flights there were from Manchester to Washington D.C.
   c How many people voted for the Green Party in the last election.

Key Points

- **Statistics** is an area of mathematics concerned with collecting and **interpreting** information.
- The statistical investigation process will usually involve a number of stages:
  - specifying the problem
  - deciding what information to collect
  - collecting the information
  - presenting and displaying the information
  - interpreting the findings
  - drawing **conclusions**.

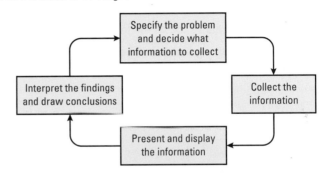

The table below shows some information about NHL ice-hockey players in the 2008–09 season.

Name	Shirt number	Team	Points scored	Age	Height (m)	Weight (kg)
Evengi Malkin	71	Pittsburgh Penguins	113	22	1.91	88.5
Alex Ovechkin	8	Washington Capitals	110	23	1.88	99.8
Sidney Crosby	87	Pittsburgh Penguins	103	21	1.80	90.7
Pavel Datsyuk	13	Detroit Red Wings	97	31	1.80	88.0
Zach Parise	9	New Jersey Devils	94	25	1.80	86.2
Ilya Kovalchuk	17	Atlanta Thrashers	91	26	1.85	104.3
Ryan Getzlaf	15	Anaheim Ducks	91	24	1.93	100.2

- **Qualitative** data is given in words such as the team name.
- **Quantitative** data is given as a numerical value such as height.
- **Discrete** data can only be a whole number, for example the shirt number.
- **Continuous** data can take any value, for example height.

Example 1 The following data has been taken from a survey of Year 11 pupils.
Decide whether each one is qualitative, discrete or continuous data.

a Method of travel to school   b Number of brothers and sisters
c Height   d Gender
e Weight

a Method of travel to school can be described in words so it is qualitative.
b Number of brothers and sisters can only take certain numerical values so it is discrete.
c Height is a number and can take any value so it is continuous.
d Gender can be described in words so it is qualitative.
e Weight is a number and can take any value so it is continuous.

Exercise 3A

Questions in this chapter are targeted at the grades indicated.

F

1 List the stages in a statistical investigation.

2 An estate agent collects the following four pieces of information about houses for sale:

Type of house   **Number of bedrooms**   **Garden size**   **Price**

a Which of these is qualitative data?   b Which is continuous data?
c Which is discrete data?

A03

3 James wants to find out how much time boys spend watching television compared to the amount of time girls spend watching television.
a Write down two pieces of information that James would need to collect.
b Describe the type of data James would need to collect.

3.2 Collecting data

◎ Objectives

- You can collect data by observation.
- You can design and use data collection sheets.
- You can understand and use a tally.
- You can group data into class intervals of equal width.

◈ Why do this?

People collect data in different ways. For example, a sales assistant might use observation to record what items customers purchase.

⬆ Get Ready

1. What are the numbers given by each set of tally marks?
a ||| b |||||| c ||||||||||

⟨🔍 **Key Points**⟩

◉ One way of collecting **data** is by observation: if you want to find out which types of vehicles travelled along a particular road, you could stand outside, observe and collect this information.

◉ A **data collection sheet** is used when **collecting data by observation**. The diagram shows a data sheet for recording the numbers of people who bought different vegetables at a supermarket.

Vegetable	Tally	Frequency
Beans	ЖⱮ |||	8
Cabbage	ЖⱮ ЖⱮ ||	12
Carrots	ЖⱮ ||||	9
Cauliflower	ЖⱮ	5
Onion	ЖⱮ ЖⱮ	10
Potato	ЖⱮ ЖⱮ ЖⱮ		16

◉ A **tally** mark is put next to beans each time a person buys them.
◉ The total number of people who bought beans is known as the **frequency**.
◉ The marks are grouped into fives with the fifth tally mark drawn through the other four.
◉ Putting tally marks in fives makes totalling up easier.

◉ If data is numerical, and widely spread, you can put it into groups of equal width known as **class intervals**. For example, class intervals for continuous data such as height will be of the form:

$150 \leqslant h < 160$ This means the **variable** h is greater than or equal to 150 but less than 160.

or $160 < h \leqslant 170$ This means the variable h is greater than 160 but less than or equal to 170.

◉ **Experiment**: if you wish to find out whether a coin lands an equal number of times on heads or tails, you could flip a coin a number of times and record the results.

◉ **Data logging**: if you want to monitor the temperature in a greenhouse, you could set up equipment to take temperature readings at set time **intervals**.

⟨🔍 **Example 2**⟩   The tally chart below shows the types of houses in which pupils in Class 10W live.

a Fill in the frequency column.
b Write down the type of house in which most of the pupils live.
c Work out how many pupils there are in Class 10W.

Type of house	Tally	Frequency
Detached	ЖⱮ ЖⱮ	
Semi-detached	ЖⱮ ЖⱮ |||	
Terraced	ЖⱮ		

a

Type of house	Tally	Frequency
Detached	ЖⱮ ЖⱮ	10
Semi-detached	ЖⱮ ЖⱮ |||	13
Terraced	ЖⱮ		6

Just add together the tallies 5 + 5 + 3 = 13.

b Semi-detached   ←   Find the Type with the highest frequency.

c 29   ←   Add together the frequencies 10 + 13 + 6 = 29.

class intervals   variable   experiment   data logging   intervals   **47**

Example 3 The examination marks for 30 students are shown below.
Draw and fill in a data collection sheet to show this information.

40	46	35	27	35	18
22	35	48	24	12	18
40	41	32	39	25	15
26	44	33	16	40	45
12	32	23	43	34	19

Mark	Tally	Frequency
11–20	|||| ||	7
21–30	|||| |	6
31–40	|||| |||| |	11
41–50	|||| |	6

The marks have been grouped together into four equal groups. The first group includes all the numbers between 11 and 20 inclusive. The groups do not overlap.

Exercise 3B

G

1 A car showroom has 30 cars. The colour of each car is shown below.

silver	white	blue	silver	black	red	red	black	silver	blue
silver	red	blue	black	red	silver	silver	red	black	blue
blue	silver	red	silver	red	red	silver	silver	silver	red

a Copy and complete the frequency table to show the colours of the cars.

b Write down the most popular colour.

Colour	Tally	Frequency
Silver		
White		
Blue		
Black		
Red		
Total		

2 A junior chess club has 30 members. Their ages are shown below.

14	15	16	13	14	16	13	14	16	16
13	14	16	16	15	15	14	15	13	14
15	15	15	14	15	16	16	16	16	15

a Copy and complete the frequency table to show the members' ages.

Age (years)	Tally	Frequency
13		
14		
15		
16		
Total		

b Write down the number of members that are 14 years old or less.

3 A scientist measured 24 worms. Their lengths, in centimetres, are shown below.

5	8	6	10	11	6
7	8	10	11	7	9
5	8	10	11	5	10
9	6	8	10	11	10

a Draw a table or chart to show the lengths of the worms.

b Write down the number of worms that are 5–6 centimetres long.

A02

3.3 Questionnaires

Objectives

- You can collect data using a questionnaire.
- You can criticise questions in a questionnaire.

Why do this?

Many organisations use questionnaires to collect data. For example, a youth centre may use one to find out whether it is providing the type of facilities that young people want.

Get Ready

1. Describe a good way to record data collected by observation.

Key Points

- A **questionnaire** is a list of questions designed to collect data. There are two types of question used on questionnaires.
  - An **open question** is one that has no suggested answers.
  - A **closed question** is one that has a set of answers to choose from. It is easier to summarise the data from this type of question.
- When designing a questionnaire, you need to make sure that possible answers are clear, do not overlap and cover all possibilities.

Example 4

Here is an example of part of a well-designed questionnaire.

1. Tick one box to indicate your age group.

☐ Under 21 ☐ 21 to 30 ☐ 31 to 40 ☐ 41 to 50 ☐ Over 50

These are response boxes. The categories do not overlap.

2. How often have you visited the dentist in the last 4 years? Tick one box.

☐ Never ☐ 1 or 2 times ☐ 3 or 4 times ☐ 5 or 6 times ☐ More than 6 times

This allows for other answers.

3. Do you agree or disagree that people who visit a dentist regularly have fewer fillings in their teeth?

☐ Agree ☐ Disagree

'Agree' or 'Disagree' makes the question unbiased.

Example 5
Here are the same questions but with a number of common errors.

1. Tick one box to indicate your age group.

☐ ☐ ☐ ☐
Under 20   20 to 30   31 to 40   40 to 50

> The categories overlap so 40-year-olds could tick two boxes.
> Other answers are not allowed for.
> Where does a 60-year-old tick?

2. How often have you visited the dentist in the last 4 years? Tick one box.

☐ ☐ ☐ ☐ ☐
Never   Seldom   Sometimes   Often   Very often

> It is difficult to decide what these words mean.

3. Do you agree that people who visit the dentist regularly have fewer fillings?

☐ ☐
Agree   Disagree

> By asking 'Do you agree...' you are inviting the answer 'Agree'. This is called a *biased* question.

Exercise 3C

1 Jenni is doing a survey on golf.
She writes the following question for a questionnaire.
'How often do you watch golf on TV each month?'

☐ 0–1 time   ☐ 1–2 times   ☐ 3–4 times

Write down one reason why this is a poor question.

2 A market research company intends to put the following question on a questionnaire.
'How old are you?'
Write down one reason why this is a poor question.

3 A town council asks the following question in a survey about council offices.
'Do you agree that the council should have new council offices?'

☐ Yes   ☐ Not sure

Write down two reasons why this is a poor question.

3.4 Sampling

Objectives

- You can collect information about a population by using a sample.
- You understand how different sample sizes may affect the reliability of conclusions drawn.
- You can identify possible sources of bias.

Why do this?

Market research companies always try to survey a representative sample of the population in order to ensure the accuracy of their data.

Get Ready

1. It takes 15 seconds to get an answer from one student.
How long should it take for you to ask everyone in your class?

Key Points

- Asking a select number of people their view is called taking a **sample**. It would be difficult to ask every person their view, so a sample is used to give information about the **population** as a whole.
- The sample must be unbiased.
- **Bias** occurs where:
  - the sample picked does not truly represent the population
  - the sample is too small.
- You need to make sure every member of the population has an equal **chance** of being picked. This is called **random sampling** and the sample will then not be biased.
- The size of a sample may vary. The larger the sample the more **representative** it is and the more accurate the information collected.

Example 6

A pollster wanted to find out who people would vote for in an election. He stopped the first 100 people he met in a shopping centre on a Saturday afternoon and asked them who they would vote for.
Explain what is wrong with this way of sampling.

This is biased because:
- Choosing the first 100 people is unlikely to give good representation of the population; for example, not all people shop in shopping centres.
- Sampling on a Saturday afternoon only may not include people who shop on other days.
- Choosing people who are shopping on Saturday afternoon may leave out people who watch or play sport at this time.

Exercise 3D

1 A supermarket manager is conducting a survey to find out how far people travel to the supermarket.
She is going to ask a sample of shoppers.
She decides to ask the first 10 people who enter the shop one Saturday morning.
What is wrong with this sample?

A03 — F

2 A head teacher decides to conduct a survey to find out how students feel about school uniform.
He asks the students in Class 2X.
What is wrong with using these students as a sample?

A03

* 3 A magazine editor wants to find out people's views about the magazine.
He organises a poll where 30 people are telephoned and asked their opinions.
Give reasons why you think this would not give a true picture of people's views.

A03 — E

---

sample   population   bias   chance   random sample   representative   **51**

3.5 Two-way and other tables

◎ Objectives

○ You can design and use two-way tables.
○ You can get data from lists and tables.
○ You can round numbers to a given degree of accuracy.

❷ Why do this?

Most organisations use databases to store and retrieve vital information.

❶ Get Ready

1. Each row has the same total and each column has the same total. Work out the values of A and B.

4	9	8	3	16
2	9	A	11	B
18	6	4	10	2

◉ Key Points

◉ A **two-way table** can be used to show how data falls into two different categories.

◉ For example, you may collect students' sex and whether they are right- or left-handed. The diagram below shows how you could record this in a two-way table.

This is the number of boys who are left-handed.

	Left-handed	Right-handed	Total
Boys	6	14	20
Girls	4	16	20
Total	10	30	40

This is the total number of children.

This is the total number of left-handed children.

This is the number of girls who are right-handed.

◉ If data has been collected by the person who is going to use it then it is called **primary data**.

◉ Data that has been collected by somebody else is known as **secondary data**. Secondary data is usually obtained from a **database**. This is data that has been collected and put together so that information can be quickly found.

🔍 Example 7

A teacher conducted a survey to find out what colour uniform students would prefer.
He gave students three possible colour choices.
The information below shows the results for girls and boys.

Girls

Green	Red	Green	Blue	Blue
Red	Green	Green	Red	Blue
Blue	Green	Red	Blue	Green
Green	Blue	Green	Green	Blue

Boys

Red	Red	Blue	Green	Blue
Blue	Blue	Green	Red	Red
Blue	Red	Blue	Red	Red
Green	Blue	Blue	Red	Red

a Display this information in a table.
b Write down the girls' top choice of colour.
c Write down the boys' top choice of colour.
d Write down the colour that was chosen by most of the students.

a

	Red	Blue	Green	Total
Girls	4	7	9	20
Boys	9	8	3	20
Total	13	15	12	40

The most suitable table is a two-way table.
Count up the number of girls who chose red and enter it here.
Do the same for the other colours and the boys' colours.

Total the rows and columns.

b Green ← Look for the highest number in the girls' row.

c Red ← Look for the highest number in the boys' row.

d Blue ← Look for the colour that has the highest total.

Example 8 The following two-way table gives information about types of housing on a new housing estate.

	Detached	Semi-detached	Terraced	Total
2 bedrooms	4	4		16
3 bedrooms	3		4	
4 bedrooms		1	1	4
Total	9	8		30

a Complete the table.

b Which type of house was the most common on the estate?

c Which types of housing were the least common?

ResultsPlus
Examiner's Tip

When completing a two-way table look for rows with only one number missing and fill these in first.
The numbers in each row must add up to the row total and the same goes for columns.

a

	Detached	Semi-detached	Terraced	Total
2 bedrooms	4	4	8	16
3 bedrooms	3	3	4	10
4 bedrooms	2	1	1	4
Total	9	8	13	30

The total number of 2-bedroom terraced houses = 16 − 4 − 4 = 8
The total number of 4-bedroom detached houses = 4 − 1 − 1 = 2
The total number of 3-bedroom semi-detached houses = 8 − 4 − 1 = 3
The total number 3-bedroom houses = 3 + 3 + 4 = 10
The total number of terraced houses = 30 − 8 − 9 = 13

b 2-bedroom terraced houses

c There was one 4-bedroom semi-detached house and one 4-bedroom terraced house.

Exercise 3E

1 A teacher is working out a timetable for Class 10B. Of the 30 students:

seven want to do Art and Music

twelve want to do Drama and PE

five want to do Music and PE

six want to do Drama and Art.

Copy and complete the two-way table below to show these data.

	Music	Drama	Total
Art			
PE			
Total			30

2 The two-way table gives some information about the numbers of different ice cream cornets sold at an ice cream van in one hour.

	Large	Small	Total
Vanilla	8	10	
Chocolate	6	4	
Total			

Copy and complete the table.

3 The following two-way table gives information about the numbers of different types of membership at a small health club.

	Junior	Senior	Family	Total
Full week	14	36	24	
Weekends	28	56	20	
Total				

a Copy and complete the table.

b Write down the number of weekend junior members.

4 The two-way table below gives information about the meals chosen by people visiting a restaurant.

	Pizza	Salad	Pasta	Total
Gateau	12	10		25
Ice Cream	10		20	40
Fruit	4	2		
Total				72

a Copy and complete the table.

b Write down the number of people who chose pasta and ice cream.

c Write down the total number of people represented by the table.

Example 9 The database below contains information about past population figures, in millions, for the United Kingdom. It also shows predicted figures for later years.

	1981	1991	2001	2005	2011	2021
England	46.8	47.9	49.4	50.4	52.0	54.6
Wales	2.8	2.9	2.9	3.0	3.0	3.2
Scotland	5.1	5.1	5.1	5.1	5.1	5.1
N. Ireland	1.5	1.6	1.7	1.7	1.8	1.8

a Write down the population of England in 2005.

b Write down what the population of Wales is expected to be in 2021.

c What happened to the population of Scotland between 1981 and 2005?

d What is expected to happen to the population of Northern Ireland between 2005 and 2021?

a 50.4 million ← Read off the figure from the intersection of the England row and the 2005 column.

b 3.2 million ← This is the intersection of the Wales row and the 2021 column.

c It stayed at 5.1 million. ← Look along the Scotland row between 1981 and 2005.

d It is expected to increase from 1.7 to 1.8 million. ← Look at what the figure for Northern Ireland was in 2005 and then at what it is predicted to be in 2021.

Exercise 3F

1 The following table provides information about the weather in Aspatria.

	Maximum temperature (degrees C)	Minimum temperature (degrees C)	Sunshine (hours)	Rainfall (mm)
January	6.4	1.2	44.3	101.9
April	11.4	3.5	155.1	50.7
July	19.2	11.2	195.9	68.2
October	12.9	6.6	98.6	110.9

a Write down the amount of rainfall in April.

b Write down the minimum temperature in January.

c Write down the month that has the most rainfall.

G

G

2 The table below provides information about planets.

Name	Number of moons	Rings	Temperature (°C)	Day length (hours)
Mercury	0	No	167	4222.6
Venus	0	No	457	2802.0
Earth	1	No	15	24.0
Mars	2	No	−63	24.6
Jupiter	63	Yes	−110	9.9
Saturn	60	Yes	−140	10.7
Uranus	27	Yes	−195	17.2
Neptune	13	Yes	−200	16.1

a How many moons has Jupiter?

b Write down the temperature of Neptune.

c Write down the planet that has the longest day.

d Write down the planet that has the highest temperature.

e How many planets have rings?

3 The following table shows information about the distance and travelling times of major cities from London.

From London to:	Distance (miles)	Air (minutes)	Train (minutes)	Coach (minutes)
Edinburgh	393	90	250	520
Glasgow	402	90	270	480
Inverness	568	105	430	720
Newcastle	270	70	170	360
Birmingham	114	50	95	160
Manchester	184	60	120	250

a Write down how long it takes to fly from London to Edinburgh.

b Write down the place that is 270 miles from London.

c What journey takes the least time if travelling by train from London?

d What journey takes the longest if travelling from London by coach?

Chapter review

- **Qualitative** data can be described in words.
- **Quantitative** data can be described using numerical values.
- **Discrete** data can only take certain numerical values.
- **Continuous** data can take any numerical value.
- When data is grouped the groups are known as **class intervals**.
- If **data** is numerical, and widely spread, you can group it into class intervals of equal width.

- A **questionnaire** is a list of questions designed to collect data.
- An **open question** is one that has no suggested answers.
- A **closed question** is one that has a set of answers to choose from.
- When designing a questionnaire, you need to make sure that possible answers are clear, do not overlap and cover all possibilities.
- A **sample** is part of a population that is used to give information about the **population** as a whole.
- A sample should be unbiased.
- **Bias** occurs where:
  - the sample picked does not truly represent the population
  - the sample is too small.
- A **two-way table** shows the frequency with which data falls into two different categories.
- **Primary data** is data collected by the person who is going to use it.
- **Secondary data** is data collected by somebody else.
- The data obtained from a **database** is secondary data.

Review exercise

1 Leanne asked each of her friends which one country they would most like to visit.
Here are her results.

| USA | France | Italy | USA | France | Australia | USA | Spain | France | Italy |
| Italy | USA | France | Italy | USA | USA | | Spain | USA | Spain | Italy |

a Copy and complete the frequency table.

Country	Tally	Frequency
Australia		
France		
Italy		
Spain		
USA		

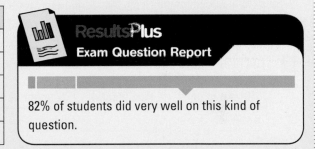

ResultsPlus
Exam Question Report

82% of students did very well on this kind of question.

b How many friends did Leanne ask?

June 2008

2 Tariq measured the lengths, in cm, of 18 books.
Here are his results.

14 13 16 15 14 14 17 13 14 16 14 15 15 17 13 15 14 16

a Complete the table to show Tariq's results.

Length (cm)	Tally	Frequency
13		
14		
15		
16		
17		

b Write down the number of books with a length of 16 cm.

March 2008

G

F

* **3** James wants to find out how many text messages people send.
He asks 10 students in his class to complete his questionnaire.
He uses this question on his questionnaire.
'How many text messages do you send?'

| 1 to 10 ☐ | 11 to 20 ☐ | 21 to 30 ☐ | more than 30 ☐ |

Write down what is wrong with this question.
Give reasons why this may not be a suitable sample.

March 2009 adapted

4 Nick has 6 coins.
Each coin comes from a different country.
Here is some information about these coins.

Coin	Country	Shape	Weight (g)
20 pence	United Kingdom	7-sided	5
500 yen	Japan	circular	7
10 centime	Switzerland	circular	3
1 dollar	Canada	11-sided	7
2 rupee	India	11-sided	6
5 cent	United States	circular	5

a Which coin comes from Switzerland?

b Which coin has the same weight as the 500 yen coin?

March 2009

5 The table shows some information about six cars.

Make of car	Age (years)	Number of doors	Engine size (litres)
BMW	7	4	2.2
Ford	5	3	1.4
Mazda	8	4	1.8
Skoda	5	5	1.4
Rover	8	5	1.4
Volvo	9	2	2.4

One of these cars has an engine size of 2.4 litres.
a Write down the make of this car.
One of these cars is 8 years old and has 4 doors.
b Write down the make of this car.

Nov 2008

6 Jason is collecting data about his school. He collects data about the following:
Number of school meals sold   Heights of the students
Colour of the walls in each classroom   Cost of school outings
Which of these is   i qualitative data   ii discrete data   iii continuous data?

7 Poppy wants to find out for how much time people use their computer.
She uses this question on a questionnaire and gives it to all the students in her class.

For how much time do you use your computer?	
0–1 hours ☐	3–4 hours ☐
1–2 hours ☐	4–5 hours ☐
2–3 hours ☐	5–6 hours ☐

Write down what is wrong with this question.
Is her sample biased? Explain why.   *Nov 2008 adapted*

8 Naomi wants to find out how often adults go to the cinema.
She uses this question on a questionnaire.

How many times do you go to the cinema?

☐   ☐   ☐
Not very often   Sometimes   A lot

Write down what is wrong with this question.
Use your answer to design a better question for her questionnaire.
You should include some response boxes.   *Nov 2008 adapted*

9 Valerie is the manager of a supermarket.
She wants to find out how often people shop at her supermarket.
She will use a questionnaire.
Design a suitable question for Valerie to use on her questionnaire.
You must include some response boxes.   *June 2008*

10 Yolande wants to collect information about the number of e-mails the students in her class send.
Design a suitable question she could use on a questionnaire.
You must include some response boxes.   *March 2008*

11 Melanie wants to find out how often people go to the cinema.
She gives a questionnaire to all the women leaving a cinema.
Her sample is biased.
Give two possible reasons why.   *March 2008*

12 Amberish is going to carry out a survey about zoo animals.
He decides to ask some people whether they prefer lions, tigers, elephants, monkeys or giraffes.
Design a data collection sheet that he can use to carry out his survey.   *March 2006*

13 Angela asked 20 people in which country they spent their last holiday.
Here are their answers.

France	Spain	Italy	England	Spain
England	France	Spain	Italy	France
England	Spain	Spain	Italy	Spain
France	England	Spain	France	Italy

Design and complete a suitable data collection sheet that Angela could have used to show this information.   *March 2004*

E

14 The table shows some information about the cost, in £s, of all-inclusive holidays to Dubai.

Hotel	Economy Class		Business Class	
	3 nights	5 nights	3 nights	5 nights
Metro	469	595	1219	1345
Habtoor	505	655	1255	1405
Hilton	509	659	1259	1409
Atlantis	659	925	1469	1735

Wing and his wife plan to go to the Hotel Habtoor for 3 nights travelling Business Class.

a How much will Wing have to pay?

A03

b How much could Wing save by travelling Economy Class?

D

15 A factory makes three sizes of bookcase.
The sizes are small, medium and large.
Each bookcase can be made from pine or oak or yew.
The two-way table shows some information about the number of bookcases the factory makes in one week.

	Small	Medium	Large	Total
Pine	7			23
Oak		16		34
Yew	3	8	2	13
Total	20		14	

Copy and complete the two-way table.

ResultsPlus
Exam Question Report

93% of students did very well on this kind of question.

Nov 2008

16 Ollie is standing in the school election, where all 240 students in the year vote for a student to join the school council.
He asks all the students in his class how they are going to vote.

a Why is this sample biased?

He then counts the first 30 votes and he has 10 votes.

b Estimate how many votes he will receive.

17 The Wildlife Trust are doing a survey into the number of field mice on a farm of size 240 acres.
They look at one field of size 6 acres.
In this field they count 35 field mice.

a Estimate how many field mice there are on the whole farm.

b Why might this be an unreliable estimate?

One of the most famous formulae you may come across is Einstein's $e = mc^2$ from his theory of relativity. Einstein's brain was removed after his death and researchers in Canada compared it with the brains of 91 people of average intelligence to try to discover the secret of his outstanding intelligence. They found that the area of Einstein's brain that is responsible for mathematical thought and spatial awareness was much larger and his brain was 15% wider than the others.

◎ Objectives

In this chapter you will:
- use letters instead of numbers
- learn the difference between variables, terms and expressions
- collect like terms
- multiply and divide variables and write them in their simplest form
- expand and factorise brackets
- learn the difference between expressions, equations and formulae.

◑ Before you start

You need to know:
- $d + d = 2d$
- $2g = 2 \times g$
- $5p - 2p = 3p$
- $4p - p = 3p$

4.1 Using letters to represent numbers

◎ Objective

○ You can use letters instead of numbers.

❓ Why do this?

You can use algebra to make a rule that will work for lots of different situations. The rule for time taken to get to a place when travelling on a motorway works for all towns and all motorways.

⬆ Get Ready

1. The total cost of  3 apples and 2 apples is the same as the cost of 5 apples.

If a is the cost of one apple, how could you write the cost of 5 apples?

$a + a + a + a + a =$

🌐 Key Point

◉ Letters can be used instead of numbers to fit all situations.

🔍 Example 1

Find these.

a $3a + 5a =$ b $6b - b =$

a $3a + 5a = 8a$

b $6b - b = 5b$

> a is the same as $1a$ so
> $a + a = 2a$ and $a + a + a = 3a$
> Remember that 6 bananas take away 1 banana would be 5 bananas!

⚙ Exercise 4A

Questions in this chapter are targeted at the grades indicated.

Find these.

1 $a + a + a + a =$

2 $a + a + a + a + a + a =$

3 $p + p + p =$

4 $6x - x =$

5 $9j - 3j =$

F

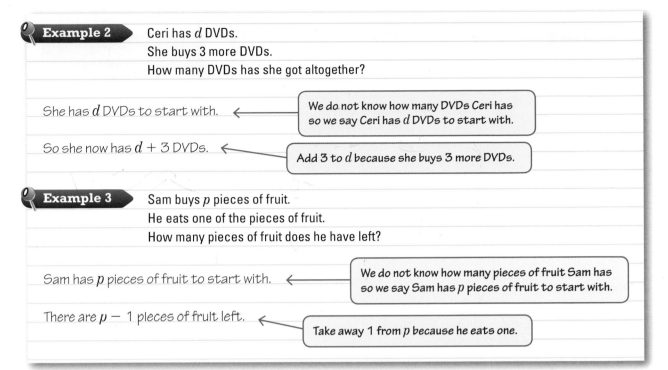

Example 2

Ceri has *d* DVDs.
She buys 3 more DVDs.
How many DVDs has she got altogether?

She has *d* DVDs to start with. ← We do not know how many DVDs Ceri has so we say Ceri has *d* DVDs to start with.

So she now has *d* + 3 DVDs. ← Add 3 to *d* because she buys 3 more DVDs.

Example 3

Sam buys *p* pieces of fruit.
He eats one of the pieces of fruit.
How many pieces of fruit does he have left?

Sam has *p* pieces of fruit to start with. ← We do not know how many pieces of fruit Sam has so we say Sam has *p* pieces of fruit to start with.

There are *p* − 1 pieces of fruit left. ← Take away 1 from *p* because he eats one.

Exercise 4B

1   Use algebra to write:
    a   3 more than *p*
    b   *x* plus 4
    c   *q* take away 5
    d   5 less than *g*
    e   4 more than *h*
    f   *k* minus 6
    g   *j* with 6 taken away
    h   *a* plus 3
    i   *y* minus 4
    j   *m* with 3 taken away
    k   *p* with 6 added
    l   7 together with *h*.

2   James had *c* CDs. He buys 12 more. How many CDs does James have now?

3   Terri had *a* apples. She eats 3 of them. How many apples does she have now?

4   Rashmi had *d* downloads on his MP3 player. He downloads 12 more.
    How many downloads has he altogether?

5   Hajra has *g* computer games. She sells 7 on the internet. How many computer games has she got now?

6   Helen and Robin go shopping. Robin buys *x* T-shirts and Helen buys *y* T-shirts.
    How many T-shirts do they have altogether?

Example 4

Rachel sells eggs.
She sells eggs in boxes of 6 or in boxes of 12.
One day she sells *s* boxes of 6 eggs and *t* boxes of 12 eggs.
How many eggs does she sell altogether?

You can write this as 6*s* and 12*t* so Rachel sold 6*s* + 12*t* eggs altogether. ← If Rachel sells *s* boxes with 6 eggs in them there will be 6 × *s* eggs in those *s* boxes and 12 × *t* eggs in the *t* boxes of 12 eggs.

F

Example 5

Rebecca sold c ice creams at 99p each and l lollipops at 75p each.

How much money, in pence, did she receive?

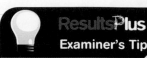

Examiner's Tip

If a question asks for the answer in pence you do not need to write p or pence and mess up your algebra.

The total is $99c + 75l$. ←

Rebecca gets $99 \times c$ for her ice creams and $75 \times l$ for her lollipops.
You can write this as $99c$ and $75l$.

Exercise 4C

1. Batteries are sold in packs of 4 and packs of 12.
   Harry buys f packs of 4 batteries and George buys t packs of 12 batteries.
   How many batteries do they buy altogether?

2. Naomi buys eggs in boxes of 4 and boxes of 10.
   One day she bought f boxes of 4 eggs and t boxes of 10 eggs.
   How many eggs did she buy altogether?

3. Moshe collects stickers in packs of 4 and packs of 9.
   One month he collected a packs of 4 stickers and b packs of 9 stickers.
   How many stickers did he collect altogether?

4. Richard sells rings. The gold rings cost £50 each and the silver rings cost £20 each.
   One day he sold g gold rings and s silver rings. How much money, in £, did he receive?

5. Jane packs boxes of chocolates.
   Plain chocolate boxes take 5 minutes to pack, milk chocolate boxes take 3 minutes and mixed boxes take 4 minutes.
   One day she packed p plain boxes, m milk boxes and n mixed boxes of chocolates.
   How much time did she spend packing chocolates on that day?

6. Nimer bakes ordinary cakes and iced cakes.
   Iced cakes are baked in batches of 6 and ordinary cakes are baked in batches of 12.
   One day, he baked x batches of iced cakes and y batches of ordinary cakes.
   Write down the total number of cakes he baked.

7. Lucy packs pencils in boxes of 12 and pens in boxes of 6.
   One week she packed x boxes of pencils and y boxes of pens.
   How many pens and pencils did she pack in total?

8. Cheryl sells flowers. She makes a profit of 90p on roses and a profit of 10p on daffodils.
   One day she sold r roses and d daffodils. How much profit, in pence, did she make on that day?

4.2 Understanding variables, terms and expressions

⊙ Objective

⊙ You know the difference between a variable, a term and an expression.

❓ Why do this?

You need to understand mathematical words to be able to understand questions in exams.

⬆ Get Ready

1. There were s tickets for the stands at a rugby match and b tickets for the boxes. How many tickets were there altogether?
2. There were h members of a hockey club at the start of the season. Twelve left at the end of the season. How many were still in the club at the end of the season?
3. Helena has r rabbits and g guinea pigs. How many pets does she have altogether?

🔑 Key Points

⊙ Variables, **expressions** and **terms** are the building blocks of **algebra**.
⊙ A variable is something that can change, e.g. speed, and is shown using a letter, e.g. a, b or c.
⊙ A term is a multiple of a letter that denotes a variable, for example $5a$, $6b$, c.
⊙ An expression is a collection of terms or variables, e.g. $5a + 6b - c$.

Example 6

Write down **a** the letters that are variables **b** the terms in this expression $2c + 5d$.

a c and d are the variables. ← | c and d can change value. |

b $2c$ and $5d$ are the terms.

⚙ Exercise 4D

1 Write down the letters that are the variables in these expressions.

a $3a + b$	**b** $x + 4y$	**c** $5a - 4t$	**d** $x - y$	**e** $2t - 5d$
f $2a - 5s$	**g** $4b - 6$	**h** $9g + 6$	**i** $5t + 7$	**j** $2a + 5b$

2 Write down the terms in these expressions.

a $3a + 4b$	**b** $x + 4y$	**c** $5a - 4t$	**d** $x - y$	**e** $2t - 5d$
f $2a - 5s + 8$	**g** $4b - 6h$	**h** $9g + 6r + 4$	**i** $5t + 7s - 3$	**j** $2a + 5b$

3 Write down the variables in these terms and expressions.

a $3a$	**b** $4y$	**c** $4t$	**d** x	**e** $5d$
f $2a - 5s$	**g** $4b - 6$	**h** $9g + 6$	**i** $5t + 7$	**j** $2a + 5b$

4 Use some of the terms in question 3 to make five new expressions.

5 Use some of the variables you identified in question 1 to make five new terms.

F

4.3 Collecting like terms

◎ Objectives

○ You can collect like terms when there is only one variable.
○ You can collect like terms when there is more than one variable.
○ You can collect like terms when there are numbers as well.
○ You can collect like terms when there are powers and numbers.

◈ Why do this?

Collecting like terms makes them easier to deal with.

◈ Get Ready

1. Find these.

   a 5 apples + 3 apples   b 10 bananas − 5 bananas − 1 pear   c 2 apples + 5 bananas + 4 apples

◉ Key Points

◉ Terms that use the same variable or letter or arrangement of letters are called **like terms**.
  a is the same as $1a$ so $a + a = 2a$ and $a + a + a = 3a$.
  Don't forget $2a - a$ is $2a - 1a$ or a.
  ◦ a and $3a$ are like terms.
  ◦ $5p$ and $8p$ are also like terms.

◉ You can add and subtract like terms to **simplify** expressions.

◉ Sometimes algebraic expressions have more than one term. You can make them simpler by collecting like terms together.

◉ Numbers will often be included as well as variables and terms. You treat these in exactly the same way as any other term.

◉ There may also be terms where variables are combined such as x^2, x^3 and ab.

◉ When you collect like terms you have to keep these more complicated terms together as well.

◉ It is also possible to have negative values when you collect like terms.

Example 7

Add or subtract these like terms to simplify the expressions.

     a $5p + 7p + p$ b $7a - 3a$

a $5p + 7p + p = 13p$
b $7a - 3a = 4a$

ResultsPlus
Examiner's Tip

'Simplify' means collect like terms.

⚙ Exercise 4E

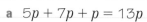

Simplify

F

1	$t + t + t + t + t$	2	$c + c$
3	$x + x + x + x$	4	$a + a + a$
5	$y + y + y + y + y + y$	6	$a + a + a + a + a + a + a + a$

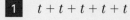

Exercise 4F

Simplify

1. $3a + 2a$
2. $5p - 3p$
3. $6s + 2s$
4. $6x - 3x$
5. $4b + 2b$
6. $8k - 3k$
7. $4a + 2a + 3a$
8. $5x + x$
9. $5b + 3b + b$
10. $2p - p$
11. $2n + 3n + 5n$
12. $2p + 5p - 3p$
13. $7x + 2x - 5x$

Example 8

Simplify $2a + 7b + 5a - 2b$.

$$2a + 7b + 5a - 2b$$

$2a + 5a + 7b - 2b$

Collect the terms in a together and collect the terms in b together.

$= 7a + 5b$

Combine the a terms together and the b terms together.

Watch Out!

Do not try to combine the as and the bs together when you are adding and taking away. They are different variables.

Exercise 4G

Simplify

1. $3a + 4b + 4a + 5b$
2. $6m + 5n + 3m + 2n$
3. $2p + 5q + 5p + 3q$
4. $7e + 2f - 5e$
5. $6g + 2h - 4g + 5h$
6. $8p - 6p + 7r - 2r$
7. $6j + 5k - 3j - 2k$
8. $7m + 8n - m$
9. $3a + 7b + 5a - 2b$
10. $7a + 8b - 4a - 5b$
11. $6m + 5n - 3m - 2n$
12. $2p + 5q + 5p - 3q$
13. $7e + 2f - e - f$
14. $6g + 8h - 4g - 5h$
15. $8p - 6r + 7r - 2p$
16. $6j + 5k + 3j - 2k$
17. $m + 8n - m$
18. $8a + 7b - 5a - 5b$
19. $8p - 2j - 5p + 5j$
20. $6p + 5t - 6p - t$
21. $6x + 4x - 3x - x$
22. $5y + 4b - 5y - 4b$
23. $k + 5g - k - g$
24. $4m - 6m + 2m + m$

Example 9 Simplify $2a + 7b + 9 + 5a - 2b - 6$. ← $2a + 7b + 9 + 5a - 2b - 6$

$2a + 5a + 7b - 2b + 9 - 6$

Collect the terms in a and in b together as before, then deal with the numbers on their own.

$= 7a + 5b + 3$

Combine the a terms together and the b terms together.

Exercise 4H

Simplify

1. $3a + 4 + 4a + 5$
2. $6m + 5 + 3m + 2$
3. $2 + 5p + 5 + 3q$
4. $7e + 2 - 5e$
5. $6 + 2h - 4 + 5h$
6. $6g + 8 - 4 - 5g$
7. $6j + 5 - 3j - 2$
8. $7m + 8 - m - 7$
9. $3a + 7b - 6 + 5a - 2b + 7$
10. $7a + 8 - 4a - 5 + c$
11. $6m + 5n - 3 + 12$
12. $2p + 9q - 8 + p - 3q + 10$
13. $5e + 2f - e - f + 2$
14. $8p + 6 - 5p + 7r - 3r - 2$
15. $8p - 6r - 7 + 7r - 7p + 7$

Example 10 Simplify $2a^2 + 5ab + 3a^2 - 3ab$.

$2a^2 + 3a^2 + 5ab - 3ab$ ← The like terms here are those in a^2 and those in ab.
$= 5a^2 + 2ab$

Exercise 4I

Simplify

1. $x^2 + x^2$
2. $3y^2 + 2y^2$
3. $7a^2 - 5a^2$
4. $3a^2 + 4b^3 + 4a^2 + 5b^3$
5. $6m^2 + 5n + 3m^2 + 2n$
6. $2p^3 + 5pq + 5p^3 - 3pq$
7. $7ef + 2ef - 5ef$
8. $6g^2 + 2h^3 - 4g^2 + 5h^3$
9. $8pq - 6pq + 7r^3 - 2r^2$
10. $6jk + 5jk - 3jk - 2jk$
11. $7m^3 + 8n - m^3$
12. $3a^2 + 7b^2 + 4a^2 - 3b^2$
13. $7a^3 + 8b^2 - 5a^3 - 5b^2$
14. $6m + 5n^2 - 4m - 2n^2$
15. $2pq + 5pq + 5p^3 - 3q^2$
16. $7e^3 + 2f^2 - e^3 - f^2$
17. $6gh + 8h^3 - 5gh - 6h^3$
18. $8pqr - 6pqr + 7pqr - 2pqr$

Example 11 Simplify $2a + 7 - 5a - 2$.

$$2a + 7 - 5a - 2$$

$2a - 5a + 7 - 2 \longleftarrow$ **Collect the terms in a together and collect the number terms together.**

$= -3a + 5 \longleftarrow$ **Combine the a terms together and the number terms together.**

Don't forget $2 - 5$ is -3. So $2a - 5a = -3a$.

Exercise 4J

Simplify

1 $3a - 6a$

2 $6m + 3n - 8m + 2n$

3 $2p + 2q - 5p - 6q$

4 $2e - 7e + 4 - 5$

5 $6g + 2h - 4g - 5h$

6 $8p^2 - 6p^2 + 3r^3 - 8r^3$

7 $3j + 5k - 3j - 8k$

8 $7m^3 - 8n - 8m^3$

9 $3a - 7b - 5a + 2b$

10 $7ab + ab - 9ab$

11 $6m + 5 - 8m - 7$

12 $2p + 5 - 5p - 11$

13 $7e + f - 6e - 2f$

14 $4g^2 + 3 - 6g^2 - 5$

15 $8p - 6r + 2 + 8r - 12p$

16 $6j + 5k \quad 8j - 2k$

17 $m - 8n - m - 5$

18 $8a - 3b - 5a - 5b$

19 $2p - 2j - 5p - 5j$

20 $6p - 5 - 6p - 3$

21 $x + 4 - 3 - 5x$

22 $5y^3 + 4 - 5y^3 - 4$

23 $k^2 + 5g^3 - k^2 - 8g^3$

24 $4mn - 6m^2 - 12mn + 8m^2$

D

4.4 Multiplying with numbers and letters

◉ Objective

○ You can multiply variables and write them in their simplest form.

◈ Why do this?

Multiplying with letters and numbers means you can solve problems such as how much carpet you would need for your bedroom.

◈ Get Ready

1. Simplify

a $2 \times 6 \times 3 \times 4$

b $4 \times 2 \times 5 \times 2$

c $2 \times 5 \times 2 \times 5 \times 8$

Key Points

- $2a$ can also be written as 2 lots of a or $2 \times a$.
- When you multiply terms and variables in algebra you can combine them by writing them next to each other.

So	$p \times q$	is written as	pq		
and	$a \times a$	is written as	aa	or	a^2 (a squared)
while	$a \times a \times a$	is written as	aaa	or	a^3 (a cubed).

Example 12   Simplify $p \times q \times r$.

$= pqr$ ←   Combine the p, q and r into pqr

Exercise 4K

Simplify

1 $a \times b$ **2** $x \times y$ **3** $b \times b$

4 $d \times d \times d$ **5** $r \times s \times t$ **6** $a \times b \times c$

7 $g \times g \times g$ **8** $2 \times e \times f$ **9** $3 \times j \times k$

10 $h \times h$ **11** $5 \times s \times s$ **12** $6 \times t \times t \times t$

13 $r \times t$ **14** $x \times y \times t$ **15** $3 \times m \times n$

16 $7 \times a \times b \times c$

Example 13   Simplify $2p \times 5t$.

$= 2 \times 5 \times p \times t$ ←   Combine the numbers 2×5.   Combine the variables $p \times t$.

$= 10pt$ ← $2 \times 5 = 10$ $p \times t = pt$.

ResultsPlus
Examiner's Tip

You combine the letters when you multiply.

Example 14   Simplify $2a \times 3a \times 4a$.

$= 2 \times 3 \times 4 \times a \times a \times a$ ←   Combine the numbers $2 \times 3 \times 4$.   Combine the variables $a \times a \times a$.

$= 24 \times aaa$ ← $2 \times 3 \times 4 = 24$

$= 24a^3$ ← $a \times a \times a = aaa = a^3$

Exercise 4L

Simplify

1	$2a \times 3b$	2	$4x \times 5y$	3	$2b \times 3b$
4	$2d \times 3d \times 2d$	5	$5r \times 7s$	6	$6b \times 2c$
7	$5g \times 3g$	8	$2e \times 7f$	9	$3j \times 8k$
10	$4h \times 5h$	11	$5s \times 5s$	12	$3t \times 2t \times 2t$
13	$6r \times 2t$	14	$5x \times 7y$	15	$3m \times 6n$
16	$5a \times 2b \times 3c$	17	$2g \times 2g$	18	$7h \times 7h$
19	$2x \times 2x \times 2x$	20	$5n \times 5n$	21	$2f \times 3g \times 5h$
22	$4j \times 6k$	23	$6h \times 6i$	24	$2a \times 2a \times b$

E

D

4.5 Dividing with numbers and letters

Objective

● You can divide variables and write them in their simplest form.

Why do this?

Simplifying expressions makes them easier to work with.

Get Ready

1. Write these fractions in their simplest form.

a $\frac{12}{4}$ b $\frac{12}{8}$ c $\frac{9}{12}$

Key Point

● Dividing algebraic expressions is like **cancelling** fractions.

Example 15   Simplify   a $10a \div 2$ b $\frac{20ab}{5b}$

a $10a \div 2$ ←

$10a \div 2$ is the same as $\frac{{}^{5}10a}{2_{1}}$

You can cancel the 2 into the 10 just as you would with fractions.

$= 5a$ ←

This leaves you with the answer $5a$ on the top.

b $\frac{20ab}{5b}$ ←

$\frac{20ab}{5b}$ is the same as $\frac{{}^{4}20 \times a \times b}{{}_{1}5 \times b}$

You can cancel the 5 into the 20 and cancel the b.

$= 4a$ ←

This leaves you with the answer $4a$ on the top.

Exercise 4M

Simplify

1	$12pq \div 3q$	**2**	$3p \div 3$	**3**	$4h \div h$	**4**	$12n \div 3$
5	$4t \div 2$	**6**	$15x \div x$	**7**	$24k \div 4$	**8**	$12ab \div 6b$
9	$\dfrac{8x}{4}$	**10**	$\dfrac{12p}{4p}$	**11**	$\dfrac{30xy}{6y}$	**12**	$\dfrac{8pq}{pq}$
13	$\dfrac{8pqr}{4qr}$	**14**	$\dfrac{8xy}{4xy}$	**15**	$\dfrac{24abc}{6ab}$	**16**	$\dfrac{8xy}{4xy}$

4.6 Expanding single brackets

Objective

○ You can expand a single bracket.

Get Ready

How many small blocks are there in these rectangles?

1.

2.

Key Point

◉ To **expand** a bracket, multiply everything inside the bracket by what is outside.

In question 1 above, the number of blocks could be written as either $2(a + 3)$ or $2 \times a + 2 \times 3$.

This means that $2(a + 3) = 2 \times a + 2 \times 3$.

The brackets have been expanded.

This section gives similar examples.

Example 16   Expand $2(a + 5)$.

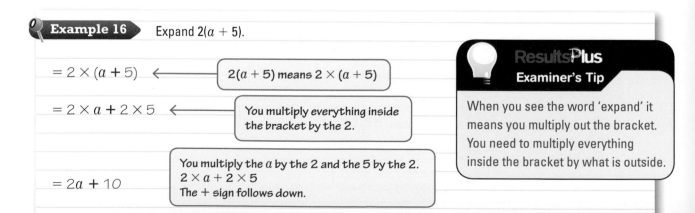

$= 2 \times (a + 5)$ ← $2(a + 5)$ means $2 \times (a + 5)$

$= 2 \times a + 2 \times 5$ ← You multiply everything inside the bracket by the 2.

$= 2a + 10$

You multiply the a by the 2 and the 5 by the 2.
$2 \times a + 2 \times 5$
The + sign follows down.

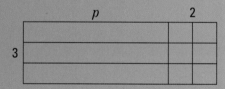

Results**Plus**
Examiner's Tip

When you see the word 'expand' it means you multiply out the bracket. You need to multiply everything inside the bracket by what is outside.

Exercise 4N

Expand

1	$2(a + 4)$	**2**	$3(b + 2)$	**3**	$4(c + 6)$	**4**	$5(a - 4)$
5	$3(b - 5)$	**6**	$5(x + 3)$	**7**	$2(y - 2)$	**8**	$6(n + 2)$
9	$3(5 + g)$	**10**	$2(5 - x)$	**11**	$3(2 - y)$	**12**	$5(4 - h)$
13	$10(a + 5)$	**14**	$3(g + 7)$	**15**	$4(s - 5)$	**16**	$3(7 - w)$

D

Example 17 Expand $2(3a + 6)$.

$= 2 \times (3a + 6)$ ← $2(3a + 6)$ means $2 \times (3a + 6)$

$= 2 \times 3a + 2 \times 6$ ← You multiply everything inside the bracket by the 2.

$= 6a + 12$ ← You multiply the $3a$ by the 2 and the 6 by the 2.
$2 \times 3a + 2 \times 6$
The $+$ sign follows down.

Exercise 4O

Expand

1	$2(3a + 4)$	**2**	$3(5b + 4)$	**3**	$4(5c + 6)$	**4**	$3(2a - 5)$
5	$3(5b - 7)$	**6**	$5(2x + 5)$	**7**	$2(3y - 4)$	**8**	$6(2n + 7)$
9	$3(5 + 3g)$	**10**	$2(5 - 2x)$	**11**	$3(2 - 5y)$	**12**	$5(4 - 3h)$
13	$10(4a + 3)$	**14**	$3(5g + 7)$	**15**	$4(3s - 5)$	**16**	$3(7 - 4w)$

D

Example 18 Expand $a(2a + 5)$.

$a \times (2a + 5)$ ← $a(2a + 5)$ means $a \times (2a + 5)$.
You multiply everything inside the bracket by the a.

$= a \times 2a + a \times 5$ ← You multiply the $2a$ by the a and the 5 by the a.

$= 2aa + 5a$ ← $2a \times a + a \times 5$

$= 2a^2 + 5a$ ← Don't forget $a \times a = aa = a^2$
The $+$ sign follows down.

D

Exercise 4P

Expand

1 $a(a + 4)$		**2** $b(b + 2)$		**3** $a(c + 6)$	
4 $a(2a - 4)$		**5** $b(b - 5)$		**6** $x(x + 3)$	
7 $y(2y - 2)$		**8** $n(3n + 2)$		**9** $g(5 + g)$	
10 $x(5 - 2x)$		**11** $y(2 - 3y)$		**12** $h(4 - 5h)$	
13 $2a(a + b)$		**14** $2g(g + 7)$		**15** $4s(s + t)$	
16 $3w(7 - w)$		**17** $5p(2p + 3)$		**18** $5x(3x - y)$	
19 $2h(g + 3h)$		**20** $5p(4 - 2p)$			

4.7 Factorising

◎ Objective

○ You can factorise an expression using a single bracket.

⬆ Get Ready

1. Expand

    a $3(7p - 4) + 5(10 - 3p)$ **b** $2(3 - 3s) + 4(5 - 2s)$ **c** $10(4 + 2g) + 3(5 - 5g)$

🌐 Key Points

◉ Multiplying out a bracket is called expanding a bracket.

◉ Putting in a bracket is called **factorising**.

📍 Example 19

Factorise $2a + 10$.

'Factorise' means take out factors that are in both terms.

ResultsPlus
Examiner's Tip

When you see the word 'factorise' it means you put the bracket back in.

$2 \times a + 2 \times 5$ ⟵

$2a = 2 \times a$ 2 and a are factors of $2a$.
$10 = 2 \times 5$ 2 and 5 are factors of 10.

$2(a + 5)$ ⟵

You take the 2 and put it outside the bracket. That leaves the a and the $+ 5$ inside the bracket.

D

Exercise 4Q

Factorise

1	$2a + 6$	2	$2n + 8$	3	$2a - 12$	4	$3k + 6$
5	$3f - 9$	6	$5p - 10$	7	$5r + 20$	8	$3x - 12$
9	$7w + 14$	10	$3m - 15$	11	$4q + 8$	12	$2s + 2$
13	$5a - 25$	14	$6x + 30$	15	$8p - 40$	16	$5y - 5$

Example 20   Factorise $x^2 - 3x$.

$= x \times x - 3 \times x$ ← $x^2 = x \times x$ $3x = 3 \times x$

$= x(x - 3)$ ← The x is in both terms so you take the x and put it outside the bracket. That leaves the other x and the $- 3$ inside the bracket.

Example 21   Factorise $y^2 + y$.

$= y \times y + 1 \times y$ ← $y^2 = y \times y$ $y = 1 \times y$

$= y(y + 1)$ ← The y is in both terms so you take the y and put it outside the bracket. That leaves the other y and the $+ 1$ inside the bracket.

Exercise 4R

Factorise

1	$a^2 + 2a$	2	$a^2 + 8a$	3	$y^3 + 2y$	4	$j^3 - 3j$
5	$s^2 - 9s$	6	$x^3 - 5x$	7	$p^2 + 6p$	8	$a^2 - a$
9	$p^3 + p$	10	$m^3 - m^2$	11	$c^3 + 8c$	12	$2a + a^2$
13	$x^3 - 2x^2$	14	$x^2 + 7x$	15	$p^3 - p$	16	$y^2 - 5y$

D

Example 22   Factorise completely $3x^2 - 6x$.

$= 3 \times x \times x - 3 \times 2 \times x$ ← $3x^2 = 3 \times x \times x$
$6x = 3 \times 2 \times x$

$= 3x(x - 2)$ ← The 3 and the x are in both terms so you take the $3x$ and put it outside the bracket. That leaves the other x and the $- 2$ inside the bracket.

ResultsPlus
Examiner's Tip

When you see 'factorise completely' it means there is more than one factor to take outside the bracket.

Exercise 4S

Factorise completely

1 $6a^2 + 2a$	**2** $3a^2 + 9a$	**3** $4y^3 + 2y$	**4** $6j^3 - 3j$
5 $3s^2 - 9s$	**6** $10x^3 - 5x$	**7** $3p^2 + 6p$	**8** $4a^2 - 2a$
9 $5p^3 + 10p$	**10** $6m^3 - 3m^2$	**11** $4c^3 + 8c$	**12** $12a + 6a^2$
13 $6x^3 - 2x^2$	**14** $5x^2 + 30x$	**15** $8p^3 - 4p$	**16** $25y^2 - 5y$

4.8 Understanding expressions, equations and formulae

Objective

○ You know the difference between an expression, an equation and a formula.

Why do this?

A formula will allow you to work out how long a journey will take at a given speed, or how much your savings will earn in interest over a given number of years.

Get Ready

1. Write three examples of expressions and identify the terms and variables in your expressions.

Key Points

◉ An expression is a collection of terms or variables.
$2x + 2y$ is an expression.

◉ **Equations** and **formulae** have equals signs in them.

◉ An equation is where it is possible to find one or more numerical values for a variable.
$2x + 1 = 7$ is an example of an equation. The value of x is 3.

◉ A formula is where one variable is equal to an expression in a different variable.
$P = 2l + 2w$ gives the perimeter of a rectangle of length l and width w. This is an example of a formula.

Example 23 Draw arrows to link the equation, expression and formula in this list.

Equation $A = \pi r^2$

Expression $3x + 4 = 10$

Formula $5r + 6$

Exercise 4T

State whether each of the following is an equation, expression or formula.

1 $y = mx + c$

2 $mx + c$

3 $3y + 1 = 10$

4 $2p + 6 = 5$

5 $F = ma$

6 $F + 3 = 10$

7 $6P + 2G$

8 $C = \pi D$

9 $a^2 + b^2 = c^2$

10 $C = \frac{9}{5}(F - 32)$

D

4.9 Replacing letters with numbers

◎ Objective

○ You can replace letters with numbers in expressions.

⦾ Why do this?

You might want to calculate the area of a square using the equation area = l^2 when $l = 3$ cm.

⬥ Get Ready

1. Work out

 a $(2 \times 3) + 5$

 b $(5 \times 2) - (3 \times 3)$

 c $(2 \times 5) + (4 \times 3) - 8$

Example 24   Find the value of
 a $3p$ when $p = 5$
 b $5x$ when $x = 4$
 c $2a + b$ when $a = 3$ and $b = 5$
 d $3p - 2q$ when $p = 2$ and $q = 5$
 e $2(x + y)$ when $x = 4$ and $y = 3$.

a $3p = 3 \times 5 = 15$ ⟵

> $3p$ means $3 \times p$ or $p + p + p$
> If $p = 5$ then $3p = 5 + 5 + 5$ or 3×5

b $5x = 5 \times 4 = 20$ ⟵

> $5x$ means $5 \times x$ or $x + x + x + x + x$
> If $x = 4$ then $5x = 5 \times 4$

c $2a + b = 2 \times 3 + 5 = 6 + 5 = 11$

d $3p - 2q = 3 \times 2 - 2 \times 5 = 6 - 10 = -4$

e $2(x + y) = 2(4 + 3) = 2 \times 7 = 14$

Exercise 4U

Find the value of these expressions when $a = 2$, $b = 5$ and $c = 3$.

1	$a + a$	2	$b + b + b$	3	$c + c + c + c$	4	$b - a$
5	$3a$	6	$5c$	7	$4b$	8	$5a$
9	$2a + b$	10	$3b + c$	11	$5c - a$	12	$2b - a$
13	$5c + 2a$	14	$2b - c$	15	$4b - 5a$	16	$4c - 3a$
17	$6a + 3b$	18	$a + b + c$	19	$2a + 4b - 8c$	20	$5c - 4c$

Exercise 4V

Find the value of these expressions when $p = 5$, $q = 3$ and $r = -2$.

1	$p + p$	2	$r + r + r$	3	$q + q + q + q$	4	$p - r$
5	$3r$	6	$5q$	7	$4p$	8	$5r + p$
9	$5p + q$	10	$3q + r$	11	$5q - p$	12	$2r - q$
13	$5p + 2q$	14	$2q - r$	15	$2p - 5r$	16	$r - 3q$
17	$2p - 3r$	18	$p + q + r$	19	$2p + 4q - r$	20	$2p - 5q$

Exercise 4W

Find the value of these expressions when $a = 2$, $b = 5$ and $c = 3$.

1	$2(a + b)$	2	$3(b + c)$	3	$4(a + c)$	4	$5(b - a)$
5	$3(a + 2b)$	6	$5(a + 2c)$	7	$4(c - b)$	8	$5(2a - b)$
9	$2(a + 2b)$	10	$3(4b + 2c)$	11	$5(c - a)$	12	$2(3b - 2a)$
13	$5(c + 2a)$	14	$2(2b - 3c)$	15	$4(5b - 3a)$	16	$4(2b - 3c)$
17	$6(a + 3b)$	18	$2(a + b + c)$	19	$2(a - 4b)$	20	$5(c - a - b)$

Chapter review

- Letters can be used instead of numbers to fit all situations.
- Variables, **expressions** and **terms** are the building blocks of **algebra**.
- A variable is something that can change, e.g. speed, and is shown using a letter, e.g. a, b or c.
- A term is a multiple of a letter that denotes a variable, for example $5a$, $6b$, c.
- An expression is a collection of terms or variables, e.g. $5a + 6b - c$.
- Terms that use the same variable or letter or arrangement of letters are called **like terms**.

- You can add and subtract like terms to **simplify** expressions.
- Sometimes algebraic expressions have more than one term. You can make them simpler by collecting like terms together.
- Numbers will often be included as well as variables and terms. You treat these in exactly the same way as any other term.
- There may also be terms where variables are combined such as x^2, x^3 and ab.
- When you collect like terms you have to keep these more complicated terms together as well.
- It is also possible to have negative values when you collect like terms.
- $2a$ can also be written as 2 lots of a or $2 \times a$.
- When you multiply terms and variables in algebra you can combine them by writing them next to each other.
- When you divide algebraic expressions you do it like **cancelling** fractions.
- To expand a bracket, multiply everything inside the bracket by what is outside.
- Multiplying out a bracket is called **expanding** a bracket.
- Putting in a bracket is called **factorising**.
- An expression is a collection of terms or variables.
- **Equations** and **formulae** have equals signs in them.
- An equation is where it is possible to find one or more numerical values for a variable.
- A formula is where one variable is equal to an expression in a different variable.

Review exercise

1 Callum has £3 more than Luke. Becky has twice as much as Callum.
Write down an expression for the total amount in pounds Callum, Luke and Becky have altogether.

A03   E

2 Here is a rod. Its length is x and its width is y.

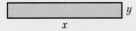

a Two of these rods are put with their widths alongside.

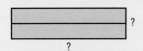

Copy and complete the diagram. Give your answer in its simplest form.

b Three of these rods are put together as shown in the diagram.

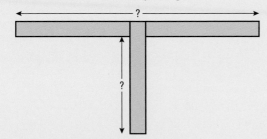

Copy and complete the diagram.
Give your answer in its simplest form.

A03   D

D

A03

3 In this set of rectangles, each expression in a rectangle is obtained by adding the two expressions immediately underneath.

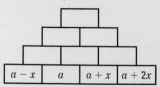

$$a - x \quad a \quad a + x \quad a + 2x$$

What expression should go in the top rectangle?

A03

4 **a** Simplify $13x - 17y - 3x - 3y$.
b Work out the value of $13 \times 99 - 17 \times 39 - 3 \times 99 - 3 \times 39$.

5 Simplify
a $c + c + c$ **b** $e + f + e + f + e$ **c** $2a + 3a$
d $2xy + 3xy - xy$ **e** $3a + 5b - a + 2b + 8$ *June 2006*

6 Simplify
a $5bc + 2bc - 4bc$ **b** $4x + 3y - 2x + 2y$ **c** $m \times m \times m$ **d** $3n \times 2p$ *Nov 2008*

7 Factorise $x^2 + 4x$ *June 2006*

8 Expand $4(3a - 7)$ *May 2008*

5 DECIMALS AND ROUNDING

In slalom racing only one skier can be on the slope at any time, so each racer is timed, as the winner can't be decided by watching them cross the finish line. The timing has to be very accurate because only fractions of a second separate competitors, so results are given to 2 decimal places.

◉ Objectives

In this chapter you will:
- ◉ work with decimals
- ◉ give values to a suitable degree of accuracy
- ◉ work out an approximate answer to a calculation quickly in your head.

◈ Before you start

You need to:
- ◉ know about digits and place value
- ◉ know what a digit is
- ◉ understand decimal places
- ◉ be able to do simple arithmetic in your head.

5.1 Understanding place value

Objective

○ You can use decimals to achieve greater accuracy than whole numbers can give.

Why do this?

You need to understand place value, including decimals, in order to carry out tasks such as shopping, measuring and timing.

Get Ready

1. Write out these numbers in words:
   a 273   b 4076   c 3753

Key Point

◉ In a decimal number, the decimal point separates the whole number from the part that is smaller than 1.

Example 1

A Formula One Grand Prix driver has his lap time recorded as 123.398 seconds.

Put 123.398 in a place value diagram.

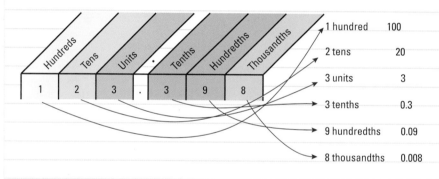

1 hundred	100
2 tens	20
3 units	3
3 tenths	0.3
9 hundredths	0.09
8 thousandths	0.008

You can better understand what 123.398 seconds means by drawing a decimal place value diagram.

Read the whole number and then read the decimal digits in order: one hundred and twenty-three point three nine eight.

Example 2

Write down the place value of the underlined digit in each number.

a 3<u>2</u>.8   b 0.38<u>5</u>   c 10.0<u>3</u>   d 4.2<u>9</u>0

a 2 units   b 5 thousandths   c 0 tenths   d 9 hundredths

Exercise 5A

Questions in this chapter are targeted at the grades indicated.

G

1 Draw a place value diagram like the one in Example 1 and write in these numbers.
   a 41.6   b 4.16   c 734.6   d 1.463
   e 0.643   f 1.005   g 5.01   h 0.086

2 What is the place value of the underlined digit in each number?
   a 2<u>5</u>.4   b 2.<u>5</u>4   c 25.4<u>6</u>   d 3.5<u>4</u>6
   e 1<u>8</u>.07   f 9.66<u>9</u>   g <u>2</u>16.031   h 2.135<u>7</u>
   i 9.1<u>0</u>2   j 3.<u>3</u>36   k 2.59<u>1</u>   l 0.02<u>7</u>

5.2 Writing decimal numbers in order of size

◎ Objective

◉ You understand the importance of place value in determining size.

❓ Why do this?

You need to order decimal numbers in order to identify the winner of a race.

◈ Get Ready

1. Write down the value of the 1 in each of these.
   a 0.102   b 0.418   c 1.002   d 10.07   e 42.001

🔍 Key Point

◉ To put decimal numbers in order of size, first compare the whole number parts, then the digits in the tenths place, then the digits in the hundredths place, and so on.

Example 3   Write these numbers in order of size, starting with the largest: 3.069, 5.2, 3.4, 3.08, 3.0901.

Step 1: Look at the whole-number parts.
        5 is bigger than 3, so 5.2 is the biggest number. | 3.069, 3.4, 3.08, 3.0901 remain unordered.
Step 2: Look at the tenths place.
        4 is bigger than 0 so 3.4 comes next. | 3.069, 3.08, 3.0901 remain unordered.
Step 3: Look at the hundredths place.
        Here the digits are 6, 8 and 9.
        The order is 5.2, 3.4, 3.0901, 3.08, 3.069.

⚙ Exercise 5B

1 The table gives the price of packets of dried apricots in different shops.

Shop	Stall	Corner	Market	Main	Store	Super
Price	£1.29	£1.18	£1.09	£1.31	£1.20	£1.13

Write the list of prices in order. Start with the lowest price.

2 Rearrange these decimal numbers in order of size. Start with the largest.
   a 0.62, 0.71, 0.68, 0.76, 0.9   b 3.4, 3.12, 3.75, 2.13, 2.09
   c 0.42, 0.065, 0.407, 0.3, 0.09   d 3.0, 6.52, 6.08, 3.58, 3.7
   e 0.06, 0.13, 0.009, 0.105, 0.024   f 2.09, 1.08, 2.2, 1.3, 1.16, 1.1087

3 The fastest lap times, in seconds, of six drivers were:
   Ascarina   53.072   Bertollini   53.207
   Rascini   52.037   Alloway   54.320
   Silverman   53.702   Killim   53.027
   Write down the drivers' times in order. Start with the fastest.

G

5.3 Adding and subtracting decimals

◉ Objective

○ You can add and subtract decimals in the same way as whole numbers.

❓ Why do this?

You add and subtract decimals when paying for your shopping.

⬆ Get Ready

1. Work out
   a 24 + 37   b 109 − 64

🔍 Key Point

◉ When adding and subtracting decimals you need the decimal points in line so that the place values match.

🔍 Example 4

Two children weigh 24.5 kg and 35.75 kg. What is their combined weight?

Combined weight is 24.5 kg + 35.75 kg.

Keep digits in their columns as in a place value diagram.

```
 24.5
35.75
```

← Put the decimal points under each other.

Then add:

```
  24.5
+ 35.75
  60.25
  1 1
```

← Decimal point in the answer should be in line.

⚙ Exercise 5C

Work these out, showing all your working.

1 1.5 + 4.6	**2** 3 + 0.25	**3** 26.7 + 42.2
4 25.7 + 0.32	**5** 0.1 + 0.9	**6** 16.1 + 2.625
7 9.9 + 9.9	**8** 10 + 1.001	**9** 0.005 + 1.909
10 117 + 1.17	**11** 6.3 + 17.2 + 8.47	**12** 13.08 + 9.3 + 6.33
13 0.612 + 3.81 + 14.7	**14** 8.6 + 3.66 + 6.066	**15** 7 + 3.842 + 0.222
16 23.43 + 5.36 + 2.216	**17** 3.07 + 12 + 0.0276	**18** 5.02 + 31.5 + 142.065

F

Example 5   Fiona buys a kettle costing £12.55. She pays with a £20 note.
How much change should she receive?

£20 − £12.55

$$\begin{array}{r} {}^{1}\cancel{2}{}^{9}\cancel{0}. {}^{9}\cancel{0}{}^{1}0 \\ -\ 1\ 2.\ 5\ 5 \\ \hline 7.\ 4\ 5 \end{array}$$

> You need to write 20 as 20.00.

She receives £7.45 in change.

> Shopkeepers often give change by counting on:
> £12.55 + £0.05 = £12.60
> £12.60 + £0.40 = £13.00
> £13.00 + £7   = £20.00
> Change is
> £7 + 40p + 5p = £7.45

Example 6   Bill earns £124.65 per week but needs to pay £33.40 in tax and national insurance.
How much does he take home?

£124.65 − £33.40

$$\begin{array}{r} \cancel{1}{}^{1}24.65 \\ -\ \ 33.40 \\ \hline 91.25 \end{array}$$

> Remember to put the decimal points under each other.

Bill takes home £91.25.

Exercise 5D

1   Work out these money calculations, showing all your working.

a £19.90 − £13.70	**b** £5.84 − £1.70	**c** £23.50 − £9.40
d £100.70 − £3.40	**e** £0.59 − £0.48	**f** £1 − £0.65
g £16.90 − £10.71	**h** £21.64 − £10.50	**i** £2.50 − £1.60
j £5.84 − £1.77	**k** £23.50 − £9.47	**l** £14 − £0.75

2   Work out these calculations, showing all your working.

a 6.125 − 4.9	**b** 14.01 − 2.361
c 3.29 − 1.036	**d** 204.06 − 35.48

F

5.4 **Multiplying decimals**

◉ **Objective**

● You can use the rule about the total number of decimal places in the answer.

◈ **Why do this?**

In the supermarket meat is weighed in kilograms and priced in pounds and pence. To work out the price of 1.5 kg you would use decimal multiplication.

◈ **Get Ready**

1. Work out
   a 464 × 4   **b** 857 × 25   **c** 68 × 42

 **Key Point**

◉ When multiplying, the total number of decimal places in the answer is the same as the total number of the decimal places in the question.

Example 7 ▶ Find the cost of 5 books at £4.64 each.

```
    464
×     5
  2320
   3 2
```

Multiply the numbers together, ignoring the decimals.

5 × 4.64

0 d.p. + 2 d.p. = 2 d.p.

Count the total number of decimal places (d.p.) in the numbers you are multiplying.

The answer must have 2 d.p.
So the cost is £23.20.

The answer must have the same number of decimal places.

Example 8 ▶ Work out 7.59 × 3.8.

```
    759
×    38
  6072
   4 7
 22,770
 28 842
    1
```

7.59 × 3.8

2 d.p. + 1 d.p. = 3 d.p.

The answer must have 3 d.p. so it is 28.842.

 Exercise 5E

Work these out, showing all your working.

G

1 Find the cost of:
  a 6 books at £2.25 each
  b 4 tins of biscuits at £1.37 each
  c 8 ice creams at £0.65 each
  d 1.5 kilos of pears at £0.80 per kilo.

2 Work out
  a 0.045 × 100
  b 0.45 × 100
  c 4.5 × 100
  d 0.0203 × 100
  e 0.203 × 100
  f 2.03 × 100

  What do you notice about your answers to question 2?

F

3 a 7.6 × 4
  b 0.76 × 4
  c 0.76 × 0.4
  d 2.25 × 5
  e 2.25 × 0.5
  f 0.225 × 0.5
  g 22.5 × 0.05
  h 2.25 × 0.005
  i 0.225 × 0.005

4
 a 6.42 × 10   b 64.2 × 10   c 0.642 × 10
 d 56.23 × 10   e 5.623 × 10   f 0.056 23 × 10
 Look carefully at your answers to question 4.
 What do you notice?

5 Work out
 a 24.6 × 7 kg   b 3.15 × 0.03 seconds   c 0.12 × 0.12 m
 d 0.2 × 0.2 miles   e 1.5 × 0.6 *l*   f 0.03 × 0.04 hours

6 A book costs £4.65. Work out the cost of buying:
 a 25 copies   b 36 copies   c 55 copies.

7 It costs £7.85 for one person to enter the Fun Beach. How much does it cost for:
 a 15 people   b 25 people   c 43 people?

8 A bucket holds 4.55 litres of water. How much water is contained in:
 a 15 buckets   b 25 buckets   c 65 buckets?

5.5 Squares and square roots, cubes and cube roots

◎ Objectives

⦿ You can extend your understanding of square and cube numbers.

⦿ You understand that finding the square root of a number is the opposite of squaring.

⦿ You understand that finding the cube root is the opposite of cubing.

⦵ Why do this?

Surveyors, engineers and architects use square and cube numbers in their jobs.

⬦ Get Ready

1. Work out:
 a 3 × 3   b 6 × 6   c 11 × 11   d 4 × 4 × 4

⬤ Key Points

⦿ **Squares** are the result of multiplying any number by itself.

⦿ 4 × 4 = 16, so we say that 4 is a **square root** of 16; it is a number which multiplied by itself gives 16. You can write the square root of 16 as $\sqrt{16}$.

⦿ Notice that −4 × −4 = 16, so −4 is also a square root of 16.

⦿ Finding a square root of a number is the opposite (inverse) of squaring. A square root of 64 (written $\sqrt{64}$) is 8, since $8^2 = 64$.

⦿ You need to know the squares of numbers up to 15 × 15 and their corresponding square roots.

⦿ **Cubes** come from multiplying any number by itself and then multiplying the result by the original number again.

⦿ You need to know the cubes of 2, 3, 4, 5 and 10.

⦿ 2 × 2 × 2 = 8, so we say that 2 is a **cube root** of 8; it is a number which multiplied by itself, then multiplied by itself again, gives 8. You can write the cube root of 8 as $\sqrt[3]{8}$.

⦿ Finding a cube root is the opposite (inverse) of cubing.

Example 9 Find: **a** the square of 2.4 **b** 1.2^3

a The square of 2.4 = 2.4 × 2.4 = 5.76

b 1.2^3 = 1.2 × 1.2 × 1.2 = 1.728

Exercise 5F

1 Work out the square of the following numbers.
  a 3.1 **b** 4.2 **c** 5.3 **d** 2.03 **e** 0.4

2 Work out the cube of the following numbers.
  a 1.5 **b** 2.5 **c** 3.2 **d** 0.2 **e** 0.5

5.6 Dividing decimals

Objective

• You can adjust decimal division so that you divide by a whole number.

Why do this?

You divide decimals whenever you work out how many calls at a certain price you can make for a £20 top up on your mobile phone.

Get Ready

1. Work out
  a 123 ÷ 3 **b** 585 ÷ 9 **c** 162 ÷ 27

Key Point

• To divide decimals, multiply both numbers by 10, 100, 1000 etc. until you are dividing with a whole number.

Example 10 Five friends win £216.35 in a charity lottery. They share the money equally. How much do they each get?

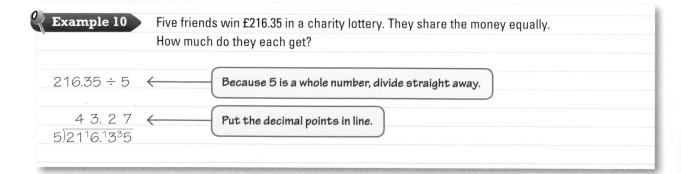

216.35 ÷ 5 ← Because 5 is a whole number, divide straight away.

$$\begin{array}{r} 4\ 3.\ 2\ 7 \\ 5\overline{)21^{1}6.^{1}3^{3}5} \end{array}$$ ← Put the decimal points in line.

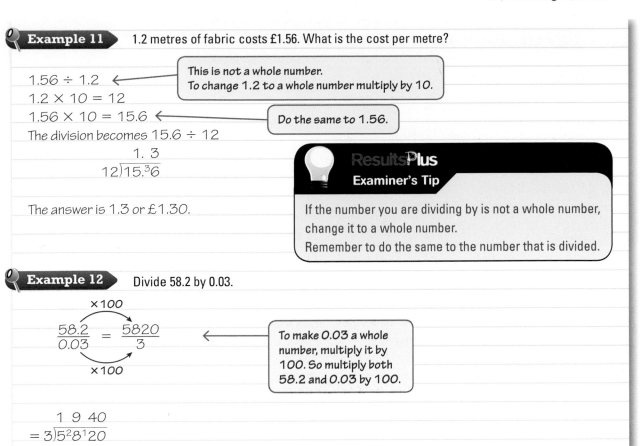

Example 11   1.2 metres of fabric costs £1.56. What is the cost per metre?

1.56 ÷ 1.2 ←

This is not a whole number.
To change 1.2 to a whole number multiply by 10.

1.2 × 10 = 12
1.56 × 10 = 15.6 ←

Do the same to 1.56.

The division becomes 15.6 ÷ 12

$$\begin{array}{r} 1.\,3 \\ 12\overline{)15.\!^36} \end{array}$$

The answer is 1.3 or £1.30.

ResultsPlus
Examiner's Tip

If the number you are dividing by is not a whole number, change it to a whole number.
Remember to do the same to the number that is divided.

Example 12   Divide 58.2 by 0.03.

×100

$$\frac{58.2}{0.03} = \frac{5820}{3}$$

×100

To make 0.03 a whole number, multiply it by 100. So multiply both 58.2 and 0.03 by 100.

$$= \begin{array}{r} 1\ 9\ 40 \\ 3\overline{)5\!^28\!^120} \end{array}$$

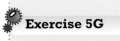

Exercise 5G

Work out questions 1–3.

1
   a 34.5 ÷ 10   b 3.45 ÷ 10   c 0.345 ÷ 10   d 2071 ÷ 10
   e 2.701 ÷ 10   f 0.2071 ÷ 10   g 65 ÷ 10   h 65 ÷ 100
   i 65 ÷ 1000

2
   a 64.48 ÷ 4   b 3.165 ÷ 5   c 133.56 ÷ 9   d 205.326 ÷ 6
   e 35.189 ÷ 7   f 0.0368 ÷ 8

3
   a 15 ÷ 2   b 23 ÷ 4   c 9 ÷ 8   d 3.5 ÷ 2
   e 14.4 ÷ 12   f 17 ÷ 20   g 310 ÷ 50   h 16.2 ÷ 9

4   Seven people share £107.80 equally. How much does each person get?

5   A 5 kilogram cheese is cut into 8 equal pieces. How much does each piece weigh?

6   Work out
   a 7.75 ÷ 0.5   b 7.92 ÷ 0.6   c 0.84 ÷ 0.04   d 7.7 ÷ 2.2
   e 6.634 ÷ 6.2   f 15.5 ÷ 2.5   g 1.242 ÷ 0.03   h 51.2 ÷ 1.6

G

F

E

5.7 Rounding decimal numbers

⊙ Objective

○ You can write numbers to a suitable degree of accuracy.

⑦ Why do this?

You can use rounding when adding up a bill in a restaurant to check that it is roughly correct.

⬆ Get Ready

1. How many decimal places are there in 2.0106?
2. How many digits are there in 2.0106?
3. Work out in your head: **a** 7 × 8   **b** 15 − 3 × 4

◔ Key Points

◉ To round a decimal to the nearest whole number, look at the digit in the tenths column (the first decimal place). If it is 5 or more, round the whole number up.

◉ To round a decimal to one decimal place (1 d.p.), look at the second decimal place. If it is 5 or more, round up the first decimal place. If it is less than 5, leave it and any further decimal places out.

◉ To round (or correct) to a given number of decimal places (d.p.), count that number of decimal places from the decimal point. Look at the next digit on. If it is 5 or more, you need to round up. Otherwise, leave off this digit and any that follow it.

Example 13   Round 7.815 to the nearest whole number.

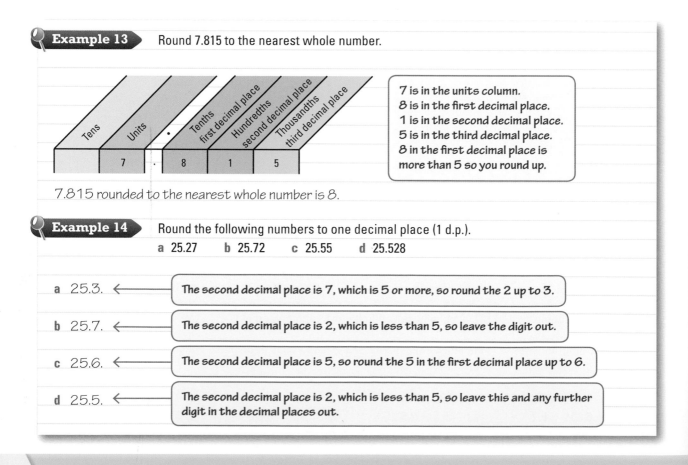

7 is in the units column.
8 is in the first decimal place.
1 is in the second decimal place.
5 is in the third decimal place.
8 in the first decimal place is more than 5 so you round up.

7.815 rounded to the nearest whole number is 8.

Example 14   Round the following numbers to one decimal place (1 d.p.).
a 25.27   **b** 25.72   **c** 25.55   **d** 25.528

a 25.3. ⟵   The second decimal place is 7, which is 5 or more, so round the 2 up to 3.

b 25.7. ⟵   The second decimal place is 2, which is less than 5, so leave the digit out.

c 25.6. ⟵   The second decimal place is 5, so round the 5 in the first decimal place up to 6.

d 25.5. ⟵   The second decimal place is 2, which is less than 5, so leave this and any further digit in the decimal places out.

Exercise 5H

1 Round these numbers to the nearest whole number.

a 7.8	b 13.29	c 14.361	d 5.802
e 10.59	f 19.62	g 0.771	h 20.499
i 0.89	j 100.09	k 19.55	l 1.99

2 Round these numbers to one decimal place (1 d.p.).

a 3.6061	b 5.3391	c 0.0901	d 9.347
e 10.6515	f 7.989	g 2.0616	h 0.4999
i 2.45	j 125.67	k 0.05	l 9.890

3 Round:

a 13.6 mm to the nearest mm	b 80.09 m to the nearest m
c 0.907 kg to the nearest kg	d £204.49 to the nearest £
e 3.601 lb to the nearest lb	f 0.299 tonne to the nearest tonne
g 10.5001 g to the nearest gram	h 8.066 min to the nearest minute.

Example 15 Round these numbers i to 3 d.p. ii to 2 d.p.

a 4.4315 b 7.3962

a i 4.4315 rounded to 3 d.p. is 4.432.

In the number 4.4315 the next digit after the 3rd d.p. is 5, so round up and the 1 becomes 2.

ii 4.4315 rounded to 2 d.p. is 4.43.

In the number 4.4315 the next digit after the 2nd d.p. is 1, so you can round down and the 3 remains the same.

b i 7.3962 to 3 d.p. is 7.396

ii 7.3962 to 2 d.p. is 7.40

The 6 makes the 9 round up to 10 and this changes the 3 to a 4.

ResultsPlus

Examiner's Tip

The final zero is important because 2 d.p. means that two decimal digits need to be shown.
e.g. The number 5.4926 would be written 5.50 to 2 d.p.

Exercise 5I

In questions 1 to 4 round the numbers: i to 3 d.p. ii to 2 d.p.

1	a 4.2264	b 9.7868	c 0.4157	d 0.058 38
2	a 10.5167	b 7.5034	c 21.7295	d 9.088 95
3	a 15.5978	b 0.4081	c 7.2466	d 6.050 77
4	a 29.1582 cm	b 0.054 86 kg	c 13.3785 km	d £5.9976

5 Round each number to the number of decimal places given in brackets.

a 5.6166 (3 d.p.)	b 0.0112 (1 d.p.)	c 0.923 98 (4 d.p.)
d 0.8639 (1 d.p.)	e 9.6619 (1 d.p.)	f 1.0076 (2 d.p.)

5.8 Rounding to 1 significant figure

Objective

You can write amounts to a suitable degree of accuracy using significant figures.

Why do this?

You can understand an approximate answer more easily. A local newspaper reported that 50 000 people had attended a rock concert. The true number was 47 231 but 50 000 made a better story and was easier to understand.

Get Ready

1. **a** Is 22 nearer to 20 or 30? **b** Is 46 nearer to 40 or 50? **c** Is 155 nearer to 100 or 200?

Key Point

To write a number to 1 **significant figure** (1 s.f.) look at the place value of the first (highest valued) non-zero digit and round the number to this place value.

Example 16 Write these numbers to 1 significant figure (1 s.f.).

a 32 **b** 452 **c** 0.0878

a 32 to 1 significant figure is 30.

The first digit is in the tens column, so you need to round to the nearest ten. 32 to the nearest ten is 30.

b 452 to 1 significant figure is 500.

The first digit is in the hundreds column, so round to the nearest hundred.

c 0.0878 to 1 significant figure is 0.09.

The first digit is in the hundredths column, so round to the nearest hundredth.

Exercise 5J

1 Write down these numbers to 1 significant figure (1 s.f.).

a 41 **b** 709 **c** 287 **d** 0.348 **e** 21 899

f 0.007 41 **g** 973 **h** 4.6 **i** 13.309 **j** 19.07

2 England won 110 medals in the 2006 Commonwealth Games. Write this number to 1 significant figure (1 s.f.).

3 The number of spectators at a rugby match final was 34 862. Write this number to 1 s.f.

5.9 Rounding to a given number of significant figures

◎ Objective

○ You can write numbers to the degree of accuracy that is sensible.

❓ Why do this?

The exact value is not always suitable for ordinary use. For example, at 3 pm on 22 April 2010 the exchange rates were £1 = $1.53889

£1 = 1.15075 euros

In real money you cannot have more accuracy than $1.54 which is 1 dollar and 54 cents.

◈ Get Ready

1. Write these average attendances for football clubs (2007/8 season) to 1 s.f.

   Manchester United 75 690   Chelsea 41 395   Real Madrid 76 200
   Newcastle United 51 320   Liverpool 43 530   Barcelona 67 560

Key Point

○ To round numbers to a given number of significant figures, you count that number of digits from the first non-zero digit. If the next digit is 5 or more then you round up.

Example 17   Round 642.803   **a** to 1 s.f.   **b** to 2 s.f.   **c** to 3 s.f.   **d** to 4 s.f.   **e** to 5 s.f.

You need the zeros to show the place value of the 6.

1 s.f.	2 s.f.	3 s.f.	.	4 s.f.	5 s.f.	6 s.f.								
6	4	2	.	8	0	3	=	6	0	0				(to 1 s.f.)
6	4	2	.	8	0	3	=	6	4	0				(to 2 s.f.)
6	4	2	.	8	0	3	=	6	4	3				(to 3 s.f.)
6	4	2	.	8	0	3	=	6	4	2	.	8		(to 4 s.f.)
6	4	2	.	8	0	3	=	6	4	2	.	8	0	(to 5 s.f.)

You need the zero to show 5 significant figures.

⚙ Exercise 5K

In questions **1** to **5** round the numbers to: **i** 2 significant figures   **ii** 3 significant figures.

1	**a** 0.061 78	**b** 0.1649	**c** 96.303	**d** 41.475
2	**a** 734.56	**b** 0.079 47	**c** 5.6853	**d** 586.47
3	**a** 0.014 84	**b** 2222.8	**c** 76.249	**d** 0.3798
4	**a** 8.3846	**b** 35.959	**c** 187.418	**d** 0.066 63
5	**a** 218 736	**b** 3 989 375	**c** 307 096	**d** 25 555

6 The exchange rate was £1 = $1.6071 on 29 May 2009. Write £1.6071 to 3 significant figures.

7 The fastest lap time in a motor race Grand Prix was 83.345 seconds. Write this time to 3 significant figures.

E

5.10 Estimating

Objective

○ You can get a rough idea of the answer by working with each of the numbers to 1 significant figure.

Why do this?

When you shop for food, having a rough idea of how much you are spending means you can keep to a budget.

Get Ready

For each of the following, round the numbers to 1 s.f., then add them.

1. $338 + 286$ **2.** $711 + 479$ **3.** $0.543 + 0.265$

Key Point

○ To get an **estimate** for an answer, you first round each number to 1 significant figure. Then you can usually do the calculation in your head.

Example 18   Estimate the answer to $289 \times \dfrac{96}{184}$

Rounding each of the numbers to 1 significant figure gives

$300 \times \dfrac{100}{200}$

This works out as 150, which is a suitable estimate.

Exercise 5L

D

1   Showing your rounding, work out estimates for:

   a $65 \times \dfrac{57}{31}$ **b** $\dfrac{206 \times 311}{154}$ **c** $\dfrac{9 \times 31 \times 97}{304}$

   d $\dfrac{200}{12 \times 99}$ **e** $\dfrac{498}{11 \times 51}$ **f** $\dfrac{103 \times 87}{21 \times 32}$

2   A football grandstand has 48 rows of seats.
Each row has 102 seats.
Work out an estimate for the total number of seats.

3   A carton of tinned peaches contains 48 tins.
Work out an estimate for the total number of tins in 73 cartons.

4   Hazel is buying 28 paving stones.
Each paving stone costs £4.85.
Work out an estimate for her total cost.

C

5   Work out estimates for each of the following calculations.

   a $17.3 \times \dfrac{0.21}{4.1}$ **b** $5.67 \times \dfrac{27.8}{0.86}$ **c** $\dfrac{873}{23.1} \times 0.476$

5.11 Manipulating decimals

⊙ Objective

⊙ You can use one calculation to find the answer to another.

⍰ Why do this?

If you knew that $1.50 was worth £1 then you could use this to work out how much 10p was worth.

⬥ Get Ready

1. Work out **a** 20×3 **b** 200×3 **c** 2000×3
2. Work out **a** $300 \div 10$ **b** $30 \div 10$ **c** $3 \div 10$

🔍 Key Point

⊙ You can use the answer from one calculation to help you find the answer to a second calculation.

Example 19

Given that $3.8 \times 5.2 = 19.76$, find the values of each of the following.

   a 38×5.2 **b** 380×0.52

a 38×5.2
$= 3.8 \times 10 \times 5.2$ ← $38 = 3.8 \times 10$
$= (3.8 \times 5.2) \times 10$ ← Rearrange the terms and substitute the known answer.
$= 19.76 \times 10$
$= 197.6$ ←   You can check the answer by estimating. 38×5.2 is roughly 40×5, which is 200.

b $380 \times 0.52 = 3.8 \times 100 \times 5.2 \div 10$ ← $380 = 3.8 \times 100$ and $0.52 = 5.2 \div 10$
$= (3.8 \times 5.2) \times 100 \div 10$ ←   Rearrange the terms and substitute the known answer.
$= 19.76 \times 10$
$= 197.6$ ←   You can check the answer by estimating. 380×0.52 is roughly 400×0.5, which is 200.

Example 20

Given that $\dfrac{40.8}{8.5} = 4.8$, find the value of each of the following.

   a $\dfrac{408}{8.5}$ **b** $\dfrac{40.8}{85}$

a $\dfrac{408}{8.5}$
$= \dfrac{40.8 \times 10}{8.5}$ ← $408 = 40.8 \times 10$
     ← Rearrange the terms and substitute the known answer.
$= \left(\dfrac{40.8}{8.5}\right) \times 10$ ←   You can check the answer by estimating. $408 \div 8.5$ is roughly $400 \div 8$, which is 50.
$= 4.8 \times 10$
$= 48$ ← $85 = 8.5 \times 10$

b $\dfrac{40.8}{85}$
$= \dfrac{40.8}{8.5 \times 10}$ ←   Rearrange the terms and substitute the known answer. Multiplying the bottom number by 10 is the same as dividing the top number by 10.
$= \left(\dfrac{40.8}{8.5}\right) \div 10$ ←   You can check the answer by estimating. $40.8 \div 85$ is roughly $40 \div 80$, which is 0.5.
$= 4.8 \div 10$
$= 0.48$ ←

Exercise 5M

1 Given that $6.4 \times 2.8 = 17.92$, work out

   a 64×28 **b** 640×2.8 **c** 0.64×28

2 Given that $\dfrac{18.3}{1.25} = 14.64$, work out

   a $\dfrac{183}{1.25}$ **b** $\dfrac{1.83}{1.25}$ **c** $\dfrac{0.183}{1.25}$

3 Given that $13.2 \times 5.5 = 72.6$, work out

   a 132×5.5 **b** 1.32×0.55 **c** 0.132×55

4 Given that $\dfrac{30.4}{4.75} = 6.4$, work out

   a $\dfrac{30.4}{47.5}$ **b** $\dfrac{3.04}{4.75}$ **c** $\dfrac{304}{4.75}$

Chapter review

- In a decimal number, the decimal point separates the whole number from the part that is smaller than 1.
- To put decimal numbers in order of size, first compare the whole number parts, then the digits in the tenths place, then the digits in the hundredths place, and so on.
- When adding and subtracting decimals, you need the decimal points in line so that the place values match.
- When multiplying, the total number of decimal places in the answer is the same as the sum of the decimal places in the question.
- **Squares** are the result of multiplying any number by itself.
- Finding a **square root** of a number is the opposite (inverse) of squaring.
- **Cubes** come from multiplying any number by itself and then multiplying the result by the original number again.
- Finding a **cube root** is the inverse of cubing.
- To divide decimals, multiply both numbers by 10, 100, 1000 etc. until you are dividing with a whole number.
- To round a decimal to the nearest whole number, look at the digit in the tenths column (the first decimal place). If it is 5 or more, round the whole number up.
- To round a decimal to one decimal place (1 d.p.), look at the second decimal place. If it is 5 or more, round up the first decimal place. If it is less than 5, leave it and any further decimal places out.
- To round (or correct) to a given number of decimal places (d.p.), count that number of decimal places from the decimal point. Look at the next digit on. If it is 5 or more, you need to round up. Otherwise, leave off this digit and any that follow it.
- To write a number to 1 **significant figure**, look at the place value of the first (highest valued) non-zero digit and round the number to this place value.
- To round numbers to a given number of significant figures, you count that number of digits from the first non-zero digit. If the next digit is 5 or more then you round up.
- To get an **estimate** for an answer, you first round each number to 1 significant figure. Then you can usually do the calculation in your head.
- You can use the answer from one calculation to help you find the answer to a second calculation.

Review exercise

1 Put these decimal numbers in order of size. Start with the smallest.
  a 4.85, 5.9, 5.16, 4.09, 5.23
  b 0.34, 0.07, 0.37, 0.021, 0.4
  c 5, 7.23, 5.01, 7.07, 5.007
  d 1.001, 0.23, 1.08, 1.14, 0.06

2 A new cereal gives these weights of vitamins and minerals per 100 g.

Fibre	1.5 g	Iron	0.014 g	Vitamin B6	0.002 g
Thiamin B1	0.0014 g	Riboflavin	0.0015 g	Sodium	0.02 g

Write down these weights in order. Start with the lowest.

3 Work out
  a 3.4 + 5.1   b 12.3 + 6.27   c 0.046 + 0.0712
  d 5.68 + 3.093 + 2.3702   e 75.3 − 16.9   f 20.3 − 4.72
  g 50 − 3.6   h 0.03 − 0.0182

4 Diana is packing to go on holiday. The baggage allowance is 20 kg.
Her suitcase weighs 2.6 kg. In it she packs her clothes weighing 11.3 kg, her shoes weighing 3.7 kg and her toiletries weighing 2.3 kg. Is her packed suitcase within the 20 kg limit?

5 An empty container weighs 27.1 kg.
When filled, it weighs 238.7 kg.
What is the weight of the contents?

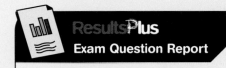

ResultsPlus
Exam Question Report

86% of students answered part **d** of this question poorly. The most common error using a long multiplication method was to add 1536 and 256.

6 Work out:
  a 2.34 × 5   b 0.24 × 6
  c 0.3 × 0.4   d 25.6 × 1.6
  e 15.3 ÷ 3   f 81.4 ÷ 4

7 In 2008, 355 024 candidates took Edexcel GCSE Mathematics.
Write the number of candidates to 2 significant figures.

8
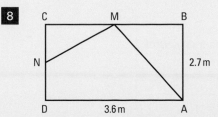

N is the midpoint of CD. M is the midpoint of CB.
Work out the difference in length between N to C to M and M to B to A.

9 Copy and complete the numbers in the boxes.
  a 6.24 × 100 = 0.624 × ☐   b 5.08 ÷ 10 = 0.508 × ☐
  c 0.455 ÷ 100 = 4.55 ÷ ☐   d 1.52 × 1000 = 152 ÷ ☐

F

10

```
          Rob's Café Price List
Cup of tea      75p   Roll        £1.70
Cup of coffee   85p   Sandwich    £1.35
Can of cola     75p
```

Joe buys a can of cola and a roll at Rob's café.

a Work out the total cost.

Susan buys two cups of tea and one sandwich.

b Work out the total cost.

Kim buys a cup of coffee and a roll.
She pays with a £5 note.

c How much change should she get?

ResultsPlus
Exam Question Report

82% of students did very well on this type of question. They were awarded method marks for explaining clearly how their answer had been achieved.

Specimen paper 2009

A03

11 Josh has two parcels of weights 2.8 kg and 1.35 kg on a trolley. The greatest total weight the trolley can carry is 5 kg. Work out the largest weight of parcel that Josh could add to the trolley.

E

12 A car travels 17.2 kilometres on 1 litre of fuel. How far will it travel on 8.5 litres of fuel?

A03

13 Angela has £15.76.
She buys as many bottles of drink costing £1.20 each as she can.
How many does she buy and how much money does she have left?

14 a $51.3 \div 0.9$ b $0.0412 \div 0.4$ c $30 \div 0.05$

15 Round these numbers to 1 decimal place.
a 23.48   b 1.7502   c 0.3479   d 150.03

16 Round these numbers to the number of decimal places given in brackets.
a 7.263 (2)   b 73.0448 (2)   c 0.041 68 (3)   d 0.7208 (3)

17 Round these numbers to the number of significant figures given in brackets.
a 8317 (2)   b 20 056 (3)   c 0.546 72 (1)   d 20.873 (3)

A02

18 Work out an estimate for the total cost of 36 books costing £7.97 each.

A02

19 A packet of 18 slices of bacon costs £5.80.
Work out an estimate for the cost of each slice of bacon.

A02

20 Here are the rates of pay in a company.

Grade	Basic Pay for an hour's work	Overtime pay for an hour's work
Operative	£5.40	£8.10
Technician	£7.50	£11.25
Supervisor	£9.00	£13.50
Driver	£7.20	£10.80

Lily has a part-time job as an operative.
Last week Lily earned basic pay for 24 hours and overtime pay for 3 hours.

a Work out Lily's total pay for last week.

b If Lily had been paid as a technician, work out how much extra pay she would have received.

June 2008 adapted

21 Sam earns £5.95 for each hour that he works. The table shows the hours he worked one week. Work out an estimate for the amount of money that Sam earned that week.

Day	1	2	3	4	5
Hours worked	5	6	6	5	7

22 For each of these calculations, work out an estimated answer.

a $\dfrac{823 \times 4872}{3261}$

b $\dfrac{3.6 \times 4.5}{9.8}$

c $\dfrac{2.4 \times 7.9}{3.9 \times 2.3}$

Exam Question Report

89% of students answered this type of question poorly.

23 Using the information that $4.8 \times 34 = 163.2$ write down the value of

a 48×34 b 4.8×3.4 c $163.2 \div 48$ *June 2008*

24 Use the information that $322 \times 48 = 15\,456$ to find the value of

a 3.22×4.8 b 0.322×0.48 c $15\,456 \div 4.8$

25 Sasha works for a company. She gets paid expenses of 40p for each mile she drives during work. Last year she worked for 48 weeks. Her total expenses for driving for the year were £2116.80. Work out an estimate for the average number of miles Sasha drove during work each week last year.

June 2009 Specimen Paper

26 Rashid and his 3 friends order 3 pizzas and 2 bottles of cola. They split the cost equally. How much do they each pay?

Pizza	£2.70
Cola	£1.56

27 Compare the labels from the Henry Turbowash and the Henry Ecowash.

a Why do you think that the Turbowash has a better energy rating?

b If you do 200 washes a year, how much energy (in kWh) do you save by buying the Turbowash rather than the Ecowash? If the electricity charges are 12p per kWh, how much do you save per year?

c The Turbowash is £50 more expensive. Is it better value for money?

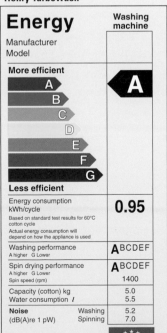

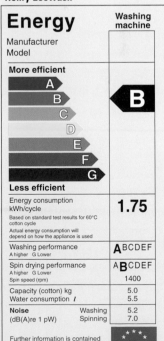

Fashion designers may get the fame and fortune, but it is their pattern makers' painstaking work that is responsible for how well the clothes fit the models. They take the designer's sketches and make a pattern out of card. This pattern is then used to cut the right shapes out of the fabric and it is these shapes that are put together to make the creations that we see on the catwalk.

◉ Objectives

In this chapter you will:
- look at the properties of triangles
- learn about congruent shapes
- look at the symmetry of various shapes
- look at the properties of quadrilaterals
- identify and name parts of circles
- construct circles.

◈ Before you start

You need to know:
- one-dimensional (1D) means it only has length
- two-dimensional (2D) shapes have area
- three-dimensional (3D) shapes have volume
- two lines are parallel if they can never meet
- two lines are perpendicular if they are at right angles to each other
- how to use a pair of compasses
- how to draw and measure angles and lines.

6.1 **Triangles**

◉ Objective

● You can identify different types of triangles.

❓ Why do this?

People sometimes need to identify or describe the shape of different objects. For example, pool balls are racked in a triangle and some road signs are triangular.

◈ Get Ready

1. Write down if each of these angles is acute, right-angled or obtuse.

a   b 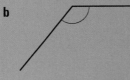 c

🌐 Key Points

● A **triangle** is any three-sided shape. Some triangles have special names.

Any triangle will have two special names: one that describes its sides and another that describes its angles.

Equilateral triangle

All three sides are the same length.
All three angles are 60°.

Isosceles triangle

Two of the sides are the same length.
Two of the angles are equal.

Scalene triangle

None of the sides or angles are equal.

Right-angled triangle

One of the angles is 90°.

Obtuse-angled triangle

One of the angles is obtuse (more than 90°).

Acute-angled triangle

All of the angles are acute (less than 90°).

⚙ Exercise 6A

Questions in this chapter are targeted at the grades indicated.

1 Match each triangle (A to F) with its mathematical names (1 to 6).
Each shape can be matched with two names.

A   B   C   D   E   F

(1) Scalene (2) Isosceles (3) Right-angled (4) Equilateral (5) Obtuse-angled (6) Acute-angled

G

G

2

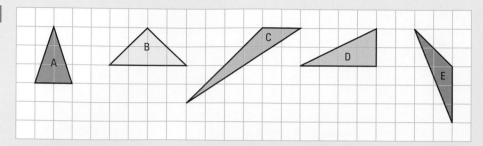

a Two of these triangles are right-angled triangles. Which two?

b Which of these triangles are isosceles triangles?

c Two of these triangles are obtuse-angled triangles. Which two?

d Which triangles are scalene triangles?

3 Draw a sketch of an obtuse-angled triangle that is also isosceles.

F

A03

4 Jemma says this triangle is an isosceles triangle.
Jemma is wrong. Explain why.

A03

5 Clare draws an equilateral triangle.
Gerry says that the triangle she has drawn is an acute-angled triangle.
Gerry is correct. Explain why.

6.2 Quadrilaterals

◉ Objective

● You know the properties of quadrilaterals.

⍰ Why do this?

Regular quadrilaterals are useful shapes. We see them in sports fields, windows and furniture.

◈ Get Ready

Sketch the following triangles, marking on any special angles.

1. Isosceles triangle   2. Right-angled triangle   3. Equilateral triangle

🌐 Key Points

● A **quadrilateral** is any four-sided shape. There are some examples of common quadrilaterals on the following page.

● The two **diagonals** of a quadrilateral go from one corner to the opposite corner.

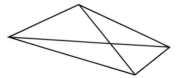

Square
- All sides are the same length.
- All angles are 90°.
- The diagonals are equal in length and **bisect** each other at right angles.
- It has 4 lines of reflection symmetry.
- It has rotational symmetry of order 4.

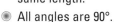

Rectangle
- Opposite sides are the same length.
- All angles are 90°.
- The diagonals are equal in length and bisect each other.
- It has 2 lines of reflection symmetry.
- It has rotational symmetry of order 2.

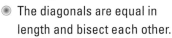

Parallelogram
- Opposite sides are parallel and are the same length.
- The diagonals bisect each other.
- It has no lines of reflection symmetry.
- It has rotational symmetry of order 2.

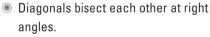

Trapezium
- One pair of opposite sides are parallel.

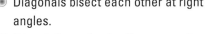

Kite
- Two pairs of **adjacent** sides are equal.
- One pair of opposite angles are equal.
- Diagonals cross each other at right angles.
- It has 1 line of reflection symmetry.

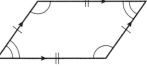

Rhombus
- All sides are the same length.
- Opposite angles are equal.
- Diagonals bisect each other at right angles.
- It has 2 lines of reflection symmetry.
- It has rotational symmetry of order 2.

⚙ Exercise 6B

1   Match each shape (A to H) with its mathematical name (1 to 6).
Some shapes will have the same name.

A   B   C   D   E   F   G   H

1 Rhombus	2 Rectangle	3 Kite
4 Trapezium	5 Square	6 Parallelogram

2   **a** Write down the mathematical name for each type of quadrilateral.
     b Which five of these shapes have two pairs of parallel sides?
     c Which shape has just one pair of opposite angles the same size?

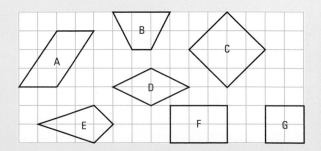

F

3   Name each of the following shapes.
  a   This shape has four equal sides and its diagonals are the same length.
  b   This shape has twice the number of sides as a quadrilateral.
  c   This shape has one pair of parallel sides and its diagonals are not the same length.
  d   This shape has half the number of sides as a hexagon and has one angle of 90°.

4   ABCD is a rhombus.
  Which of the following statements are true and which are false?
  a   AB = DC   b   angle BAD = angle ADC
  c   BC ∥ AD   d   AC is perpendicular to BD

5   EFGH is a kite.
  FH and EG meet at J.
  a   Name a side equal in length to HG.
  b   Is angle EJH the same size as angle GJH?
  c   Name an angle equal to angle EFG.
  d   Is triangle EFG a scalene triangle?
  e   Write down the mathematical name for triangle FGH.

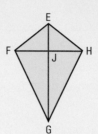

A03

6   Which one of the following pairs of lines could be the diagonals of a parallelogram?
  a   b   c

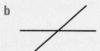

E

A02
A03

7   Pietro designs a company badge.
  He is told that the badge must have 5 sides and must be made up of two different quadrilaterals
  and the longest side must be the same length as the total length of two different sides.
  The diagram shows his design.
  Copy the diagram and show the two different quadrilaterals.
  Write down the mathematical names for the two different
  types of quadrilateral Pietro has used.

6.3 Congruent and similar shapes

◉ Objectives

- ◉ You can identify congruent shapes.
- ◉ You can identify similar shapes.

⬥ Why do this?

Being able to identify if two shapes are an exact match
can be a vital skill, particularly in the manufacturing
industry where you need to remove defective products.

⬥ Get Ready

1. Two of these shapes are exactly the same size. Which two?
  a   b   c

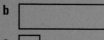

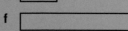

  d   e   f

Key Point

◉ **Congruent** shapes are shapes that are exactly the same size and exactly the same shape.

◉ When one shape is an enlargement of another the shapes are called **similar** shapes.

Example 1

a Which of these shapes are congruent to shape **A**?

b Which of these shapes are similar to shape **A**?

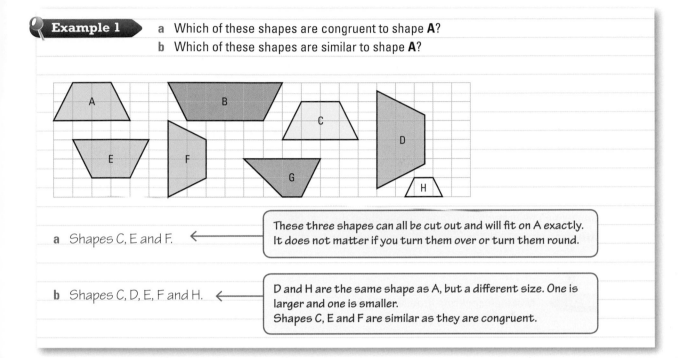

a Shapes C, E and F. ⟵

> These three shapes can all be cut out and will fit on A exactly. It does not matter if you turn them over or turn them round.

b Shapes C, D, E, F and H. ⟵

> D and H are the same shape as A, but a different size. One is larger and one is smaller.
> Shapes C, E and F are similar as they are congruent.

Exercise 6C

E

1 Write down the letters of three pairs of shapes that are congruent.

2 In each of the following, there is one pair of congruent shapes.

  i Write down the letters of each pair.   ii Which shapes are similar?

3 On the grid, draw a shape that is congruent to the shaded shape but has been turned so it is not the same way up.

6.4 Accurate drawings

Objectives

- You can make accurate drawings of triangles and quadrilaterals.
- You can draw parallel lines.

Why do this?

Architects need to make accurate drawings of their structures.

Get Ready

1. Draw the following lines accurately using a pencil and ruler.
   a 6.8 cm   b 9.2 cm   c 76 mm   d 12.7 cm   e 83 mm

2. Draw the following angles accurately using a pencil, ruler and protractor.
   a 76°   b 42°   c 118°   d 55°   e 107°

Key Points

- Most triangles can be drawn using three details about the triangle.
- To **construct** a triangle with the lengths of the sides given you should use a compass only.

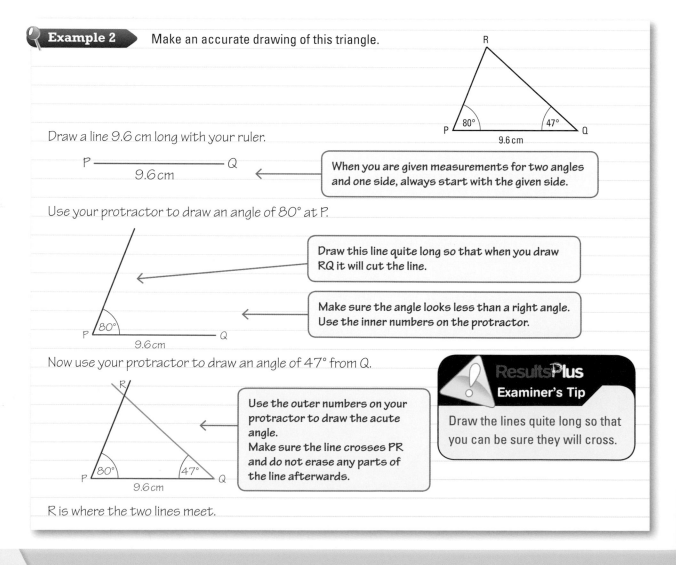

Example 2   Make an accurate drawing of this triangle.

Draw a line 9.6 cm long with your ruler.

> When you are given measurements for two angles and one side, always start with the given side.

Use your protractor to draw an angle of 80° at P.

> Draw this line quite long so that when you draw RQ it will cut the line.

> Make sure the angle looks less than a right angle. Use the inner numbers on the protractor.

Now use your protractor to draw an angle of 47° from Q.

> Use the outer numbers on your protractor to draw the acute angle.
> Make sure the line crosses PR and do not erase any parts of the line afterwards.

ResultsPlus
Examiner's Tip

Draw the lines quite long so that you can be sure they will cross.

R is where the two lines meet.

E

Exercise 6D

1   Make an accurate drawing of each of the following triangles.

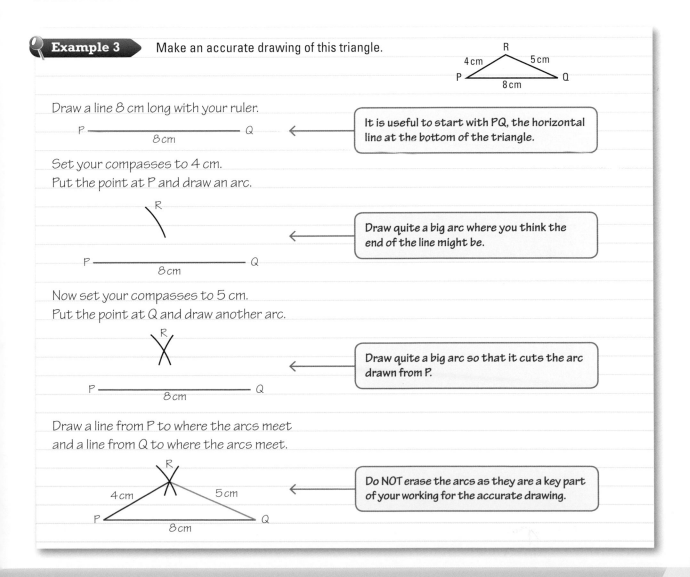

a

6 cm

58°

9 cm

b

37°   53°

11.7 cm

c

8.8 cm   8.8 cm

6.6 cm

d

12.5 cm

103°

5.4 cm

e

8.7 cm

124°

7.2 cm

f

8.6 cm

5.7 cm

Example 3   Make an accurate drawing of this triangle.

R

4 cm   5 cm

P   8 cm   Q

Draw a line 8 cm long with your ruler.

P ——————————— Q

8 cm

> It is useful to start with PQ, the horizontal line at the bottom of the triangle.

Set your compasses to 4 cm.
Put the point at P and draw an arc.

R

P ——————————— Q

8 cm

> Draw quite a big arc where you think the end of the line might be.

Now set your compasses to 5 cm.
Put the point at Q and draw another arc.

R

P ——————————— Q

8 cm

> Draw quite a big arc so that it cuts the arc drawn from P.

Draw a line from P to where the arcs meet
and a line from Q to where the arcs meet.

R

4 cm   5 cm

P   8 cm   Q

> Do NOT erase the arcs as they are a key part of your working for the accurate drawing.

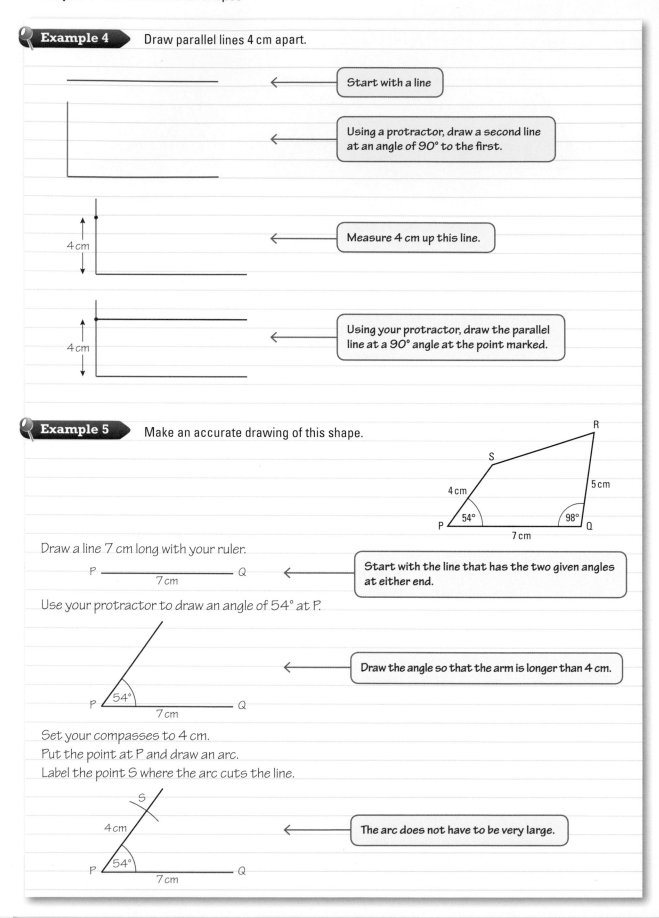

Example 4   Draw parallel lines 4 cm apart.

Start with a line

Using a protractor, draw a second line at an angle of 90° to the first.

Measure 4 cm up this line.

4 cm

Using your protractor, draw the parallel line at a 90° angle at the point marked.

4 cm

Example 5   Make an accurate drawing of this shape.

R
S
4 cm
5 cm
P
54°
98°
Q
7 cm

Draw a line 7 cm long with your ruler.

P ——— Q
7 cm

Start with the line that has the two given angles at either end.

Use your protractor to draw an angle of 54° at P.

Draw the angle so that the arm is longer than 4 cm.

P 54°
7 cm   Q

Set your compasses to 4 cm.
Put the point at P and draw an arc.
Label the point S where the arc cuts the line.

S
4 cm
P 54°
7 cm   Q

The arc does not have to be very large.

Use your protractor to draw an angle of 98° at Q.

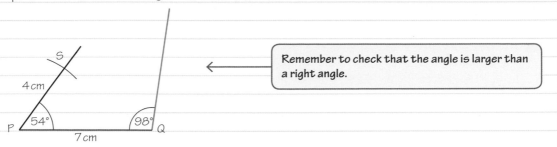

Remember to check that the angle is larger than a right angle.

Set your compasses to 5 cm. Put the point at Q and draw an arc.

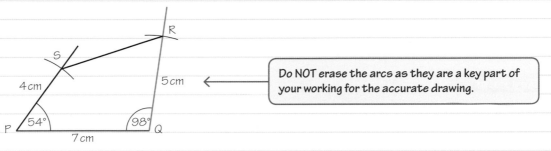

Do NOT erase the arcs as they are a key part of your working for the accurate drawing.

Label the point R where this arc cuts the line just drawn. Join S to R.

Exercise 6E

1   Make an accurate drawing of each of the following sketches of triangles.

a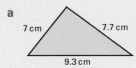

b

c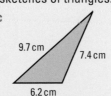

d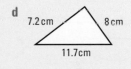

2   a   Make an accurate drawing of this quadrilateral.

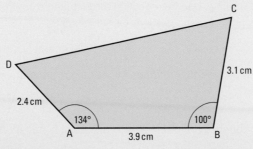

   b   Measure the length of CD.
   c   Measure the size of angle BCD.

3   In triangle EFG, EF = 9.3 cm, FG = 9.7 cm and angle FEG = 74°.
   a   Draw a sketch of triangle EFG showing all three given measurements.
   b   Make an accurate drawing of triangle EFG.
   c   Measure the length of EG.

E

4   Using compasses, make an accurate drawing of triangle HJK where HJ = 10.6 cm, HK = 4.3 cm and JK = 5.1 cm. What problems do you find? Explain why.

5   Draw pairs of parallel lines that have a distance between them of:
   a   3 cm   **b**   5 cm   **c**   3.5 cm   **d**   4.5 cm

6.5   Circles

◎ Objective

● You can identify and name parts of a circle.

❓ Why do this?

The circle is a particularly important shape. What wheels have you seen today?

🔍 Key Point

● There are several key words associated with **circles**.

Circumference	Diameter	Radius	Tangent

Chord	Arc	Segment	Sector

⚙ Exercise 6F

Copy and complete the sentences below. Use the correct word chosen from the following list:

circumference	diameter	radius	tangent
chord	arc	segment	sector

G

1   A line through the centre of a circle that touches the circumference at each end is the

    ....

2   A line outside a circle that touches the circle at only one point is called a ....

3   A line that does not pass through the centre of a circle but touches the circumference at each end is

    called a ....

4   A line from the centre of a circle that is half the length of the diameter is the ....

5   A part of a circle that has a chord and an arc as its boundary is called a ....

6   A part of a circle that has two radii and an arc as its boundary is called a ....

6.6 Drawing circles

Objectives

- You can draw accurately a circle with a given radius.
- You can draw accurately an arc of a given radius and angle.

Why do this?

Landscape gardeners use circles to represent plants and sculptures in their design plans.

Get Ready

1. Sketch 3 circles. On each of these circles draw
   a   a radius   b   a chord   c   a sector.

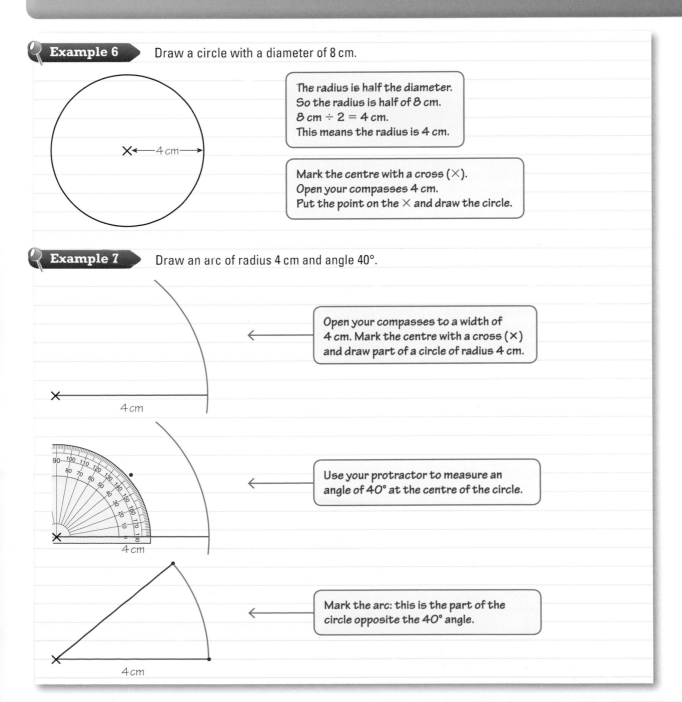

Example 6   Draw a circle with a diameter of 8 cm.

The radius is half the diameter.
So the radius is half of 8 cm.
8 cm ÷ 2 = 4 cm.
This means the radius is 4 cm.

Mark the centre with a cross (×).
Open your compasses 4 cm.
Put the point on the × and draw the circle.

Example 7   Draw an arc of radius 4 cm and angle 40°.

Open your compasses to a width of 4 cm. Mark the centre with a cross (×) and draw part of a circle of radius 4 cm.

Use your protractor to measure an angle of 40° at the centre of the circle.

Mark the arc: this is the part of the circle opposite the 40° angle.

111

Exercise 6G

G

1. Draw a circle of radius 5.3 cm.

2. Draw a circle with a diameter of 12 cm.

3. a Draw a circle of radius 6.7 cm.   b Shade a segment of your circle.

4. a Draw a circle of diameter 15.4 cm.   b Shade a sector of your circle.

5. a Draw a circle of diameter 11.6 cm.
   b On your circle, draw and label i a radius   ii a chord   iii a tangent.

6. Draw arcs with the following measurements.
   a radius 3 cm, angle 30°   b radius 4 cm, angle 80°   c radius 5 cm, angle 60°
   d radius 4 cm, angle 75°   e radius 3 cm, angle 40°   f radius 4.5 cm, angle 65°

6.7 Line symmetry

Objective

You can understand line symmetry and identify and draw lines of symmetry on a 2D shape.

Why do this?

Symmetry and balance tend to be closely related – this accounts for the symmetry of the human body.

Get Ready

1. In each of these questions, does the red line in the middle act as a mirror?

a   b   c

Key Points

● A shape is **symmetrical** if you can fold it in half and one half is the **mirror image** of the other half. The dividing line is called a **line of symmetry** or a **mirror line**.

● You can use tracing paper to help you. Trace the diagram and then fold it in half on the mirror line. You can then check if each half folds exactly onto the other half.

Example 8   Draw all the lines of symmetry for: a a kite; b a rectangle.

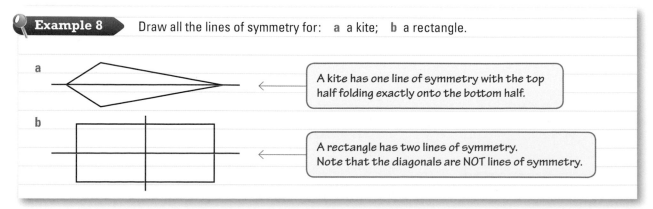

a

A kite has one line of symmetry with the top half folding exactly onto the bottom half.

b

A rectangle has two lines of symmetry.
Note that the diagonals are NOT lines of symmetry.

Example 9 — Reflect the shaded shape in the mirror line.

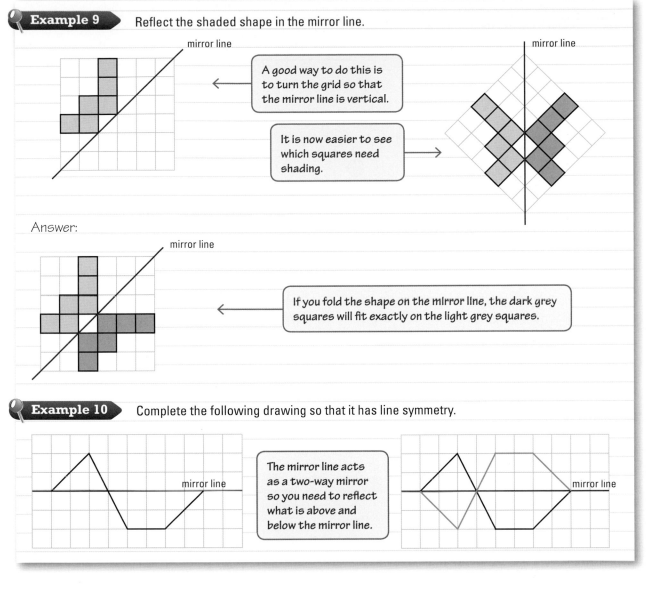

mirror line

A good way to do this is to turn the grid so that the mirror line is vertical.

It is now easier to see which squares need shading.

mirror line

Answer:

mirror line

If you fold the shape on the mirror line, the dark grey squares will fit exactly on the light grey squares.

Example 10 — Complete the following drawing so that it has line symmetry.

mirror line

The mirror line acts as a two-way mirror so you need to reflect what is above and below the mirror line.

mirror line

Exercise 6H

1 Copy the following shapes and draw all the lines of symmetry on each one.

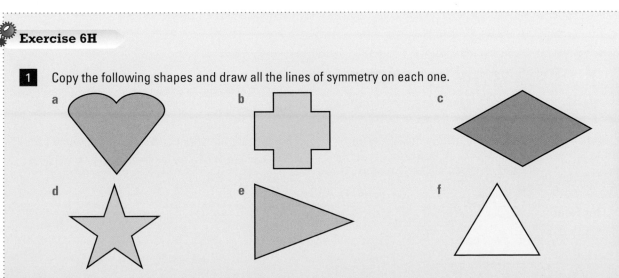

a

b

c

d

e

f

F

F

2 For each shape, state whether or not it has line symmetry.
If it does, write down how many lines of symmetry it has.

a   b    c

d 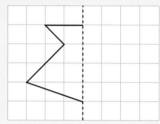   e 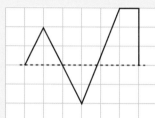   f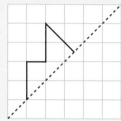

3 Copy and complete each drawing so that it has line symmetry.

a   b   c

E

4 Copy and complete each drawing so that the final pattern is symmetrical about both lines.

a 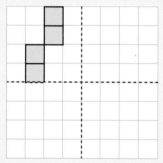   b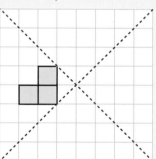

6.8 Rotational symmetry

Objective

You can understand rotational symmetry and can identify the order of rotational symmetry of a 2D shape.

Why do this?

Rotational symmetry can be found all around us. For example, it occurs in the sails of a windmill or a hubcap on a car.

Get Ready

1. How many degrees are there in
   a a full turn
   b a quarter turn
   c a half turn?

Key Points

⦿ To see if a shape has **rotational symmetry**, rotate it one full turn and see how many times along the rotation the shape still looks the same.

⦿ When a rectangle [⋅] is turned through 360° (one full turn) around its centre you can see that it looks the same twice:

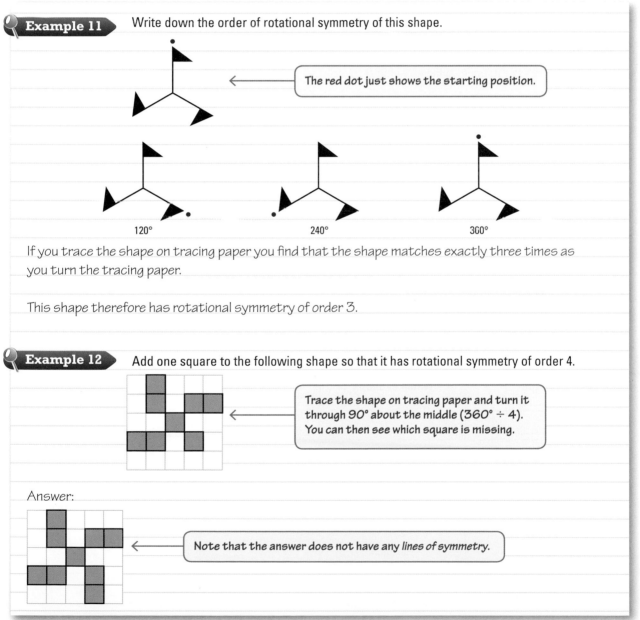

Therefore a rectangle has rotational symmetry of **order** 2.

Example 11   Write down the order of rotational symmetry of this shape.

The red dot just shows the starting position.

120°   240°   360°

If you trace the shape on tracing paper you find that the shape matches exactly three times as you turn the tracing paper.

This shape therefore has rotational symmetry of order 3.

Example 12   Add one square to the following shape so that it has rotational symmetry of order 4.

A03

Trace the shape on tracing paper and turn it through 90° about the middle (360° ÷ 4). You can then see which square is missing.

Answer:

Note that the answer does not have any lines of symmetry.

Exercise 6I

1 For each letter, write down if it has rotational symmetry or not.
If it does, write down the order of rotational symmetry.

a H   b T   c X   d Z

2 Write down the order of rotational symmetry for each of the following shapes.

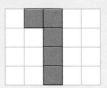

3 On a copy of this grid, add one square to the shape
so that it has rotational symmetry of order 2.

4 On a copy of this grid, add three squares to the shape
so that it has rotational symmetry of order 4.

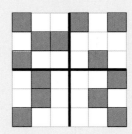

Exercise 6J

1 Copy and complete the following table.

Shape	Name of shape	Number of lines of symmetry	Order of rotational symmetry
		2	
	Equilateral triangle		
			2
	Hexagon		

Chapter review

- A **triangle** is any three-sided shape. Some triangles have special names: **equilateral**, **isosceles**, **scalene**, **right-angled**, acute-angled and obtuse-angled.
- A **quadrilateral** is any four-sided shape. Some examples of quadrilaterals are: **square**, **rectangle**, **parallelogram**, **trapezium**, **kite**, **rhombus**.
- The two **diagonals** of a quadrilateral go from one corner to the opposite corner.
- **Congruent** shapes are shapes that are exactly the same size and exactly the same shape.
- When one shape is an enlargement of another the shapes are called **similar** shapes.
- Most triangles can be drawn using three details about the triangle.
- To **construct** a triangle with the lengths of the sides given you should use a compass only.
- There are several key words associated with **circles**: **circumference**, **diameter**, **radius**, **tangent**, **chord**, **arc**, **segment** and **sector**.
- A shape is **symmetrical** if you can fold it in half and one half is the **mirror image** of the other half. The dividing line is called a **line of symmetry** or a **mirror line**.
- You can use tracing paper to help you. Trace the diagram and then fold it in half on the mirror line. You can then check if each half folds exactly onto the other half.
- To see if a shape has **rotational symmetry**, rotate it one full turn and see how many times along the rotation the shape still looks the same.

Review exercise

1. Here is a triangle.
   What type of triangle is it?

   Nov 2007

2. Here are 6 shapes drawn on a grid.

   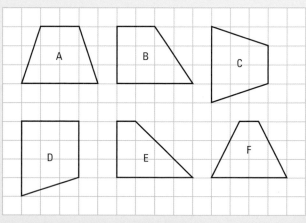

   Two of these shapes are congruent.
   a Write down the letters of these two shapes.

G

G

b On a copy of the grid below, draw a shape that is congruent to shape **P**.

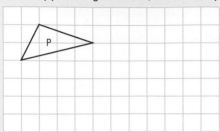

June 2009

3 Here are 5 diagrams and 5 labels.

In each diagram the centre of the circle is marked with a cross (×).

Match each diagram to its label. One has been done for you.

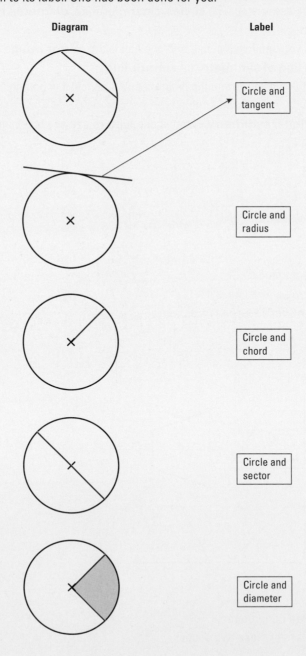

| Diagram | Label |

Circle and tangent

Circle and radius

Circle and chord

Circle and sector

Circle and diameter

June 2008

4 **a** Here is a quadrilateral.
What type of quadrilateral is it?

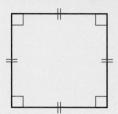

b On a copy of the grid, draw a trapezium.

Nov 2007

5 Here are some quadrilaterals.
Draw an arrow from each quadrilateral to its mathematical name.
The square has been done for you.

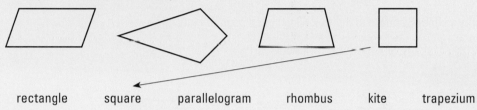

rectangle   square   parallelogram   rhombus   kite   trapezium   *June 2007*

6 Write down which of the triangles below are isosceles triangles.

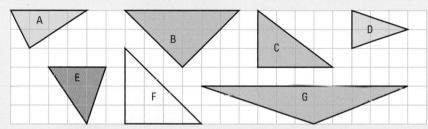

7 **a** On the diagram below, shade **one** square so that the shape has exactly **one** line of symmetry.

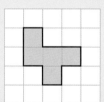

b On the diagram below, shade **one** square so that the shape has rotational symmetry of order 2.

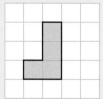

Nov 2008

8 Here are four shapes.

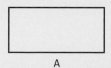

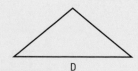

Write down the letter of the shape which has
i exactly **one** line of symmetry   **ii** **no** lines of symmetry   **iii** exactly **two** lines of symmetry.

Nov 2008

F

9 **a** Reflect the shaded shape in the mirror line. **b** Draw the lines of symmetry on this triangle.

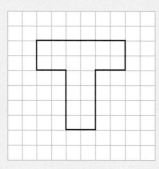

Mirror line

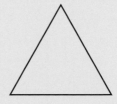

May 2008 adapted

10 Add one more shaded square to the following shape so that it has line symmetry.

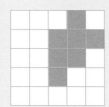

E

11 The shape below has one line of symmetry. The shape below has rotational symmetry.
a On the grid, draw this line of symmetry. **b** Write down the order of rotational symmetry.

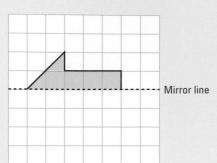

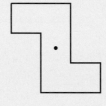

Nov 2007

12 **a** On a copy of the diagram, shade **one** more square to make a pattern with 1 line of symmetry. **b** On a copy of the diagram, shade **one** more square to make a pattern with rotational symmetry of order 2.

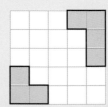

June 2007

13 The diagram shows part of a shape.
    The shape has rotational symmetry of order 3 about the point P.
    On a copy of the grid, complete the shape.

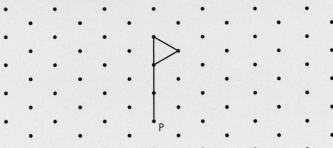

Nov 2006

14 a On a copy of this diagram, shade **one** more
       square so that the shape has exactly **one**
       line of symmetry.

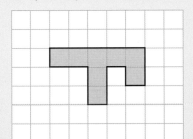

    b On a copy of this diagram, shade **one** more
       square so that the shape has rotational
       symmetry of order **2**.

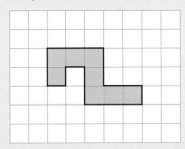

Nov 2006

15 Here are five shapes.

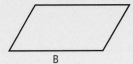

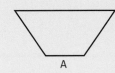

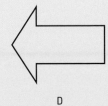

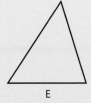

Two of these shapes have only **one** line of symmetry.
a Write down the letter of each of these two shapes.

Two of these shapes have rotational symmetry of order 2.
b Write down the letter of each of these **two** shapes.

June 2007

E

16 Add one more shaded square to the following shape so that it has rotational symmetry of order 2.

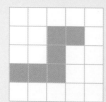

17 AB = 8 cm.   AC = 6 cm.   Angle A = 52°.
Make an accurate drawing of triangle ABC.

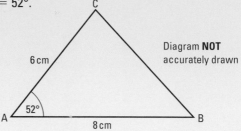

Diagram **NOT** accurately drawn

Nov 2007

18 Here is a sketch of a triangle.
The lengths of the sides of the triangle are 9 cm, 7 cm and 5 cm.
Use ruler and compasses to make an accurate scale drawing of the triangle.

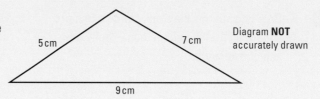

Diagram **NOT** accurately drawn

19

Diagram **NOT** accurately drawn

Make an accurate drawing of the quadrilateral ABCD.

June 2007

7 ANGLES 2

Rugby players need to think about angles when they are converting a try or taking a penalty; 45° is the optimum angle at which the player should kick the ball. Getting the ball between the posts is a trade-off between energy and distance travelled, and an angle of 45° gives enough height and force for the ball to clear the horizontal bar.

◉ Objectives

In this chapter you will:
- learn how to use angle facts to solve problems
- learn how to solve problems involving bearings and scale drawings.

◆ Before you start

You need to:
- know how to find missing angles on a straight line and in triangles
- be able to draw and measure angles using a protractor.

7.1 Angles in quadrilaterals

◎ Objective

- You can use angle facts to find missing angles in quadrilaterals.

❓ Why do this?

Carpenters need to be able to work out angles in quadrilaterals so that they can work out the angles in joints.

⬆ Get Ready

1. Find the missing angles in the diagrams below.

a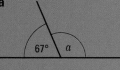

b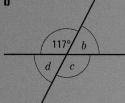

c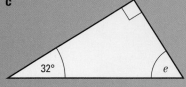

🔍 Key Points

- The interior angles of a quadrilateral (a four-sided shape) always add up to 360°.

You can see this by measuring the angles…

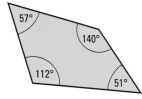

… or by dividing the quadrilateral into two triangles…

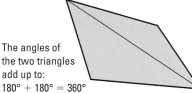

The angles of the two triangles add up to:
180° + 180° = 360°

… or by tearing off the four corners of a quadrilateral.

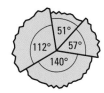

Put the angles together. They make a full turn of 360°.

A02 🔍 **Example 1**   Find the missing angle in this quadrilateral.

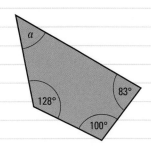

The total of the three angles marked = 311°

So a = 360° − 311°

a = 49°

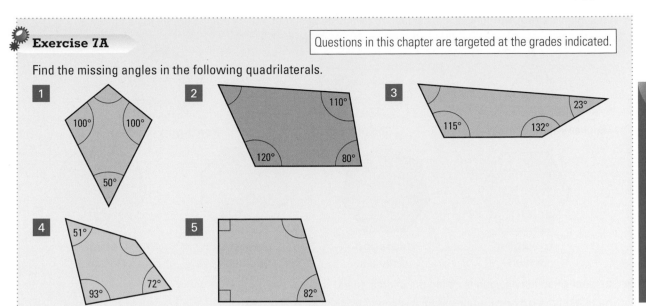

Exercise 7A

Questions in this chapter are targeted at the grades indicated.

Find the missing angles in the following quadrilaterals.

E

7.2 Polygons

◎ Objectives

○ You can recognise and name regular polygons.
○ You can work out the sum of the interior angles of polygons.

⑦ Why do this?

Polygons occur in both the natural and man-made world, for example the pentagon shape can be seen in the vegetable okra and also the US department of defence building called The Pentagon.

◈ Get Ready

1. Write down the name of the shape that has 3 equal sides and 3 equal angles.
2. Use one word to complete the following sentence.
   A square is a quadrilateral with 4 _____ sides and 4 _____ angles.

◉ Key Points

◉ A **polygon** is a 2D shape with straight sides. The following shapes are all polygons:

Not all sides of a polygon need to be the same length.

Polygons can have sides which point inwards.

Polygons are closed shapes – there are no gaps in the perimeter.

A polygon is a **regular** polygon if its sides are all the same length and its angles are all the same size. You need to know the names of the following special polygons:

Equilateral triangle (3 sides)

Square (4 sides)

Pentagon (5 sides)

Hexagon
(6 sides)

Heptagon
(7 sides)

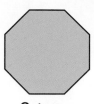

Octagon
(8 sides)

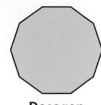

Decagon
(10 sides)

The angle inside a polygon is called the interior angle.
The total of the **interior** angles of an n-sided polygon is $(n - 2) \times 180°$.

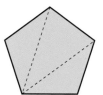

The interior angles of this pentagon can be found by dividing the shape into three triangles.
The sum of the interior angles of a triangle $= 180°$.
So the sum of the interior angles of the pentagon $= 3 \times 180° = 540°$.
If the polygon is regular, each interior angle is the sum of all the interior angles divided by the number of sides.

All regular polygons with the same number of sides, and all circles, are similar shapes.

Example 2

a Find the total of the interior angles of a decagon.

b The sum of the interior angles of a polygon is 2160°.
 Work out the number of sides of the polygon.

a A decagon has 10 sides.
 The sum of its interior angles is $(10 - 2) \times 180° = 1440°$.

b The sum of the interior angles of a polygon with n sides is $(n - 2) \times 180$.
 Therefore $(n - 2) \times 180 = 2160$
 $$n - 2 = 2160 \div 180$$
 $$n - 2 = 12$$
 $$n = 12 + 2$$
 $$n = 14$$
 The polygon has 14 sides.

Exercise 7B

F

1 Name the following polygons.

a

b

c

regular   pentagon   hexagon   heptagon   octagon   decagon   interior

2 Calculate the sum of the interior angles of:

    a a hexagon   **b** an octagon   **c** a 15-sided polygon   **d** a 20-sided polygon.

3 The totals of the interior angles for some polygons are shown below.
For each polygon calculate the number of sides.

    a 900°   **b** 1620°   **c** 3600°

D

7.3 **Exterior and interior angles**

◎ Objective

● You can use angle facts about interior and exterior angles of a polygon to solve problems.

⦾ Why do this?

To make a football you need to know how to calculate the correct angles.

⬀ Get Ready

1. Find the value of the marked angle.
2. $x + 28 = 180$. Find x.
3. $3x = 240$. Find x.

🔍 Key Points

● The angle outside a polygon is called the **exterior angle**.
The exterior angles of a polygon add up to 360°.
● The interior angle and the exterior angle for any polygon add up to 180°.

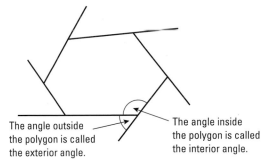

The angle outside the polygon is called the exterior angle.

The angle inside the polygon is called the interior angle.

🔍 Example 3

An n-sided regular polygon has an interior angle of 160°.

    a Calculate the exterior angle of the polygon.

    b Use your answer to part **a** to work out how many sides the polygon has.

a The interior angle and the exterior angle of the polygon add up to 180°.
Therefore 160° + exterior angle = 180°.
So the exterior angle = 20°.
b The exterior angles of a polygon add up to 360°.
So $n \times 20 = 360$
$n = 360 \div 20$
$n = 18$.   The polygon has 18 sides.

Results**Plus**

Watch Out!

The formula exterior angles = 360° ÷ n does not work for irregular polygons.

Exercise 7C

E

1. Work out the size of the marked angles.

D **A03**

2. Work out the exterior angle of:

   a a pentagon   b an equilateral triangle   c a 20-sided polygon.

3. Use your answers to question **2** to work out the interior angle of each of the polygons.

C **A03**

4. A regular polygon has an exterior angle of 12°. Work out how many sides it has.

A03

5. A regular polygon has an interior angle of 144°. Find the exterior angle and work out how many sides the polygon has.

7.4 Tessellations

◎ Objective

● You can tessellate various shapes.

❓ Why do this?

Tessellations can be found in nature. Bees tessellate hexagons when they build their honeycombs.

⬆ Get Ready

1. Is each floor covered with only one shape of tile?

   a   b   c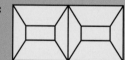

🔑 Key Points

● A **tessellation** is when a shape is drawn over and over again so that it covers an area without any gaps or overlaps. You often tessellate a shape to tile a floor.

This shape has been tessellated to cover a floor.

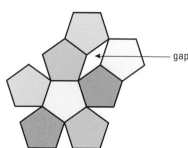

This pentagon cannot tessellate as gaps are formed.

Tessellations can also be made using more than one shape. Here is a tessellation made from squares and regular octagons.

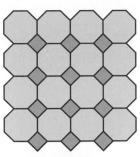

Sometimes tessellations are used to create optical illusions. Here, a rhombus has been tessellated so that it looks as if there are lots of cubes.

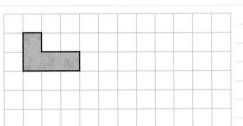

Example 4 Complete the following diagram to show how the shaded shape will tessellate. You should draw at least six more shapes.

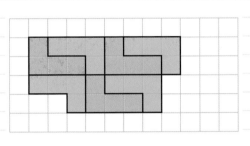

You could continue in this way to cover the whole grid with no gaps between the shapes.

Exercise 7D

1 Show how each of the following shapes will tessellate.

a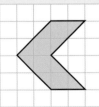

b

c

d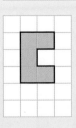

e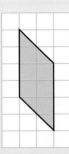

E A02 A03

2 Copy and complete the following diagram to show how the shaded shape will tessellate. You should draw at least five more shapes.

7.5 Perpendicular and parallel lines

◎ Objective

● You know what perpendicular and parallel lines are and can mark them in diagrams.

⦿ Why do this?

Parallel lines are used on running tracks to make sure that runners stay in their lanes.

Key Points

● Two lines are **perpendicular** to each other if they meet at a 90° angle.

Lines AB and CD are perpendicular to each other.

Line FO is perpendicular to EG.

● Lines that remain the same distance apart are called **parallel** lines. On diagrams this is shown by marking the parallel sides with arrows. If there is a second pair of parallel lines in one diagram these are marked with double arrows.

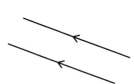

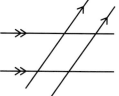

● The distance between the two edges of a ruler is the same all the way along it. Similarly, the distance between the two rails of a train track is the same wherever it is measured.

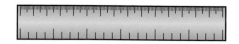

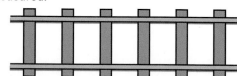

perpendicular parallel

Example 5

a Look at the diagram.
Which sides are perpendicular?

b Which sides are parallel in the diagram?

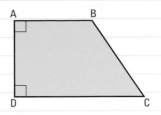

a Sides AB and AD meet at 90° so they are perpendicular to each other.
Sides AD and DC are perpendicular to each other as well.

b Sides AB and DC are parallel. They are the same distance apart all along their length.

Exercise 7E

1 a Find and name as many pairs of parallel lines as you can in the diagram.

b Now find and name as many pairs of perpendicular lines as you can in the diagram.

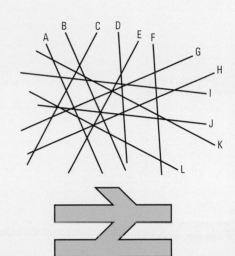

2 a Copy the diagram and mark three pairs of perpendicular lines.

b Now mark three pairs of parallel lines on your diagram.

7.6 **Corresponding and alternate angles**

◈ **Get Ready**

In each of the following questions find the value of a. Give a reason for your answer.

1.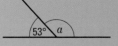

2.

3.

Key Points

- The marked angles below are equal. They are called **corresponding angles**.

 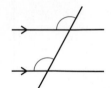 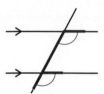

- The marked angles below are equal. They are called **alternate angles**.

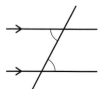 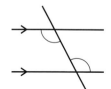 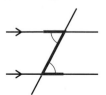

Example 6

a Find the size of angle a. Give a reason for your answer.

b Now find the size of angle b, giving a reason for your answer.

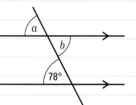

a Angle $a = 78°$ corresponding angles are equal.

b Angle $b = 78°$ alternate angles are equal.

Example 7   Find the size of the lettered angles, giving reasons for your answers.

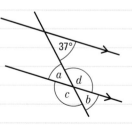

$a = 37°$ alternate angles are equal.
$b = 37°$ corresponding angles are equal.
$c = 143°$ angles on a straight line add up to 180°.
$d = 143°$ vertically opposite angles are equal.

Exercise 7F

D

1 a In the diagram, which pairs of angles are corresponding angles?

   b Which pairs of angles in the diagram are alternate angles?

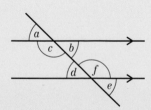

D

2 In the diagrams below, find the size of each angle marked with a letter. Give reasons for your answers.

a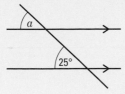

b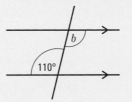

c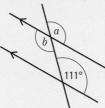

d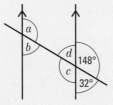

e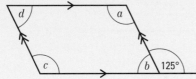

f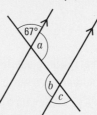

3 Find the size of the angles marked with a letter in the diagrams below.

A02

a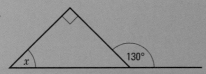

b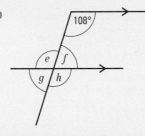

7.7 Proofs

◉ Objective

● You can demonstrate proofs about angle facts of exterior angles, angle sums of a triangle and opposite angles of a parallelogram.

◈ Why do this?

Proofs are used in science to prove that a law is true and that a conclusion is correct.

◈ Get Ready

1. Find the value of angle x.

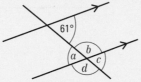

2. Write down the value of a and b. Give a reason for your answer.

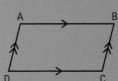

3. ABCD is a parallelogram.
   Angle ABC = 81°
   Write down the size of the other 3 angles.

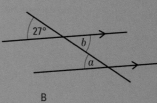

Key Points

◉ **Proof 1**

Proving the exterior angle of a triangle is equal to the sum of the two interior opposite angles.

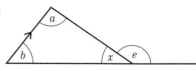

Through the point of angle e draw a line parallel to the opposite side of the triangle.

Angle e is now divided into two angles, c and d.

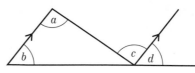

$a = c$ alternate angles
$b = d$ corresponding angles

So $a + b = c + d = e$.

◉ **Proof 2**

Proving the angle sum of a triangle is 180°.

The angles of the triangle are a, b and x.

$c + d + x = 180°$ angles on a straight line

But $c + d = a + b$ exterior angle = sum of interior opposite angles

Therefore $a + b + x = 180°$.

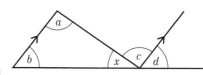

◉ **Proof 3**

Proving the opposite angles of a parallelogram are equal.

$a = b$ corresponding angles
$b = c$ alternate angles

So $a = c$.

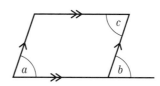

7.8 Bearings

◎ Objectives

◉ You know what a bearing is.
◉ You can solve problems involving bearings by calculation and drawing.

ⓘ Why do this?

You would use bearings when orienteering.

⬦ Get Ready

1. Copy and complete this diagram showing the 8 points of the compass.

2. Use a protractor to measure angle a.

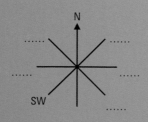

3. Draw angle ABC = 138°.

4. Write down the size of the marked angles.

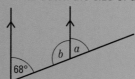

Key Points

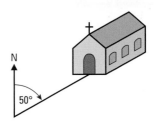

- Bearings are used to describe directions with angles.
  - If you begin facing north then turn clockwise until you face the church, you have turned through 50°.
  - The angle you turn is called the **bearing**.
  - It is always written as a three-figure number.
  - When there are less than three digits in the angle you need to add zeros to make a three-figure number. For example, 9° = 009°.
  - You write the bearing of the church as 050°.

- Bearings are always measured from north in a clockwise direction. In this diagram the bearing of London from Birmingham is 135°. The bearing of Birmingham from London is 315°.

Exercise 7G

1 Complete the following sentences using one of the eight points of the compass.

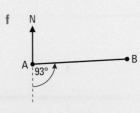

  a The ship is _____ of the lighthouse.
  b The lighthouse is _____ of the airport.
  c The lighthouse is _____ of the harbour.

2 In each of the following, write down the bearing of B from A.

a 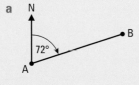 **b** 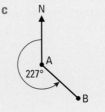 **c**

d 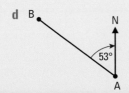 **e** 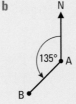 **f**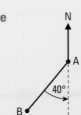

3 Write down the bearing of:
  a B from A
  b C from B
  c B from C.

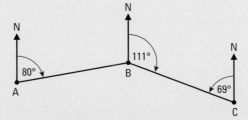

Example 8

Measure and write down the bearing of B from A.

Use a protractor to measure the angle between the north line and AB.

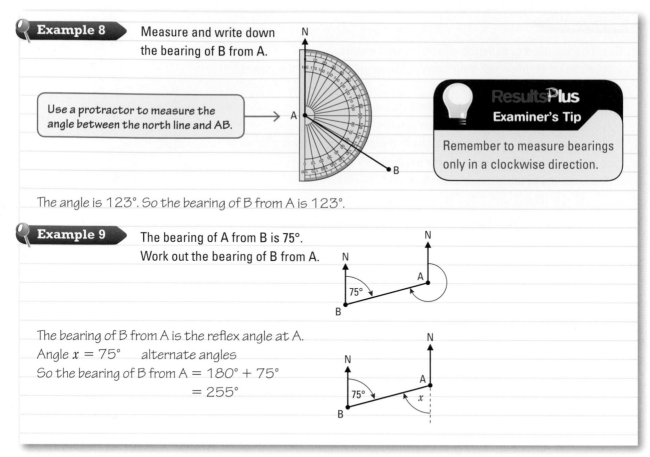

ResultsPlus
Examiner's Tip

Remember to measure bearings only in a clockwise direction.

The angle is 123°. So the bearing of B from A is 123°.

Example 9

The bearing of A from B is 75°.
Work out the bearing of B from A.

The bearing of B from A is the reflex angle at A.
Angle $x = 75°$ alternate angles
So the bearing of B from A = $180° + 75°$
                         = $255°$

Exercise 7H

D

1 In each of the following diagrams measure the bearing of T from S.

a   b   c

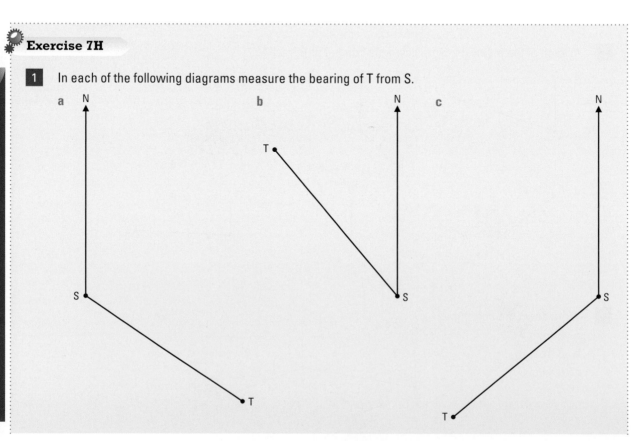

D

2 Use a protractor to draw bearings of:
   a 085°   **b** 147°   **c** 238°.

3 Use a protractor to find the bearings of:
   a Q from P
   b P from R
   c R from Q.

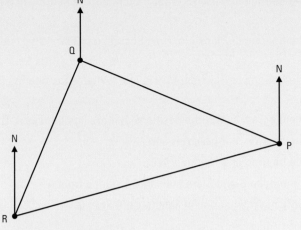

4 Paul is using the map below to find the bearing from Tavistock to:
   a Tiverton
   b Plymouth
   c Bodmin.
   Write down these bearings.

5 Three ships, A, B and C, are at sea.
The bearing of B from A is 065°.
The bearing of B from C is 125°.
The bearing of C from A is 040°.
Draw a sketch to show the positions of the three ships.

6 The bearing of P from Q is 055°.
Work out the bearing of Q from P.

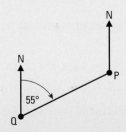

D
A02
A03

7 The bearing of T from S is 168°. Work out the bearing of S from T.

A03

8 The bearing of Blackburn from Rochdale is 303°. What is the bearing of Rochdale from Blackburn?

A02

9 The bearing of Hull from York is 115°. What is the bearing of York from Hull?

A03

10 The diagram shows the positions of three lighthouses A, B and C.

The bearing of B from A is 048°.

The bearing of C from A is 126°.

Lighthouse B and C are the same distance from A.

Work out the bearing of lighthouse C from B.

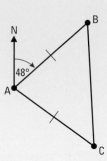

7.9 Maps and scale drawings

◎ Objectives

- ◉ You can use and interpret maps and scale drawings
- ◉ You can construct accurate scale drawings.
- ◉ You can draw lines and shapes to scale and estimate lengths on scale drawings.
- ◉ You can work out lengths using a scale factor.

? Why do this?

When planning to build an extension to a house, accurate scale drawings have to be made.

⬆ Get Ready

1. Convert **a** 56 000 mm into cm; **b** 5.6 m into cm.
2. Divide 36 g in the ratio
    a 1:2 **b** 3:1 **c** 3:6 **d** 8:4

🌐 Key Points

- ◉ Maps and plans are accurate drawings from which measurements are made.
- ◉ A **scale** is a ratio which shows the relationship between a length on a drawing and the actual length in real life. (There is more about ratios in Chapter 24).
- ◉ Ordnance Survey Pathfinders maps, used by hill walkers, are on a scale of 1 to 25 000, written 1 : 25 000.
- ◉ 1 cm on the map represents 25 000 cm in real life, which is 250 m or a quarter of a kilometre.
- ◉ Drawing diagrams to scale and using bearings means you can measure and calculate missing angles and distances.

Example 10   The scale on a road map is 1 : 200 000.
Sunderland and Newcastle are 9 cm apart on the map.

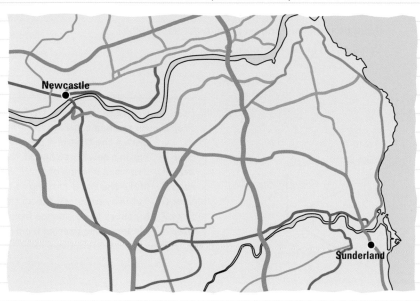

a   Work out the real distance, in km, between Sunderland and Newcastle.

b   Middlesbrough is 55 km in a straight line from Newcastle.
Work out the distance of Middlesbrough from Newcastle on the map.

a   The distance on the map is 9 cm.

> Remember:
> 100 cm = 1 m
> 1000 m = 1 km

real distance = 9 cm × 200 000 = 1 800 000 cm

real distance = 18 000 m   ← Divide by 100 to change cm to m.

The real distance between Sunderland and Newcastle is 18 km.   ← Divide by 1000 to change m to km.

b   The real distance is 55 km.

real distance = 55 × 1000 = 55 000 m   ← Multiply by 1000 to change km to m.

real distance = 55 000 × 100 = 5 500 00 cm   ← Multiply by 100 to change m to cm.

Distance on the map = $\frac{5500000}{200000}$ = 27.5 cm   ← Divide by 200 000 to find the distance on the map.

Example 11

Irie walks for 2 miles on a bearing of 060° from home.

She then walks a further 4 miles on a bearing of 300°.

How far is Irie from home?

What bearing must Irie walk on to get back home?

Use a scale of 2 cm to represent one mile.

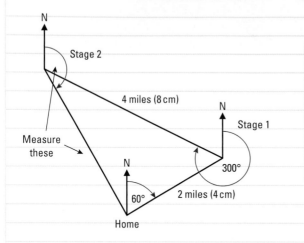

Step 1   Draw the bearing of 060° from home.
Step 2   As the scale is 2 cm for 1 mile, you need to
          draw a line that is 4 cm long.
Step 3   Put in a new north line at the end of Stage 1.
Step 4   Draw the bearing of 300° from Stage 1.
Step 5   Make the line 8 cm long (4 miles is 4 × 2 cm).
Step 6   Put in a new north line at the end of Stage 2.
Step 7   Measure the distance from Stage 2 to home.
Step 8   Measure the bearing from Stage 2 to home.

Answer   Irie is about 3.5 miles from home and must walk on a bearing of 150°.

Exercise 7I

D

1   The scale on a map is 1 cm to 2 km.
        The distance between Ashton and Dacreville is 5.5 cm on the map.
        How many kilometres apart are Ashton and Dacreville in real life?

2   A map is drawn on a scale of 3 cm to 1 km.
        a Work out the real length of a lake, which is 4.2 cm long on the map.
        b The distance between the church in Canwick and the town hall in Barnton is 5.8 km.
            Work out the distance between them on the map.

3   Jane walks for 10 miles on a bearing of 060°.
        Use a scale of 1 cm to represent 1 mile to show the journey.

4   Sam runs for 4 km on a bearing of 120°. Use a scale of 1 cm to represent 2 km to show the journey.

5   Witley is 2 km due south of Milford. The bearing of Hydestile from Milford is 125° and the distance from
        Milford to Hydestile is 2.8 km.
        a Make a scale drawing to show the three villages. Use a scale of 1 : 25 000.
        b Use your drawing to find
            i   the distance of Hydestile from Witley
            ii the bearing of Hydestile from Witley.

6 Ian sails his boat from the Isle of Wight for 20 km on a bearing of 135°.
He then sails on a bearing of 240° for 10 km.
How far is Ian from his starting point?
What bearing does he need to sail on to get back to the start?
Use a scale of 1 cm to represent 2 km.

7 Peg Leg the pirate buried his treasure 100 yards from the big tree on a bearing of 045°. One-Eyed Rick
dug up the treasure and moved it 50 yards on a bearing of 310° from where it had been buried.
How far is the treasure from the big tree now?
What bearing is the new hiding place of the treasure?
Use a scale of 1 cm to represent 10 yards.

8 Ray flew his plane on a bearing of 300° for 200 km. He then changed direction and flew on a
bearing of 150° for 100 km.
What bearing must Ray fly on to get back to the start?
How far is he away from the start?

9 This is a sketch of Arfan's bedroom. It is *not* drawn to scale.
Draw an accurate scale drawing on cm squared paper of Arfan's bedroom.
Use a scale of 1 : 50.

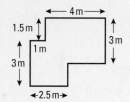

Chapter review

- The interior angles of a quadrilateral (a four-sided shape) always add up to 360°.
- A **polygon** is a 2D shape with straight sides.
- A polygon is a **regular** polygon if its sides are all the same length and its angles are all the same size.
- The angle inside a polygon is called the interior angle.
  The total of the interior angles of an n-sided polygon is $(n - 2) \times 180°$.
- The angle outside a polygon is called the **exterior angle**.
- The exterior angles of a polygon add up to 360°.
- The interior angle and the exterior angle for any polygon add up to 180°.
- All regular polygons with the same number of sides are similar shapes.
- A **tessellation** is when a shape is drawn over and over again so that it covers an area without any gaps or overlaps.
- **Perpendicular** lines meet at an angle of 90°.
- Lines which remain the same distance apart along their length are **parallel** lines.
- **Corresponding angles** are equal.
- **Alternate angles** are equal.
- The exterior angle of a triangle is equal to the sum of the two interior opposite angles.
- The angle sum of a triangle is 180°.
- The opposite angles of a parallelogram are equal.
- **Bearings** are measured clockwise from north and are always written as a three-figure number.
- Maps and plans are accurate drawings from which measurements are made.
- A **scale** is a ratio which shows the relationship between a length on a drawing and the actual length in real life.

G

⚙ **Review exercise**

1 **a** On a copy of the grid, draw a
line that is parallel to the line **L**.

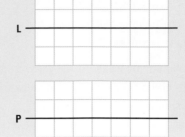

b On a copy of the grid, draw a line
perpendicular to the line **P**.

Nov 2007

2 Copy the following flag.
Mark two pairs of parallel lines
and a pair of perpendicular lines.

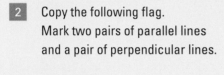

E

3 In the diagram, ABC is a triangle.
ACD is a straight line.
Angle CAB = 50°.
Angle ABC = 60°.
Work out the size of the angle marked x.

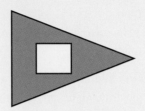

Diagram **NOT**
accurately drawn

Nov 2008

4 Work out the size of the angle marked x.

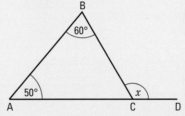

Diagram **NOT**
accurately drawn

Nov 2008

5 PQR is a straight line.
PQ = QS = QR.
Angle SPQ = 25°.
a **i** Write down the size of angle w.
ii Work out the size of angle x.
b Work out the size of angle y.

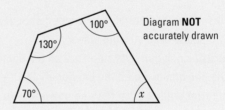

Diagram **NOT**
accurately drawn

Nov 2008

6 Work out the value of x.

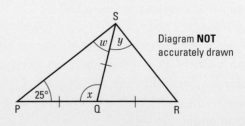

Diagram **NOT**
accurately drawn

Nov 2007

7 On the grid, show how this shape will tessellate.
You should draw at least 8 shapes.

June 2007

8 The diagram shows the position of a farm F and a bridge B on a map.
a Measure and write down the bearing of B from F.

A church C is on a bearing of 155° from the bridge B.
On the map, the church is 5 cm from B.
b Mark the church with a cross (×) and label it C.

Nov 2006

9 In the diagram, ABC is a straight line and BD = CD.
a Work out the size of angle x.
b Work out the size of angle y.

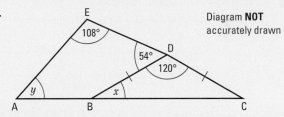

Diagram **NOT** accurately drawn

Nov 2006

10 A car is 4 m long and 1.8 m wide.
A model of the car, similar in all respects, is 5 cm long. How wide is it?

11 A model of a car is 12 cm long and 5.2 cm high.
If the real car is 3.36 m long, how high is it?

12 AB is parallel to CD.
Angle BEF = 127°.
i Write down the value of y.
ii Give a reason for your answer.

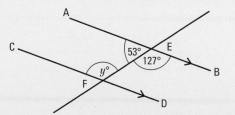

Diagram **NOT** accurately drawn

Nov 2008

13 AB is parallel to CD.
i Write down the value of y.
ii Give a reason for your answer.

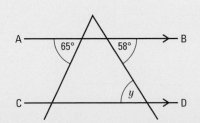

Diagram **NOT** accurately drawn

June 2008

D

14 AB is parallel to CD. EF is a straight line.

   i Write down the value of y.

   ii Give a reason for your answer.

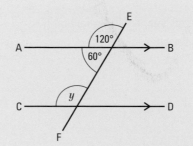

Diagram **NOT** accurately drawn

June 2008

* **15** ABCD is a straight line.

PQ is parallel to RS.

Write down the size of the angles marked x and y and give a reasons for your answers.

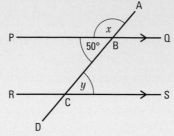

Diagram **NOT** accurately dawn

March 2008 adapted

16 This shape is a regular polygon.

   a Write down the special name for this type of regular polygon.

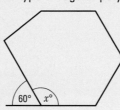

   b **i** Work out the size of the angle marked $x°$.

      ii Give a reason for your answer.

June 2007

17 ABC is an isosceles triangle.

BCD is a straight line.

AB = AC.

Angle A = 54°.

   a **i** Work out the size of the angle marked x.

      ii Give a reason for your answer.

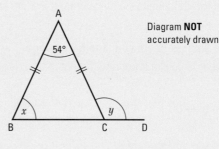

Diagram **NOT** accurately drawn

   b Work out the size of the angle marked y.

June 2007

18 **a** Find the bearing of B from A.

   b On a copy of the diagram, draw a line on a bearing of 135° from A.

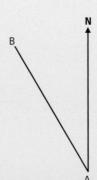

Nov 2006

19 The diagram shows a regular hexagon.
  a Work out the size of angle x.
  b Work out the size of angle y.

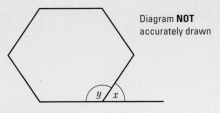

Diagram **NOT**
accurately drawn

Nov 2006

20 Find the size of the marked angles. Give reasons for your answers.

  a

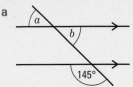

  b

21 The diagram shows the position of two airports, A and B.
  A plane flies from airport A to airport B.
  a Measure the size of the angle marked x.
  b Work out the real distance between airport A and airport B.
    Use the scale 1 cm represents 50 km.
  Airport C is 350 km on a bearing of 060° from airport B.
  c On the diagram, mark airport C with a cross (×). Label it C.

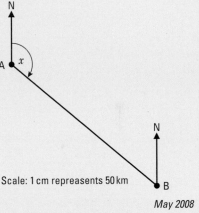

Scale: 1 cm repreasents 50 km

May 2008

22 The diagram shows part of a regular 10-sided polygon.
  Work out the size of the angle marked x.

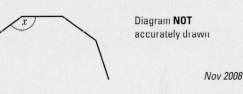

Diagram **NOT**
accurately drawn

Nov 2008

* 23 Prove that the angle sum of a triangle is 180°.

8 FRACTIONS

We use fractions to describe the phases of the moon. The moon in the photo is commonly called a half moon as it looks like half a circle. However, it is more correctly called a quarter moon as when it looks like this it has completed one quarter of an orbit around the Earth and one quarter of the moon's surface is visible from Earth.

◉ Objectives

In this chapter you will:
- learn how to recognise fractions
- learn how to work with fractions
- learn how to convert between fractions and decimals.

◈ Before you start

You need to:
- know your multiplication tables and the rules of basic arithmetic.

8.1 Understanding fractions

Objective

- You can identify fractions from diagrams or words.

Why do this?

Fractions such as halves and quarters are used in many contexts, and shops often have offers, such as '$\frac{1}{3}$ off'.

Get Ready

1. Work out the value of:
   a $12 \div 3$ **b** 5×4 **c** $28 \div 4$ **d** $54 \div 9$

Key Points

- Fractions are parts of a unit.
  - The top number is called the **numerator**.
  - The bottom number is called the **denominator**.

Example 1

Identify the fraction of the parking spaces that are occupied from this information.

The car park is divided into three spaces.

Two spaces have cars in them.

Two thirds or $\frac{2}{3}$ of the parking spaces are occupied. ←

> There are 3 spaces, so 3 is the denominator.
> 2 spaces are occupied, so 2 is the numerator.

Example 2

John has 150 marbles.

30 of the marbles are made from metal. The remainder are made from glass.

What fraction of John's marbles are

a metal   **b** glass?

a 30 of the 150 marbles are metal.
   The fraction is $\frac{30}{150}$

b $150 - 30 = 120$ marbles are glass.
   The fraction is $\frac{120}{150}$

Exercise 8A

Questions in this chapter are targeted at the grades indicated.

1 Copy these shapes into a table like the one on the right.
Complete your table.
The first shape has been done for you.

G

Shape	Fraction shaded	Fraction not shaded
⊘	$\frac{1}{2}$	$\frac{1}{2}$

G

2 Make four copies of this rectangle.
Shade them to show these fractions (one fraction on each copy).

    a $\frac{1}{16}$ b $\frac{3}{16}$ c $\frac{8}{16}$ d $\frac{16}{16}$

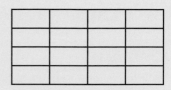

3 Make three copies of this circle.
Shade them to show these fractions.

    a $\frac{1}{6}$ b $\frac{3}{6}$ c $\frac{1}{2}$

4 There are 28 people on a martial arts course. Thirteen are female and 15 are male.
What fraction of the people are

    a male   b female?

5 28 competitors took part in a surfing competition on a Saturday, and 47 other competitors took part on Sunday.

    a How many surfers were there altogether?

    b What fraction of the surfers competed on Sunday?

    c What fraction of the surfers competed on Saturday?

8.2 Equivalent fractions

◎ Objective

● You can find families of fractions that are equal.

⍰ Why do this?

To share a pizza equally with a friend, cut it into six pieces and have three each, or cut it into four pieces and have two each.

◈ Get Ready

1. Copy and complete

    a $7 \times ? = 56$ b $3 \times ? = 27$ c $9 \times ? = 54$

🔍 Key Points

● You can make an **equivalent fraction** by multiplying or dividing both numerator and denominator by the same whole number.

● You can write a fraction in its **simplest form** by dividing numerator and denominator by the same common factor (see Section 1.10).

🔍 Example 3

Write the fraction $\frac{18}{24}$ in its simplest form.

$$\overset{\div 6}{\frac{18}{24}} = \frac{3}{4}$$
$$\underset{\div 6}{\phantom{\frac{18}{24}}}$$

> Divide the numerator and denominator by the *same* number.

$\frac{3}{4}$ is equivalent to $\frac{18}{24}$

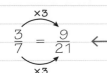 **Example 4** Convert $\frac{3}{7}$ to an equivalent fraction with a denominator of 21.

Multiply the numerator and the denominator by the *same* number.

$\frac{3}{7}$ is equivalent to $\frac{9}{21}$

 Exercise 8B

1 For each of these diagrams write down at least two equivalent fractions that describe the shaded fraction.

a   **b**

c   **d**

G

2 Copy and complete each set of equivalent fractions.

a $\frac{3}{4} = \frac{\square}{8} = \frac{\square}{12} = \frac{\square}{16} = \frac{\square}{20} = \frac{\square}{24}$ **b** $\frac{2}{7} = \frac{\square}{14} = \frac{\square}{21} = \frac{\square}{28} = \frac{\square}{35} = \frac{\square}{42}$

c $\frac{4}{5} = \frac{\square}{10} = \frac{\square}{15} = \frac{\square}{20} = \frac{\square}{25} = \frac{\square}{30}$ **d** $\frac{1}{3} = \frac{\square}{9} = \frac{\square}{18} = \frac{\square}{27} = \frac{\square}{36} = \frac{\square}{45}$

3 Copy and complete these equivalent fractions.

a $\frac{1}{6} = \frac{\square}{18}$ **b** $\frac{3}{7} = \frac{\square}{14}$ **c** $\frac{3}{8} = \frac{\square}{48}$ **d** $\frac{4}{7} = \frac{\square}{21}$

e $\frac{5}{6} = \frac{\square}{36}$ **f** $\frac{2}{3} = \frac{6}{\square}$ **g** $\frac{4}{9} = \frac{24}{\square}$ **h** $\frac{5}{7} = \frac{\square}{56}$

i $\frac{9}{10} = \frac{90}{\square}$ **j** $\frac{7}{12} = \frac{84}{\square}$ **k** $\frac{7}{8} = \frac{49}{\square}$ **l** $\frac{2}{9} = \frac{\square}{81}$

4 **a** Find a fraction equivalent to $\frac{1}{2}$ and a fraction equivalent to $\frac{1}{3}$ so that the bottom numbers of the two new fractions are the same.

b Repeat part **a** for

i $\frac{2}{5}$ and $\frac{3}{6}$ **ii** $\frac{1}{10}$ and $\frac{1}{7}$ **iii** $\frac{1}{4}$ and $\frac{5}{6}$

iv $\frac{1}{2}$ and $\frac{3}{5}$ **v** $\frac{2}{3}$ and $\frac{1}{8}$ **vi** $\frac{3}{4}$ and $\frac{3}{5}$

8.3 Ordering fractions

⊙ Objective

○ You can put fractions in order of size.

? Why do this?

To get the best deal when shopping, you need to know if half price is better than one-third off.

⬆ Get Ready

1. Find the lowest common multiple of
   a 5 and 4   b 10 and 7   c 2, 4 and 3

🔍 Key Points

◉ Fractions can be ordered by using equivalent fractions with the same denominator.
◉ To find a **common denominator**, you need to find a common multiple (see Section 1.10).

🔍 Example 5

Write these fractions in order, starting with the largest.

$\frac{1}{3}$ $\frac{2}{5}$ $\frac{3}{10}$ $\frac{1}{6}$

Step 1 Decide on a common denominator to use. Find a common multiple of 3, 5, 10 and 6. 30 is the lowest.

$\frac{1}{3} = \frac{1 \times 10}{3 \times 10} = \frac{10}{30}$ $\frac{2}{5} = \frac{12}{30}$ $\frac{3}{10} = \frac{9}{30}$ $\frac{1}{6} = \frac{5}{30}$ ←

Step 2 Re-write the fractions as their equivalent fractions using 30 as denominator.

$\frac{2}{5}$ $\frac{1}{3}$ $\frac{3}{10}$ $\frac{1}{6}$ ←

Step 3 Compare the equivalent fractions then write the original fractions in order.

⚙ Exercise 8C

G

1   By writing equivalent fractions, find the smaller fraction in each pair.
   a $\frac{2}{5}$ or $\frac{1}{4}$ b $\frac{2}{4}$ or $\frac{4}{5}$ c $\frac{2}{3}$ or $\frac{3}{4}$ d $\frac{3}{5}$ or $\frac{7}{10}$

2   Which is larger?
   a $\frac{2}{5}$ or $\frac{3}{6}$ b $\frac{1}{10}$ or $\frac{1}{7}$ c $\frac{1}{4}$ or $\frac{5}{6}$ d $\frac{1}{2}$ or $\frac{3}{5}$ e $\frac{2}{3}$ or $\frac{1}{8}$ f $\frac{3}{4}$ or $\frac{3}{5}$

3   Write these fractions in order of size. Put the smallest one first.
   a $\frac{1}{2}, \frac{3}{4}, \frac{2}{3}$ b $\frac{4}{5}, \frac{5}{6}, \frac{7}{15}$ c $\frac{3}{4}, \frac{4}{5}, \frac{1}{2}$ d $\frac{3}{7}, \frac{5}{14}, \frac{1}{2}, \frac{4}{7}$

4   Put these fractions in order of size, starting with the largest.
   $\frac{2}{5}$ $\frac{1}{2}$ $\frac{7}{8}$ $\frac{3}{4}$ $\frac{2}{10}$

8.4 Improper fractions and mixed numbers

Objective

- You can use fractions that are bigger than 1.

Why do this?

Distances on road signs are often displayed as mixed numbers, e.g. 'Oxford $7\frac{1}{2}$ miles'.

Get Ready

1. Calculate

   a $23 \div 4$ **b** $2 \times 7 + 4$ **c** $4 \times 8 + 3$

Key Points

- A fraction with a numerator that is larger than the denominator is called an **improper fraction**.
- An improper fraction can be written as a **mixed number** with a whole number part and a proper fraction part.

Example 6   Change $\frac{23}{7}$ into a mixed number.

$23 \div 7 = 3$ remainder 2 ← **Step 1** Do the division.

$= 3\frac{2}{7}$ ← **Step 2** As this is 3 whole ones with 2 left over, the mixed number is $3\frac{2}{7}$

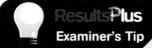

ResultsPlus

Examiner's Tip

Remember: a fraction is a division waiting to be done.

Example 7   Change $5\frac{3}{4}$ into an improper fraction.

$5\frac{3}{4} = \frac{20}{4} + \frac{3}{4} = \frac{23}{4}$ ← Each whole number is the same as $\frac{4}{4}$ (4 quarters). So, the whole number 5 is $5 \times \frac{4}{4} = \frac{20}{4}$ (20 quarters).

Exercise 8D

1   Change these improper fractions to mixed numbers.

   a $\frac{5}{2}$ **b** $\frac{7}{4}$ **c** $\frac{9}{7}$ **d** $\frac{11}{8}$ **e** $\frac{9}{8}$ **f** $\frac{16}{5}$

   g $\frac{23}{10}$ **h** $\frac{24}{5}$ **i** $\frac{16}{7}$ **j** $\frac{12}{5}$ **k** $\frac{20}{3}$ **l** $\frac{16}{9}$

   m $\frac{39}{4}$ **n** $\frac{27}{5}$ **o** $\frac{26}{9}$ **p** $\frac{17}{10}$

2   Change these mixed numbers to improper fractions.

   a $1\frac{1}{2}$ **b** $5\frac{1}{2}$ **c** $2\frac{3}{4}$ **d** $1\frac{2}{3}$ **e** $3\frac{1}{4}$ **f** $4\frac{2}{5}$

   g $3\frac{7}{10}$ **h** $5\frac{1}{5}$ **i** $7\frac{3}{4}$ **j** $2\frac{1}{4}$ **k** $1\frac{9}{10}$ **l** $9\frac{1}{3}$

   m $2\frac{5}{6}$ **n** $5\frac{3}{8}$ **o** $3\frac{5}{8}$ **p** $1\frac{9}{100}$

G

8.5 Multiplying fractions

◈ **Get Ready**

1. Write down

   a 13×6 **b** 7×19 **c** 15×7

🔧 **Key Points**

◎ To multiply two fractions, multiply the numerators together and multiply the denominators.

◎ To find a fraction of a quantity, multiply the fraction by the quantity. For example, $\frac{3}{4}$ of £60 $= \frac{3}{4} \times$ £60.

🔍 **Example 8**   Work out

      a $\frac{3}{5} \times \frac{2}{3}$ **b** $\frac{4}{7} \times 3$ **c** $2\frac{3}{5} \times 1\frac{1}{4}$

> *Step 1* Change any mixed numbers to improper fractions.
> *Step 2* Multiply numerators together and denominators together.
> *Step 3* Simplify if possible.

a $\frac{3}{5} \times \frac{2}{3} = \frac{3 \times 2}{5 \times 3} = \frac{6}{15} = \frac{2}{5}$ ←   Divide top and bottom by 3.

b $\frac{4}{7} \times 3$

   $= \frac{4}{7} \times \frac{3}{1}$ ←   First, write the whole number 3 as $\frac{3}{1}$

   $= \frac{4 \times 3}{7 \times 1} = \frac{12}{7}$

c $2\frac{3}{5} \times 1\frac{1}{4} = \left(\frac{10}{5} + \frac{3}{5}\right) \times \left(\frac{4}{4} + \frac{1}{4}\right) = \frac{13}{5} \times \frac{5}{4}$

             $= \frac{13 \times 5}{5 \times 4} = \frac{65}{20} = \frac{13}{4}$ ←   Divide top and bottom by 5.

⚙ **Exercise 8E**

E

1   Work out

   a $\frac{1}{2} \times \frac{3}{4}$ **b** $\frac{3}{8} \times \frac{1}{4}$ **c** $\frac{2}{5} \times \frac{4}{5}$ **d** $\frac{3}{8} \times \frac{3}{4}$

   e $\frac{5}{12} \times \frac{1}{3}$ **f** $\frac{7}{10} \times \frac{3}{4}$ **g** $\frac{3}{10} \times \frac{3}{5}$ **h** $\frac{2}{3} \times \frac{2}{3}$

   i $\frac{1}{2} \times \frac{3}{8}$ **j** $\frac{4}{5} \times \frac{2}{3}$ **k** $\frac{4}{7} \times \frac{1}{3}$ **l** $\frac{2}{3} \times \frac{2}{5}$

   m $\frac{2}{7} \times \frac{1}{5}$ **n** $\frac{2}{3} \times \frac{5}{7}$ **o** $\frac{3}{2} \times \frac{3}{4}$ **p** $\frac{3}{5} \times \frac{1}{3}$

2 Work out

a $\frac{1}{2} \times \frac{4}{5}$ b $\frac{3}{4} \times \frac{4}{5}$ c $\frac{5}{6} \times \frac{3}{5}$ d $\frac{4}{5} \times \frac{3}{10}$

e $\frac{5}{6} \times \frac{3}{4}$ f $\frac{7}{12} \times \frac{3}{14}$ g $\frac{8}{9} \times \frac{3}{10}$ h $\frac{3}{4} \times \frac{16}{21}$

i $\frac{1}{3} \times \frac{6}{7}$ j $\frac{6}{7} \times \frac{5}{12}$ k $\frac{1}{2} \times \frac{4}{10}$ l $\frac{2}{3} \times \frac{1}{4}$

m $\frac{3}{7} \times \frac{2}{6}$ n $\frac{6}{5} \times \frac{1}{3}$ o $5 \times \frac{7}{10}$ p $\frac{9}{10} \times \frac{13}{18}$

3 Work out

a $\frac{1}{2} \times 7$ b $\frac{2}{3} \times 5$ c $6 \times \frac{4}{5}$ d $8 \times \frac{3}{4}$

e $\frac{7}{10} \times 20$ f $9 \times \frac{2}{3}$ g $10 \times \frac{2}{5}$ h $\frac{5}{6} \times 12$

4 Work out

a $\frac{1}{2}$ of £8   b $\frac{1}{5}$ of £25   c $\frac{1}{3}$ of £21   d $\frac{1}{6}$ of £54

e $\frac{1}{4}$ of 28 cm   f $\frac{3}{4}$ of 28 cm   g $\frac{1}{10}$ of 440 kg   h $\frac{3}{5}$ of 30 kg

5 Work out

a $\frac{2}{3} \times 1\frac{1}{3}$ b $\frac{2}{5} \times 2\frac{1}{3}$ c $1\frac{1}{2} \times \frac{1}{4}$ d $3\frac{1}{4} \times \frac{1}{2}$

e $\frac{2}{3} \times 4\frac{1}{4}$ f $\frac{5}{6} \times 1\frac{1}{3}$ g $2\frac{1}{2} \times \frac{7}{10}$

6 A machine takes $5\frac{1}{2}$ minutes to produce a special type of container.
How long would the machine take to produce 15 containers at the same rate?

7 Work out

a $3\frac{1}{2} \times 1\frac{1}{2}$ b $2\frac{1}{3} \times 2\frac{3}{8}$ c $1\frac{4}{5} \times 2\frac{1}{3}$ d $3\frac{3}{4} \times 1\frac{2}{5}$

e $2\frac{1}{2} \times \frac{1}{4}$ f $1\frac{2}{5} \times 1\frac{1}{3}$ g $6 \times 2\frac{2}{3}$ h $2\frac{1}{7} \times 1\frac{2}{5}$

8.6 Dividing fractions

◉ Objective

◉ You can divide fractions by other fractions or a whole number.

❓ Why do this?

Doctors calculate the dosage of drugs for children as a fraction of the adult dose, e.g. half a half-teaspoon.

◈ Get Ready

1. Work out

a 8×3 b 2×15 c 10×7

🔍 Key Point

◉ To divide fractions, turn the dividing fraction upside down and multiply the fractions.

Example 9 Work out

$$\textbf{a} \quad \frac{1}{4} \div \frac{3}{5} \qquad\qquad \textbf{b} \quad \frac{15}{16} \div 5 \qquad\qquad \textbf{c} \quad 3\frac{1}{2} \div 4\frac{3}{4}$$

a $\frac{1}{4} \div \frac{3}{5} = \frac{1}{4} \times \frac{5}{3}$ ⟵ | Turn the second fraction upside down and multiply.

$\qquad = \frac{1 \times 5}{4 \times 3} = \frac{5}{12}$

b $\frac{15}{16} \div 5 = \frac{15}{16} \div \frac{5}{1}$

Results Plus
Watch Out!

$\qquad = \frac{15}{16} \times \frac{1}{5}$

$\qquad = \frac{15 \times 1}{16 \times 5}$

$\qquad = \frac{15}{80}$

$\qquad = \frac{3}{16}$ ⟵ | Divide top and bottom by 5.

Sometimes students turn the first fraction upside down by mistake. Make sure you turn the second fraction upside down.

c $3\frac{1}{2} \div 4\frac{3}{4} = \frac{7}{2} \div \frac{19}{4}$ ⟵ | Change mixed numbers to improper fractions.

$\qquad = \frac{7}{2} \times \frac{4}{19}$ ⟵ | Turn the second fraction upside down and multiply.

$\qquad = \frac{28}{38}$

$\qquad = \frac{14}{19}$ ⟵ | Simplify the answer.

Exercise 8F

1 Work out

a $\frac{1}{3} \div \frac{1}{4}$	**b** $\frac{1}{4} \div \frac{1}{3}$	**c** $\frac{3}{4} \div \frac{1}{2}$	**d** $\frac{1}{2} \div \frac{7}{10}$
e $\frac{2}{3} \div \frac{1}{5}$	**f** $\frac{5}{8} \div \frac{1}{3}$	**g** $\frac{5}{6} \div \frac{3}{4}$	**h** $\frac{7}{10} \div \frac{4}{5}$
i $\frac{2}{9} \div \frac{1}{2}$	**j** $\frac{2}{5} \div \frac{3}{4}$	**k** $\frac{3}{8} \div \frac{2}{3}$	**l** $\frac{1}{2} \div \frac{1}{4}$

2 Work out

a $8 \div \frac{1}{2}$	**b** $12 \div \frac{3}{4}$	**c** $6 \div \frac{3}{5}$	**d** $8 \div \frac{7}{8}$
e $4 \div \frac{4}{5}$	**f** $1 \div \frac{7}{12}$	**g** $5 \div \frac{1}{3}$	**h** $6 \div \frac{1}{4}$

3 Work out

a $2\frac{1}{2} \div \frac{1}{2}$	**b** $3\frac{1}{4} \div 2\frac{1}{2}$	**c** $3\frac{3}{4} \div 2\frac{1}{4}$	**d** $1\frac{5}{8} \div 3\frac{1}{6}$
e $3\frac{2}{3} \div 7\frac{1}{3}$	**f** $5\frac{1}{2} \div 2\frac{3}{4}$	**g** $1\frac{7}{10} \div 2\frac{7}{10}$	**h** $\frac{7}{8} \div 1\frac{2}{3}$

4 Work out

a $\frac{3}{4} \div 8$	**b** $\frac{5}{6} \div 2$	**c** $\frac{3}{5} \div 6$	**d** $\frac{4}{5} \div 5$
e $1\frac{1}{3} \div 4$	**f** $3\frac{1}{4} \div 6$	**g** $2\frac{5}{6} \div 10$	**h** $2\frac{1}{2} \div 15$
i $\frac{8}{9} \div 4$	**j** $\frac{2}{3} \div 6$	**k** $4\frac{2}{3} \div 4$	**l** $5\frac{1}{4} \div 3$

8.7 Adding and subtracting fractions

⊙ **Objective**

⊙ You can add and subtract fractions.

❓ **Why do this?**

You might need to add or subtract distances involving fractions.

⬆ **Get Ready**

1. Find the lowest common multiple of these.
   a 8 and 7   b 3 and 5   c 6 and 3

🔑 **Key Points**

⊙ Fractions can be added or subtracted when they have the same denominator.

⊙ To add or subtract fractions that have different denominators you need to find equivalent fractions that have the same denominator (see Section 8.3).

🔍 **Example 10**   Work out $\frac{5}{8} + \frac{3}{7}$

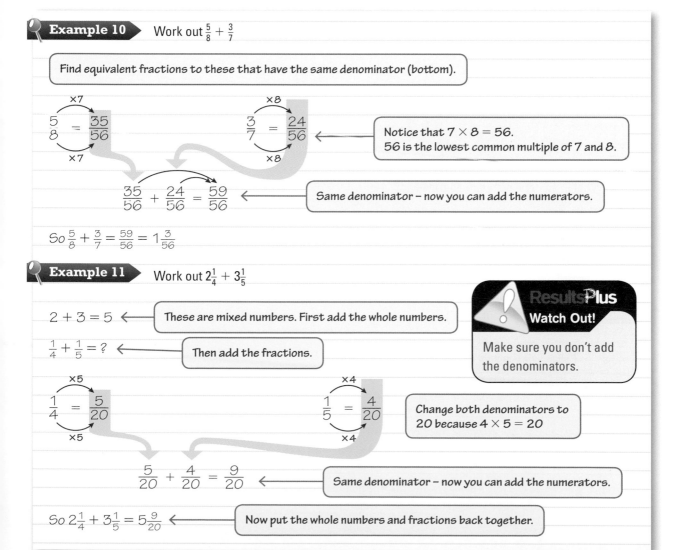

Find equivalent fractions to these that have the same denominator (bottom).

$\overset{\times 7}{\frac{5}{8}} = \frac{35}{56} \underset{\times 7}{}$ $\overset{\times 8}{\frac{3}{7}} = \frac{24}{56} \underset{\times 8}{}$

Notice that $7 \times 8 = 56$.
56 is the lowest common multiple of 7 and 8.

$\frac{35}{56} + \frac{24}{56} = \frac{59}{56}$

Same denominator – now you can add the numerators.

So $\frac{5}{8} + \frac{3}{7} = \frac{59}{56} = 1\frac{3}{56}$

🔍 **Example 11**   Work out $2\frac{1}{4} + 3\frac{1}{5}$

$2 + 3 = 5$ These are mixed numbers. First add the whole numbers.

$\frac{1}{4} + \frac{1}{5} = ?$ Then add the fractions.

ResultsPlus
Watch Out!

Make sure you don't add the denominators.

$\overset{\times 5}{\frac{1}{4}} = \frac{5}{20} \underset{\times 5}{}$ $\overset{\times 4}{\frac{1}{5}} = \frac{4}{20} \underset{\times 4}{}$

Change both denominators to 20 because $4 \times 5 = 20$

$\frac{5}{20} + \frac{4}{20} = \frac{9}{20}$

Same denominator – now you can add the numerators.

So $2\frac{1}{4} + 3\frac{1}{5} = 5\frac{9}{20}$ Now put the whole numbers and fractions back together.

Exercise 8G

1 Work out

a $\frac{3}{8} + \frac{4}{8}$ b $\frac{2}{9} + \frac{5}{9}$ c $\frac{5}{12} + \frac{7}{12}$ d $\frac{5}{18} + \frac{11}{18}$

e $\frac{1}{8} + \frac{3}{8}$ f $\frac{2}{7} + \frac{4}{7}$ g $\frac{2}{5} + \frac{3}{5}$ h $\frac{9}{10} + \frac{7}{10}$

i $\frac{7}{9} + 2\frac{4}{9}$ j $\frac{5}{6} + 1\frac{5}{6}$ k $\frac{3}{4} + \frac{3}{4} + \frac{1}{4}$ l $\frac{3}{8} + \frac{5}{8} + \frac{7}{8}$

2 Work out

a $\frac{1}{2} + \frac{1}{4}$ b $\frac{1}{4} + \frac{3}{8}$ c $\frac{1}{2} + \frac{7}{8}$ d $\frac{2}{3} + \frac{1}{6}$

e $\frac{5}{6} + \frac{1}{3}$ f $\frac{2}{5} + \frac{3}{10}$ g $\frac{7}{12} + \frac{3}{4}$ h $\frac{3}{4} + \frac{7}{20}$

3 Work out

a $\frac{1}{3} + \frac{7}{8}$ b $\frac{3}{4} + \frac{1}{10}$ c $\frac{4}{9} + \frac{5}{12}$ d $\frac{7}{8} + \frac{9}{10}$

e $\frac{3}{10} + \frac{4}{15}$ f $\frac{5}{6} + \frac{1}{4}$ g $\frac{3}{8} + \frac{7}{12}$ h $\frac{1}{6} + \frac{8}{9}$

i $\frac{5}{8} + \frac{1}{4}$ j $\frac{1}{6} + \frac{5}{8}$ k $\frac{3}{8} + \frac{11}{16}$

4 Work out

a $\frac{1}{2} + \frac{1}{3}$ b $\frac{2}{5} + \frac{1}{6}$ c $\frac{5}{8} + \frac{1}{5}$ d $\frac{3}{4} + \frac{1}{9}$

e $\frac{5}{6} + \frac{3}{7}$ f $\frac{9}{10} + \frac{2}{7}$ g $\frac{2}{3} + \frac{7}{10}$ h $\frac{3}{5} + \frac{2}{7}$

i $\frac{1}{5} + \frac{3}{8}$ j $\frac{1}{5} + \frac{1}{6}$ k $\frac{2}{3} + \frac{2}{7}$

5 Work out

a $1\frac{1}{2} + 2\frac{1}{8}$ b $2\frac{3}{4} + 3\frac{7}{8}$ c $1\frac{3}{4} + 2\frac{5}{16}$ d $\frac{3}{4} + 3\frac{5}{8}$

e $2\frac{9}{16} + 1\frac{5}{8}$ f $1\frac{3}{10} + 1\frac{2}{3}$ g $3\frac{1}{6} + \frac{2}{7}$ h $2\frac{5}{6} + 1\frac{1}{7}$

i $3\frac{2}{5} + 2\frac{7}{15}$ j $1\frac{2}{3} + 1\frac{2}{9}$

6 Jo cycled $2\frac{3}{4}$ miles to one village and then a further $4\frac{1}{3}$ miles to her home. What is the total distance Jo travelled?

7 Work out

a $3\frac{1}{4} + 2\frac{1}{2}$ b $2\frac{1}{2} + \frac{2}{3}$ c $1\frac{1}{4} + 2\frac{7}{8}$ d $3\frac{1}{3} + 5\frac{3}{4}$

e $3\frac{5}{16} + 1\frac{7}{8}$ f $2\frac{11}{12} + \frac{3}{4}$ g $\frac{5}{6} + 6\frac{1}{3}$ h $2\frac{2}{3} + 4\frac{3}{5}$

Example 12   Work out $4\frac{2}{5} - 1\frac{1}{2}$

Step 1 Change to a common denominator.

$4\frac{2}{5} - 1\frac{1}{2} = 4\frac{4}{10} - 1\frac{5}{10}$ ⟵ 10 is a common denominator for 5 and 2.

Step 2 You cannot take away the fraction part as $\frac{5}{10}$ is bigger than $\frac{4}{10}$. So use 1 from $4\frac{4}{10}$ and change it into $\frac{10}{10}$ so that $4\frac{4}{10}$ becomes $3\frac{14}{10}$

Step 3 You can do the whole numbers and the fractions separately.

$3\frac{14}{10} - 1\frac{5}{10} = 2\frac{9}{10}$ ⟵ $\left(3 - 1 = 2 \text{ and } \frac{14}{10} - \frac{5}{10} = \frac{9}{10}\right)$

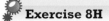 **Exercise 8H**

1 Work out

a $\frac{5}{11} - \frac{3}{11}$ b $\frac{7}{9} - \frac{5}{9}$ c $\frac{7}{8} - \frac{1}{8}$ d $\frac{7}{12} - \frac{5}{12}$

e $\frac{3}{4} - \frac{1}{4}$ f $\frac{5}{8} - \frac{3}{8}$ g $\frac{15}{16} - \frac{7}{16}$ h $\frac{6}{7} - \frac{3}{7}$

2 $\frac{2}{5}$ of the students at Hay College wear contact lenses. What fraction of the students do not wear them?

3 Work out

a $\frac{1}{2} - \frac{1}{4}$ b $\frac{7}{8} - \frac{3}{4}$ c $\frac{5}{8} - \frac{1}{2}$ d $\frac{3}{4} - \frac{1}{8}$

e $\frac{5}{6} - \frac{1}{3}$ f $\frac{7}{12} - \frac{1}{3}$ g $\frac{9}{10} - \frac{2}{5}$ h $\frac{1}{4} - \frac{1}{20}$

i $\frac{1}{2} - \frac{3}{8}$ j $\frac{7}{8} - \frac{1}{2}$ k $\frac{11}{12} - \frac{3}{4}$

4 Work out

a $\frac{2}{3} - \frac{1}{2}$ b $\frac{5}{8} - \frac{1}{3}$ c $\frac{1}{5} - \frac{1}{6}$ d $\frac{3}{5} - \frac{1}{6}$

e $\frac{4}{5} - \frac{2}{3}$ f $\frac{3}{4} - \frac{3}{5}$ g $\frac{7}{10} - \frac{1}{3}$ h $\frac{9}{10} - \frac{3}{4}$

i $5\frac{1}{4} - \frac{1}{10}$ j $7\frac{1}{2} - \frac{1}{3}$

5 In a school, $\frac{7}{16}$ of the students are girls. What fraction of the students are boys?

6 Work out

a $4\frac{5}{8} - 2\frac{1}{4}$ b $6\frac{1}{2} - 5\frac{1}{4}$ c $9\frac{1}{2} - 7\frac{3}{10}$ d $4 - 1\frac{3}{10}$

e $4\frac{4}{5} - 3\frac{9}{10}$ f $1\frac{2}{3} - \frac{11}{12}$ g $5\frac{3}{4} - 2\frac{19}{20}$ h $4\frac{7}{8} - 1\frac{2}{3}$

i $5\frac{7}{9} - 3\frac{1}{3}$ j $3\frac{4}{5} - \frac{3}{8}$ k $7\frac{4}{7} - 4\frac{2}{5}$

E

D

C

8.8 Converting between fractions and decimals

◎ Objectives

◉ You can change a fraction into a decimal and a decimal into a fraction.
◉ You can recognise that some fractions are recurring decimals.

⦾ Why do this?

Food labels show the amount of protein, fat etc in the food we eat, and what fractions these are of your recommended daily amount.

⬙ Get Ready

1. Do these divisions. Give your answers as decimals.
   a $3 \div 8$ b $1 \div 2$ c $4 \div 100$

Key Points

- All fractions can be written as decimals.
- You can change a fraction into a decimal by doing the division.
- You can change a decimal into a fraction by looking at the place value of its smallest digit.
- You should be familiar with the equivalent fractions and decimals in this table.

Decimal	0.01	0.1	0.25	0.5	0.75
Fraction	$\frac{1}{100}$	$\frac{1}{10}$	$\frac{1}{4}$	$\frac{1}{2}$	$\frac{3}{4}$

- Fractions that have an exact decimal equivalent are called **terminating decimals**.
- To change a fraction into a decimal, divide the numerator by the denominator.
- Fractions that have a decimal equivalent that repeats itself are called **recurring decimals**.

Example 13 Change these fractions into decimals. a $\frac{3}{4}$ b $\frac{1}{3}$ c $\frac{3}{11}$

a 0.75
$4\overline{)3.^30^20}$

$\frac{3}{4} = 0.75$

b 0.333
$3\overline{)1.^10^10^10}$

$\frac{1}{3} = 0.\dot{3}$

c 0.272727
$11\overline{)3.^30^80^30^80^30^80}$

$\frac{3}{11} = 0.\dot{2}\dot{7}$ ← Place dots over the 2 and 7 to show that they recur.

Example 14 Change these decimals into fractions. a 0.3 b 0.709

a The smallest digit of 0.3 is 3 tenths.
$0.3 = 3$ tenths $= \frac{3}{10}$

b The smallest digit of 0.709 is 9 thousandths.
$0.709 = 709$ thousandths $= \frac{709}{1000}$

Put the number into a place value table.

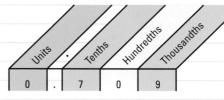

Units	.	Tenths	Hundredths	Thousandths
0	.	7	0	9

Exercise 8I

G

1 Change these fractions into decimals. Show your working.

a $\frac{3}{5}$ b $\frac{1}{2}$ c $\frac{7}{10}$ d $\frac{7}{20}$

e $\frac{4}{25}$ f $\frac{3}{50}$ g $\frac{7}{8}$ h $\frac{9}{20}$

i $\frac{19}{25}$ j $\frac{5}{16}$ k $\frac{1}{8}$ l $\frac{27}{50}$

m $\frac{9}{100}$ n $\frac{13}{200}$ o $\frac{2}{3}$ p $\frac{19}{20}$

2 Change these decimals into fractions.

a 0.3	b 0.37	c 0.93	d 0.137
e 0.293	f 0.7	g 0.59	h 0.003
i 0.00003	j 0.0013	k 0.77	l 0.077
m 0.39	n 0.0041	o 0.019	p 0.031

3 Write these fractions as decimals.

a $\frac{4}{5}$	b $\frac{3}{4}$	c $1\frac{1}{8}$	d $\frac{19}{100}$
e $3\frac{3}{5}$	f $\frac{13}{25}$	g $\frac{5}{8}$	h $3\frac{17}{40}$
i $\frac{7}{50}$	j $4\frac{3}{16}$	k $3\frac{3}{20}$	l $4\frac{5}{16}$
m $\frac{7}{1000}$	n $1\frac{7}{25}$	o $15\frac{15}{16}$	p $2\frac{7}{20}$

4 Write these decimals as fractions in their simplest form.

a 0.48	b 0.25	c 1.7	d 3.406
e 4.003	f 2.025	g 0.049	h 4.875
i 3.75	j 10.101	k 0.625	l 2.512
m 0.8125	n 14.14	o 9.1875	p 60.065

Chapter review

- Fractions are parts of a unit.
  - The top number is called the **numerator**.
  - The bottom number is called the **denominator**.
- You can make an **equivalent fraction** by multiplying or dividing both numerator and denominator by the same whole number.
- You can write a fraction in its **simplest form** by dividing numerator and denominator by the same common factor.
- You can order fractions by using equivalent fractions with the same denominator.
- To find a **common denominator**, you need to find a common multiple.
- A fraction whose numerator is larger than its denominator is called an **improper fraction**.
- An improper fraction can be written as a **mixed number** with a whole number part and a proper fraction part.
- To multiply two fractions, multiply the numerators together and multiply the denominators.
- To divide fractions, turn the dividing fraction upside down and multiply the fractions.
- Fractions can be added or subtracted when they have the same denominator.
- To add or subtract fractions that have different denominators you need to find equivalent fractions that have the same denominator.
- You can change a fraction into a decimal by doing the division.
- You can change a decimal to a fraction by looking at the place value of its smallest digit.
- Fractions that have an exact decimal equivalent are called **terminating decimals**.
- Fractions that have a decimal equivalent that repeats itself are called **recurring decimals**.

Review exercise

1 In each case write down the fraction of the shape that has been shaded.

a   b 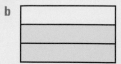   c   d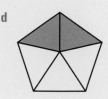

2 Write down three fractions equivalent to each of the following.

a $\frac{4}{5}$ b $\frac{2}{7}$ c $\frac{18}{24}$ d $\frac{16}{20}$ e $\frac{30}{100}$ f $\frac{18}{27}$

3 Change these fractions into decimals.

a $\frac{1}{4}$ b $\frac{3}{8}$ c $\frac{7}{10}$ d $\frac{3}{5}$ e $\frac{24}{200}$ f $\frac{37}{50}$

4 Change these decimals into fractions.

a 0.34   b 0.125   c 0.3   d 0.025   e 0.15   f 3.1

5 There are 600 counters in a bag. 90 of the counters are yellow.
Work out 90 as a fraction of 600.
Give your answer in its simplest form.

May 2009

6 On an average day, William spends 9 hours asleep, 8 hours at school, 2 hours watching television, 1 hour travelling and 4 hours on other things.
Work out the fraction of time spent

a asleep   b at school   c watching television
d travelling   e on other things.

7 In each case, write down which fraction is larger. You must show your working.

a $\frac{2}{3}$ or $\frac{3}{5}$ b $\frac{3}{4}$ or $\frac{4}{5}$ c $\frac{7}{10}$ or $\frac{3}{4}$ d $\frac{4}{9}$ or $\frac{11}{25}$

8 Put these fractions in order of size. Start with the largest.

$\frac{3}{10}$ $\frac{1}{3}$ $\frac{2}{7}$ $\frac{4}{15}$ $\frac{29}{100}$

9 Here are the fractions $\frac{3}{4}$ and $\frac{4}{5}$.
Which is the larger fraction? You must show working to explain your answer.
You may copy and use the grids to help with your explanation.

June 2007

10 Put these fractions and decimals in order. Start with the smallest.

$\frac{7}{10}$ $\frac{3}{4}$ 0.6 $\frac{450}{1000}$

11 Work out

a $\frac{3}{5} \times \frac{2}{9}$ **b** $\frac{7}{15} \times \frac{5}{21}$ **c** $\frac{9}{16} \div \frac{3}{4}$

12 Which one of these fractions is equal to 0.28?

$\frac{28}{10}$ $\frac{1}{28}$ $\frac{2}{8}$ $\frac{7}{25}$ $\frac{28}{1}$

13 Work out

a $\frac{5}{12} + \frac{1}{4}$ **b** $\frac{7}{15} + \frac{4}{45}$ **c** $\frac{3}{5} + \frac{3}{8}$ **d** $\frac{7}{12} + \frac{4}{9}$ **e** $\frac{7}{9} - \frac{1}{3}$

14 The diagram shows a rectangle.
Work out the distance all the way around the rectangle.

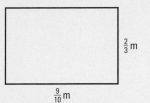

$\frac{2}{3}$ m

$\frac{9}{10}$ m

15 Carats are used to measure the purity of gold jewellery. Purity is measured in 24ths.
24 carat gold is pure gold but is not often used in jewellery as is it very expensive and very soft.
12 carat gold is $\frac{1}{2}$ gold. It is an alloy of which $\frac{1}{2}$ the weight is gold and $\frac{1}{2}$ the weight is made up of another metal.

a Anisha buys an 18 carat gold bracelet.
Explain what this means in terms of the amount of gold in the bracelet.

b Glen buys an 8 carat ring. What fraction of gold does the ring contain?
Give your answer in its simplest form.

c Kelly has a 12 carat gold bangle weighing 60 g. What weight of gold is in the bangle?

d Raj buys a 10 carat necklace that weighs 120 g. What is the weight of gold in the necklace?

e Carol's watch strap weighs 144 g. The weight of gold in the strap amounts to 60 g. What carat is this?

16   **a** $2\frac{1}{2} \times 3\frac{2}{5}$ **b** $3\frac{3}{8} \div 2\frac{4}{7}$ **c** $1\frac{2}{5} \div 1\frac{11}{14}$

17 Work out $2\frac{2}{3} \times 1\frac{1}{4}$
Give your answer in its simplest form.

ResultsPlus
Exam Question Report

82% of students answered this question poorly. Many students did turn the mixed numbers into improper fractions, but wrote them with a common denominator of 12.

June 2007

18 Work out

a $1\frac{3}{4} - \frac{7}{16}$ **b** $2\frac{1}{4} - \frac{1}{2}$ **c** $4\frac{3}{8} - 1\frac{1}{2}$ **d** $5\frac{1}{5} - 2\frac{7}{15}$

19 George won some money in a lottery.
He gave $\frac{2}{5}$ to his wife and $\frac{1}{3}$ to his daughter.
What fraction did he have left?

20 The diagram represents a part of a machine. In order to fit the machine, the part must be between $6\frac{1}{16}$ cm and $6\frac{3}{16}$ cm long.

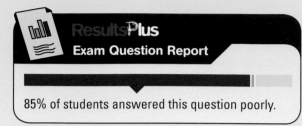

Diagram **NOT** accurately drawn

$2\frac{1}{2}$ cm $\qquad$ $3\frac{5}{8}$ cm

Will the part fit the machine? You must explain your answer.

21 A full glass of water holds $\frac{1}{6}$ of a bottle of water. How many glasses of water can be filled from $2\frac{1}{2}$ bottles of water?

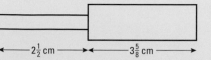

Exam Question Report

85% of students answered this question poorly.

June 2007

9 ALGEBRA 2

In this chapter, we will look at writing numbers using powers. For example, one trillion is written in long form as 1 000 000 000 000 but can be written using powers as 10^{12}. This form of writing large numbers is very useful in science. The distance from the Earth to the Sun and back is approximately 10^8 miles.

◉ Objectives

In this chapter you will:
- calculate with powers
- write expressions as a single power of the same number
- use powers in algebra
- multiply out brackets in algebra
- factorise expressions.

◈ Before you start

You need to know how to:
- use letters instead of numbers
- collect like terms
- multiply and divide variables and write them in their simplest form.

9.1 Calculating with powers

◎ Objectives

- You can work out the value of numbers raised to a power.
- You can write numbers using index notation.
- You can work out values of expressions given in index notation.

◈ Why do this?

It is easy to make mistakes when working with very large numbers with lots of zeros, or with repeated multiples. Using powers reduces mistakes.

◈ Get Ready

1. Work out the following.

   a 2×2 b $5 \times 5 \times 5$ c $8 \times 8 \times 8$ d $10 \times 10 \times 10 \times 10 \times 10 \times 10$

◐ Key Points

- The ^{2} in 5^2 is called a **power** or an **index**. It tells you how many times the given number must be multiplied by itself. The plural of index is indices.
- The 5 in 5^2 is called the **base**.
- You can solve equations such as $3^x = 81$ by working out how many times the base has to be multiplied by itself to give the answer.

Example 1   Work out the value of the following.

   a 3^2 b 2 to the power of 5   c $2^3 \times 3^2$

a $3^2 = 3 \times 3 = 9$ ← 3 to the power of 2 is usually called 3 squared.

b 2 to the power of 5 = 2^5

$2^5 = 2 \times 2 \times 2 \times 2 \times 2 = 32$ ← Write as a power. 2^5 means 2 multiplied by itself 5 times.

c $2^3 \times 3^2$

$2^3 = 2 \times 2 \times 2 = 8$ ← Work out 2^3 first.

$3^2 = 3 \times 3 = 9$ ← Work out 3^2.

$2^3 \times 3^2 = 8 \times 9$ ← Use the two values above.

$= 72$

Example 2   Rewrite these expressions using index notation.

   a $3 \times 3 \times 3 \times 3$ b $4 \times 4 \times 5 \times 5 \times 5$ c $6 \times 6 \times 6 \times 6 \times 6$

a $3 \times 3 \times 3 \times 3 = 3^4$

b $4 \times 4 \times 5 \times 5 \times 5 = 4^2 \times 5^3$ ← Replace 4×4 with 4^2 and $5 \times 5 \times 5$ with 5^3.

c $6 \times 6 \times 6 \times 6 \times 6 = 6^5$ ← The index is 5 because 5 lots of 6 are multiplied together.

Example 3 Find the value of x.

  a $5^x = 25$ b $4^x = 64$

a $5 \times 5 = 25$
  So $5^2 = 25$ ← Write as a power. Compare powers.
  $x = 2$

b $4 \times 4 \times 4 = 64$
  So $4^3 = 64$ ← Write as a power. Compare powers.
  $x = 3$

Exercise 9A

Questions in this chapter are targeted at the grades indicated.

E

1 Find the value of
   a 2^6 b 4 to the power 4   c 1^6
   d 10 to the power 4   e 5^4 f 6 to the power 5

2 Write these using index notation.
   a $2 \times 2 \times 2 \times 2$ b $4 \times 4 \times 4 \times 4 \times 4$
   c $1 \times 1 \times 1 \times 1 \times 1 \times 1$ d $8 \times 8 \times 8$
   e $3 \times 3 \times 8 \times 8 \times 8 \times 8$ f $4 \times 4 \times 4 \times 4 \times 2 \times 2 \times 2$

3 Work out the value of
   a 2^4 b 3^5 c 6^3 d 5^2
   e 8^3 f $2^4 \times 9^3$ g $2^6 \times 4^5$ h $5^3 \times 3^4$
   i $2^7 \times 3^5$ j $4^3 \times 4^1$

4 Copy and complete the table for powers of 10.

Power of 10	Index	Value	Value in words
	3		One thousand
10^2		100	
		1 000 000	One million
	1	10	
10^5	5		

5 Work out the value of
   a $4^3 \times 10^2$ b 4×10^2 c 6×10^3
   d $10^2 \div 5^2$ e $10^3 \div 2^3$ f $4^3 \div 2^2$

6 Find x when
   a $5^x = 125$ b $3^x = 81$ c $2^x = 64$ d $10^x = 10\,000$
   e $9^x = 81$ f $3^x = 27$ g $2^x = 16$ h $7^x = 49$

D

9.2 Writing expressions as a single power of the same number

Objective

- You can use rules of indices to simplify expressions.

Why do this?

Using rules of indices saves time and makes calculations easier.

Get Ready

1. Work out the value of
   a 2^2 b 4^3 c 7^4 d 10^4

Key Points

- To multiply powers of the same number, add the indices.
  e.g. $3^2 \times 3^3 = 3^{2+3} = 3^5$
- To divide powers of the same number, subtract the indices.
  e.g. $5^6 \div 5^3 = 5^{6-3} = 5^3$
- Any number raised to the power of 1 is equal to the number itself.
  e.g. $4^1 = 4$
- To raise a power of a number to a further power, multiply the powers (or indices).
  e.g. $(10^3)^2 = 10^{3 \times 2} = 10^6$

Example 4   Simplify these expressions by writing them as a single power of the number.

a $2^3 \times 2^4$ b $5^8 \div 5^3$ c $(8^2)^5$

a $2^3 \times 2^4 = 2^{3+4}$ ← [Add the powers.]
$= 2^7$

b $5^8 \div 5^3 = 5^{8-3}$ ← [Subtract the powers.]
$= 5^5$

c $(8^2)^5 = 8^{2 \times 5}$ ←
$= 8^{10}$

$(8^2)^5 = (8 \times 8) \times (8 \times 8) \times (8 \times 8) \times (8 \times 8) \times (8 \times 8)$
$= 8^{10}$
Multiply the powers.

Exercise 9B

Simplify these expressions by writing as a single power of the number.

1 a $6^8 \times 6^3$ b $8^3 \times 8^5$ c $2^4 \times 2^2$

2 a $4^3 \div 4^2$ b $6^6 \div 6^3$ c $7^5 \div 7$

3 a $4^2 \times 4^3$ b $5^3 \div 5$ c $3^9 \div 3^8$

C

C

4	a	$5^6 \times 5^4 \times 5^3$	b	$2^3 \times 2^7 \times 2$
5	a	$10^2 \times 10^2 \times 10$	b	$9^4 \div 9^4$
6	a	$6^3 \times 6^7 \times 6$	b	$5^2 \times 5^2 \times 5^2$
7	a	$3^5 \times 3 \times 3^2$	b	$4^7 \times \dfrac{4^5}{4^6}$

8 a $\dfrac{6^8}{6^2} \times 6^3$ b $5^8 \times \dfrac{5^4}{5^7}$ c $\dfrac{4^9}{4^2} \times 4^5$

9 a $(5^2)^3$ b $(7^4)^2$

9.3 Using powers in algebra to simplify expressions

Objectives

- You can multiply powers of the same letter.
- You can divide powers of the same letter.
- You can raise a power of a letter to a further power.

Why do this?

Using powers makes calculating easier.

Get Ready

1. Simplify

a $7^2 \times 7^5$ b $8^5 \div 8^3$ c $(5^3)^2$

Key Points

- In the expression x^n, the number n is called the power or index.
- $x^m \times x^n = x^{m+n}$
- $x^m \div x^n = x^{m-n}$
- $(x^m)^n = x^{m \times n}$
- Any letter raised to the power of 1 is equal to the letter itself, e.g. $x^1 = x$.
- Any letter raised to the power 0 is equal to 1, e.g. $x^0 = 1$.

Example 5 Simplify a $x^5 \times x^3$ b $y^7 \div y^4$ c $a^2 \times a^3 \times a^5$ d $(x^3)^2$

e $3x^2 \times 4x^3$ f $10x^6 \div 5x^3$ g $(3a^2)^4$

a $x^5 \times x^3 = x^{5+3}$ ⟵ [Add the powers.]

$\quad = x^8$

b $y^7 \div y^4 = y^{7-4}$ ⟵ [Subtract the powers.]

$\quad = y^3$

c $a^2 \times a^3 \times a^5 = a^{2+3+5}$

$\quad = a^{10}$

ResultsPlus
Examiner's Tip

Show your working.
The mark is scored here.

d $(x^3)^2 = x^{3 \times 2}$ ← Multiply the powers.

$\quad = x^6$

e $3x^2 \times 4x^3 = 3 \times 4 \times x^2 \times x^3$ ← Write the numbers and letters together.

$\quad = 12 \times x^{2+3}$ ← Multiply the numbers 3 and 4.

$\quad = 12x^5$ ← Add the powers 2 and 3.

f $10x^6 \div 5x^3 = \frac{10}{5} \times x^6 \div x^3$ ← Divide the numbers 10 and 5.

$\quad = 2 \times x^{6-3}$ ← Subtract the powers.

$\quad = 2x^3$

g $(3a^2)^4 = 3^4 \times (a^2)^4$ ← Both 3 and a^2 are raised to the power of 4. $3^4 = 3 \times 3 \times 3 \times 3 = 81$

$\quad = 81 \times a^{2 \times 4}$ ← Multiply the powers.

$\quad = 81a^8$

Exercise 9C

Simplify the following.

1
 a $x^8 \times x^2$
 b $y^3 \times y^8$
 c $x^9 \times x^5$

2
 a $a^5 \times a^3$
 b $b^3 \times b^3$
 c $d^7 \times d^4$

3
 a $p^5 \div p^2$
 b $q^{12} \div q^2$
 c $t^8 \div t^4$

4
 a $j^9 \div j^3$
 b $k^5 \div k^4$
 c $n^{25} \div n^{23}$

5
 a $x^5 \times x^2 \times x^2$
 b $y^2 \times y^4 \times y^3$
 c $z^3 \times z^5 \times z^2$

6
 a $3x^2 \times 2x^3$
 b $5y^9 \times 3y^{20}$
 c $6z^8 \times 4z^2$

7
 a $12p^8 \div 4p^3$
 b $15q^5 \div 3q^3$
 c $6r^5 \div 3r^2$

8
 a $(d^3)^4$
 b $(e^5)^2$
 c $(f^3)^3$
 d $(g^7)^9$

9
 a $(g^6)^4$
 b $(h^2)^2$
 c $(k^4)^0$
 d $(m^0)^{56}$

10
 a $(3d^2)^7$
 b $(4e)^3$
 c $(3f^{129})^0$

11
 a $a^4 \times \dfrac{a^5}{a^9}$
 b $b^7 \times \dfrac{b}{b^4}$
 c $c^3 \times \dfrac{c^4}{c^2} \times c^5$

12
 a $4d^9 \times 2d$
 b $8e^8 \div 4e^4$
 c $(4f^2)^2$

9.4 Understanding order of operations

Objective

● You can work out the value of numerical expressions.

Why do this?

When making a cake, you need to know what order to add the ingredients in. The same is true of a calculation such as $3 \times 4 + 2 \times 5$. It is important that the operations are carried out in the correct order or the answer will be wrong.

Get Ready

1. Work out
   a $(6 + 3) \times (2 - 1)$ b $6 + (3 \times 2) - 1$ c $((6 + 3) \times 2) - 1)$

Key Points

● **BIDMAS** gives the order in which **operations** should be carried out.

● Remember that B I D M A S stands for

B rackets   If there are brackets, work out the value of the expression inside the brackets first.

I ndices   Indices include square roots, cube roots and powers.

D ivide   If there are no brackets, do dividing and multiplying before adding and subtracting, no
M ultiply   matter where they come in the expression.

A dd
S ubtract   If an expression has only adding and subtracting then work it out from left to right.

Example 6   Work out $(3 \times 2) - 1$

$(3 \times 2) - 1 = 6 - 1$ ← Work out the Brackets first.
$\qquad\qquad = 5$

Example 7   Work out $3 + 2 \times 5 - 1$

There is no Bracket or Divide, so start with Multiply, then Add, then Subtract.

$3 + 2 \times 5 - 1 = 3 + 10 - 1$
$\qquad\qquad\quad = 13 - 1$
$\qquad\qquad\quad = 12$

Example 8   Work out $(10 + 2)^2 - 5 \times 3^2$

Brackets first, then Indices, Multiply, and finally Subtract.

$(10 + 2)^2 - 5 \times 3^2 = 12^2 - 5 \times 3^2$
$\qquad\qquad\qquad = 144 - 5 \times 9$
$\qquad\qquad\qquad = 144 - 45$
$\qquad\qquad\qquad = 99$

Exercise 9D

1 Use BIDMAS to help you find the value of these expressions.

a $5 + (3 + 1)$

b $5 - (3 + 1)$

c $5 \times (2 + 3)$

d $5 \times 2 + 3$

e $3 \times (4 + 3)$

f $5 \times 4 + 3$

g $20 \div 4 + 1$

h $20 \div (4 + 1)$

i $6 + 4 \div 2$

j $(6 + 4) \div 2$

k $24 \div (6 - 2)$

l $24 \div 6 - 2$

m $7 - (4 + 2)$

n $7 - 4 + 2$

o $((15 - 5) \times 4) \div ((2 + 3) \times 2)$

2 Make these expressions correct by replacing the • with $+$ or $-$ or $\times$ or $\div$ and using brackets if you need to. The first one is done for you.

a $4 • 5 = 9$ becomes $4 + 5 = 9$

b $4 • 5 = 20$

c $2 • 3 • 4 = 20$

d $3 • 2 • 5 = 5$

e $5 • 2 • 3 = 9$

f $4 • 2 • 8 = 10$

g $5 • 4 • 5 • 2 = 27$

h $5 • 4 • 5 • 2 = 23$

3 Work out

a $(3 + 4)^2$

b $3^2 + 4^2$

c $3 \times (4 + 5)^2$

d $3 \times 4^2 + 3 \times 5^2$

e $2 \times (4 + 2)^2$

f $2^3 + 3^2$

g $2 \times (3^2 + 2)$

h $\dfrac{(2 + 5)^2}{3^2 - 2}$

i $\dfrac{5^2 - 2^2}{3}$

j $4^2 - 2^4$

k $2^5 - 5^2$

l $4^3 - 8^2$

9.5 Multiplying out brackets in algebra

◎ Objectives

○ You can add expressions with brackets.

○ You can subtract expressions with brackets.

❓ Why do this?

Using brackets and collecting like terms helps you get the correct answer.

⬆ Get Ready

1. Expand

a $6(5 + 3)$

b $3(4 \times 4 - 5 \times 3)$

c $(5 \times 4)(2 \times 4 - 2 \times 1)$

🌐 Key Points

◉ Expanding brackets means multiplying each term inside the brackets by the term outside the brackets.

◉ To simplify an expression with brackets, expand the brackets and collect like terms.

Example 9   Simplify   **a** $2(3x - y) + 5(y - 2x)$

   b $3x(2x + y) - 2x(5y - 1)$

a $2(3x - y) + 5(y - 2x)$ ←

> 2 times $(3x - y)$ plus
> 5 times $(y - 2x)$
> Multiply each pair of terms.

$= 2 \times (3x - y) + 5 \times (y - 2x)$

> Remember $+ + = +$
> $+ - = -$
> $- + = -$
> $- - = +$

$= 2 \times 3x - 2 \times y + 5 \times y - 5 \times 2x$

$= 6x - 2y + 5y - 10x$

$= 3y - 4x$ ← Collect the terms.

b $3x(2x + y) - 2x(5y - 1)$

$= 3x \times 2x + 3x \times y - 2x \times 5y - 2x \times -1$

$= 6x^2 + 3xy - 10xy + 2x$ ← $x \times x = x^2$

$= 6x^2 - 7xy + 2x$ ← Collect the terms.

Exercise 9E

Expand and simplify.

1 $3(x + 2) + 2(x + 4)$ **2** $4(2x - 1) + 3(4x + 7)$

3 $5(3x + 2) + 4(2x + 1)$ **4** $7(3 - 2x) + 3(2x - 3)$

5 $6(4 - 2x) - 3(5 + 3x)$ **6** $4(3 - 2x) + 3(1 - 5x)$

7 $2(3x - 5y) + 3(2x - 4y)$ **8** $5(6y + 2x) - 4(3x + 2y)$

9 $3(2x - 3y) - 2(5x + 6y)$ **10** $3(2x + 3y) - 5(x + y)$

11 $4(3y - 2) - 5(y - 2)$ **12** $2(3x + 6) - 3(2x - 5)$

13 $4(3 - 2x) - 3(5 - 3x)$ **14** $2(3 - y - 2x) - 3(4x - 3y)$

15 $3(2x - 3y) + 5(3x - 2y)$ **16** $5(3y - 5x) - 2(x - 3y)$

17 $(4x - 3y) + 2(3x - 2y)$ **18** $7(3x - 5y) - (x - 3y)$

19 $x(2y + 1) + 2x(3y + 1)$ **20** $2x(3y + 1) + y(2x + 1)$

21 $2y(3x - 2) + 3x(2 - 3y)$ **22** $4x(2y - 5x) + 2y(x - y)$

C

9.6 Factorising expressions

◎ Objectives

○ You can factorise expressions by taking out a single factor.

○ You can factorise an expression by taking out multiple factors.

⟁ Why do this?

If you can see a common number or factor it can make calculations easier. Buying two burgers and two fries $(2b + 2f)$ is the same as buying $2(b + f)$.

⟁ Get Ready

1. Find the highest common factors of these.

   a 16 and 24   **b** 48 and 20   **c** ab and abc

⚲ Key Points

◉ Factorising is the reverse process to expanding brackets.

◉ To factorise an expression, find the common factor of the terms in the expression and write the common factor outside a bracket. Then complete the bracket with an expression which, when multiplied by the common factor, gives the original expression.

⚲ Example 10   Factorise   **a** $3x^2 + 5x$ **b** $10a^2 - 15ab$

a $3x^2 + 5x = x(3x + 5)$ ◂——— Take x outside the bracket. It is a common factor of $3x^2 + 5x$.

b $10a^2 - 15ab$
   $= 5a(2a - 3b)$ ◂——— The common factor may be both a number and a letter.
   Take $5a$ outside the bracket. It is a common factor of both $10a^2$ and $15ab$.

ResultsPlus
Examiner's Tip

Always check your answer by expanding.

⚙ Exercise 9F

Factorise each of the expressions in questions 1–6.

D

1 **a** $2x + 6$ **b** $6y + 2$ **c** $15b - 5$
     d $4r - 2$ **e** $3x + 5xy$ **f** $12x + 8y$
     g $12x - 16$ **h** $9 - 3x$ **i** $9 + 15g$

C

2 **a** $3x^2 + 4x$ **b** $5y^2 - 3y$ **c** $2a^2 + a$
     d $5b^2 - 2b$ **e** $7c - 3c^2$ **f** $d^2 + 3d$
     g $6m^2 - m$ **h** $4xy + 3x$ **i** $n^3 - 8n^2$

3 a $8x^2 + 4x$
   d $3b^2 - 9b$
   g $21x^4 + 14x^3$

   b $6p^2 + 3p$
   e $12a + 3a^2$
   h $16y^3 - 12y^2$

   c $6x^2 - 3x$
   f $15c - 10c^2$
   i $6d^4 - 4d^2$

4 a $ax^2 + ax$
   d $qr^2 + q^2$
   g $6a^3 - 9a^2$

   b $pr^2 - pr$
   e $a^2x + ax^2$
   h $8x^3 - 4x^4$

   c $ab^2 - ab$
   f $b^2y - by^2$
   i $18x^3 + 12x^5$

5 a $12a^2b + 18ab^2$
   d $4x^2y + 6xy^2 - 2xy$

   b $4x^2y - 2xy^2$
   e $12ax^2 + 6a^2x - 3ax$

   c $4a^2b + 8ab^2 + 12ab$
   f $a^2bc + ab^2c + abc^2$

6 a $5x + 20$
   d $4y - 3y^2$
   g $cy^2 + cy$

   b $12y - 10$
   e $8a + 6a^2$
   h $3dx^2 - 6dx$

   c $3x^2 + 5x$
   f $12b^2 - 8b$
   i $9c^2d + 15cd^2$

Chapter review

- The ^{2} in 5^2 is called a **power** or an **index**. It tells you how many times the given number must be multiplied by itself. The plural of index is indices.
- The 5 in 5^2 is called the **base**.
- You can solve equations such as $3^x = 81$ by working out how many times the base has to be multiplied by itself to give the answer.
- To multiply powers of the same number, add the indices.
- To divide powers of the same number, subtract the indices.
- Any number raised to the power of 1 is equal to the number itself.
- To raise a power of a number to a further power, multiply the powers (or indices).
- In the expression x^n, the number n is called the power or index.
- $x^m \times x^n = x^{m+n}$
- $x^m \div x^n = x^{m-n}$
- $(x^m)^n = x^{m \times n}$
- Any letter raised to the power of 1 is equal to the letter itself, e.g. $x^1 = x$.
- **BIDMAS** gives the order in which **operations** should be carried out.
- Remember that B I D M A S stands for

B rackets   If there are brackets, work out the value of the expression inside the brackets first.

I ndices   Indices include square roots, cube roots and powers.

D ivide   If there are no brackets, do dividing and multiplying before adding and subtracting, no
M ultiply   matter where they come in the expression.

A dd

S ubtract   If an expression has only adding and subtracting then work it out from left to right.

- Expanding brackets means multiplying each term inside the brackets by the term outside the brackets.
- To simplify an expression with brackets, expand the brackets and collect like terms.
- Factorising is the reverse process to expanding brackets.
- To factorise an expression, find the common factor of the terms in the expression and write the common factor outside a bracket. Then complete the bracket with an expression which, when multiplied by the common factor, gives the original expression.

Review exercise

1 Write down these using index notation.
  a $6 \times 6 \times 6$ **b** 11×11 **c** $2 \times 2 \times 2 \times 2 \times 2 \times 2$

2 Write down the value of
  a 5^4 **b** 2^7 **c** 10^3 **d** 10^5

3 Work out the value of
  a $3^2 \times 4^2$ **b** $2^4 \times 7^2$ **c** 4×10^2 **d** 3×10^4

4 Rewrite these expressions using index notation.
  a $2 \times 2 \times 3 \times 3 \times 3$ **b** $5 \times 5 \times 7 \times 7$ **c** $4 \times 4 \times 8 \times 8 \times 8 \times 8$ **d** $6 \times 6 \times 6 \times 2 \times 2 \times 2$

5 Work out
  a 8^3 **b** 10^4 **c** 5^3 **d** $2^4 \times 3^2$ **e** $5^2 \times 2^5$

6 Find x when
  a $3^x = 243$ **b** $2^x = 32$ **c** $10^x = 1000$

7 Factorise
  a $5x + 15y$ **b** $15p - 9q$ **c** $cd + ce$

8 Jake thinks of a number, squares it, multiplies his answer by 2 and gets 72.
  What number did Jake think of?

9 Work out $6^5 \div (2^4 \times 3^5)$.

10 In this set of squares, each number in a square is obtained by multiplying the two numbers immediately underneath.
  What number should go in the top square?
  Give your answer as a power of 2.

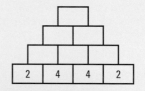

| 2 | 4 | 4 | 2 |

11 **a** Factorise $5p - 20$
   b Solve $3(x - 2) = x + 1$

Results**Plus**
Exam Question Report

83% of students answered part **b** of this question poorly.

March 2007

12 Simplify
  a $2^3 \times 2^4$ **b** $5^3 \times 5^2$ **c** 3×3^4 **d** $7^5 \div 7^2$
  e $9^8 \div 9^4$ **f** $8^3 \div 8$ **g** $7^2 \times \dfrac{7^4}{7^3}$ **h** $6^4 \div \dfrac{6}{6^2}$

13 Simplify

   a $x^6 \times x^3$ **b** $x^8 \div x^5$ **c** $(x^3)^5$ **d** $x^5 \div x^4$

   e $x \times x^4$ **f** $(x^6)^2$ **g** $x^8 \div x^8$ **h** $x^7 \div x$

   i $x^2 \times x^6 \times x^3$ **j** $x^8 \times x$ **k** $x^6 \times \dfrac{x^4}{x^7}$ **l** $x^3 \times \dfrac{x^7}{x^4} \times x^5$

14 Simplify

   a $4x^3 \times x^5$ **b** $3x^2 \times 5x^6$ **c** $7x \times 3x^4$ **d** $8x^9 \div 2x^5$

   e $24x^6 \div 3x$ **f** $36x^9 \div 4x^8$ **g** $(x^5)^2$ **h** $(x^3)^3$

15 Simplify

   a $a^3 \times a^4$ **b** $3x^2y \times 5xy^3$

16 Expand and simplify

   a $3a(b - 2a) + 2b(3a - 2b)$ **b** $4p(2q + 3p) + 3p(2p + q)$

   c $5c(3c + 2d) - 2c(c - d)$ **d** $a(a + b) + b(a + b)$

   e $3a(b + c) + 2b(a + c) - c(2a + 3b)$ **f** $2a(b - 2c) - 3b(2a + 3c)$

17 Factorise

   a $x^2 - 7x$ **b** $t^2 + at$ **c** $bx^2 - x$ **d** $3p^2 + py$ **e** $aq^2 - at$

18 **a** Factorise $x^2 - 5x$.

    b Work out the value of $105^2 - 5 \times 105$.

Nov 2007 adapted

A03

C

Charles Babbage was a 19th century inventor who was frustrated by the errors found in mathematical and astronomical tables calculated by hand. He designed the 'Difference Engine', a machine that would do the calculations correctly. Many people thought the task impossible and he died before he was able to complete it. 150 years later a team at the Science Museum in London finally built the machine.

◉ Objectives

In this chapter you will:
- work out reciprocals, powers, square roots and cube roots using a calculator
- use a calculator to work out complex calculations.

◈ Before you start

You need to be able to:
- write a fraction as a decimal
- work out powers, including squares and cubes
- work out square roots and cube roots
- round to one decimal place
- use the correct order of operations when carrying out a calculation.

10.1 Recognising terminating and recurring decimals

⊙ Objectives

⊙ You can change fractions to decimals.
⊙ You can recognise that some fractions are recurring decimals.

⊘ Why do this?

You need to recognise recurring and terminating decimals when trying to share a third of £1; you can't share it exactly as the numbers keep recurring so you would have to share it approximately.

⬦ Get Ready

1. Write $\frac{1}{4}$ as a decimal. **2.** Write $\frac{1}{10}$ as a decimal. **3.** Write $\frac{1}{100}$ as a decimal.

🔑 Key Points

⊙ To change a fraction into a decimal, divide the numerator by the denominator.
For example, $\frac{7}{25} = 7 \div 25 = 0.28$
0.28 is a terminating decimal.

⊙ Not all fractions can be written as terminating decimals.
$\frac{1}{3} = 1 \div 3 = 0.3333\ldots$
This is called a recurring decimal since one of the figures recurs (repeats).
You put a dot over the figure that repeats.
So 0.3333… is written as $0.\dot{3}$

⊙ Sometimes, a recurring decimal has more than one figure that repeats.
$\frac{5}{11} = 0.454545\ldots = 0.\dot{4}\dot{5}$
$\frac{1}{7} = 0.142857142857\ldots = 0.\dot{1}4285\dot{7}$
When more than one figure repeats, put a dot above the first figure in the group that repeats and a dot above the last figure in the group.

🔍 Example 1

Write $\frac{27}{40}$ as a decimal.

$\frac{27}{40}$ means $27 \div 40$ ← numerator ÷ denominator

$\boxed{2}\boxed{7}\boxed{\div}\boxed{4}\boxed{0}\boxed{=}$ ← Key in on your calculator.

$\frac{27}{40} = 0.675$

⚙ Exercise 10A

Questions in this chapter are targeted at the grades indicated.

1 Change these fractions to decimals.

a $\frac{3}{4}$ b $\frac{1}{8}$ c $\frac{3}{8}$ d $\frac{1}{16}$ e $\frac{9}{16}$

2 Change these fractions to decimals.

a $\frac{16}{25}$ b $\frac{7}{40}$ c $\frac{13}{20}$ d $\frac{23}{50}$ e $\frac{47}{80}$

G

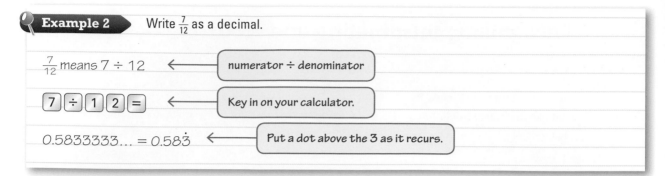

Example 2 Write $\frac{7}{12}$ as a decimal.

$\frac{7}{12}$ means $7 \div 12$ ← numerator ÷ denominator

$7 \div 1 2 =$ ← Key in on your calculator.

$0.5833333... = 0.58\dot{3}$ ← Put a dot above the 3 as it recurs.

Exercise 10B

1 Change these fractions to decimals.

a $\frac{2}{3}$ b $\frac{2}{9}$ c $\frac{4}{9}$ d $\frac{7}{9}$ e $\frac{8}{9}$

2 Change these fractions to decimals.

a $\frac{1}{6}$ b $\frac{5}{11}$ c $\frac{9}{11}$ d $\frac{5}{12}$ e $\frac{17}{22}$

3 a Change each of these fractions to a recurring decimal.

$\frac{1}{7}$ $\frac{2}{7}$ $\frac{3}{7}$ $\frac{4}{7}$ $\frac{5}{7}$ $\frac{6}{7}$

b What do you notice about your recurring decimals in part **a**?

10.2 Finding reciprocals

Objective

● You can work out the reciprocal of a number.

Why do this?

Reciprocals are used in engineering and applied maths.

Get Ready

Work out

1. $1 \div 4$ **2.** $1 \div 0.4$ **3.** $1 \div 25$

Key Points

● The **reciprocal** of a number is 1 divided by the number.

  The reciprocal of 2 is $\frac{1}{2}$ (or 0.5). The reciprocal of 3 is $\frac{1}{3}$ (or $0.\dot{3}$).

● To find the reciprocal of a fraction, turn it upside down.

  The reciprocal of $\frac{3}{4}$ is $\frac{4}{3}$ (or $1\frac{1}{3}$). The reciprocal of $\frac{1}{3}$ is $\frac{3}{1}$ (or 3).

● To work out reciprocals you can use the reciprocal key on a calculator.

  It is usually shown by $\boxed{1/x}$ or $\boxed{x^{-1}}$

● Any number multiplied by its reciprocal gives the answer 1.

● Zero has no reciprocal, because you cannot divide a number by zero.

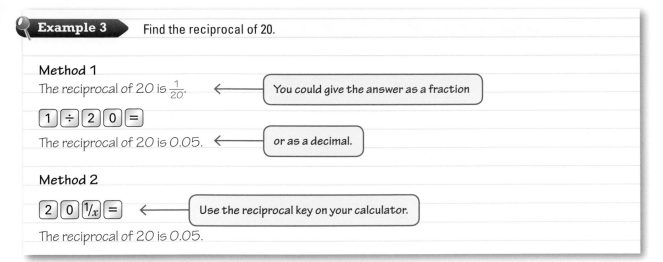

Example 3   Find the reciprocal of 20.

Method 1

The reciprocal of 20 is $\frac{1}{20}$.   ← You could give the answer as a fraction

$1 \div 2 0 =$

The reciprocal of 20 is 0.05.   ← or as a decimal.

Method 2

$2 0 \; ^1/_x \; =$ ← Use the reciprocal key on your calculator.

The reciprocal of 20 is 0.05.

Exercise 10C

1   Find the reciprocals of these numbers.

     a   10   b   4   c   8   d   5   e   9

2   Find the reciprocals of these fractions.

     a $\frac{1}{3}$ b $\frac{1}{4}$ c $\frac{2}{3}$ d $\frac{5}{6}$ e $\frac{3}{10}$

3   Use your calculator to find the reciprocals of these numbers.

     a   2.5   b   50   c   16   d   80   e   0.2

     f   0.5   g   0.05   h   0.125   i   0.04   j   0.01

4   The reciprocal of 1000 is 0.001. What is the reciprocal of 0.001?

5   a   Find the reciprocal of 40.   b   Multiply 40 by its reciprocal.

6   a   Find the reciprocal of 100.   b   Multiply 100 by its reciprocal.

D

Mixed exercise 10D

1   Change these fractions to decimals.

     a $\frac{5}{8}$ b $\frac{7}{16}$ c $\frac{5}{9}$ d $\frac{7}{11}$ e $\frac{1}{12}$

2   Which of these fractions can be written as a recurring decimal?

     $\frac{1}{2}$ $\frac{1}{4}$ $\frac{1}{5}$ $\frac{1}{9}$

     Explain your answer.

3   Lauren says that $\frac{1}{3}$ is equal to 0.3.

     Is Lauren correct? Explain your answer.

4   Find the reciprocal of 25.

5   Find the reciprocal of 0.8.

D

10.3 Interpreting a calculator display

◎ Objective

○ You can interpret the answer on a calculator display.

⊘ Why do this?

If you are checking a shopping bill you need to be able to interpret the calculator display correctly.

⬦ Get Ready

1. Work out the total cost of 4 magazines costing £1.67 each.
2. £375 is shared equally between 5 people. How much does each person get?
3. Work out the total cost of a notebook costing £2.79 and 3 pens costing 65p each.

🔑 Key Points

◉ You need to take care when writing down the answer from a calculator display.

If you are working in pounds, the calculator display 3.4 means £3.40.

Answers which are in pounds and pence should always be written with two figures after the decimal point.

◉ You must make sure that your answer makes sense in the context of the question.

Sometimes the answer to a problem must be a whole number and if the calculator display shows a decimal you will need to think carefully about whether to round it up or round it down.

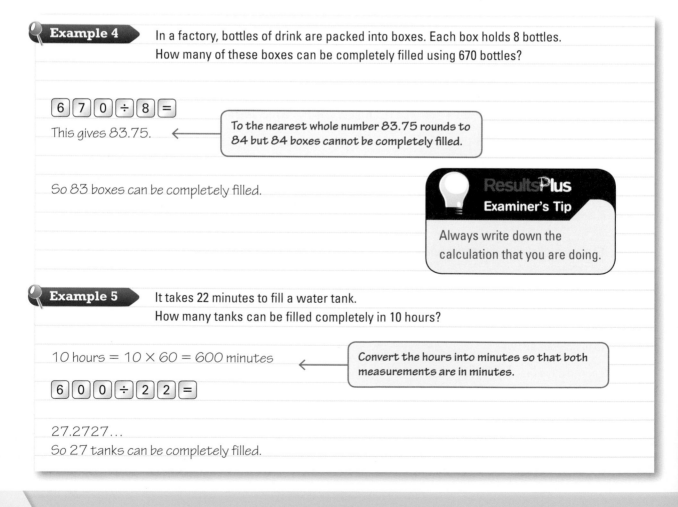

Example 4   In a factory, bottles of drink are packed into boxes. Each box holds 8 bottles.
How many of these boxes can be completely filled using 670 bottles?

6 7 0 ÷ 8 =

This gives 83.75.

To the nearest whole number 83.75 rounds to 84 but 84 boxes cannot be completely filled.

So 83 boxes can be completely filled.

ResultsPlus
Examiner's Tip

Always write down the calculation that you are doing.

Example 5   It takes 22 minutes to fill a water tank.
How many tanks can be filled completely in 10 hours?

10 hours = 10 × 60 = 600 minutes

Convert the hours into minutes so that both measurements are in minutes.

6 0 0 ÷ 2 2 =

27.2727...
So 27 tanks can be completely filled.

E

Exercise 10E

1   The total cost of five adult cinema tickets is £44.50. Work out the cost of one adult cinema ticket.

2   Ryan buys four bars of chocolate costing £1.45 each and one packet of sweets costing £1.30.
    Work out the total cost.

3   Hannah buys two magazines costing £3.85 each.
    Work out how much change she should get from £10.

4   A garden centre sells plants for £1.90 each. Lee buys 14 plants. Work out the total cost.

5   Tom's company pays him 45p for each mile that he drives his car.
    Work out how much money Tom's company pays him when he drives 126 miles.

6   Petrol costs 115.9 pence per litre. Richard buys 38 litres of petrol. How much should Richard pay?

7   Colin needs 160 tiles for a room.
    Tiles are sold in boxes. There are 12 tiles in each box.
    Work out the least number of boxes of tiles that Colin needs.

8   458 students and teachers are going on a coach trip.
    Each coach holds 54 passengers.
    Work out the smallest number of coaches needed.

9   A beaker holds 225 m*l* of orange squash.
    How many of these beakers can be completely filled using 2000 m*l* of orange squash?

10 It takes 35 seconds to fill a bucket. How many buckets can be completely filled in 20 minutes?

11 The battery life of a calculator is 420 hours. Work out the battery life in days and hours.

12 A pen costs 38p. Sam has £5. He buys as many pens as he can.
    Work out how much change Sam should get from £5.

* 13 Raja sees this new monthly plan for a mobile phone.
    Raja's current plan gives him 200 minutes and unlimited
    texts for £25 per month. He wants to find out if he should
    switch to the new monthly plan.
    In September, Raja used 140 minutes and 230 texts.
    In October, he used 145 minutes and 190 texts.
    In November, he used 135 minutes and 260 texts.
    Should Raja switch to the new monthly plan? Explain your answer.

YOU PAY per month	YOU GET per month
£15	**FREE** - 100 minutes **FREE** - 200 texts

Extra minutes: 20p each
Extra texts: 12p each

10.4 Working out powers and roots

Objectives

- You can work out powers using a calculator.
- You can work out square roots and cube roots using a calculator.

Why do this?

Scientists, financial analysts and economists make use of powers and roots.

Get Ready

1. Work out 6^2.
2. Work out 2^3.
3. Work out $\sqrt{25}$.

Key Points

- With a scientific calculator you can work out squares using the $\boxed{x^2}$ key. (You met square numbers in Section 5.5.)
- Some scientific calculators have an $\boxed{x^3}$ key for working out cubes.

  To work out 3.5^2, key in $\boxed{3}\ \boxed{\cdot}\ \boxed{5}\ \boxed{x^2}\ \boxed{=}$

  To work out 2.7^3, key in $\boxed{2}\ \boxed{\cdot}\ \boxed{7}\ \boxed{x^3}\ \boxed{=}$

- Scientific calculators have a power (or index) key.

  It can be shown by $\boxed{x^y}$ or $\boxed{y^x}$ or $\boxed{x^\blacksquare}$ or $\boxed{\wedge}$.

  To work out 2.7^3, key in $\boxed{2}\ \boxed{\cdot}\ \boxed{7}\ \boxed{x^y}\ \boxed{3}\ \boxed{=}$

- To work out square roots on a calculator use the $\boxed{\sqrt{}}$ key.

- To work out cube roots on a calculator use the $\boxed{\sqrt[3]{}}$ key.

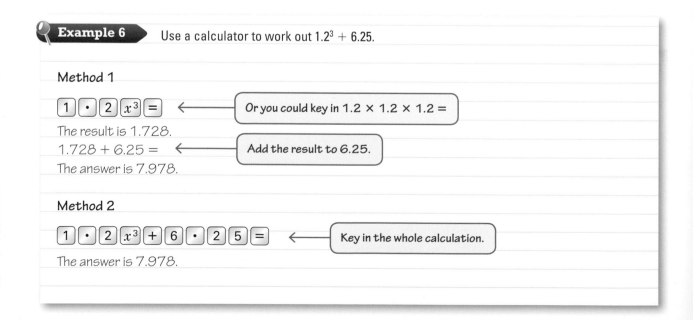

Example 6

Use a calculator to work out $1.2^3 + 6.25$.

Method 1

$\boxed{1}\ \boxed{\cdot}\ \boxed{2}\ \boxed{x^3}\ \boxed{=}$ ⟵ Or you could key in 1.2 × 1.2 × 1.2 =

The result is 1.728.

1.728 + 6.25 = ⟵ Add the result to 6.25.

The answer is 7.978.

Method 2

$\boxed{1}\ \boxed{\cdot}\ \boxed{2}\ \boxed{x^3}\ \boxed{+}\ \boxed{6}\ \boxed{\cdot}\ \boxed{2}\ \boxed{5}\ \boxed{=}$ ⟵ Key in the whole calculation.

The answer is 7.978.

Exercise 10F

1 Work out
   a 2.5^2 **b** 3.2^2 **c** 47^2 **d** 1.8^2 **e** 17.9^2

2 Work out
   a 6^3 **b** 22^3 **c** 2.1^3 **d** 3.4^3 **e** 1.5^3

3 Work out
   a 2^5 **b** 11^4 **c** 3^5 **d** 12^4 **e** 5^6

4 Work out
   a $17^2 + 10$ **b** $2.4^2 + 15$ **c** $19^2 - 322$ **d** $2.7^2 + 5.42$

5 Work out
   a $2.1^3 + 1.96$ **b** $1.9^3 + 5.29$ **c** $2.5^3 - 7.29$ **d** $1.4^3 - 1.544$

E

Example 7   Use a calculator to work out $\sqrt{14.44}$.

> On some calculators you first key in 14.44 and then press the square root key.

$\sqrt{14.44} = 13.8$ ←

> Calculators always give the positive square root.

Exercise 10G

1 Work out
   a $\sqrt{529}$ **b** $\sqrt{289}$ **c** $\sqrt{1156}$ **d** $\sqrt{625}$

2 Work out
   a $\sqrt{2.56}$ **b** $\sqrt{22.09}$ **c** $\sqrt{13.69}$ **d** $\sqrt{88.36}$

3 Work these out, giving your answers correct to one decimal place.
   a $\sqrt{150}$ **b** $\sqrt{80}$
   c $\sqrt{124}$ **d** $\sqrt{240}$

4 Work out
   a $\sqrt[3]{216}$ **b** $\sqrt[3]{729}$
   c $\sqrt[3]{343}$ **d** $\sqrt[3]{1728}$

E

ResultsPlus
Examiner's Tip

Write down all the figures on your calculator display before you round your answer to one decimal place. (See Section 5.7 on Rounding.)

E

5 Work these out, giving your answers correct to one decimal place.

    a $\sqrt[3]{40}$ **b** $\sqrt[3]{200}$ **c** $\sqrt[3]{120}$ **d** $\sqrt[3]{84}$

6 Work out

    a $\sqrt{51.84} + 4.8$ **b** $\sqrt{841} - 21.3$ **c** $\sqrt[3]{9.261} - 1.9$ **d** $\sqrt[3]{1.728} + 1.8$

D

A02
A03

7 How far can you see?

To work out the distance, in kilometres, you can see:

1 Find the height, in metres, of your eyes above sea level.

2 Multiply this height by 13.

3 Find the square root of the answer.

James is standing on a cliff, at J. His eyes are 20 metres above sea level.

Matthew is standing in a lighthouse, M. His eyes are 50 metres above sea level.

Which of the three boats, A, B and C, should James be able to see?

Which of the three boats should Matthew be able to see?

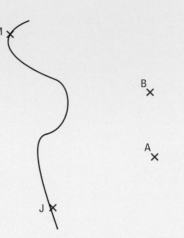

Scale: 1 cm represents 5 km.

10.5 Using a calculator to work out complex calculations

◎ Objective

○ You can use a calculator to work out complex calculations.

⊘ Why do this?

Engineers, architects, computer programmers and accountants all need to be able to work out complex calculations using a calculator.

◈ Get Ready

1. Work out 1.5×4^2.   **2.** Work out $3.5 + 2.5 \times 4$.   **3.** Work out $\dfrac{20 - 10}{5}$.

◉ Key Points

◉ All scientific calculators carry out mathematical operations in the same order.
This order is sometimes abbreviated to BIDMAS (see Section 9.4).
You need to know about BIDMAS in order to use a scientific calculator properly.

◉ On most scientific calculators you can key in calculations in the order in which they are written down.
To work out $(1.27 + 3.8)^2 \times 3.5$, for example, you would key in

$$(\quad 1 \quad \cdot \quad 2 \quad 7 \quad + \quad 3 \quad \cdot \quad 8 \quad) \quad x^2 \quad \times \quad 3 \quad \cdot \quad 5 \quad =$$

◉ When you are doing a division calculation you must remember to divide by ALL of the denominator.
To work out $\dfrac{14.5}{1.32 + 1.28}$, for example, you must divide 14.5 by the result of $1.32 + 1.28$.

◉ Most scientific calculators have the negative sign $(-)$. To enter the number -6, for example, key in $(-)$ 6.

Example 8 Work out the value of $\dfrac{16.3 + 7.82}{7.2 - 4.7}$.

Method 1

[1] [6] [·] [3] [+] [7] [·] [8] [2] [=] ← Key in the numerator.

The result is 24.12.

[7] [·] [2] [−] [4] [·] [7] [=] ← Key in the denominator.

The result is 2.5.

[2] [4] [·] [1] [2] [÷] [2] [·] [5] [=] ← Divide the first result by the second result.

The value is 9.648.

> **ResultsPlus**
> **Examiner's Tip**
>
> If you work out the numerator and the denominator separately, make sure you write down the value of each.

Method 2

[(] [1] [6] [·] [3] [+] [7] [·] [8] [2] [)] [÷] [(] [7] [·] [2] [−] [4] [·] [7] [)] [=]

Put brackets around the numerator and around the denominator.

 Exercise 10H

1   Work out
   a $(5.2 + 2.7)^2$ **b** $(12.4 - 9.71)^2$ **c** $(2.43 + 1.87)^3$ **d** $(5.1 - 3.7)^3$

2   Work out
   a $12^2 + 13^2$ **b** $34^2 + 6^3$ **c** $12^3 - 23^2$ **d** $38^2 - 18^2$

3   Work out
   a $\sqrt{17.8 + 13.56}$ **b** $\sqrt{415 - 159}$ **c** $\sqrt[3]{129 - 65}$ **d** $\sqrt[3]{1.85 + 1.525}$

4   Work out the value of each of these.
   Write down all the figures on your calculator display.
   a $\dfrac{4.78 - 1.42}{0.84}$ **b** $\dfrac{48.88}{3.62 + 5.78}$
   c $\dfrac{12.24 \times 2.5}{6.8}$ **d** $\dfrac{35.36}{12.6 - 5.8}$

5   Work out the value of each of these.
   Write down all the figures on your calculator display.
   a $\dfrac{13.2 - 6.84}{2.8 + 3.41}$ **b** $\dfrac{5.6 \times 8.1}{12.5 - 3.9}$
   c $\dfrac{4.37 \times 6.52}{2.8 + 7.19}$ **d** $\dfrac{17.6 + 9.82}{23.6 - 5.94}$

E

D

Chapter review

- You can change a fraction into a decimal by doing the division.
- Fractions that have an exact decimal equivalent are called terminating decimals.
- Fractions that have a decimal equivalent that repeats itself are called recurring decimals.
- The **reciprocal** of a number is 1 divided by the number.
- To find the reciprocal of a fraction, turn it upside down.
- To work out reciprocals you can use the reciprocal key on a calculator.
  It is usually shown by $\boxed{1/x}$ or $\boxed{x^{-1}}$.
- Any number multiplied by its reciprocal gives the answer 1.
- Zero has no reciprocal, because you cannot divide a number by zero.
- You need to take care when writing down the answer from a calculator display.
- Answers which are in pounds and pence should always be written with two figures after the decimal point.
- You must make sure that your answer makes sense in the context of the question.
- Sometimes the answer to a problem must be a whole number and if the calculator display shows a decimal you will need to think carefully about whether to round it up or round it down.
- With a scientific calculator you can work out squares using the $\boxed{x^2}$ key.
- Some scientific calculators have an $\boxed{x^3}$ key for working out cubes.
- Scientific calculators have a power (or index) key.
  It can be shown by $\boxed{x^y}$ or $\boxed{y^x}$ or $\boxed{x^\blacksquare}$ or $\boxed{\wedge}$.
- To work out square roots on a calculator use the $\boxed{\sqrt{}}$ key.
- To work out cube roots on a calculator use the $\boxed{\sqrt[3]{}}$ key.
- All scientific calculators carry out mathematical operations in the same order. This order is sometimes abbreviated to BIDMAS.
- You need to know about BIDMAS in order to use a scientific calculator properly.
- On most scientific calculators you can key in calculations in the order in which they are written down.
- When you are doing a division calculation you must remember to divide by ALL of the denominator.
- Most scientific calculators have the negative sign $\boxed{(-)}$.

Review exercise

E

1   Christine buys a calculator costing £3.99, a pencil case costing £1.65 and two rulers costing 28p each. She pays with a £10 note. How much change should she get from her £10 note?

2   Shares in a company cost £6.23 each. Tauqeer has £500. He buys as many shares as he can. Work out how many shares Tauqeer can buy.

3   A milk crate holds 24 bottles. Amraiz has 357 bottles of milk. Work out how many milk crates he can fill completely.

4 The table below shows the cost of each of three calculators.

Quicksum	£2.30
Basic	£2.15
Easycalc	£2.90

a Emily buys one Quicksum calculator and two Easycalc calculators.
She pays with a £10 note. How much change should she get?

b Mrs Windsor wants to buy some Basic calculators. She has £60 to spend.
Work out the greatest number of Basic calculators she can buy.

5 Work out
a 2.9^2 **b** 12^3 **c** 3.7^2 **d** 2.2^3

6 Work out
a $\sqrt{51.84}$ **b** $\sqrt{784}$ **c** $\sqrt[3]{512}$ **d** $\sqrt[3]{15.625}$

7 Work these out, giving your answers correct to one decimal place.
a $2.8^2 + \sqrt{34}$ **b** $\sqrt{56} - 2.3^2$ **c** $4.7^2 - \sqrt{28}$ **d** $3.8^2 - \sqrt{50}$

8 Which of the following fractions will be recurring when written as decimals?
$$\frac{3}{4} \qquad \frac{5}{6} \qquad \frac{2}{5} \qquad \frac{1}{7} \qquad \frac{15}{24}$$

9 Jonathan buys a can of cola and a roll.
  a Work out the total cost.

Sachin buys a cup of tea, a cup of coffee and 2 sandwiches.
  b Work out the total cost.

Kim buys a can of cola, a cup of coffee and a sandwich.
She pays with a £5 note.
  c Work out how much change she should get.

Joe's Café

Prices

Cup of tea	70p
Cup of coffee	85p
Can of cola	75p
Roll	£1.60
Sandwich	£1.35

June 2007

10 Cans of drink are put into packs of 24. How many packs can be filled from 750 cans of drink?

11 There are 1230 students in a school. All the students go on a trip.
Each bus can take 48 students. How many buses are needed?

12 Plain tiles cost 28p each.
Patterned tiles cost £9.51 each.
Julie buys 450 plain tiles and 15 patterned tiles.
Work out the total cost of the tiles.

Nov 2007

13 Use a calculator to work out
$\sqrt{2.56} + 8.4$

Nov 2008

14 Work out
a $(3.7 + 2.64)^2$ **b** $\sqrt{17 + 25.25}$ **c** $(2.1 + 2.8)^2 \times 1.2$

D

15 Work out the value of each of these.
Write down all the figures on your calculator display.

a $\dfrac{5.68 - 1.52^2}{0.83}$ b $\dfrac{1}{3.58^2 - 2.87}$ c $\dfrac{8.7 + 5.92}{16.3 - 4.56}$

16 Work out the value of each of these.
Write down all the figures on your calculator display.

a $\dfrac{\sqrt{3.96} + 1.8}{7.625 - 3.48}$ b $\sqrt{\dfrac{4.92 + 3.48}{9.2 - 3.75}}$

* **17** To work out a person's daily calorie requirement you can use one of these rules.

Gender	Daily calorie requirement
Female	655 + (9.6 × weight in kg) + (1.8 × height in cm) − (4.7 × age in years)
Male	66 + (13.7 × weight in kg) + (5 × height in cm) − (6.8 × age in years)

The table below shows some information about four people.

Name	Gender	Age (years)	Weight (kg)	Height (cm)
Sophie	F	32	68	165
Chelsea	F	47	55	175
Kenny	M	27	98	191
Hassan	M	38	117	182

Work out the recommended daily calorie intake for each person.
Which person has the greatest daily calorie requirement?
Which person has the smallest daily calorie requirement?

18 Find the reciprocal of   a $\frac{5}{8}$ b 2.5   c $\frac{1}{4}$

19 Change the following fractions to decimals.   a $\frac{17}{20}$ b $\frac{3}{40}$

20 Use your calculator to work out $\dfrac{22.4 \times 14.5}{8.5 \times 3.2}$

Write down all the figures on your calculator display.

June 2007

A03 * **21**

A large tub of popcorn costs £3.80 and holds 200 g.
A regular tub of popcorn costs £3.50 and holds 175 g.
Rob says that the 200 g large tub is the better value for money.
Linda says that the 175 g regular tub is the better value for money.
Who is correct?
Explain the reasons for your answer.
You must show all your working.

June 2006

22 Work out $\dfrac{4.6 + 3.85}{3.2^2 - 6.51}$

Write down all the numbers on your calculator display.

June 2009

23 Salma has £1.55.

She wants to buy a burger and fries.

a What are the different combinations that can she buy?

Mark buys 2 double burgers with cheese, 1 large fries and 1 large cola.

He pays with a £10 note

b He gets the best price. What change should he get?

A02
A03

Ben's Burger Bar

Burgers

Single Burger	£0.85
Single Burger with Cheese	£0.95
Double Burger	£1.55
Double Burger with Cheese	£1.70

Fries		**Cola**	
Regular	£0.65	Regular	£0.85
Large	£0.99	Large	£1.10

Meal Deals

Regular

Single Burger with Cheese regular Fries and regular Cola	£2.09

Large

Double Burger with Cheese large Fries and large Cola	£3.49

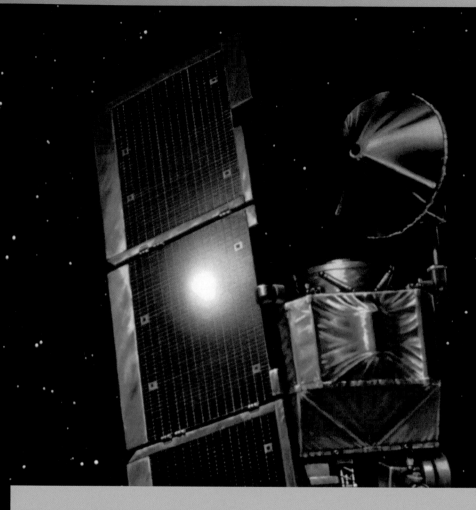

In 1999, NASA spent $125 million on a space probe designed to orbit Mars. The mission ended in disaster after the probe steered too close to Mars and burned up whilst skimming the planet's thin atmosphere. Apparently navigation commands to the probe's engines were provided in imperial units rather then the metric ones that NASA had been using since at least 1990.

◉ Objectives

In this chapter you will:
- learn to take and estimate readings from dials and scales
- learn how to choose the most appropriate unit for taking a measurement and convert between metric units of measurement
- work with time and solve problems relating to speed
- learn how to convert between imperial units and between metric and imperial units.

◈ Before you start

You need to be able to:
- estimate simple measurements and larger measurements
- carry out simple measurements using rulers and weighing scales
- tell the time using clock faces or digital clocks.

11.1 Reading scales

Objectives

- You can take readings from dials and scales.
- You can estimate readings from dials and scales.

Why do this?

Diabetics need to accurately monitor their blood sugar levels and asthmatics use a peak flow meter to monitor their asthma. In both cases this will help doctors treat them effectively.

Get Ready

1. Measure the length and width of this book, using suitable units of measure.

Key Points

- Lines can be measured using a ruler. The following ruler is marked off in centimetres (cm) but the smaller marks show you millimetres (mm) (10 mm = 1 cm).

This line can be measured as 8.5 cm or 85 mm.

- Some dials have a scale. You need to interpret the scale to take an accurate reading from the dial.
  On this scale there are 5 spaces between 10 and 20, so each mark shows 2 units.
  The reading is 18 kg.

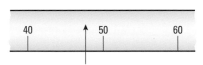

- Sometimes you will need to estimate a reading.
  The arrow on this scale is more than halfway between 40 and 50, so it is more than 45 but less than 50. It is also some distance from 50, so it is not 49.

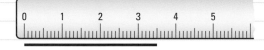

A good estimate is 47 or 48.

Example 1

Measure and write down the reading from this ruler. Give your answer in cm and mm.

The reading from this ruler is 3.5, which is 3.5 cm.
In millimetres it is 35 mm.

ResultsPlus
Examiner's Tip

Always remember to show the units with your answer.

Exercise 11A

G

1 Measure and write down the lengths of these lines in cm and mm.

a _____

b _____

c _____

d _____

e _____

f _____

g _____

h _____

i _____

j _____

k _____

l _____

2 Draw and label lines with the following lengths.

a	6 cm	b	3 cm	c	4 cm	d	6.3 cm
e	5.1 cm	f	6.8 cm	g	3.4 cm	h	7.9 cm
i	3.5 cm	j	4.2 cm	k	5.7 cm	l	7.6 cm

3 Draw and label lines with the following lengths.

a	40 mm	b	70 mm	c	14 mm	d	35 mm
e	12 mm	f	31 mm	g	48 mm	h	26 mm
i	27 mm	j	59 mm	k	73 mm	l	66 mm

4 Write down the readings shown on the following scales.

F

5 Write down the readings shown on the following scales.

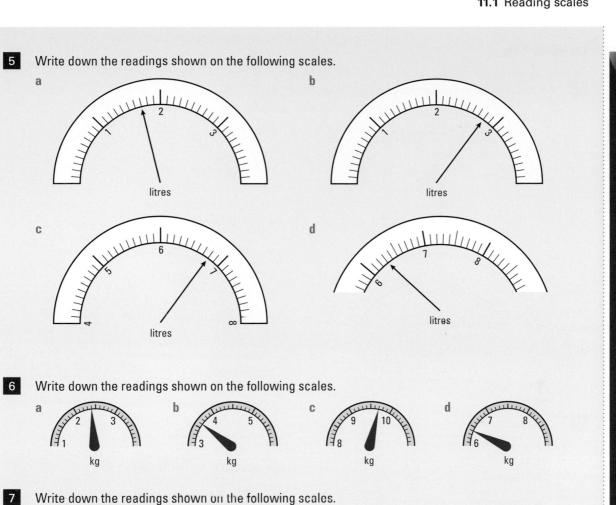

6 Write down the readings shown on the following scales.

7 Write down the readings on the following scales.

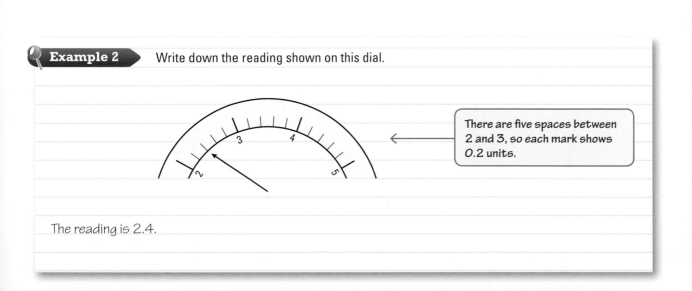

Example 2 Write down the reading shown on this dial.

> There are five spaces between 2 and 3, so each mark shows 0.2 units.

The reading is 2.4.

Exercise 11B

G

1 Write down the readings shown on the following scales.

a

amps

b

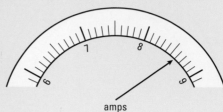

amps

c

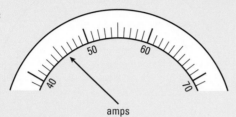

amps

d

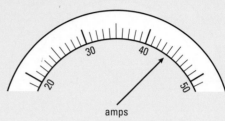

amps

F

2 Write down the readings shown on the following scales.

a

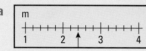

b

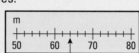

c

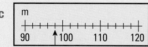

d

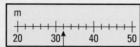

e

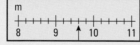

f

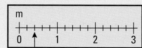

3 Write down the readings shown on the following scales.

a  b  c  d  e f

4 Write down the readings shown on the following scales.

a 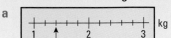 kg

b 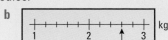 kg

c 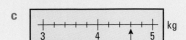 kg

d 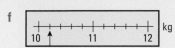 kg

e 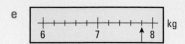 kg

f 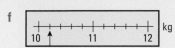 kg

g 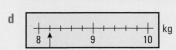 kg

h 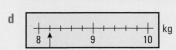 kg

---

Example 3   Estimate the reading shown on this dial.

mph

The speed is more than 25 and less than 30.
It is nearer to 25 than 30.
A good estimate is 26 or 27 mph.

Exercise 11C

1 Estimate the reading shown on the following scales.

a

b

c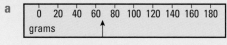

d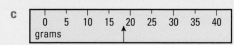

2 Estimate the reading shown on the following scales.

a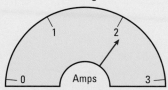

b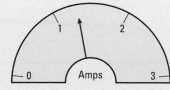

c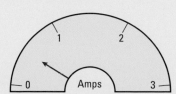

d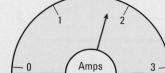

3 Estimate the readings shown on the following scales. Give the readings in both °C and °F.

a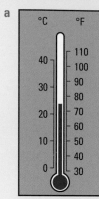
b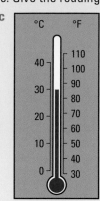
c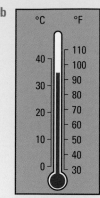
d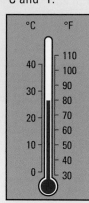

e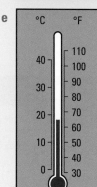
f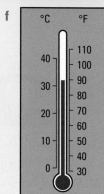
g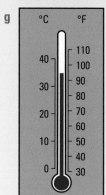
h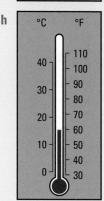

4 Estimate the readings shown on the following scales.

a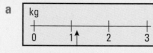
b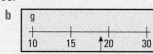

c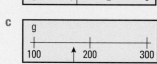
d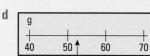

5 Estimate the readings shown on the following scales. Give the readings in both kilometres per hour and miles per hour.

a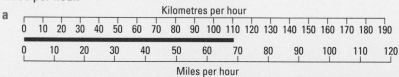

b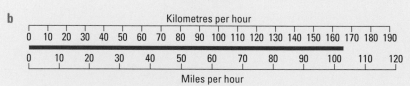

c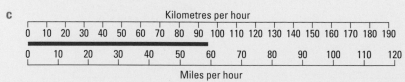

11.2 Time

◎ Objectives

- ◉ You can read time from a variety of clocks.
- ◉ You can convert between 12-hour time and 24-hour time.
- ◉ You can work out durations of time.
- ◉ You can solve problems involving time and timetables.

◈ Why do this?

You need to use time in order to read a bus or train timetable or to plan how long to spend on each question in an exam.

◈ Get Ready

1. Write down the times at which you have your meals during the day, and find the duration of time between each one.

◈ Key Points

- ◉ A 12-hour clock goes from 0 to 12, using am for times before noon, and pm for times after noon.
- ◉ A 24-hour clock goes from 0 to 24, where 24 is midnight.

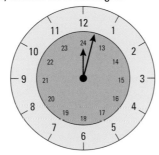

- ◉ You will need to know the following **conversions** involving time:

  60 seconds = 1 minute
  60 minutes = 1 hour
  24 hours = 1 day
  7 days = 1 week
  365 days = 1 year

  366 days = 1 leap year
  3 months = 1 quarter
  12 months = 1 year
  52 weeks = 1 year

◈ Example 4

The time shown on this clock is in the afternoon.
Write the time using both 12-hour time and 24-hour time.

ResultsPlus
Examiner's Tip

Make sure you write time using the correct notation.

The time is quarter past five, or 5.15 pm, in 12-hour time.
In 24-hour time, this time is 17:15 hours.

Exercise 11D

G

1 These clock faces show times in the morning.

a  b 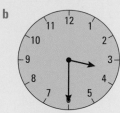 c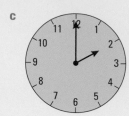

d 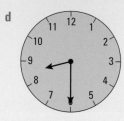 e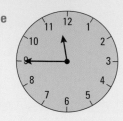

Write each of the times in both 12-hour time and in 24-hour time.

2 These clock faces show times in the afternoon.

a  b 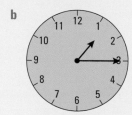 c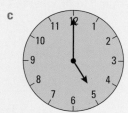

d 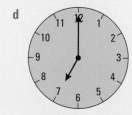 e

Write each of the times in both 12-hour time and 24-hour time.

3 Change the following 12-hour times to 24-hour times.

a 8 am	b 11.15 am	c 3.40 pm	d 8.20 am
e 8.55 pm	f 3.25 pm	g 2.30 am	h 5.25 pm
i 10.15 pm	j 7.20 am	k 9.45 am	l 1.15 pm
m 11.25 pm	n 2.50 am	o 1.50 pm	p 12.20 pm

4 Change the following 24-hour times to 12-hour times.

a 11:10 h	b 08:20 h	c 07:40 h	d 23:35 h
e 14:17 h	f 09:35 h	g 18:16 h	h 17:25 h
i 13:20 h	j 13:10 h	k 08:30 h	l 13:35 h
m 03:42 h	n 22:16 h	o 09:17 h	p 13:37 h

G

5 Write down the correct time in the following questions.

a 4 hours before 3.15 pm

b $4\frac{1}{2}$ hours before 16:40 h

c $3\frac{1}{4}$ hours after 11.15 am

d $2\frac{1}{4}$ hours before 1 pm

e $1\frac{1}{2}$ hours before 13:00 h

f 3 hours after 2.40 am

g $2\frac{1}{2}$ hours after 11.20 am

h $4\frac{1}{4}$ hours before 14:45 h

i $5\frac{1}{2}$ hours after 10:55 h

j $6\frac{1}{2}$ hours before 15:10 h

k $5\frac{1}{4}$ hours before 3 am

l $2\frac{3}{4}$ hours after 22:00 h

ResultsPlus
Examiner's Tip

It may be useful to use a clock face to help you answer this type of question. There will usually be a clock in the exam room that you can use.

6 Change the units of time in the following questions.

a 3 years into weeks

b $3\frac{1}{2}$ hours into minutes

c 5 minutes into seconds

d 5 years into months

e 4 days into hours

f 540 minutes into hours

g 312 weeks into years

h $2\frac{1}{2}$ minutes into seconds

i 90 seconds into minutes

j 4 years into weeks

k 8 hours into minutes

l $2\frac{1}{2}$ days into hours

m $3\frac{1}{2}$ years into weeks

n 2 years into days

7 a A man buys three packets of crisps each week. How many does he buy in a year?

b It takes 3 minutes for Bill to make a toy. How many can he make in 1 hour?

c Sally pays her gas bill every month. How many gas bills does she pay in a year?

d A clock chimes every hour. How many times does the clock chime in a week?

F

A02

Example 5

ResultsPlus
Watch Out!

Do not use a calculator to solve time problems.

These two clocks both show times in the morning.
What is the time difference between the clocks?

3.30 am to 4.00 am is 30 minutes ← Start with the earliest time and count on to the next hour

4.00 am to 6.00 am is 2 hours ← Count the number of full hours.

6.00 am to 6.15 am is 15 minutes ← Count the minutes to the end time.

Total time: 30 minutes + 2 hours + 15 minutes = 2 hours 45 minutes.

Exercise 11E

1 Work out the time difference between each of the following times:

a 9.50 am to 10.15 am   b 07:30 h to 08:20 h

c 12.15 pm to 2.25 pm   d 13:45 h to 00:15 h

e 10 pm Monday to 7 am Tuesday   f 09:17 h to 13:27 h

g 9.37 am to 1.15 pm   h 03:42 h to 22:14 h

i 8.30 am to 11.15 pm   j 9.15 am to 1.05 pm

k 09:15 Tue to 08:05 Wed   l 18:35 h to 21:50 h

2 Saima leaves her house at 6.15 am. She travels by aeroplane to her parents' home, arriving at 10.35 pm. How long does her journey take?

3 A man arrives at work at 08:55 h and leaves at 05:05 h. How long is he at work?

4 A ferry sets sail from Portsmouth at 08:50 h and arrives in France at 15:40 h. How long does the crossing take?

5 The following table shows information about flight times to Aberdeen. Ella wants to fly to Aberdeen. What time does her flight depart and arrive if she wants to take the fastest flight?

Flight number	Departure time	Arrival time	Flight time
BA52	22:20	04:45	
XA160	05:42	09:14	
FC492	14:15		4 h 40 min
TC223	10:02		4 h 23 min
AL517		07:59	5 h 37 min
AB614	19:17	05:21	
FX910	02:43		3 h 51 min
BI451		12:17	2 h 32 min
AE105		02:25	5 h 29 min
DA452	15:39		6 h 48 min

Example 6

Part of a bus timetable is shown.

a Ali gets the 09:28 bus from Burton. How long does it take him to get to Didcom?

b Chelsea arrives at her bus stop in Camberley at 09:40. She is going to Earlstown. What time will she get there?

Aldwich	09:16	10:05
Burton	09:28	10:17
Camberley	09:35	10:34
Didcom	09:55	10:54
Earlstown	10:08	11:07

a *The 09:28 bus arrives in Didcom at 09:55. 09:28 to 09:55 is 27 minutes.*

b *Chelsea has just missed a bus! The next bus to Earlstown leaves Camberley at 10:34 and arrives at 11:07.*

Exercise 11F

1   Use the bus timetable to answer the questions below.

Bus timetable: Ordsall to Bury

Ordsall, Salford Quays			07:30		08:30		18:30	19:00	20:00	21:00	22:00
Trafford Rd			07:35		08:35		18:35	19:05	20:05	21:05	22:05
Pendleton Precinct arr.			07:41		08:41		18:41				
Pendleton Precinct dep.	06:43	07:13	07:43	08:13	08:43	and	18:43	19:10	20:10	21:10	22:10
Lower Kersal	06:54	07:24	07:54	08:24	08:54	every	18:54	19:19	20:19	21:19	22:19
Agecroft	06:57	07:27	07:57	08:27	08:57	30	18:57	19:22	20:22	21:22	22:22
Butterstile Lane	07:04	07:34	08:04	08:34	09:04	mins	19:04				
Prestwich	07:10	07:40	08:10	08:40	09:10	until	19:10				
Besses o'th' Barn	07:14	07:44	08:14	08:44	09:14		19:14				
Unsworth	07:24	07:54	08:24	08:54	09:24		19:24				
Bury, Interchange	07:40	08:10	08:40	09:10	09:40		19:40				

  a   How long does it take the 06:43 Pendleton bus to get to Besses o'th' Barn?

  b   How long do buses wait at Pendleton?

  c   At what time does the last bus arrive in Bury?

  d   At what time does the first bus call at Trafford Road?

  e   How long does it take to travel from Lower Kersal to Unsworth?

  f   How long does it take to travel from Ordsall to Prestwich?

  g   How many buses call at Trafford Road before 10:00?

  h   How many buses call at Butterstile Lane during the day?

  i   Shaun arrives at his bus stop at Prestwich at 7.30 am.
      How long will he have to wait for a bus to Unsworth?

  j   Jane arrives at her bus stop in Trafford Road at 7.50 am.
      How long will she have to wait for a bus to Agecroft?

  k   Umar lives 5 minutes away from his bus stop in Agecroft.
      What is the latest time he can leave his house to get to Bury by 9 am?

  l   Eko wants to catch a bus from Pendleton to Prestwich, to arrive in Prestwich no later than 12 noon.
      What is the departure time of the latest bus he can catch from Pendleton?

F

2 The train route diagram shows the times it takes to travel from Manchester Victoria to all stations on the line. Use the information in the diagram to answer the questions below.

a How long does it take to travel between:
  i Manchester Victoria and Hindley
  ii Swinton and Daisy Hill
  iii Wigan and Farnworth
  iv Bolton and Salford Crescent?

b James is planning a trip from Swinton to Bolton. He will have to wait 12 minutes at Salford Crescent to change trains. What will be his total journey time from Swinton to Bolton?

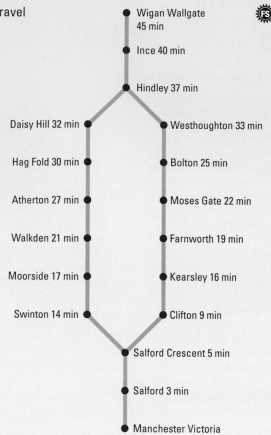

c Copy and complete the following timetables.

Manchester Victoria	09:05	11:35	Wigan Wallgate	08:30	11:55
Salford			Ince		
Salford Crescent			Hindley		
Swinton			Westhoughton		
Moorside			Bolton		
Walkden			Moses Gate		
Atherton			Farnworth		
Hag Fold			Kearsley		
Daisy Hill			Clifton		
Hindley			Salford Crescent		
Ince			Salford		
Wigan Wallgate			Manchester Victoria		

11.3 Metric units

⊙ Objectives

⦿ You can choose an appropriate unit to use for measurement.

⦿ You can convert between metric units.

⊘ Why do this?

It would be useful to understand metric units when you go on holiday as many European countries use kilometres rather than miles on their road signs.

⬥ Get Ready

1. Write down a list of objects you could measure using the following metric units of measure: centimetres, metres, grams, kilograms and litres.

Key Points

⦿ The following facts show how to change from one **metric unit** to another metric unit.

Length	Weight	Capacity
10 mm = 1 cm	1000 mg = 1 g	100 cl = 1 litre
100 cm = 1 m	1000 g = 1 kg	1000 ml = 1 litre
1000 mm = 1 m	1000 kg = 1 tonne	1000 l = 1 cubic metre
1000 m = 1 km		1000 cm^3 = 1 litre

The pictures below show examples of everyday items along with appropriate units of measurement.

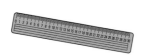

| A ruler is about 30 cm long. | A door has a height of about 2 m. | A can of cola is about 300 ml. | A bag of sugar is about 1 kg. |

⬥ Example 7

Write down the metric unit you would choose to measure the following:

a the length of your classroom

b the weight of a pen

c the amount of water in a bucket.

A02

a *metres*   b *grams*   c *litres*

G

⚙ **Exercise 11G**

Write down the metric unit you would use to take the measurements listed below:

1 an amount of medicine

2 the length of a finger nail

3 the weight of a house brick

4 the weight of a lorry

5 the amount of petrol in a car's petrol tank

6 the length of a bus

7 the length of a ballpoint pen

8 the weight of 30 of these books

9 the weight of a £1 coin

10 the length of an ant

11 the capacity of a kettle

12 the weight of a human hair

13 the distance from home to school

14 the weight of a box of cornflakes

15 the height of a person.

Example 8

a Change 3 kilometres to metres.

b Change 450 mm to cm.

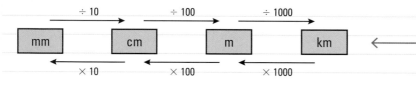

When you divide by 10, 100 or 1000 everything moves to the right by 1, 2 or 3 places.

a $3\,km = 3 \times 1000 = 3000\,m.$

Kilometres are longer than metres so you should expect to get more metres than kilometres. $1\,km = 1000\,m$

b $450\,mm = 450 \div 10 = 45\,cm.$

Millimetres are smaller than centimetres so you should expect to get fewer centimetres than millimetres. $10\,mm = 1\,cm$

G

⚙ **Exercise 11H**

1 Change the following lengths to centimetres.

   a 4 m   **b** 50 mm   **c** 8 m   **d** 13 m

   e 200 mm   **f** 35 m   **g** 74 mm   **h** 122 mm

2 Change the following lengths to millimetres.

   a 3 cm   **b** 6 cm   **c** 22 cm   **d** 40 cm

   e 200 cm   **f** 5.4 cm   **g** 13.7 cm   **h** 5.15 cm

3 Change the following lengths to metres.

a 6 km b 500 cm c 20 km d 3000 cm
e 45 km f 0.8 km g 1.4 km h 2.45 km

4 Change the following weights to grams.

a 2 kg b 30 kg c 400 kg d 250 kg
e 2000 kg f 55 kg g 0.12 kg h 4.2 kg

5 Change the following capacities to litres.

a 4000 m*l* b 7000 m*l* c 20 000 m*l* d 45 000 m*l*
e 2500 m*l* f 3700 m*l* g 6520 m*l* h 3130 m*l*

6 Change the following capacities to millilitres.

a 3 *l* b 20 *l* c 200 *l* d 450 *l*
e 35 *l* f 7.5 *l* g 0.4 *l* h 1.43 *l*

7 Change the following lengths to kilometres.

a 3000 m b 8000 m c 30 000 m d 68 000 m
e 4200 m f 5600 m g 5410 m h 2140 m

8 Change the following weights to tonnes.

a 5000 kg b 6000 kg c 40 000 kg d 57 000 kg
e 3600 kg f 4500 kg g 7630 kg h 4250 kg

9 Change the following weights to kilograms.

a 4000 g b 2 tonnes c 20 000 g d 15 tonnes
e 200 000 g f 3.7 tonnes g 6400 g h 1230 g

10 A large box contains 3 kg of plastic cubes. Each cube weighs 6 grams.
How many cubes are there in the box? AO2 F

11 A jar contains 450 m*l* of water. A tank can fill 200 of these jars. What is the capacity, in litres, of the tank? AO2

12 One lap of a running track is 400 m. How many laps must be run to complete a 10 km race? AO2

13 Neejal estimates that a car weighs $\frac{3}{4}$ tonne. How many kilograms is this? AO2

14 Jessica has a 60 c*l* bottle of medicine. How many 30 m*l* doses can be poured from it? AO2

15 A group of men can lay 300 m of pipe in one day. Working at the same rate, how long should it take them to lay the pipe for a length of 15 km? AO2 E

Example 9 Put the following weights in order, with the largest first.

250 g   25 g   2 kg   250 kg   3000 g

250 g, 25 g, 2000 g, 250 000 g, 3000 g ← First, change all of the weights to grams.

250 000 g, 3000 g, 2000 g, 250 g, 25 g ← Then order, with the largest first.

250 kg, 3000 g, 2 kg, 250 g, 25 g ← Finally, change back to the original units.

⚙ Exercise 11I

F

1 Put these lengths in order, with the smallest first.

4 m   6 mm   3 cm   4 km   30 cm   60 mm

2 Put these capacities in order, with the smallest first.

400 m*l*   6 *l*   700 m*l*   3000 m*l*   1*l*

3 Put these weights in order, with the smallest first.

600 g   450 g   0.5 kg   0.62 kg

4 Put these lengths in order, with the smallest first.

40 cm   0.6 cm   370 mm   1.4 m   600 mm   55 cm

5 Put these lengths in order, with the smallest first.

6 cm   55 mm   46 cm   0.2 cm   77 mm   0.4 cm   9 mm

6 Put these capacities in order, with the smallest first.

600 m*l*   450 m*l*   0.3 *l*   260 m*l*   0.08 *l*   75 m*l*

11.4 Imperial units

◎ Objectives

- You can convert between imperial units.
- You know the common metric–imperial conversions.
- You can convert between metric and imperial units.

◈ Why do this?

The UK is still making things using imperial units, for parts of the world that still use imperial units (parts of Africa, Asia and the US). It is also important that we understand our mathematical heritage.

⬆ Get Ready

1. Measure the heights of a number of objects in feet and inches. Also measure the same heights in centimetres.

🔍 Key Points

Imperial unit conversions

| 12 inches = 1 foot |
| 3 feet = 1 yard |
| 16 ounces = 1 pound |
| 14 pounds = 1 stone |
| 8 pints = 1 gallon |

Metric–imperial approximate equivalent conversions

Metric	Imperial	Metric	Imperial
8 km →	5 miles	1 kg →	2.2 pounds
1 m →	39 inches	25 g →	1 ounce
30 cm →	1 foot	4.5 litres →	1 gallon
2.5 cm →	1 inch	1 litre →	1.75 pints

Example 10   A man weighs $10\frac{1}{2}$ stones. Change this weight to pounds.

Use 14 pounds = 1 stone
$10\frac{1}{2}$ stones = 10.5 × 14 = 147 pounds.

Exercise 11J

1   Change the following imperial measurements.
   a 36 inches into feet   **b** 32 pints into gallons
   c 2 ft 4 in into inches   **d** 4 pounds into ounces
   e $5\frac{1}{4}$ feet into inches   **f** 21 feet into yards
   g 4 ft 4 in into inches   **h** 6 gallons into pints
   i $3\frac{1}{2}$ stones into pounds   **j** $2\frac{1}{4}$ gallons into pints
   k 300 pounds into stones and pounds   **l** 4 yards into feet
   m $\frac{1}{4}$ yard into inches   **n** 22 pints into gallons and pints
   o 1 stone into ounces

2   Ben was 4 foot $10\frac{1}{2}$ inches tall. He has grown another $4\frac{1}{2}$ inches. What is his height now?   **A03**

3   Anna weighs 9 stone 10 pounds. She wants to lose $\frac{1}{2}$ stone. What will she then weigh?

4   A curtain measures 5 foot 8 inches. It has a folded hem of 6 inches. What is its total length of the curtain material?   **A03**

F

E

Example 11   Change 20 km into miles.

Use 8 km = 5 miles
20 km = 20 ÷ 8 × 5 miles = 12.5 miles.

ResultsPlus
Examiner's Tip

In the exam, you will be expected to know the metric–imperial conversions.

Exercise 11K

1   Change the following measurements.
   a 15 miles into kilometres   **b** 10 kg into pounds   **c** 4 litres into pints
   d 6 inches into cm   **e** 3 yards into cm   **f** 48 km into miles
   g 11 pounds into kg   **h** 2.5 m into inches   **i** 75 miles into km

2   A house is estimated to be 24 feet high. What is this in metres?

3   A shirt collar measures 14 inches. What is this in centimetres?

F

F

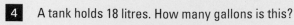

A02
A03

4 A tank holds 18 litres. How many gallons is this?

5 A baby needs 1 pint of milk a day. His bottle contains 0.2 litres.
How many bottles will his carer have to make for him in one day?

6 Yasmin travels from London to Scotland. The distance is 400 miles. What is the distance in kilometres?

7 A container has a capacity of 5 litres. What is this in pints?

8 A family on holiday in Majorca travel 150 km while touring the island. How many miles do they travel?

E **A02**
A03

9 A scuba diver weighs 110 lb. She needs to weigh a total of 59 kg to dive. Her airtank weighs 4 kg.
How many pound weights does she need to wear?

A02

10 Shop A is selling 4 kg of potatoes for £1.60. Shop B is selling a 5-pound bag of potatoes for £1.60.
Which is the better buy?

11.5 Speed

◉ Objectives

○ You can calculate speed given distance and time.
○ You can solve problems involving speed.

◈ Why do this?

You need to be able to calculate speed if you are training to improve your times in sports such as running, swimming or athletics.

◈ Get Ready

1. Find out how speed is measured on the speedometer of a car.

◉ Key Points

◉ **Speed** $= \dfrac{\text{distance}}{\text{time}}$

◉ Time $= \dfrac{\text{distance}}{\text{speed}}$

◉ Distance $=$ speed $\times$ time

◉ **Average speed** $= \dfrac{\text{total distance travelled}}{\text{total time taken}}$

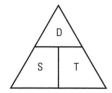

◉ Units of speed are usually miles per hour (mph), kilometres per hour (km/h) or metres per second (m/s).

🔍 Example 12

A car covers a distance of 160 miles in 4 hours. What is the car's average speed?

Average speed $= \dfrac{\text{total distance travelled}}{\text{total time taken}} = \dfrac{160}{4} = 40$ mph. ←

> Units are miles and hours, so speed is given as miles per hour (mph).

Exercise 11L

D

1   What is the average speed of a car that takes 3 hours to travel 90 miles?

2   Giles ran for 2 hours and covered 16 miles. At what average speed was he running?

3   Gareth was travelling by canal boat. He went 32 miles in 8 hours.
    At what average speed was he travelling?

C

4   Mandy swam 5 miles. It took her $2\frac{1}{2}$ hours. What was her average speed?

5   Ahmed travelled 200 miles on a business trip. He left home at 9 am and arrived at his destination at 1 pm.
    What was his average speed?

6   What is the average speed of a train that takes $2\frac{1}{2}$ hours to travel 80 km?

7   After a $3\frac{1}{2}$ hour journey a car had travelled 210 miles. What was its average speed?

8   Rami drives 125 miles in $2\frac{1}{2}$ hours. Work out his average speed.

9   Jon travels 40 km in 30 minutes. Calculate his average speed in km/h.

10 An aeroplane flies 1400 km in $3\frac{1}{2}$ hours. What is its average speed?

Example 13   Jill is a motorcycle courier. She travels at an average speed of 5 mph in the city.
How long will it take her to travel 25 miles?

A02

$$\text{Time} = \frac{distance}{speed} = \frac{25 \text{ miles}}{5 \text{ mph}} = 5 \text{ hours.}$$

Exercise 11M

E

1   A car travels for 2 hours at 40 mph. How far will the car have gone?

2   Find the time taken to travel 12 km at 3 km/h.

3   A delivery van takes 3 hours to complete a journey at a speed of 35 mph.
    What distance will it have covered?

D

4   How long does it take to travel 60 miles at an average speed of 40 mph?

5   How long does it take to travel 50 km at 20 km/h?

6   Find the time taken to walk 10 miles at an average walking speed of $2\frac{1}{2}$ mph.

7   A man walks at an average speed of 5 km/h. What distance will he cover in 2 hours 30 minutes?

D

8   Find the time taken to travel 10 miles at a speed of 4 mph.

9   An aeroplane flew at 400 mph. How far did it travel in $3\frac{1}{2}$ hours?

10   Rajesh travels for $2\frac{1}{2}$ hours at 64 mph. Calculate the distance he travelled.

Example 14   Martin cycles for 2 hours 18 minutes at a speed of 12 mph.
How far will he have travelled in this time?

2 h 18 min = $2\frac{18}{60}$ hour = 2.3 h   ← *The time needs to be converted into hours.*

ResultsPlus
Examiner's Tip

Remember:
0.1 h = 6 minutes
$\frac{1}{10}$ h = 6 minutes

Distance = speed × time = 12 × 2.3 = 27.6 miles.

Exercise 11N

C

1   How far will you have gone if you travel for 1 hour 45 minutes at 50 mph?

2   Mark runs at an average speed of 5 mph. He goes running for 1 hour and 18 minutes.
What distance does he run in that time?

3   A train travelled a distance of 208 km in 2 hours 36 minutes. What was the average speed of the train?

4   A cyclist rode at a speed of 10 mph. She covered a distance of 32 miles. For how long was she cycling?
(Give your answer in hours and minutes.)

5   A speedboat has an average speed of 20 km/h. It travels around the coast for 4 hours 6 minutes.
What distance does it cover in that time?

A02   6   A train travelled from London to Scotland. It left London at 08:00 hours and arrived at its destination at
13:24 hours. It travelled 324 miles. What was the average speed of the train?

7   A lorry made a journey of 390 km at an average speed of 50 km/h. How long did it take?
Give your answer in hours and minutes.

8   A train was travelling on a track at a speed of 90 mph. It was travelling at this speed for
3 hours 18 minutes. How far did it go in this time?

9   A horse takes 6 minutes to gallop 5 km. What is the average speed of the horse?

10   a   A racing car travels at 85 m/s. Work out the distance the car travels in 0.4 seconds.
b   Change a speed of 85 m/s into km/h.

11.6 Accuracy of measurements

◎ Objective

◉ You can recognise that measurements given to the nearest whole unit may be inaccurate by up to one half in either direction.

◈ Why do this?

Small differences in the measurement of an object can make a big difference if multiplied over 10 objects.

◈ Get Ready

1. Round the following numbers to the nearest whole number.
   a 6.8   b 52.49   c 13.5
2. Change the following lengths to cm.
   a 3 m   b 70 mm
3. Change the following lengths to mm.
   a 22 cm   b 4.7 cm

◈ Key Points

◉ Measured with a centimetre stick, the length of a piece of A4 paper is 30 cm.

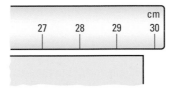

◉ This does not mean that the length of the A4 is exactly 30 cm. It is 30 cm correct to the nearest centimetre.

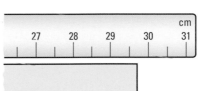

◉ The exact length of the piece of A4 paper is somewhere between 29.5 and 30.5 cm.

◉ Measured with a ruler, the length of a piece of A4 paper is 297 mm.

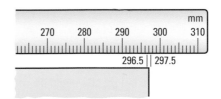

◉ The exact length of the piece of A4 paper is somewhere between 296.5 mm and 297.5 mm.

◉ So 297 mm to the nearest millimetre means that the minimum (least) possible length is 296.5 mm, and the maximum (greatest) possible length is 297.5 mm.

◉ Measurements given to the nearest whole unit may be inaccurate by up to one half of a unit below and one half of a unit above.

Example 15

1. The weight of a cocker spaniel is 14 kg correct to the nearest kilogram.
   Write down   **a**   the smallest possible weight   **b**   the greatest possible weight.

2. A small bowl holds 244 millilitres of water correct to the nearest millilitre.
   Write down   **a**   the smallest possible volume   **b**   the greatest possible volume.

1. a 13.5 kg   **b** 14.5 kg
2. a 243.5 ml   **b** 244.5 ml

Example 16   The length of a calculator is 12.8 cm correct to the nearest millimetre.
Write down   **a**   the minimum possible length   **b**   the maximum possible length.

12.8 cm = 128 mm ← *Convert the length from centimetres to millimetres. 1 cm = 10 mm.*

a   127.5 mm or 12.75 cm
b   128.5 mm or 12.85 cm

Exercise 11O

C

1 The length of a pencil is 12 cm correct to the nearest centimetre.
Write down the maximum length it could be.

2 The weight of an envelope is 45 grams correct to the nearest gram.
Write down the minimum weight it could be.

3 The capacity of a jug is 4 litres correct to the nearest litre.
Write down the minimum capacity of the jug.

4 The radius of a plate is 9.7 cm correct to the nearest millimetre.
Write down
   a   the least possible length it could be   **b**   the greatest possible length it could be.

5 Magda's height is 1.59 m correct to the nearest centimetre. Write down in metres
   a   the minimum possible height she could be
   b   the maximum possible height she could be.

6 The length of a pencil is 10 cm correct to the nearest cm.
The length of a pencil case is 102 mm correct to the nearest mm.
Explain why the pen might **not** fit in the case.

7 The width of a cupboard is measured to be 82 cm correct to the nearest centimetre.
There is a gap of 817 mm correct to the nearest mm in the wall.
Explain how the cupboard might fit in the wall.

Chapter review

- Lines can be measured using a ruler.
- Some dials have a scale. You need to interpret the scale to take an accurate reading from the dial.
- Sometimes you will need to estimate a reading.
- A 12-hour clock goes from 0 to 12, using am for times before noon, and pm for times after noon.
- A 24-hour clock goes from 0 to 24, where 24 is midnight.
- You will need to know the following **conversions** involving time:

60 seconds = 1 minute	366 days = 1 leap year
60 minutes = 1 hour	3 months = 1 quarter
24 hours = 1 day	12 months = 1 year
7 days = 1 week	52 weeks = 1 year
365 days = 1 year	

- The following facts show how to change from one **metric unit** to another metric unit.

Length	Weight	Capacity
10 mm = 1 cm	1000 mg = 1 g	100 cl = 1 litre
100 cm = 1 m	1000 g = 1 kg	1000 ml = 1 litre
1000 mm = 1 m	1000 kg = 1 tonne	1000 l = 1 cubic metre
1000 m = 1 km		1000 cm^3 = 1 litre

- **Imperial unit** conversions   Metric–imperial approximate equivalent conversions

12 inches = 1 foot	
3 feet = 1 yard	
16 ounces = 1 pound	
14 pounds = 1 stone	
8 pints = 1 gallon	

Metric	Imperial	Metric	Imperial
8 km ⟶	5 miles	1 kg ⟶	2.2 pounds
1 m ⟶	39 inches	25 g ⟶	1 ounce
30 cm ⟶	1 foot	4.5 litres ⟶	1 gallon
2.5 cm ⟶	1 inch	1 litre ⟶	1.75 pints

- **Speed** $= \dfrac{\text{distance}}{\text{time}}$
- Time $= \dfrac{\text{distance}}{\text{speed}}$
- Distance = speed × time
- **Average speed** $= \dfrac{\text{total distance travelled}}{\text{total time taken}}$
- Units of speed are usually miles per hour (mph), kilometres per hour (km/h) or metres per second (m/s).
- Measurements given to the nearest whole unit may be inaccurate by up to one half of a unit below and one half of a unit above.

Review exercise

1 **a** Complete the table by writing a sensible **metric** unit for each measurement.
   The first one has been done for you.

The length of the river Nile	6700 kilometres
The height of the world's tallest tree	110
The weight of a chicken's egg	70
The amount of petrol in a full petrol tank of a car	40

   b Change 4 metres to centimetres.
   c Change 1500 grams to kilograms.

June 2008

G

2 This is part of a ruler.

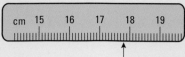

a Write down the length marked with an arrow.

This is a thermometer.

b Write down the temperature shown.

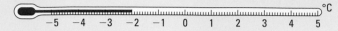

This is a parcel on some scales.

c Write down the weight of the parcel.

3 a Write down a sensible metric unit that can be used to measure
  i the height of a tree ii the weight of a person.

b Change 2 centimetres to millimetres.

Nov 2008

4 a Write down a sensible **metric** unit for measuring
  i the distance from London to Paris ii the amount of water in a swimming pool.

b i Change 5 centimetres to millimetres. ii Change 4000 grams to kilograms.

Nov 2008

5 a

Write down the number marked by the arrow.

b
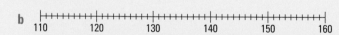

Find the number 127 on the number line.
Mark it with an arrow (↑).

c

Write down the number marked by the arrow.

d
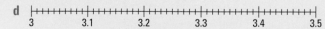

Find the number 3.18 on the number line.
Mark it with an arrow (↑).

Nov 2008

6

The diagram shows a building and a man.

The man is of normal height.

The man and the building are drawn to the same scale.

a Write down an estimate for the height of the man.

b Write down an estimate for the height of the building.

Nov 2008

7

The picture shows a man standing next to a flagpole.

The man is of normal height.

The man and the flagpole are drawn to the same scale.

a Write down an estimate for the height, in metres, of the man.

b Work out an estimate for the height, in metres, of the flagpole.

June 2008

8 Here is a picture of a woman opening a door that is 2 m high.

Estimate the height of the woman.

Nov 2007

9 Shalim says 1.5 km is less than 1400 m.

Is he right?

Explain your answer.

A03

June 2007

F

10 Here is part of a train timetable from Peterborough to London.

a Which station should the train leave at 09:01?

The train arrives in Sandy at 09:12.

b How many minutes should the train wait in Sandy?

The train should take 41 minutes to travel from Arlesey to London.

c What time should the train arrive in London?

Station	Time of leaving
Peterborough	08:44
Huntingdon	09:01
St Neots	09:08
Sandy	09:15
Biggleswade	09:19
Arlesey	09:24

Nov 2008

11 a Write down the weight in kg shown on this scale.

b i How many pounds are there in 1 kg?

The weight of a baby is 5 kg.

ii Change 5 kg to pounds.

Nov 2008

12 Here is part of a railway timetable.

A train leaves from Bristol Temple Meads at 09:00.

a At what time should the train arrive at Swindon?

Jambaya gets to the station in Chippenham at 08:45.
She waits for the next train to Didcot.

b i How long should she have to wait?

ii At what time should she arrive at Didcot?

Bristol Temple Meads	08:00	08:30	09:00
Bath	08:15	08:45	09:15
Chippenham	08:30	09:00	09:30
Swindon	08:50	09:20	09:50
Didcot	09:15	09:45	10:15
Reading	09:35	10:05	10:35
London Paddington	09:55	10:25	10:55

All the trains should take the same time to travel from Bath to Reading.

c How long, in minutes, should it take to travel from Bath to Reading?

June 2008

13 Here is part of a bus timetable.

Bus Station	07:00	07:30	08:00
Castle Street	07:10	07:40	08:15
High Street	07:25	07:55	08:25
Station Road	07:37	08:07	08:37
Church Street	07:50	08:20	08:50
Wharf Inn	07:55	08:25	08:55

A bus leaves the Bus Station at 07:00.

a At what time should the 07:00 bus arrive at Station Road?

Jill arrives at High Street at 07:45.
She wants to catch a bus to Wharf Inn.

b How long should she have to wait for the next bus?

A bus leaves Station Road at 08:37.

c How long should this bus take to travel from Station Road to Wharf Inn?

Nov 2007

14 Complete this table.

Write a sensible unit for each measurement.

Three have been done for you.

	Metric	Imperial
Distance from London to Cardiff	km	
Weight of a bag of potatoes		pounds
Volume of fuel in a car's fuel tank		gallons

Nov 2007

15 Zoe is planning a trip to Palma from Sa Pobla on the local train.

This is part of the train timetable.

Sa Pobla	09:23	10:23	11:23	12:23	13:23
Inca	09:41	10:41	11:41	12:41	13:41
Santa Maria	09:57	10:57	11:57	12:57	13:57
Marratxi	10:05	11:05	12:05	13:05	14:05
Palma	10:20	11:20	12:20	13:20	14:20

a How long, in minutes, is the train journey from Sa Pobla to Palma?

The drive from Zoe's hotel to the station will take 30 minutes.

She wants to travel by train from Sa Pobla to Palma, visiting Inca for $1\frac{1}{2}$ hours on the way.

b Complete the table.

	Time
Zoe leaves the hotel	
Zoe leaves Sa Pobla on a train	
Train arrives at Inca (Zoe gets off)	
Zoe leaves Inca on another train	
Zoe arrives at Palma	

16 The distance from London to New York is 3456 miles.

A plane takes 8 hours to fly from London to New York.

Work out the average speed of the plane.

June 2008

17 Car P and Car Q travel from Amfield to Barton.

Car P averages 10 km per litre of petrol.

It needs 48 litres of petrol for this journey.

Car Q averages 4 km per litre of petrol.

Work out the number of litres of petrol Car Q needs for the same journey.

18 There are 14 pounds in a stone.

There are 2.2 pounds in a kilogram.

A man weighs 13 stone 6 pounds.

Work out his weight in kilograms, giving your answer to the nearest kilogram.

You must show all your working.

C

19 Stuart drives 180 km in 2 hours 15 minutes.
Work out Stuart's average speed.   *Nov 2008*

20 A gold necklace has a mass of 127 grams, correct to the nearest gram.
a Write down the least possible mass of the necklace.
b Write down the greatest possible mass of the necklace.   *June 2006*

12 PROCESSING, REPRESENTING AND INTERPRETING DATA

Look at the bar chart of Bangladesh's annual rainfall. It receives the majority of its annual rainfall during the rainy season (June to September). These rains flood fields and homes, leaving 70% of Bangladesh's land underwater.

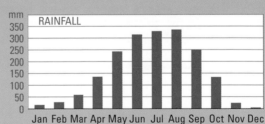

Objectives

In this chapter you will:
- learn how to produce and interpret pictograms, pie charts and bar charts for various data types
- learn how to produce and interpret vertical line graphs for discrete data, comparative bar charts, composite bar charts, histograms for continuous data and frequency polygons.

Before you start

You need to know:
- how to add and subtract numbers to 360
- how to measure angles
- what a sector of a circle is.

12.1 Pictograms

⬆ Get Ready

Describe each of the following as either qualitative data, discrete quantitative data or continuous quantitative data.

1. Can be given as a whole number only
2. Can be described in words
3. Can take any numerical value

🔍 Key Points

◉ Data are often easier to understand if they are presented in the form of a diagram.
   When **representing** data, the method chosen depends on the type of data.
◉ A **pictogram** can be used to represent qualitative data. It uses a number of symbols or pictures to represent a number of items.
◉ A **key** tells you the number of items represented by a single symbol or picture.

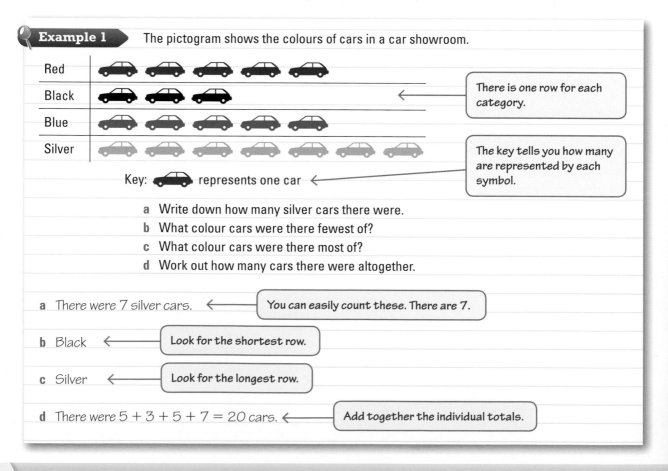

Example 1 The pictogram shows the colours of cars in a car showroom.

Red
Black ← There is one row for each category.
Blue
Silver

Key: 🚗 represents one car ← The key tells you how many are represented by each symbol.

a Write down how many silver cars there were.
b What colour cars were there fewest of?
c What colour cars were there most of?
d Work out how many cars there were altogether.

a There were 7 silver cars. ← You can easily count these. There are 7.

b Black ← Look for the shortest row.

c Silver ← Look for the longest row.

d There were 5 + 3 + 5 + 7 = 20 cars. ← Add together the individual totals.

Example 2

Atul asked some of his friends which sport they liked best.
The table shows the results.

Sport	Number of friends
Football	16
Rugby	6
Cricket	11
Swimming	8

Draw a pictogram to represent this information.
Use ⊕ to represent four friends.

Sport	Number of friends	Number of symbols
Football	16	$16 \div 4 = 4$
Rugby	6	$6 \div 4 = \frac{6}{4} = \frac{3}{2} = 1\frac{1}{2}$
Cricket	11	$11 \div 4 = \frac{11}{4} = 2\frac{3}{4}$
Swimming	8	$8 \div 4 = 2$

Favourite sports

Football	⊕ ⊕ ⊕ ⊕
Rugby	⊕ ⊖
Cricket	⊕ ⊕ ⊖
Swimming	⊕ ⊕

$1\frac{1}{2} \times 4 = 6$

Key
⊕ represents four friends

$2\frac{3}{4} \times 4 = 11$

Don't forget the key.

Exercise 12A

Questions in this chapter are targeted at the grades indicated.

G

1   The pictogram shows the number of emails sent from an office in one week.

Number of emails

Monday	✉ ✉ ✉ ✉
Tuesday	✉ ✉ ✉ ◗
Wednesday	✉ ✉ ✉ ✉ ◖
Thursday	✉ ✉ ✉ ✉ ✉
Friday	✉ ✉ ✉

Key
 represents
20 emails

a   On which day were the greatest number of emails sent from the office?
b   How many emails were sent from the office on Thursday?
c   How many emails were sent from the office on Monday?
d   How many emails were sent from the office on Tuesday?

G

2 The pictogram shows the number of drinks sold in one day from a machine.

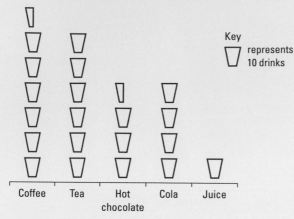

Key
represents
10 drinks

Coffee   Tea   Hot chocolate   Cola   Juice

a Write down the drink that was the least popular.

b How many drinks of tea were sold?

c How many drinks of hot chocolate were sold?

d Work out the total number of drinks sold.

3 The pictogram shows the numbers of three different types of pizza sold in one day.

Pizzas sold

Margherita	⊕ ⊕ ◁
BBQ chicken	⊕ ⊕ ⊕ ⊕
Hawaiian	⊕ ⊕ ⊕
Meat feast	

Key
⊕ represents
eight pizzas

20 meat feast pizzas were also sold.

a Copy and complete the pictogram.

b How many of each type of pizza were sold?

F

A02 **4** 40 students were asked which of five subjects was their favourite.

Represent this information in a suitable chart.

Subject	Number of students
English	6
Mathematics	8
Spanish	4
Science	9
Technology	3

A02 **5** The table shows how 60 students travel to school.

Form of travel	Number of students
Walk	6
Bus	36
Train	2
Cycle	14
Car	2

Represent this information in a suitable chart.

12.2 Pie charts

◉ Objectives

- ◉ You can represent the proportions of different categories of data using a pie chart.
- ◉ You can use pie charts to find the frequency for each category.
- ◉ You can find the total population from a pie chart.
- ◉ You can find the greatest and least values from a pie chart.

◈ Why do this?

A pie chart would be a good way to display a company's market share, or a school could use one to show the relative popularity of GSCE or A level subjects.

◈ Get Ready

1. A sector of a circle takes up $\frac{3}{4}$ of the circle. What is the angle of the sector?
2. A sector of a circle has an angle of 18°. What fraction of a circle does it take up?
3. A sector of a circle takes up 290°. What angle is taken up by the remainder of the circle?

◈ Key Points

- ◉ A **pie chart** is a circle that is divided into sectors and shows how the total is split up between the different categories.

- ◉ In a pie chart the area of the whole circle represents the total number of items.

- ◉ The area of each sector represents the number of items in that category.
  This pie chart shows people's favourite pets.

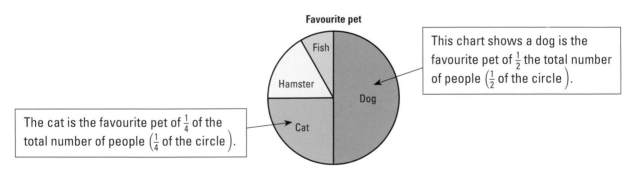

Favourite pet

This chart shows a dog is the favourite pet of $\frac{1}{2}$ the total number of people ($\frac{1}{2}$ of the circle).

The cat is the favourite pet of $\frac{1}{4}$ of the total number of people ($\frac{1}{4}$ of the circle).

- ◉ The angles at the centre of a pie chart must add up to 360°.

- ◉ The angle for a particular sector is found as follows:

  $$\text{Sector angle} = \frac{\text{frequency} \times 360°}{\text{total frequency}}$$

- ◉ To read frequencies from a pie chart use the formula:

  $$\text{Frequency} = \frac{\text{sector angle} \times \text{total frequency}}{360°}$$

- ◉ The frequency represented by corresponding sectors of two pie charts might have the same sector angle but the frequency they represent depends on the total frequency.

Example 3 The table shows the favourite colours of a sample of 30 students.

Colour	Blue	Red	Green	Black
Frequency	10	15	3	2

Draw a pie chart to represent this information.

Method 1

$360 \div 30 = 12$

12° represents 1 student.

> There are 360° in a full circle.
> There are 30 students.
> To work out the angle for
> 1 student divide 360 by 30.

Blue $10 \times 12 = 120°$

> If 1 student is represented by
> 12°, then 10 students will need
> $10 \times 12° = 120°$.

Red $15 \times 12 = 180°$
Green $3 \times 12 = 36°$
Black $2 \times 12 = 24°$

> The angle for each of the other
> colours is found by multiplying the
> frequency of the colour by 12°.

ResultsPlus
Examiner's Tip

Make sure the angles in a pie
chart add up to 360°.

Method 2

Blue $\frac{10}{30} \times 360° = 120°$

> 10 out of 30 students chose blue as their
> favourite colour so $\frac{10}{30}$ of the whole circle is
> needed to represent blue. $\frac{10}{30}$ of 360° = 120°.

Red $\frac{15}{30} \times 360° = 180°$

Green $\frac{3}{30} \times 360° = 36°$

Black $\frac{2}{30} \times 360° = 24°$

> The angles for the other colours
> can be found in the same way.

Draw a circle. Draw a line OA
from its centre to its
circumference.

Use your protractor to measure
the 120° angle.
Mark it and draw line OB.

Place your protractor on OB.
Measure the angle 180°.
Mark it and draw line OC.

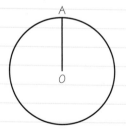

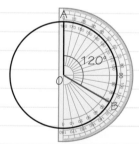

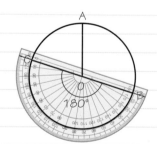

Continue in this way to draw a 36° angle.
The finished pie chart looks like this:

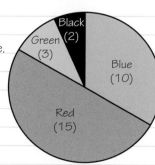

The frequencies have been
included to make it easy to
compare them.

Exercise 12B

1 40 passengers at a train station were asked which city they were travelling to.
The table shows this information.

City	Reading	Swindon	Bristol	Cardiff
Frequency	6	5	9	20

Work out the sector angle for each city. Draw an accurate pie chart to show this information.

2 Ryan asked 60 friends to name their favourite fruit. The table shows his results.

Fruit	Strawberry	Apple	Banana	Grape	Orange
Frequency	25	15	6	9	5

Work out the sector angle for each flavour. Draw an accurate pie chart to show this information.

3 The table shows the time it takes each of 30 people to travel to work.
Draw an accurate chart to show this information.

Time in minutes	Frequency
Less than 10	3
Between 10 and <15	4
Between 15 and 30	12
More than 30	11

4 The table shows information about the 540 books in a small library.

Category	Thriller	Classic	Romance	Non-fiction
Frequency	120	60	270	90

Draw an accurate chart to show this information.

5 The table shows the numbers of the different types of cars in a car park.

Type of car	Ford	Nissan	Toyota	Renault
Frequency	24	30	27	9

Draw an accurate chart to show this information.

6 The table shows the number of students absent in a year group on each of five days.

Day	Monday	Tuesday	Wednesday	Thursday	Friday
Number of students absent	7	8	10	6	5

Draw an accurate chart to show this information.

Example 4 The pie chart shows the numbers of each type of pet seen in a vet's surgery one morning.
There were 20 pets altogether.

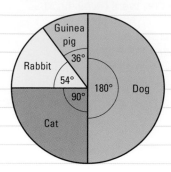

a Which type of pet was most common?

b Work out the number of each type of pet.

ResultsPlus
Examiner's Tip

In an exam 'work out' means calculate the frequency, so don't just measure the angle.

a Dog is the most common pet. | This has the largest sector. |

b **Method 1**

$$\text{Dog frequency} = \frac{\text{sector angle} \times \text{total frequency}}{360°}$$

$$= \frac{180° \times 20}{360°} = 10$$

$$\text{Cat frequency} = \frac{90° \times 20}{360°} = 5$$

$$\text{Rabbit frequency} = \frac{54° \times 20}{360°} = 3$$

$$\text{Guinea pig frequency} = \frac{36° \times 20}{360°} = 2$$

ResultsPlus
Examiner's Tip

Always add up the frequencies for each sector to make sure they total to the right number.

Method 2

If 20 pets are represented by 360°, then 1 pet is represented by $\frac{360}{20} = 18°$.
Divide each angle by 18°.

$$\text{Dog frequency} = \frac{180°}{18°} = 10 \qquad \text{Cat frequency} = \frac{90°}{18°} = 5$$

$$\text{Rabbit frequency} = \frac{54°}{18°} = 3 \qquad \text{Guinea pig frequency} = \frac{36°}{18°} = 2$$

Exercise 12C

E

1 120 adults were asked to name their favourite pastime.
   The pie chart shows the results of this survey.
   a Which was the most popular pastime?
   b Which pastime was the least popular?
   c What fraction of the adults say going out with friends is their
      favourite pastime? Give your fraction in its simplest form.
   d How many of the adults prefer going out with friends?

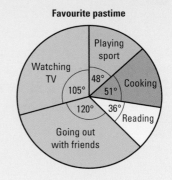

Favourite pastime

2 The pie chart shows the holiday destinations of 90 girls.
   a Which destination is the most popular?
   b How many degrees represent one person on the pie chart?
   c How many girls said Great Britain was their holiday destination?
   d What angle represents Greece on the pie chart?
   e How many girls said Greece was their holiday destination?

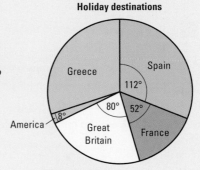

Holiday destinations

3 The pie chart shows information about how Letitia spends
   her time in one 24-hour day.
   Copy and complete the table. You will need to measure the
   angles in the pie chart.

Activity	Angle (degrees)	Number of hours
Sleep		
School		
Play		
Watch TV		
Eat		
Homework		

Letitia's day

4 The pie chart shows information about the makes of car driven by
   1200 people.
   a What fraction of the 1200 people drive a Nissan?
      Give your fraction in its simplest form.
   b How many people drive a Nissan?
   c How many people are represented by 1 degree in the pie chart?
   d How many people drive a Toyota?

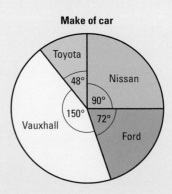

Make of car

E

5 In a survey, some adults were asked to name their favourite ice cream flavour. The results are shown in the pie chart.

30 adults said that mint was their favourite ice cream flavour.

a How many degrees represent one person in the pie chart?

b How many adults took part in the survey?

A corner shop wishes to stock 2 varieties of ice cream.

c Suggest, with reasons, what they should stock.

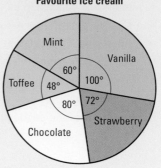

Favourite ice cream

6 In a survey, some students were asked to name their favourite animal. The pie chart shows information about their answers.

a Write down the fraction of the students who answered horse.

Write your fraction in its simplest form.

12 students answered horse.

b Work out the number of students that took part in the survey.

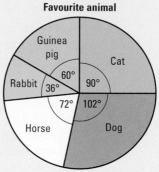

Favourite animal

12.3 **Bar charts**

⊙ Objectives

○ You can represent qualitative data as a bar chart.

○ You can represent grouped discrete data as a bar chart.

○ You can read off frequency values from a bar chart, and find the greatest and least values.

○ You can find total populations from a bar chart.

⊘ Why do this?

It is easy to see data shown on a bar chart and read off individual values by using the heights of the bars. They can be used to show how many inches of snow fall per month or how many hours of sunshine per day you can expect in a certain destination.

⬦ Get Ready

Work out

1. $8 + 6 + 10 + 1$

2. $12 + 21 + 15 + 36$

3. 9×80

4. 35×4

Key Points

● A **bar chart** can be used to display qualitative data.

The **frequency table** shows the eye colours of 40 adults.

Eye colour	Brown	Hazel	Blue	Green
Frequency	15	8	12	5

The two bar charts also show this information.

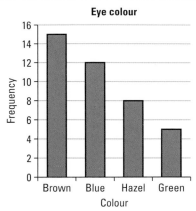

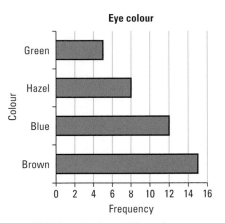

This is a vertical bar chart.   This is a horizontal bar chart.

● In a bar chart:
  ● all the bars are the same width   ● both the vertical and horizontal axes have labels
  ● there is a gap between the bars   ● the bars can be drawn horizontally or vertically.

● Bar charts may also be used for grouped discrete data.

Example 5   The bar chart shows the shoe sizes of a number of people.

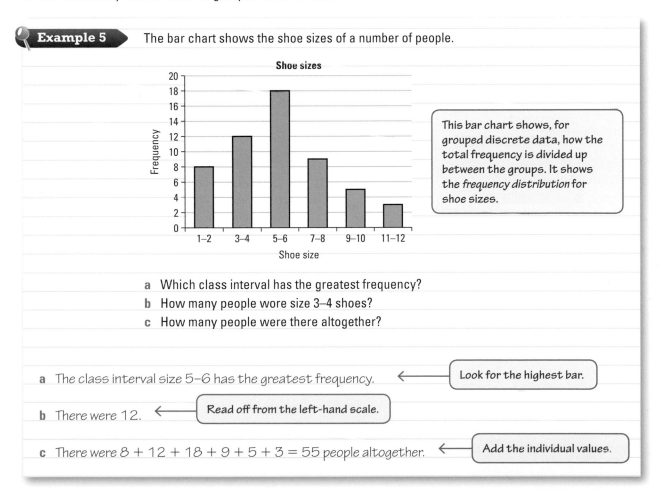

This bar chart shows, for grouped discrete data, how the total frequency is divided up between the groups. It shows the frequency distribution for shoe sizes.

  a Which class interval has the greatest frequency?
  b How many people wore size 3–4 shoes?
  c How many people were there altogether?

  a The class interval size 5–6 has the greatest frequency. ← Look for the highest bar.

  b There were 12. ← Read off from the left-hand scale.

  c There were 8 + 12 + 18 + 9 + 5 + 3 = 55 people altogether. ← Add the individual values.

Exercise 12D

G

1 The bar chart shows information about the hair colour of all the students in a class.

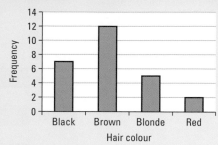

  a How many students have black hair?

  b How many students are in the class?

2 The bar chart shows the number of second-hand cars sold by different garages in one week in September.

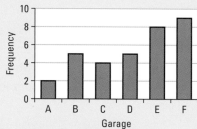

  a Write down the number of cars sold by garage B.

  b Write down the garage at which most cars were sold.

  c Work out how many cars were sold altogether.

3 The bar chart shows the number of hours that some people spent watching television in one week.

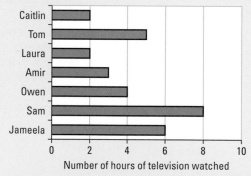

  a Who watched the most hours of television?

  b Write down the number of hours of television watched by Amir.

  c Write down the two people who watched the same number of hours of television.

  d Who watched 4 hours of television?

4 The table gives information about the meals served in a canteen.
Draw a vertical bar chart to show this information.

Meal	Fish and chips	Lasagne	Pie and chips	Cheese salad	Sausage and mash
Frequency	55	27	34	12	43

12.4 Comparative and composite bar charts

Objectives

● You can draw comparative bar charts.
● You can draw composite bar charts.

Why do this?

You might want to compare two people's results in different exams, or show the numbers of people employed in different industries in two countries.

Get Ready

1. This picture shows three boxes stacked on top of each other.
   a Which box is the tallest?
   b Which box is the smallest?

Key Points

● Comparative or **dual bar charts** can be drawn to compare data.
● In a comparative bar chart:
   ● two (or more) bars are drawn side by side for each category
   ● the bars can be horizontal or vertical
   ● the heights of the bars can be compared category by category.
● A composite bar chart shows the size of individual categories split into their separate parts.

Example 6 ▶ The dual bar chart shows the number of houses sold by two agents in four months.

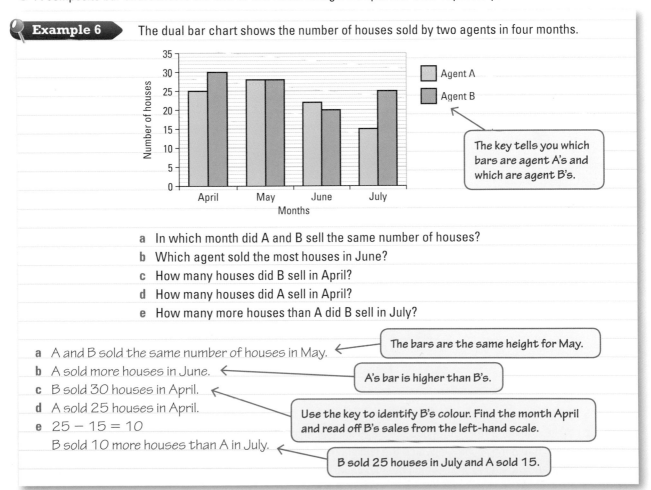

The key tells you which bars are agent A's and which are agent B's.

a In which month did A and B sell the same number of houses?
b Which agent sold the most houses in June?
c How many houses did B sell in April?
d How many houses did A sell in April?
e How many more houses than A did B sell in July?

a A and B sold the same number of houses in May. ← The bars are the same height for May.
b A sold more houses in June. ← A's bar is higher than B's.
c B sold 30 houses in April. ←
d A sold 25 houses in April. ← Use the key to identify B's colour. Find the month April and read off B's sales from the left-hand scale.
e 25 − 15 = 10
   B sold 10 more houses than A in July. ← B sold 25 houses in July and A sold 15.

Example 7 The composite bar chart shows the types of land use on three different farms.

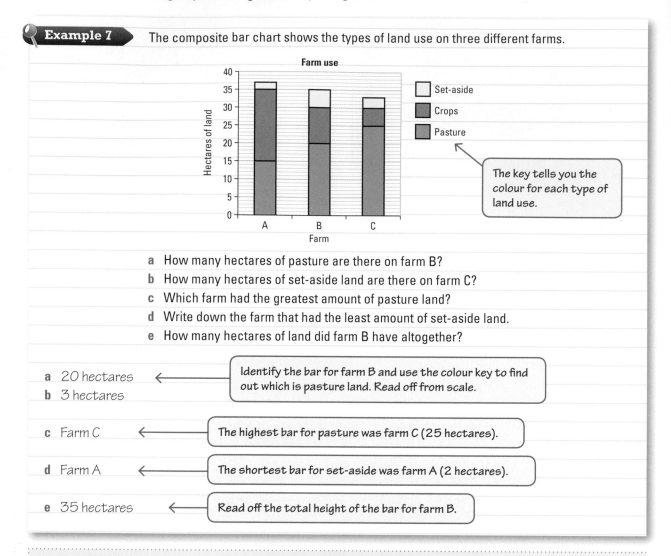

a How many hectares of pasture are there on farm B?

b How many hectares of set-aside land are there on farm C?

c Which farm had the greatest amount of pasture land?

d Write down the farm that had the least amount of set-aside land.

e How many hectares of land did farm B have altogether?

a 20 hectares ← Identify the bar for farm B and use the colour key to find out which is pasture land. Read off from scale.

b 3 hectares

c Farm C ← The highest bar for pasture was farm C (25 hectares).

d Farm A ← The shortest bar for set-aside was farm A (2 hectares).

e 35 hectares ← Read off the total height of the bar for farm B.

Exercise 12E

1 The comparative bar chart shows the temperature in a number of resorts in April and October.

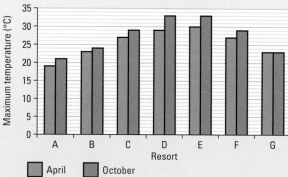

a Write down the maximum temperature in April.

b Write down the maximum temperature in October.

c Write down the resort that had the same maximum temperature in both months.

d Write down the resorts in which the maximum temperature in October was 29°C.

e Write down the resort in which the maximum temperature in April was 19°C.

2   Two factories making bolts employ male and female workers.
The numbers of males and females are shown on the
composite bar chart.

a   Write down the factory that employed the most people.

b   Write down the number of males employed in Factory A.

c   Write down the number of people employed by Factory B.

d   Work out how many people were employed by both
factories altogether.

One factory has a bigger wage bill than the other.

e   Which do you think this is and why?

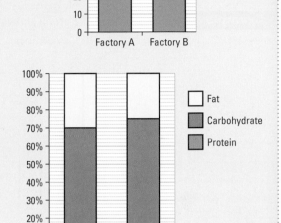

3   The composite bar chart gives information about
the nutritional content of chocolate and shortbread
biscuits.

a   Write down the name of the biscuit that has the
most carbohydrates.

b   Which of the nutritional contents is the same in
both biscuits?

c   Work out the percentage of fat in a shortbread
biscuit.

* **4**   The table gives information about the numbers of
males and females in each of the five classes in Year 11.

Draw a comparative bar chart to represent these data.

Class	Males	Females
P	14	13
Q	15	12
R	12	15
S	11	12
T	8	13

12.5 Line diagrams for discrete data and histograms for continuous data

Objectives

- You can represent discrete data
using a vertical line graph.
- You can represent grouped
continuous data as a histogram.

Why do this?

When quantitative data are shown as a line graph or histogram, it
is easy to see and read off the individual values using the heights
of the lines or bars, and to find the greatest and least values by
comparing the heights of the bars.

Get Ready

1. This diagram shows five lines A, B, C, D and E.

a   Which is the shortest line?

b   Which is the longest line?

c   Which two lines are the same length?

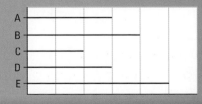

Key Points

- Quantitative data can be either discrete or continuous.
- A vertical **line graph** is used to display ungrouped discrete data.
- Because the data are numerical, the horizontal axis is a numerical scale.
- The discrete data can only take certain values on this scale.
- A **histogram** can be used to display grouped continuous data.
- A histogram is similar to a bar chart but, because it represents continuous data, no gap is left between the bars.

Example 8 The vertical line graph shows the number of children in a sample of families.

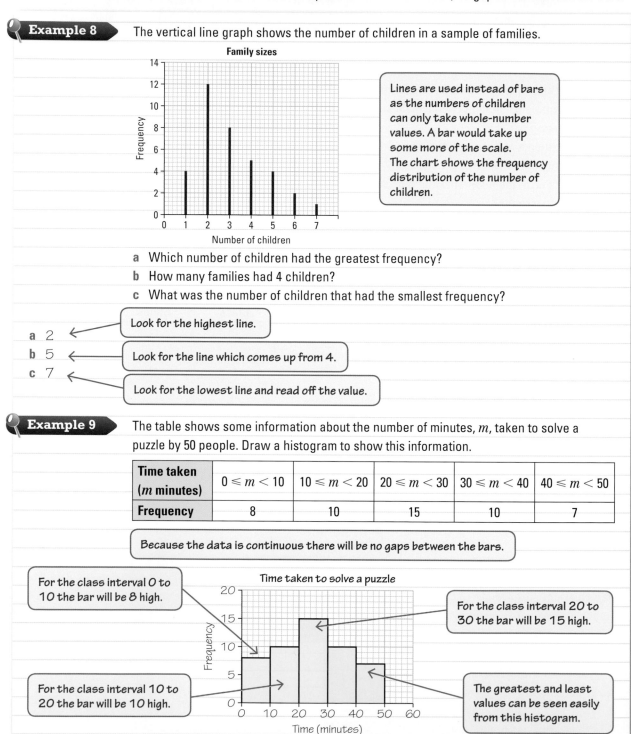

Family sizes

Lines are used instead of bars as the numbers of children can only take whole-number values. A bar would take up some more of the scale.
The chart shows the frequency distribution of the number of children.

a Which number of children had the greatest frequency?

b How many families had 4 children?

c What was the number of children that had the smallest frequency?

Look for the highest line.

a 2

b 5   Look for the line which comes up from 4.

c 7

Look for the lowest line and read off the value.

Example 9 The table shows some information about the number of minutes, m, taken to solve a puzzle by 50 people. Draw a histogram to show this information.

Time taken (m minutes)	$0 \leqslant m < 10$	$10 \leqslant m < 20$	$20 \leqslant m < 30$	$30 \leqslant m < 40$	$40 \leqslant m < 50$
Frequency	8	10	15	10	7

Because the data is continuous there will be no gaps between the bars.

For the class interval 0 to 10 the bar will be 8 high.

Time taken to solve a puzzle

For the class interval 20 to 30 the bar will be 15 high.

For the class interval 10 to 20 the bar will be 10 high.

The greatest and least values can be seen easily from this histogram.

Exercise 12F

G

1 The vertical line graph shows how often each number on a dice came up when it was thrown a number of times.

  a How many times was the number 3 thrown?

  b Which number was thrown five times?

  c How many times was the dice thrown altogether?

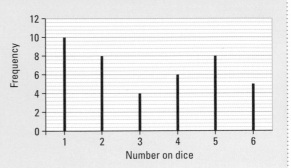

2 A survey was done to see the number of people travelling in each car entering a factory car park.

The results are shown on the vertical line graph.

  a Write down the number of cars that had four people in them.

  b Write down the most common number of people.

  c Work out the total number of cars in the survey.

The factory wish to encourage car sharing.

They decide to ban all cars carrying less than 3 people from using the car park.

  d How many people are going to have to change their travel arrangements?

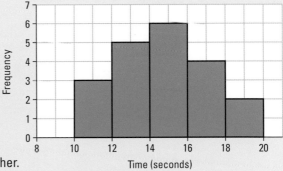

3 The histogram shows information about the times taken by some runners to run 100 m.

  a Write down the reason why there are no gaps between the bars.

  b Write down the number of runners that took between 12 and 14 seconds.

  c Work out the number of runners that took more than 16 seconds.

  d Work out how many runners there were altogether.

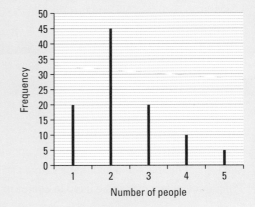

4 A speed monitor recorded the speed of cars travelling below a bridge on a motorway during a one-hour period. The table shows this information.

Speed (s mph)	Frequency
$20 \leqslant s < 30$	0
$30 \leqslant s < 40$	8
$40 \leqslant s < 50$	12
$50 \leqslant s < 60$	28
$60 \leqslant s < 70$	38
$70 \leqslant s < 80$	14

Draw a histogram to show these data.

12.6 Frequency polygons

◎ Objectives

○ You can draw frequency polygons.
○ You can recognise simple trends from a frequency polygon.

⊘ Why do this?

In a sample of families, as the number of children in a family increases how does the frequency change? By drawing a frequency polygon the trend for the frequency to increase, decrease or stay the same can be recognised.

⊕ Get Ready

1. Draw a histogram for the following data.

Size	0 to <10	10 to <20	20 to <30	30 to <40
Frequency	5	10	8	6

⚲ Key Points

⦿ When drawing a **frequency polygon** for discrete data, you draw straight lines to connect the tops of the lines on a vertical line chart.
⦿ When drawing a frequency polygon for continuous data, you draw a histogram then mark the midpoints of the tops of the bars and join these with a straight line.
⦿ Frequency polygons can be used to compare the **frequency distributions** of two (or more) sets of data.

🔍 Example 10

The table shows the number of children in a sample of families.

Number in family	1	2	3	4	5	6	7
Frequency	4	12	8	5	4	2	1

a Draw a frequency polygon for these data.
b Describe the trend of these data.

You can draw a vertical line chart for the discrete data, and then join the tops of the lines to get the frequency polygon.

a

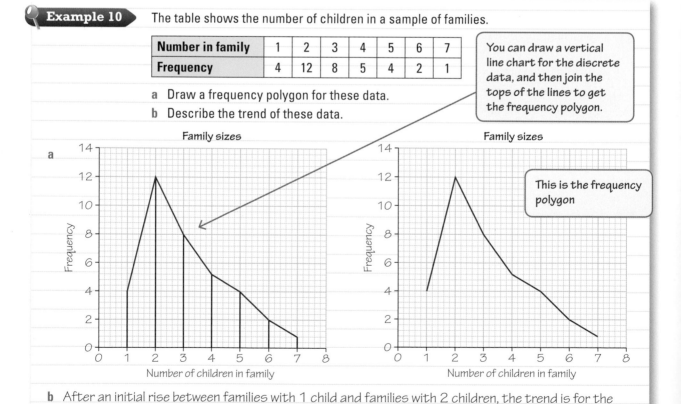

This is the frequency polygon

b After an initial rise between families with 1 child and families with 2 children, the trend is for the frequency to decrease as the family size increases.

Example 11 Draw a frequency polygon for the data in Example 9.

This is the frequency polygon.

You could also plot the midpoints and draw the polygon without the bars

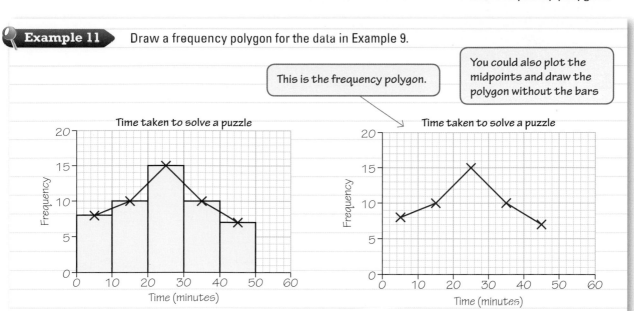

Example 12 The two frequency polygons show the heights of a group of girls and the heights of a group of boys.

Compare the heights of the two groups. Give reasons for your answers.

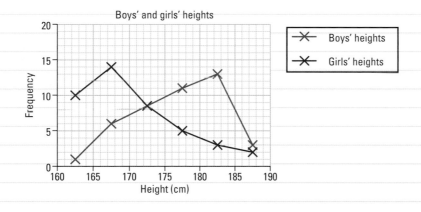

The boys are generally taller than the girls. ← The line showing the boys' heights is above the line showing the girls' heights to the right of the graph.

There were more boys above 172.5 cm in height. ← The lines cross at 172.5 cm and the boys' line is above the girls' line above this height.

There were more girls below 172.5 cm in height. ← The lines cross at 172.5 cm and the girls' line is above the boys' line below this height.

Exercise 12G

E

1 The table shows information about the shoe sizes of the members of a gym club.

Shoe size	Frequency
3	3
4	5
5	7
6	10
7	10
8	6
9	1

a Draw a vertical line graph for these data.

b Use your answer to part **a** to draw a frequency polygon for these data.

D

2 The table shows the distance the members of a sports centre threw a cricket ball in a competition.

Distance (d metres)	Frequency
$10 \leqslant d < 20$	2
$20 \leqslant d < 30$	6
$30 \leqslant d < 40$	15
$40 \leqslant d < 50$	20
$50 \leqslant d < 60$	4

a Draw a histogram for these data.

b Use your answer to part **a** to draw a frequency polygon for these data.

C

3 In three months Zainab travelled by train 20 times and by bus 20 times.
The frequency polygons show information about the amount of time Zainab spent waiting for the train and for the bus.

a How many times did Zainab wait for between 15 and 20 minutes for the bus?

b How many times did Zainab wait for between 5 and 10 minutes for the train?

c For what fraction of the times Zainab went by train did she wait for less than 10 minutes?
Give your fraction in its simplest form.

A03

d For which transport did Zainab generally have to wait the longest time, the train or the bus?
You must give a reason for your answer.

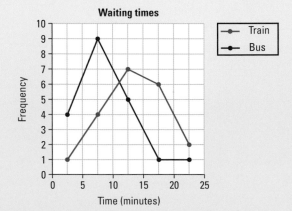

Chapter review

- A **pictogram** uses symbols or pictures to **represent** a number of items.
- A **key** tells you the number of items represented by a single symbol or picture.
- In a **pie chart** the area of the whole circle represents the total number of items.
- The area of each sector represents the number of items in that category.

$$\text{Sector angle} = \frac{\text{frequency} \times 360°}{\text{total frequency}} \qquad \text{Frequency} = \frac{\text{sector angle} \times \text{total frequency}}{360°}$$

- The frequency represented by corresponding sectors of two pie charts might have the same sector angle but the frequency they represent depends on the total population.
- A **bar chart** can be used to display qualitative data or grouped discrete data.
- In a bar chart:
  - all the bars are the same width
  - both the vertical and horizontal axes have labels
  - there is a gap between the bars
  - the bars can be drawn horizontally or vertically.
- In a comparative bar chart two (or more) bars are drawn side by side for each category.
- A composite bar chart shows the size of individual categories split into their separate parts.
- A **line graph** is used to display ungrouped discrete data.
- A **histogram** can be used to display grouped continuous data.
- A histogram is similar to a bar chart but, because it represents continuous data, no gap is left between the bars.
- When drawing a **frequency polygon** for discrete data, you draw straight lines to join the tops of the lines on a vertical line chart.
- When drawing a frequency polygon for continuous data, you draw a histogram then mark the midpoints of the tops of the bars and join these with a straight line.
- Frequency polygons can be used to compare the **frequency distributions** of two (or more) sets of data.

Review exercise

1 Here is a pictogram. It shows the number of books read by Asad, by Betty, and by Chris.

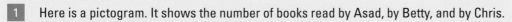

Asad	⊞ ⊞
Betty	⊞ ⊞ ⊞ ▫
Chris	⊞ ⊞ ⊟
Diana	
Erikas	

Key

⊞ represents 4 books

  a Write down the number of books read by
     i Asad   ii Chris.

  Diana read 12 books.

  Erikas read 9 books.

  b Show this information on a copy of the pictogram.

March 2009

G

G

2   The bar chart shows the number of TVs sold by
    a shop six days last week.

    a How many TVs were sold on Friday?

    b On which day was the **least** number of TVs sold?

    c On which two days were the same number of
       TVs sold?

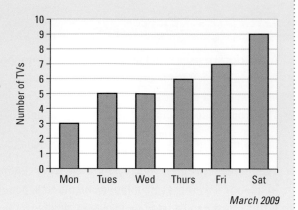

March 2009

3   Jessica asked some students to tell her their
    favourite pet.
    She used the information to draw this bar chart.

    a How many students said a rabbit?

    b Which pet did most students say?

    c Work out the number of students that Jessica
       asked.

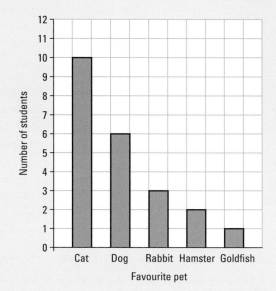

Nov 2008

4   The pictogram shows the number of plates sold by a shop on Monday, Tuesday, Wednesday and
    Thursday of one week.

Monday	◯ ◯
Tuesday	◯ ◖
Wednesday	◯ ◯ ◯
Thursday	◯
Friday	
Saturday	

Key: ◯ represents 10 plates

    a Work out the number of plates sold on Monday.

    b Work out the number of plates sold on Tuesday.

    The shop sold 40 plates on Friday.
    The shop sold 25 plates on Saturday.

    c Use this information to complete the pictogram.

Nov 2008

5 Steve asked his friends to tell him their favourite colour.
Here are his results.

Favourite colour	Tally	Frequency
Red	IIII I	6
Blue	IIII III	8
Green	IIII	5
Yellow	III	3

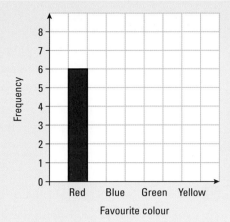

a Copy and complete the bar chart to show his results.

b Which colour did most of his friends say?

May 2008

6 The bar chart shows information
about the amount of time, in
minutes, that Andrew and Karen
spent watching television on
four days last week.

Karen spent more time watching
television than Andrew on two
of these four days.

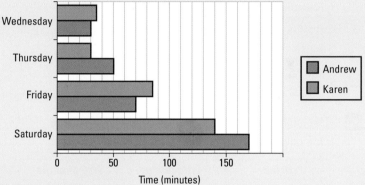

a Write down these two days.

b Work out the total amount of
time Andrew spent watching television on these four days.

June 2008

7 Mr White recorded the number of students absent one week.
The dual bar chart shows this information for the first four days.

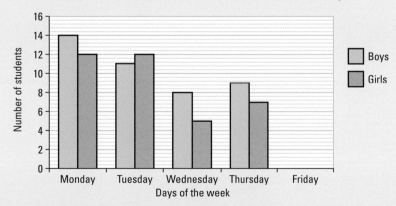

a How many boys were absent on Monday?

b How many girls were absent on Wednesday?

On Friday, 9 boys were absent and 6 girls were absent.

c Use this information to complete the bar chart.

On only one day more girls were absent than boys.

d Which day?

March 2008

241

E

8 Colin carried out a survey. He asked some students in Year 10 which type of film they liked best.
He used the results to draw this pie chart.

a What fraction of the students said "Comedy"?

20 students said "Horror".

b Work out the total number of students Colin asked.

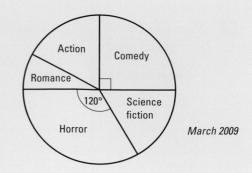

March 2009

A02

9 60 students were asked to choose one of four subjects.
The table gives information about their choices.

Subject	Number of students
Art	12
French	10
History	20
Music	18

Draw an accurate chart to show this information.

10 The table gives information about the drinks sold in a café one day.

Drink	Frequency	Size of angle
Hot chocolate	20	80°
Soup	15	
Coffee	25	
Tea	30	

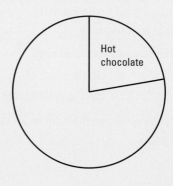

Copy and complete the pie chart to show this information.

Nov 2008

11 José is in hospital.
Here is his temperature chart during one day.

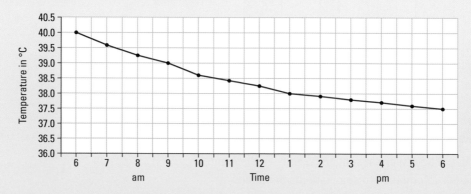

a At what time was José's temperature 39.0°C?

b What can you say about José's temperature from 6 am to 6 pm?

Nov 2007

12 The pie chart below shows government spending for 2008/9

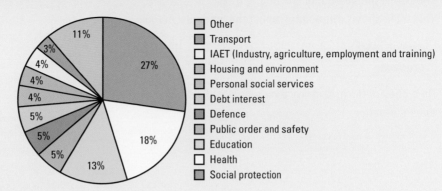

□ Other
□ Transport
□ IAET (Industry, agriculture, employment and training)
□ Housing and environment
□ Personal social services
□ Debt interest
□ Defence
□ Public order and safety
□ Education
□ Health
□ Social protection

The government spent £21 billion on transport.

a How much did it spend on health?

b How much did it spend on public order and safety?

c What is the total amount of money that the government spent?

13 The table shows some information about the weights (w grams) of 60 apples.

Weight (w grams)	Frequency
$100 \leqslant w < 110$	5
$110 \leqslant w < 120$	9
$120 \leqslant w < 130$	14
$130 \leqslant w < 140$	24
$140 \leqslant w < 150$	8

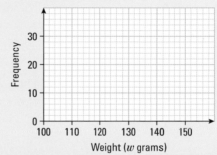

On a copy of the grid, draw a frequency polygon to show this information. *March 2009*

14 The pie chart gives information about the mathematics exam grades of some students.

a What fraction of the students got grade D?

8 of the students got grade C.

b i How many of the students got grade F?

ii How many students took the exam?

This accurate pie chart gives information about the English exam grades for a different set of students.

Sean says "More students got a grade D in English than in mathematics."

c Sean could be **wrong**.
Explain why.

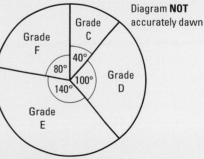

Diagram **NOT** accurately dawn

Mathematics exam grades

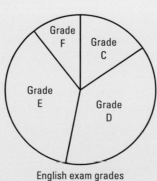

English exam grades

June 2008

D

15 60 students take a science test.
The test is marked out of 50.
This table shows information about the students' marks.

Science mark	0–10	11–20	21–30	31–40	41–50
Frequency	4	13	17	19	7

On a copy of the grid, draw a frequency polygon to show
this information.

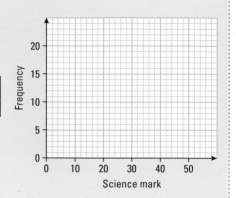

June 2008

A02

* **16** The table below gives some information about the nutritional content of 120 g of baked beans.

Protein	Carbohydrate	Fibre	Other
6 g	16 g	5 g	

Copy and complete the table.
Draw a chart for these data.

17 The pie chart shows the sources of the UK's energy production
in the early part of the 21st century.
Estimate the percentages of each type of energy and use your
answers to draw a composite bar chart.
Draw a new composite bar chart showing how you think it will
have changed by 2050, stating reasons for your answers.

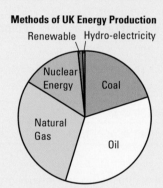

Methods of UK Energy Production

13 SEQUENCES

Neptune was the first planet to be discovered by mathematical prediction. It was found by looking at the number patterns of the other planets in the Solar System, and its position was correctly predicted to within a degree. Two scientists were eventually jointly credited with the discovery, one British and one French.

Objectives

In this chapter you will:
- continue number patterns using the four rules of number
- continue patterns using pictures and give the rule for continuing the pattern
- use number machines to produce a number pattern and write down the rule
- complete a table of values using a number machine and write down the rule
- use the first difference to find the nth term and use the nth term to find any number in a sequence
- identify whether or not a number is in a sequence.

Before you start

You need to:
- recognise simple patterns and simple number differences
- know about odd numbers
- know about even numbers
- know that multiples are members of a multiplication table, e.g. 3, 6, 9, 12 are multiples of 3
- know the properties of simple shapes.

13.1 Sequences

○ You can continue number patterns by adding or subtracting, or multiplying or dividing by a number.

○ You can give the term to term rules for continuing number patterns.

○ You can continue patterns using pictures and give the rule for continuing the pattern.

Why do this?

You may use sequences when learning a dance routine or a note sequence when playing a musical instrument.

Get Ready

Even numbers form a pattern   2, 4, 6, 8, 10, 12, ... They go up in twos.

Odd numbers also form a pattern   1, 3, 5, 7, 9, 11, ... These also go up in twos.

1. Write down all the even numbers up to 20.
2. Write down all the odd numbers up to 20.
3. Check that you have written all the numbers from 1 to 20.

Key Points

⊚ A **sequence** is a pattern of numbers or shapes that follows a rule.

⊚ Number patterns can be continued by adding, subtracting, multiplying and dividing.

⊚ Patterns with pictures can be continued by finding the rule for continuing the pattern.

⊚ The numbers in a number pattern are called terms.

⊚ The **term to term rule** for a number pattern means you can say how you find a term from the one before it.

Continuing patterns by adding

Example 1

a Write down the next two numbers in this number pattern.

    2   6   10   14

b What is the rule you use to find the next number in the number pattern?

c Find the 10th number in this pattern.

a $14 + 4 = 18$
   $18 + 4 = 22$

    2   6   10   14   18
     +4   +4   +4   +4

b To get the next number you add 4 each time.

c The 10th number in the pattern is 38.

2 6 10 14 18 22 26 30 34 38
Carry on the number pattern until you get to the 10th number in the pattern.

Exercise 13A

Questions in this chapter are targeted at the grades indicated.

1 Find the two missing numbers in these number patterns.
For each pattern, write down the term to term rule.

   a 3, 6, 9, __, __, 18, 21   b 3, 7, 11, __, __, 23, 27

   c 5, 10, 15, 20, __, __, 35, 40   d 2, 7, 12, 17, __, __, 32, 37

   e 1, 4, 7, 10, __, __, 19, 22   f 5, 7, 9, 11, __, __, 17, 19

   g 3, 8, 13, 18, __, __, 33, 38   h 4, 7, 10, 13, __, __, 22, 25

   i 2, 6, 10, 14, __, __, 26, 30   j 10, 20, 30, __, __, 60, 70

2 a Write down the next two numbers in these sequences.

     i 1, 5, 9, 13, 17, …   ii 2, 5, 8, 11, 14, …

   iii 3, 7, 11, 15, 19, …   iv 4, 8, 12, 16, 20, …

    v 5, 8, 11, 14, 17, …   vi 5, 11, 17, 23, …

  vii 2, 6, 10, 14, 18, …   viii 1, 7, 13, 19, 25, …

   ix 3, 11, 19, 27, 35, …   x 5, 9, 13, 17, 21, …

  b Write down the rule you used to find the missing numbers in each sequence.

3 Find the 10th number of each of the number patterns in questions 1 and 2.

4 Jenny saves £2 each week in her piggy bank.
Here is the pattern of how her money grows.

Week	1	2	3	4	5
Money in piggy bank (£)	2	4	6		

  a Copy and complete the table.

  b Jenny is saving for a present for her Mum's birthday that costs £20.
    How many weeks will this take?

> **ResultsPlus**
> **Examiner's Tip**
>
> … means that the sequence carries on.

A03

Continuing patterns by subtracting

Key Point

● Number patterns can be continued by subtracting the same number from each term.

Example 2   a Write down the next two numbers in this number pattern.

        60   54   48   42   36

  b What is the rule you use to find the next number in the number pattern?

  c Find the 8th number in this pattern.

a $36 - 6 = 30$
   $30 - 6 = 24$

> To get to the next number you take away 6.
> Take away 6 from 36 to get 30 then
> take away 6 from 30 to get 24.

b To get the next number you subtract 6 each time.

c The 8th number in the pattern is 18.

> 60 54 48 42 36 30 24 18
> Carry on the number pattern until you get to the 8th number in the pattern.

Exercise 13B

1 Find the two missing numbers in these number patterns.
Write down the rule for each number pattern.
a 20, 18, 16, 14, __, __, 8
b 17, 15, 13, 11, __, __, 5
c 55, 50, 45, 40, __, __, 25
d 42, 37, 32, 27, __, __, 12
e 22, 19, 16, 13, __, __, 4
f 19, 17, 15, 13, __, __, 7
g 45, 38, 31, 24, __, __, 3
h 25, 22, 19, 16, __, __, 7
i 29, 25, 21, 17, __, __, 5
j 80, 70, 60, __, __, 30

2 a Write down the next two numbers in these sequences.
i 41, 37, 33, 29, …   ii 27, 24, 21, 18, …
iii 59, 55, 51, 47, …   iv 34, 31, 28, 25, …
v 30, 27, 24, 21, …   vi 61, 55, 49, 43, …
vii 22, 20, 18, 16, …   viii 51, 46, 41, 36, …
ix 64, 57, 50, 43, …   x 8, 6, 4, 2, 0, -2, …

b Write down the rule you used to find the missing numbers
in each sequence.

3 Find the 10th number of each of the number patterns in questions 1 and 2.

4 Abdul's mother gives him £20 each week to buy his school lunch.
His lunch costs him £3 each day.
Here is the pattern of how he spends his money.

Day	M	Tu	W	Th	F
Money left at end of day (£)	17	14			

How much money will Abdul have left at the end of the week?

Continuing number patterns by multiplying

Example 3

a Write down the next two numbers in this number pattern.

1   3   9   27   81

b What is the rule you use to find the next number in the number pattern?

c Find the 8th number in this pattern.

a $81 \times 3 = 243$ 1   3   9   27   81
$243 \times 3 = 729$

×3   ×3   ×3   ×3

> To get to the next number you multiply by 3. Multiply 81 by 3 to get 243, then multiply 243 by 3 to get 729.

b To get the next number you multiply by 3 each time.

c The 8th number in the pattern is 2187.

> 1   3   9   27   81   243   729   2187
> Carry on the number pattern until you get to the 8th number in the pattern.

Exercise 13C

1 Find the missing numbers in these number patterns.
For each pattern, write down the rule.

a 1, 2, 4, 8, ___, ___, 64   **b** 1, 4, 16, 64, ___, 1024

c 1, 5, ___, 125, ___, 3125   **d** 1, 10, 100, ___, ___, 100 000

e 3, 6, 12, 24, ___, ___, 192   **f** 2, 6, 18, ___, ___, 486

g 2, 8, 32, ___, ___, 2048   **h** 2, 20, 200, 2000, ___, ___, 2 000 000

i 2, 10, 50, ___, ___, 6250   **j** 3, 15, 75, ___, 1875

2 **a** Write down the next two numbers in these sequences.

   i 2, 4, 8, 16, …   **ii** 3, 9, 27, 81, …

  iii 4, 16, 64, 256, …   **iv** 5, 25, 125, 625, …

   v 5, 10, 20, 40, …   **vi** 4, 12, 36, 108, …

 vii 10, 30, 90, 270, …   **viii** 5, 50, 500, 5000, …

  ix 10, 20, 40, 80, …   **x** 6, 36, 216, 1296, …

b Write down the rule you used to find the missing numbers in each sequence.

3 Find the 10th number of each of the number patterns in questions 1 and 2.

4 The number of rabbits in a particular colony doubled every month for 10 months.
The table shows the beginning of the pattern.

Month	1	2	3	4	5
Number of rabbits	2	4	8		

a Copy and complete the table.

b How many rabbits were in the colony in month 10?

F

E

A02

Continuing number patterns by dividing

Example 4

a Write down the next two numbers in this number pattern.

729   243   81   27

b What is the rule you use to find the next number in the number pattern?

c Find the 8th number in this pattern.

a $27 \div 3 = 9$
   $9 \div 3 = 3$

729 243   81   27

    $\div 3$ $\div 3$ $\div 3$

> To get to the next number you divide by 3. Divide 27 by 3 to get 9 then divide 9 by 3 to get 3.

b To get the next number you divide by 3 each time.

c The 8th number in the pattern is $1 \div 3 = \frac{1}{3}$

> 729   243   81   27   9   3   1 $\frac{1}{3}$
> Carry on the number pattern until you get to the 8th number in the pattern.
> $1 \div 3 = \frac{1}{3}$

Exercise 13D

1 Find the missing numbers in these number patterns.
Write down the rule for each number pattern.

a 64, 32, 16, 8, __, __, 1

b 1024, 256, 64, __, 4

c 3125, 625, 125, __, __, 1

d 100 000, 10 000, 1000, __, __, 1

e 192, 96, 48, 24, __, __, 3

f 486, 162, 54, 18, __, 2

g 1024, 512, 256, 128, __, __, 16

h 300 000, 30 000, 3000, __, __, 3

i 6250, 1250, 250, __, __, 2

j 2000, 200, 20, __, __, 0.02

2 a Write down the next two numbers in these sequences.

   i   16, 8, 4, 2, …

   ii   243, 81, 27, 9, …

   iii   128, 64, 32, 16, …

   iv   625, 125, 25, 5, …

   v   80, 40, 20, 10, …

   vi   972, 324, 108, 36, …

   vii   2430, 810, 270, 90, …

   viii   50 000, 5000, 500, 50, …

   ix   160, 80, 40, 20, …

   x   1296, 216, 36, 6, …

b Write down the rule you used to find the missing number in each sequence.

3 Find the 8th number of each of the number patterns in questions 1 and 2.

4 The number of radioactive atoms in a radioactive isotope halves every 10 years.
The table shows the beginning of the pattern.

Years	0	10	20	30	40
Number of atoms	2560	1280	640		

a Copy and complete the table.

b How many radioactive atoms were in the isotope in year 100?

Continuing patterns in pictures

Example 5

a Copy and complete the table for the number of matches used to make each member of the pattern.

Pattern number	1	2	3	4	5	6	7
Number of matches used	4	7	10				

b Write down the rule to get the next number in the pattern.

c How many matches are there in pattern number 10?

a

Pattern number	1	2	3	4	5	6	7
Number of matches used	4	7	10	13	16	19	22

> Count the number of matches in each pattern and write down the number of matches used.

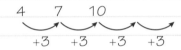

$$10 + 3 = 13 \qquad 13 + 3 = 16$$
$$16 + 3 = 19 \qquad 19 + 3 = 22$$

b Add 3 to the previous number.

c Pattern number 10 has 31 matches. ←

> Continue the patterns.
> 22   25   28   31

E

Exercise 13E

1 For these patterns:
  i draw the next two patterns
  ii write down the rule in words to find the next pattern
  iii use your rule to find the 10th term.

a
```
X     XX    XXX
X     XX    XXX
```

b
```
X      XXX     XXXXX
XX     XXXX    XXXXXX
```

c

d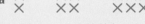

e
```
      X     XX     XXX     XXXX
X     XX    XXX    XXXX    XXXXX
```

E

2 **a** Write down the number of matches in each of these patterns.

Pattern 1   Pattern 2   Pattern 3

b Draw the next two patterns.
c Write down the rule in words to continue the pattern.
d Use your rule to find the number of matches needed for pattern number 10.

3 Repeat question 2 with the hexagon shape shown below.

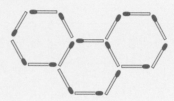

13.2 Using input and output machines to investigate number patterns

◎ Objectives

- ○ You can use number machines to produce a number pattern and write down the rule (term number to term).
- ○ You can complete a table of values using a number machine and write down the rule (term number to term).
- ○ You can find missing values in a table of values and use the term number to term rule.

⊘ Why do this?

When baking, you take your ingredients, mix them together and bake them in the oven, and you end up with a cake.

Input ⟶ Action ⟶ Output

This process applies to anything from baking a cake to making a motor car.

⬆ Get Ready

1. Put the following numbers into this number machine and write down the answers.
   a 5   **b** 3   **c** 11   **d** 17

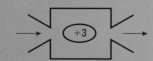

2. Draw a number machine for the process −6 and use it to find the answer when the following numbers are put into it.
   a 10   **b** 6   **c** 2

Key Points

◉ In this number pattern

3	7	11	15	19	23	...
Term 1	Term 2	Term 3	Term 4			

3, 7, 11, 15, 19, 23, ... are the terms.

◉ The term number tells you the position of each term in the pattern.
In the sequence 3, 7, 11, 15, 19, ... term 1 is 3, term 2 is 7, etc.

◉ You can use number machines to produce a number pattern and write down the rule (term number to term rule).

◉ You can complete a table of values using a number machine and write down the rule (term number to term).

◉ Sometimes you can put two number machines together to make a sequence.

One-stage input and output machines

Example 6   This number machine has been used to produce the terms of a pattern.

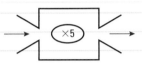

a Complete the term numbers and terms in this table of values for the number machine.

Term number	Term
1	5
2	
3	
4	

b What is the rule for working out the term from the term number?

c Write down the rule for finding the next term from the term before it.

a

Term number	Term
1	5
2	10
3	15
4	20

Term number	Term
1	5
2	10
3	15
4	20

+5
+5
+5

> The rule for the number machine is multiply the term number by 5 so the terms will be 5, 10, 15, 20.

b Multiply the term number by 5. ← To get to the term from the term number you multiply by 5.

c Add 5. ← To get to the next term from the term before it you have to add 5 since the pattern is 5, 10, 15, 20, ...

Exercise 13F

For each of these questions:

a copy and complete the table of values for the number machine

b write down the rule for finding the term from the term number

c write down the rule for finding the next term from the term before it.

1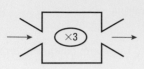

Term number	Term
1	3
2	6
3	
4	

2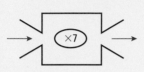

Term number	Term
1	7
2	14
3	
4	

3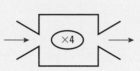

Term number	Term
1	4
2	8
3	
4	

4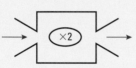

Term number	Term
1	2
2	4
3	
4	

5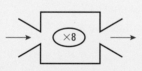

Term number	Term
1	8
2	16
3	
4	

6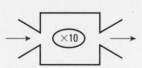

Term number	Term
1	10
2	20
3	
4	

7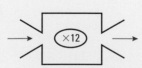

Term number	Term
1	12
2	24
3	
4	

8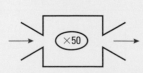

Term number	Term
1	50
2	100
3	
4	

Two-stage input and output machines

 Example 7

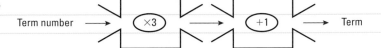

Term number → ×3 → +1 → Term

This number machine has been used to produce the terms of a pattern.

Results Plus
Examiner's Tip

Use the rule from the number machine on the term number to get to the term. You feed the result of the first machine into the second machine.

a Complete the terms in a table of values for the number machine.

Term number	Term
1	4
2	
3	
4	

b What is the rule for working out the term from the term number?

c Write down the rule for finding the next term from the term before it.

a

Term number	Term
1	4
2	7
3	10
4	13

Term number	Term	Working
1	4	1 × 3 + 1 = 4
2	7	2 × 3 + 1 = 7
3	10	3 × 3 + 1 = 10
4	13	4 × 3 + 1 = 13

+3
+3
+3

b Multiply by 3 and add 1. ←

To get to the term from the term number you ×3 and +1.

c Add 3. ←

To get to the next term from the term before it you have to add 3 since the pattern is 4, 7, 10, 13, …

Exercise 13G

For each of these questions:

a   copy and complete the table of values for the number machine

b   write down the rule for finding the term from the term number

c   write down the rule for finding the next term from the term before it.

1
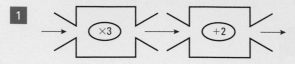

Term number	1	2	3	4	5
Term					

2
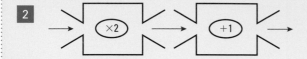

Term number	1	2	3	4	5
Term					

3
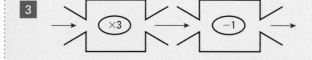

Term number	1	2	3	4	5
Term					

4
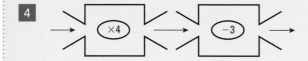

Term number	1	2	3	4	5
Term					

5
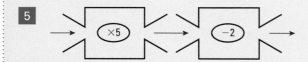

Term number	1	2	3	4	5
Term					

6
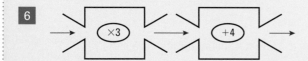

Term number	1	2	3	4	5
Term					

7
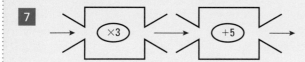

Term number	1	2	3	4	5
Term					

8

Term number	1	2	3	4	5
Term					

9

Term number	1	2	3	4	5
Term					

10

Term number	1	2	3	4	5
Term					

Example 8 Complete this table of values for the number pattern with term number to term rule 'Multiply by 4 and subtract 2'.

×4 → −2

Term number	Term
1	2
2	6
3	
4	
5	
↓	↓
8	
↓	↓
	38

Term number	Term	Working
1	2	1 × 4 − 2 = 2
2	6	2 × 4 − 2 = 6
3	10	3 × 4 − 2 = 10
4	14	4 × 4 − 2 = 14
5	18	5 × 4 − 2 = 18
↓		↓
8	30	8 × 4 − 2 = 30
↓		↓
10	38	10 × 4 − 2 = 38

You can find these terms by using the rule ×4 then −2.

ResultsPlus
Examiner's Tip

Don't forget Bidmas: you do the × before the −.
You met Bidmas in Chapter 9.

 Exercise 13H

Copy and complete these tables of values.

1 ×3 → +1

Term number	Term
1	4
2	
3	
4	
5	
↓	↓
10	
↓	↓
	34

2 ×2 → −1

Term number	Term
1	1
2	
3	
4	
5	
↓	↓
10	
↓	↓
	25

3 ×5 → +3

Term number	Term
1	8
2	
3	
4	
5	
↓	↓
10	
↓	↓
	78

E

E

4

×4 → −3	
Term number	**Term**
1	1
2	
3	
4	
5	
↓	↓
10	
↓	↓
	45

5

×10 → +1	
Term number	**Term**
1	11
2	
3	
4	
5	
↓	↓
10	
↓	↓
	151

6

×5 → −3	
Term number	**Term**
1	2
2	
3	
4	
5	
↓	↓
10	
↓	↓
	67

7 a Find the 10th number in this number pattern.   3, 7, 11, 15, …
   b What is the term number for the term that is 47?

8 a Find the 10th number in this number pattern.   4, 9, 14, 19, …
   b What is the term number for the term that is 69?

9 a Find the 10th number in this number pattern.   8, 11, 14, 17, …
   b What is the term number for the term that is 50?

13.3 Finding the nth term of a number pattern

◉ Objective

● You can use the first difference to find the nth term of a number pattern and use the nth term to find any number in a number pattern or sequence.

❓ Why do this?

This may be useful when your teacher is dividing the class into groups, so that you can work out which group you are going to be in, or make sure you will be in a group with your friends.

⬆ Get Ready

1. Write down the difference between each term in these number patterns.
   a 5, 10, 15, 20, 25, 30, …   b 40, 35, 30, 25, 20, …   c 4, 7, 10, 13, 16, …
   d 7, 11, 15, 19, 21, …   e 50, 47, 44, 41, 37, …
2. Find the 10th term in each of the number patterns in question **1**.

🌐 Key Point

◉ The first difference can be used to find the nth term of a number pattern and then the nth term can be used to find any number in a sequence.

Example 9   Here is a number pattern   4, 7, 10, 13, 16, …
a Find the *n*th term in this pattern.
b Find the 20th term in this number pattern.

a

Term number	Term	Difference
1	4	
2	7	+3
3	10	+3
4	13	+3
5	16	+3
n	3*n* + 1	

Step 1
Put the number pattern into a table of values.

Step 2
Find the difference between the terms in the number pattern. In this case it is +3.

Step 3
Multiply each term number by the difference to get a new pattern.
3, 6, 9, 12, 15 …

b The 20th term is 61.

Step 4
Compare your new pattern with the original one and see what number you need to add or subtract to/from each term to get the original number pattern. In this case it is +1.
The *n*th term is 3*n* + 1.
You replace the *n* by 20 in the *n*th term to find the 20th term. It is 3 × 20 + 1 = 61

Exercise 13I

1 For questions 1, 2 and 3 in Exercise 13H, find the *n*th term of each of the number patterns.

2 Write each pattern in a table and use the table to find the *n*th term of these number patterns.
Use your *n*th term to find the 20th term in each of these number patterns.

a 1, 3, 5, 7, 9, 11, …

b 3, 5, 7, 9, 11, 13, …

c 2, 5, 8, 11, 14, 17, …

d 5, 8, 11, 14, 17, 20, …

e 1, 5, 9, 13, 17, 21, …

f 2, 6, 10, 14, 18, 22, …

g 2, 7, 12, 17, 22, 27, …

h 4, 9, 14, 19, 24, 29, …

i 8, 13, 18, 23, 28, …

j 5, 7, 9, 11, 13, …

k 40, 35, 30, 25, 20, …

l 38, 36, 34, 32, 30, …

m 35, 32, 29, 26, 23, …

n 20, 18, 16, 14, 12, …

o 19, 17, 15, 13, 11, …

p 190, 180, 160, 150, …

ResultsPlus
Examiner's Tip

To find the *n*th term of a sequence that gets smaller you subtract a multiple of *n* from a fixed number.
e.g. 15 − 2*n* is the *n*th term of 13, 11, 9, 7, …

C

C

3 Here is a pattern made from sticks.

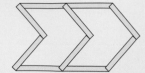

Pattern number 1   Pattern number 2   Pattern number 3

a Draw pattern number 4.

b Copy and complete this table of values for the number of sticks used to make the patterns.

Pattern number	1	2	3	4	5	6
Number of sticks	6	10				

c Write, in terms of n, the number of sticks needed for pattern number n.

d How many sticks would be needed for pattern number 20?

13.4 Deciding whether or not a number is in a number pattern

◎ Objective

• You can use number patterns or use the nth term to identify whether a number is in the pattern.

❓ Why do this?

This is useful when you want to work out what will happen in the future, for example, you could work out whether next year will be a leap year as this happens every four years.

◆ Get Ready

1. Write each of these patterns in a table and use the table to find the nth term.
   Use your nth term to find the 20th term in each pattern.

   a 4, 7, 10, 13, …   b 3, 8, 13, 18, …   c 13, 15, 17, 19, …

🔑 Key Points

◉ Number patterns or the nth term can be used to identify whether a number is in the pattern.

◉ Sometimes you will be asked how you know if a number is part of a sequence. You would then have to explain why the number is in the sequence or, even, why it is not in the sequence.

Example 10 Here is a number pattern.

3   8   13   18   23

a   Explain why 423 is in the pattern.

b   Explain why 325 is not in the pattern.

a   423 is in the number pattern.
Every odd term ends in 3 and goes up 3, 13, 23, etc,
so 423 will be a member as it ends in a 3.

> There are other ways of answering questions like these. For example, you could identify the nth term.
> The nth term is $5n - 2$ if
> $$5n - 2 = 423$$
> $$5n = 425 \text{ so } n = 85$$
> so 423 is the 85th term.

b   325 is not in the number pattern.
325 ends in a 5 and every member
of the pattern ends in either a 3 or
an 8 so 325 cannot be in the pattern.

> The nth term is $5n - 2$ so if 325 is in the pattern.
> $$5n - 2 = 325$$
> $$5n = 327 \quad \text{so} \quad n = 65.4$$
> If 325 is in the pattern n must be a whole number.
> 65.4 is not a whole number so 325 is not in the pattern.

Exercise 13J

For each of these number patterns, explain whether each of the numbers in brackets are members of the number pattern or not.

1   1, 3, 5, 7, 9, 11, …   (21, 34)

2   3, 5, 7, 9, 11, 13, …   (63, 86)

3   2, 5, 8, 11, 14, 17, …   (50, 66)

4   5, 8, 11, 14, 17, 20, …   (50, 62)

5   1, 5, 9, 13, 17, 21, …   (101, 150)

6   2, 6, 10, 14, 18, 22, …   (101, 98)

7   2, 7, 12, 17, 22, 27, …   (97, 120)

8   4, 9, 14, 19, 24, 29, …   (168, 169)

9   40, 35, 30, 25, 20, …   (85, 4)

10 38, 36, 34, 32, 30, …   (71, 82)

11 3, 7, 11, 15, 19, 21, …   (46, 79)

12 5, 11, 17, 23, 29, …   (119, 72)

A03 C

Chapter review

- A **sequence** is a number or shape pattern which follows a rule.
- Number patterns can be continued by adding, subtracting, multiplying and dividing.
- Patterns using pictures can be continued by finding the rule for continuing the pattern.
- The numbers in a number pattern are called terms.
- The **term to term rules** for continuing number patterns can be given.
- The term number tells you the position of each term in the pattern.
- You can use number machines to produce a number pattern and write down the rule (term number to term rule)
- You can complete a table of values using a number machine and write down the rule (term number to term).
- The first difference can be used to find the nth term of a number pattern and then the nth term can be used to find any number in a sequence.
- Number patterns or the nth term can be used to identify whether a number is in the pattern.

Review exercise

F

1 Here are some patterns made of squares.

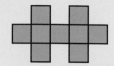

Pattern number 1   Pattern number 2   Pattern number 3

The diagram below shows part of Pattern number 4.

a Copy and complete Pattern number 4.

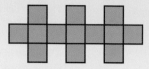

Pattern number 4

A02

b Find the number of squares used for Pattern number 10.   *Nov 2008 adapted*

D

2 Here are the first 4 terms in a number sequence.

   124   122   120   118

a Write down the next term in this number sequence.

b Write down the 7th term in this number sequence.

A03

c Can 9 be a term in this number sequence? You must give a reason for your answer.   *May 2009*

3 The diagram shows a mathematical rule.

input ⟶ ×3 ⟶ −3 ⟶ output

It multiplies a number by 3 and then subtracts 3

Copy and complete the diagram in each case.

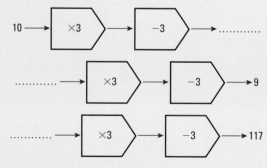

10 ⟶ ×3 ⟶ −3 ⟶ ....

.... ⟶ ×3 ⟶ −3 ⟶ 9

.... ⟶ ×3 ⟶ −3 ⟶ 117

Nov 2008

4 Here is a table for a 2-stage number machine.

It subtracts 5 and then multiplies by 2

a Copy and complete the table.

b The input is n.

   Write down an expression in terms of n for the output.

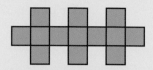

− 5 then × 2	
Input	**Output**
4	−2
2	....
−3	....

Nov 2006

5 The nth term of a sequence is $n^2 + 4$.
Alex says 'The nth term of the sequence is always a prime number when n is an odd number'.
Is Alex correct? You must give a reason for your answer.   *Nov 2008 adapted*

6 Here are the first 5 terms of a sequence.

    1   1   2   3   5

The rule for the sequence is 'The first two terms are 1 and 1. To get the next term add the two previous terms'.
a Find the 6th term and the 7th term.
b Find the 10th term.

7 Dylan and Evie are studying a number pattern.
The first three numbers in the number pattern are 1, 2, 4.
Dylan says that the next number is 8.
Evie says that the next number is 7.
Explain how both Dylan and Evie could be right.

8 Here are the first four terms of an arithmetic sequence.

    5   8   11   14

Is 140 a term in the sequence? You must give a reason for your answer.

9 The first term of a sequence is x. To get the next term, multiply the previous term by 2 and add 1.
The third term of the sequence is 21. Find the value of x.

10 Here are the first five terms of a number sequence.

    3   7   11   15   19

a Work out the 8th term of the number sequence.
b Write down an expression, in terms of n, for the nth term of the number sequence.   *Nov 2006*

14 PERIMETER AND AREA OF 2D SHAPES

Many people need to be able to calculate both the perimeter and area of various shapes in their day-to-day work. For example, the organisers at Glastonbury need to know the area and perimeter of the festival fields so they can hire enough safety barriers.

◎ **Objectives**

In this chapter you will:
- learn the difference between the perimeter and area of a shape
- find the perimeter and area of two-dimensional shapes and solve problems involving area.

◇ **Before you start**

You should be able to:
- measure the length of a line
- change measurements between millimetres (mm), centimetres (cm), metres (m) and kilometres (km).

14.1 Perimeter

14.1 Perimeter

◉ Objectives

- ◉ You can find the perimeter of simple shapes.
- ◉ You can find the perimeter of shapes made from squares, rectangles and triangles.

❓ Why do this?

A gardener will need to know the distance around a boundary edge in order to calculate the length of fence needed to surround a garden.

◆ Get Ready

1. What is the length and width of the shaded rectangle drawn on the centimetre grid?

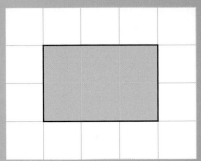

2. Measure the lengths of the sides of the following triangle.

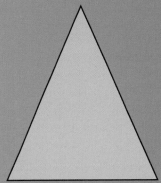

🔍 Key Points

- ◉ The **perimeter** of a **two-dimensional (2D)** shape is the total distance around the edge of the shape. The examples show you how to work out the perimeter of a variety of shapes.
- ◉ To work out the perimeter of a rectangle you can use the following formula.

  Perimeter of a rectangle $= l + w + l + w$
  $$= 2l + 2w$$

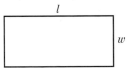

$l =$ the length of the rectangle
$w =$ the width of the rectangle

🔍 Example 1

What is the perimeter of the following shape?

Perimeter $= 3 + 2 + 4 + 5$
$= 14\,cm$ ← Add the lengths of all four sides to find the perimeter.

3 cm
2 cm
4 cm
5 cm

ResultsPlus
Examiner's Tip

Always remember to include the units in your answer.

Example 2 The diagram shows a rectangle drawn on a centimetre grid.

Work out the perimeter of the rectangle.

Method 1

Perimeter = 5 + 3 + 5 + 3

= 16 cm ← Work out the length of each side and add them together.

Method 2

Perimeter = $2l + 2w$

= $(2 \times 5) + (2 \times 3)$

= $10 + 6$

= 16 cm ← Use the formula for the perimeter of a rectangle.

Exercise 14A

Questions in this chapter are targeted at the grades indicated.

G

1   Here are three shapes drawn on a centimetre grid.

Work out the perimeter of each shape.

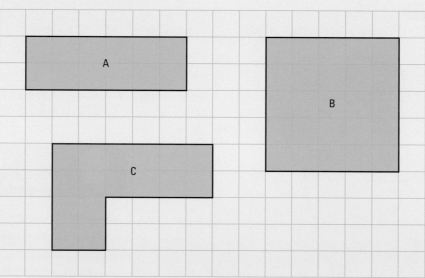

2 Work out the perimeters of the following shapes.

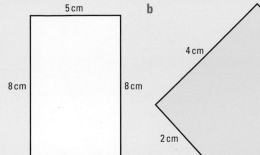
a

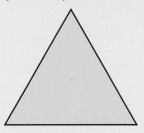
b

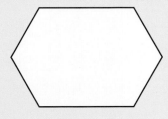
c

3 These shapes are drawn accurately. Find the perimeter of each shape.

a

b

c

4 Jenny has a rectangular pond in her garden.
The length of the pond is 3 m.
The width of the pond is half the length of the pond.
She wants to put a low fence around the edge of the pond.
What length of fencing does Jenny need?

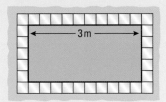

A02

A03

F

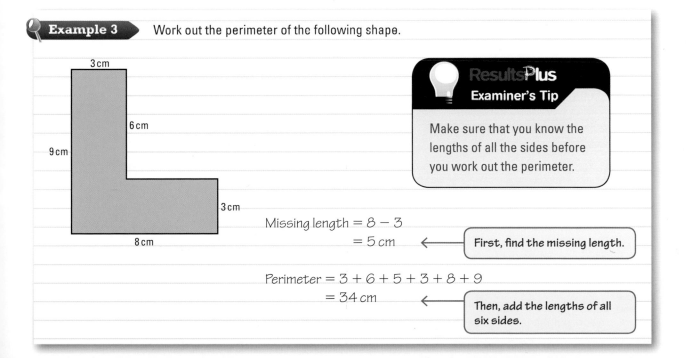

Example 3 Work out the perimeter of the following shape.

ResultsPlus
Examiner's Tip

Make sure that you know the lengths of all the sides before you work out the perimeter.

Missing length = 8 − 3
= 5 cm ← First, find the missing length.

Perimeter = 3 + 6 + 5 + 3 + 8 + 9
= 34 cm ← Then, add the lengths of all six sides.

Exercise 14B

F

1   Each side of a regular pentagon has length 8 cm.
    Work out the perimeter of the pentagon.

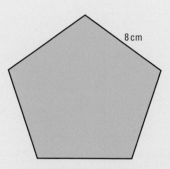

E

2   Work out the perimeters of the following three shapes.

    a
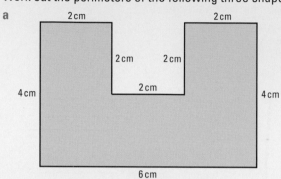

    b
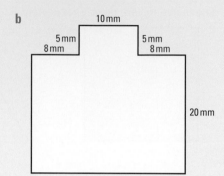

    c
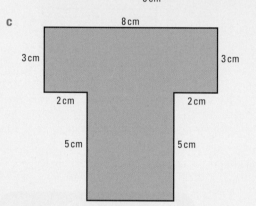

3   Work out the perimeter of this trapezium.

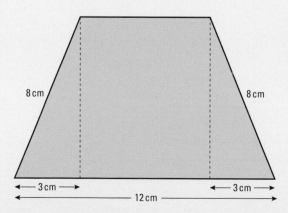

4   The perimeter of an equilateral triangle is 21.6 cm.
    Work out the length of each side.

14.2 Area

⊙ Objectives

- ⊙ You can find the area of simple shapes by counting squares.
- ⊙ You can find the area of more complicated shapes by counting squares.

⊘ Why do this?

It would be useful to know the area of a wall you were going to paint, so that you could buy the right amount of paint for the job.

⬥ Get Ready

1. How many squares do you need to cover each of the shaded shapes?

a

b
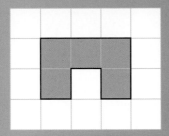

c
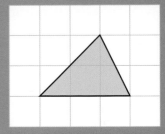

🔍 Key Points

- ⊙ The **area** of a two-dimensional (2D) shape is a measure of the amount of space inside the shape.
- ⊙ The area of a square with sides of length 1 cm is 1 square centimetre.

1 cm

1 cm

This is written as 1 cm^2.

- ⊙ Other units of area include square millimetres (mm^2), square metres (m^2) and square kilometres (km^2).

🔍 Example 4

The following diagram shows a rectangle drawn on a centimetre grid.
Find the area of the rectangle.

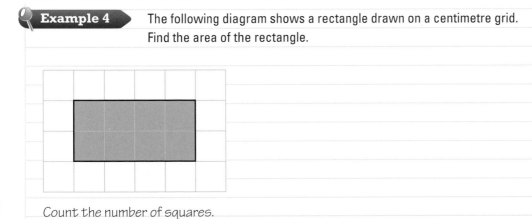

Count the number of squares.
Each square has area of 1 cm^2.
Area = 8 squares = 8 cm^2.

Example 5 Estimate the area of the shaded shape drawn on the centimetre grid.

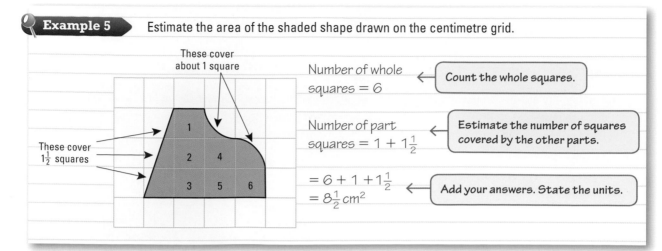

These cover about 1 square

These cover $1\frac{1}{2}$ squares

Number of whole squares = 6 ← Count the whole squares.

Number of part squares = $1 + 1\frac{1}{2}$ ← Estimate the number of squares covered by the other parts.

$= 6 + 1 + 1\frac{1}{2}$
$= 8\frac{1}{2}\ cm^2$ ← Add your answers. State the units.

Exercise 14C

G

1 Find the area of the shape shown on the centimetre grid.
Give the units with your answer.

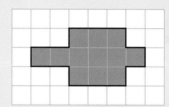

2 The diagram shows four shapes drawn on a centimetre grid.
Find the area of each shape.

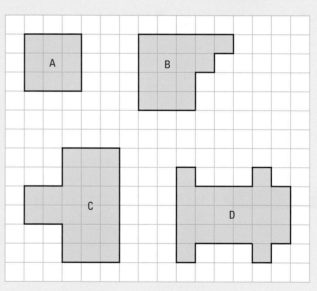

3 Find the area of each of the four triangles, T_1, T_2, T_3 and T_4, drawn on the centimetre grid.

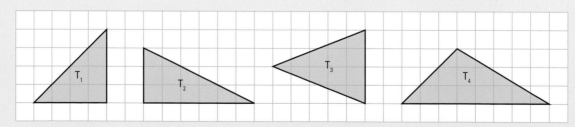

G

4 Find the area of each of the three shapes, F, G and H, drawn on the centimetre grid.

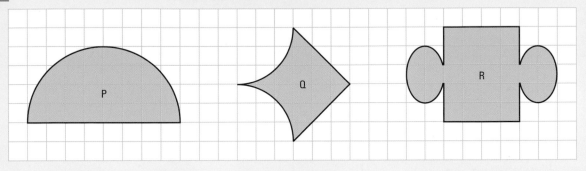

5 Estimate the area of each of the three shapes, P, Q and R, drawn on the centimetre grid.

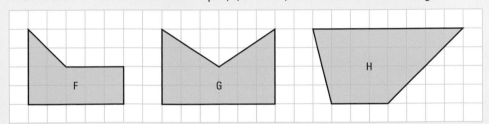

Mixed exercise 14D

G

1 This shape has been drawn on a centimetre grid.
  a Find the perimeter of the shape.
  b Find the area of the shape.

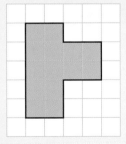

2 Estimate the area of this shape that has been drawn on a centimetre grid.

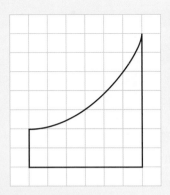

E

3 Work out the perimeter of the shape below.

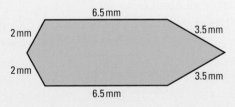

E

4 The perimeter of triangle ABC is 52 m.
What is the length of side AB?

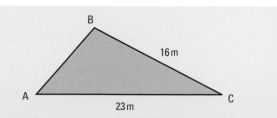

14.3 Finding areas using formulae

◉ Objectives

○ You can find the areas of rectangles, triangles and parallelograms.
○ You can use the formula to find the area of a trapezium.

❓ Why do this?

It might be necessary for your school to find out the area of a field, so that they can work out if it is big enough for an athletics track.

⬥ Get Ready

The diagram shows a triangle drawn on a centimetre grid.
The length of the base is 5 cm.
The vertical height is 3 cm.

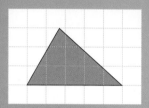

State the length of the base and the vertical height of these triangles.

1. 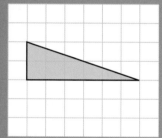 **2.** 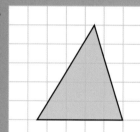 **3.**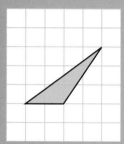

🔧 Key Points

◉ To work out the area of a rectangle, square, triangle and parallelogram use the following formulae.

Rectangle
Area of a rectangle
= length × width
= $l \times w$

Parallelogram
You can cut a triangle off a parallelogram
and put it on the other side to make a rectangle,
so:
Area of a parallelogram
= base × vertical height
= $b \times h$

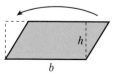

Triangle
The area of a triangle is half the area
of a rectangle that surrounds it.
Area of a triangle
= $\frac{1}{2} \times$ base × vertical height
= $\frac{1}{2} \times b \times h$

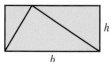

Square
Area of a square
= length × length
= $l \times l$
= l^2

The area of a trapezium is worked out by finding the average of the lengths of the parallel sides and multiplying by the distance between them.

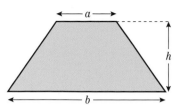

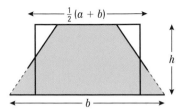

A formula is used to work out the area of a trapezium:

Area $= \frac{1}{2} \times$ sum of the lengths of the parallel sides $\times$ distance between the parallel sides

$A = \frac{1}{2}(a + b)h$

Example 6 Work out the area of this rectangle.

Area of rectangle ← *Choose the formula to use.*
$= $ length $\times$ width
$= l \times w$ ← *Put in the values of l and w.*
$= 6 \times 4$
$= 24 \text{ cm}^2$ ← *Remember to include the units.*

6 cm

4 cm

Exercise 14E

1 Find the area of the following rectangles.
Remember to give the units with your answer.

a   5 cm
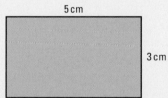
3 cm

b   2 mm
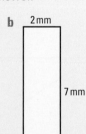
7 mm

c   8 m
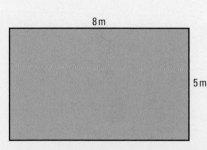
5 m

2 The following three squares are accurately drawn. Find the area of each one.

a

b
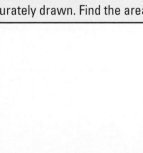

c

E

E

3 A decorator wants to paint two rectangular walls.
The walls are 4.5 m by 2.6 m and 3.1 m by 2.6 m.
What is the total area of the two walls to be painted?

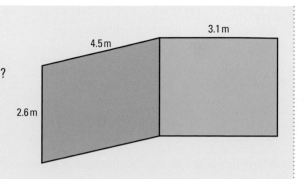

Example 7 Work out the area of this triangle.

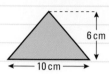

Area of triangle

$= \frac{1}{2} \times$ base $\times$ vertical height ← Choose the formula to use.

$= \frac{1}{2} \times b \times h$

$= \frac{1}{2} \times 10 \times 6$ ← Put in the values of b and h.

$= 30 \text{ cm}^2$ ← Remember to include the units.

Exercise 14F

D

1 Find the area of the following triangles.

a

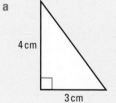

b

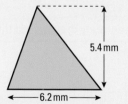

c

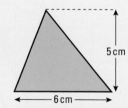

d

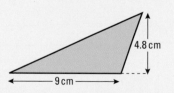

e

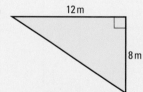

2 Find the area of this triangle.

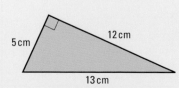

3 A company makes flags in this shape.
It makes **50** identical flags.
Work out the area of fabric used to make these flags.

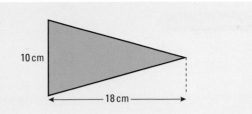

10 cm

18 cm

D

Example 8   Find the shaded area in the company logo.

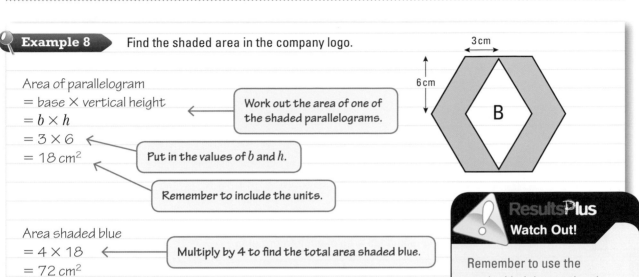

3 cm

6 cm

B

Area of parallelogram
= base × vertical height
= $b \times h$
= 3×6
= 18 cm^2

Work out the area of one of
the shaded parallelograms.

Put in the values of b and h.

Remember to include the units.

Area shaded blue
= 4×18
= 72 cm^2

Multiply by 4 to find the total area shaded blue.

Results Plus
Watch Out!

Remember to use the
vertical height, not the slant
height, when working out
the area of a parallelogram.

Exercise 14G

1 Find the areas of the parallelograms drawn on the centimetre grid.

a   b   c

D

2 A tiler creates the following pattern using
parallelogram-shaped tiles.
Work out the total area covered by the red tiles.

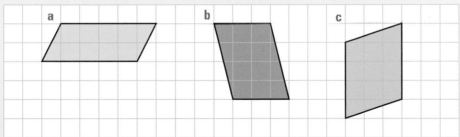

240 cm

200 cm

A03

Example 9   Work out the area of this trapezium.

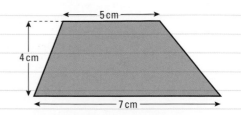

Area of trapezium

$= \frac{1}{2} \times$ sum of parallel sides $\times$ distance between
   parallel sides

$= \frac{1}{2}(a + b)h$ ← Choose the formula to use.

$= \frac{1}{2} \times (5 + 7) \times 4$ ← Put in the values of a, b and h.

$= \frac{1}{2} \times 12 \times 4$

$= 24\,cm^2$

Exercise 14H

1   Copy and complete the table to find the area of each trapezium.

	a	b	h	Area
Trapezium 1	4 cm	6 cm	3 cm	
Trapezium 2	10 cm	12 cm	5 cm	
Trapezium 3	9 m	7 m	6 m	
Trapezium 4	5 m	10 m	4 m	

2   Work out the area of each trapezium.

a

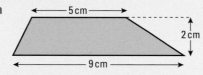

b

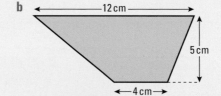

c

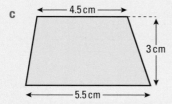

d

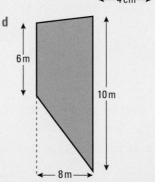

3   A trapezium has an area of 40 cm².
    The two parallel sides have lengths 7 cm and 13 cm.
    The distance between the two parallel sides is h cm.
    Work out the value of h.

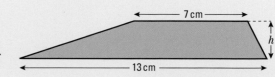

14.4 **Problems involving areas**

Objective

- You can find the area of a more complicated shape by splitting it up into simple shapes.

Why do this?

Garden patios are not always simple shapes. A gardener will need to work out the area of a new patio to know how many tiles to buy.

Get Ready

1. Here are some shapes. Copy them and draw lines to show how each one can be split up into squares, rectangles and triangles.

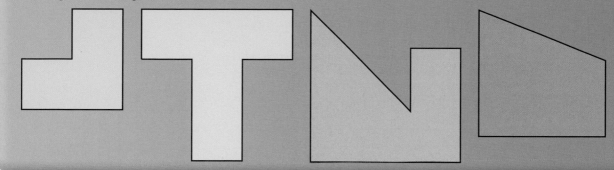

Key Point

- To find the area of more complicated shapes you will need to split the shape into a number of simpler shapes such as rectangles, squares, triangles or parallelograms. You can then find the area of each part and add these areas together to find the total area.

  The following examples will show you how to do this.

Example 10

Find the area of this shape.

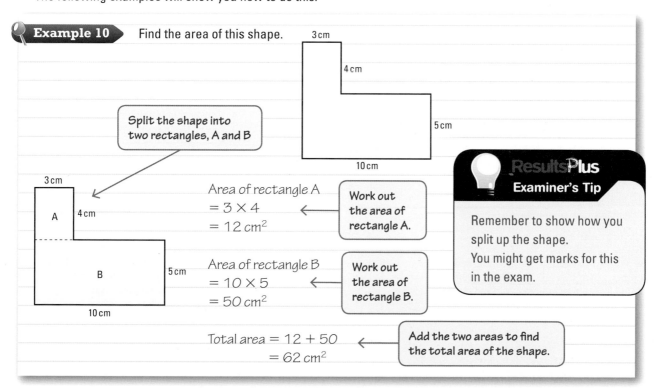

Split the shape into two rectangles, A and B

Area of rectangle A
$= 3 \times 4$
$= 12 \text{ cm}^2$

Work out the area of rectangle A.

Area of rectangle B
$= 10 \times 5$
$= 50 \text{ cm}^2$

Work out the area of rectangle B.

Total area $= 12 + 50$
$= 62 \text{ cm}^2$

Add the two areas to find the total area of the shape.

ResultsPlus
Examiner's Tip

Remember to show how you split up the shape.
You might get marks for this in the exam.

Example 11 Work out the area of this shape.

Area of shape
= area of triangle
+ area of square

> Split into simpler shapes.

Area of triangle
$= \frac{1}{2} \times b \times h$
$= \frac{1}{2} \times 5 \times 4$
$= 10 \, cm^2$

> Work out any unknown lengths you need and mark them on the diagram. Then calculate the area of the triangle.

Area of square
$= l^2$
$= 6^2$
$= 36 \, cm^2$

> Calculate the area of the square.

Total area $= 10 + 36$
$\qquad\quad = 46 \, cm^2$

> Add the two areas to find the total area of the shape.

A03

Example 12 Julie wants to make a rectangular patio in her garden.

She needs to cover an area which measures 5 m by 3.5 m with square paving stones.

Each paving stone measures 50 cm by 50 cm and costs £3.99.

Work out the cost of the paving stones Julie will need.

Method 1

500 cm

50 cm

50 cm

Paving stone

350 cm

Patio

> Draw a diagram to help you understand the question. Convert your units so they are all the same.

> Work out the area of the patio.

Area of patio $\qquad\qquad = 500 \times 350 = 175\,000 \, cm^2$
Area of a paving stone $\quad = 50 \times 50 = 2500 \, cm^2$

Number of paving stones needed $= \dfrac{175\,000}{2500}$
$\qquad\qquad\qquad\qquad\qquad\qquad = 70$

> Work out the area of a paving stone. Divide the area of the patio by the area of a paving stone to calculate the required number of paving stones.

Total cost of paving stones $= £3.99 \times 70$
$\qquad\qquad\qquad\qquad\qquad = £279.30$

> Multiply the cost of 1 paving stone by the number needed.

Method 2

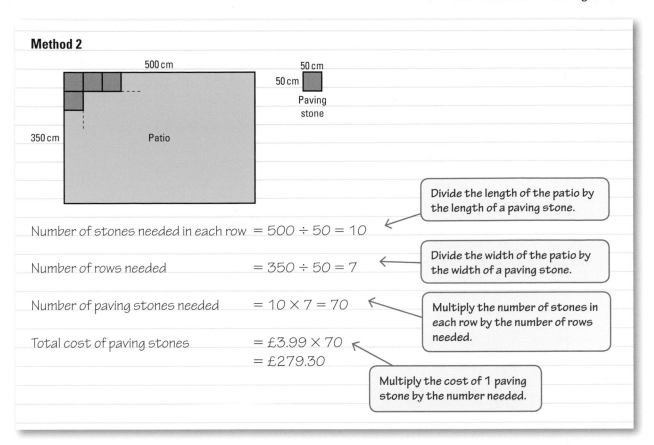

Number of stones needed in each row $= 500 \div 50 = 10$

> Divide the length of the patio by the length of a paving stone.

Number of rows needed $= 350 \div 50 = 7$

> Divide the width of the patio by the width of a paving stone.

Number of paving stones needed $= 10 \times 7 = 70$

> Multiply the number of stones in each row by the number of rows needed.

Total cost of paving stones $= £3.99 \times 70$
$= £279.30$

> Multiply the cost of 1 paving stone by the number needed.

Exercise 14I

1 A square piece of card has sides of length 10 cm.
A hole is cut from the card.
The hole is a square of side 6 cm.
  a Work out the area of the large square.
  b Work out the area of the small square.
  c Work out the area of the card left.

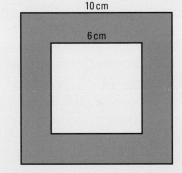

E

2 The floor of the hall in a house is a 225 cm by 150 cm rectangle. Tiles which are squares of side 15 cm are used to tile the floor. Work out how many tiles are needed.

A02 **D**

3 Liam wants to replace the carpet in his room. The floor of the room is a rectangle measuring 4 metres by 3 metres. The carpet he wants to buy costs £8.65 per square metre.
Work out how much it will cost Liam to buy enough carpet to cover the floor.

A02

4 Libby wants to buy some grass seed so that she can sow a new lawn in her garden. She wants the lawn to be a rectangle measuring 3.2 metres by 2.5 metres. She needs 35 grams of lawn seed for every square metre of lawn. One box of lawn seed contains 250 g.
  a How many boxes of lawn seed will Libby need to buy?
  b How much lawn seed will be left over?

A03

5 Find the area of the following shapes.

a
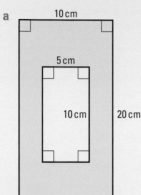
4 cm
10 cm
8 cm
4 cm

b
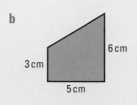
6 cm
3 cm
5 cm

c
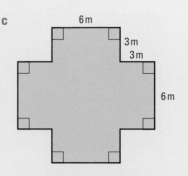
6 m
3 m
3 m
6 m

6 Work out the shaded area in each diagram.

a
10 cm
5 cm
10 cm   20 cm

b
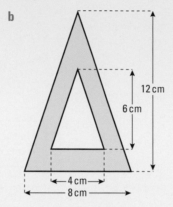
12 cm
6 cm
4 cm
8 cm

c
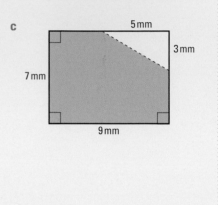
5 mm
3 mm
7 mm
9 mm

Chapter review

- The **perimeter** of a **2D shape** is the total distance around the edge of the shape.
- The perimeter of a rectangle can be found using the formula
  Perimeter of a rectangle $= l + w + l + w$
  $$= 2l + 2w$$
- The **area** of a 2D shape is the amount of space inside the shape.
- The area of a 2D shape can be found by counting squares or by using the formulae:
  - Area of rectangle $=$ length $\times$ width
    $$= l \times w$$
  - Area of square $=$ length $\times$ length
    $$= l^2$$
  - Area of triangle $= \frac{1}{2} \times$ base $\times$ vertical height
    $$= \frac{1}{2} \times b \times h$$
  - Area of parallelogram $=$ base $\times$ vertical height
    $$= b \times h$$
  - Area of trapezium $= \frac{1}{2} \times$ sum of the lengths of the parallel sides $\times$ distance between the parallel sides
    $$= \frac{1}{2}(a + b)h$$
- To find the area of more complicated shapes you will need to split the shape into a number of simpler shapes such as rectangles, squares, triangles or parallelograms. You can then find the area of each part and add these areas together to find the total area.

Review exercise

1 Here is a shaded shape on a grid of centimetre squares.
  a Find the perimeter of the shaded shape.
  b Find the area of the shaded shape.

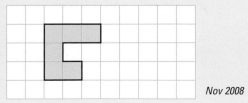

Nov 2008

2 Here is a rectangle.
  a Work out the perimeter of the rectangle.
  b Work out the area of the rectangle.

10 cm

Diagram **NOT** accurately drawn

20 cm

Nov 2008

3 A carpet 60 cm wide is to be used to cover a rectangular floor measuring 4 metres by 9 metres. Calculate the length of carpet needed.

4 Office regulations say the gap between the desks should be 900 mm to allow for wheelchair users. A03
A desk has a length of 2 m and a width of 1 m.

2 m

1 m

Diagram **NOT** accurately drawn

Six of these desks are arranged as shown in the diagram below.
The gap between each desk is 900 mm.

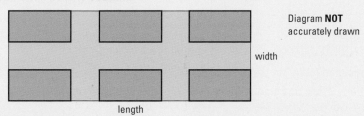

Diagram **NOT** accurately drawn

width

length

What is the total length and the total width of the office space needed for these desks?

5 ABC is a right-angled triangle.
AB = 7 cm,
BC = 8 cm.
Work out the area of the triangle.

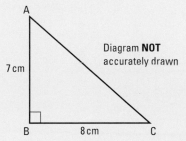

A

7 cm

Diagram **NOT** accurately drawn

B   8 cm   C

June 2008

6 **a** Work out the area of this rectangle.

4.5 cm

2.5 cm

Diagram **NOT** accurately drawn

A square has an area of 324 cm^2.
  b Work out the length of one side of the square.

Area
324 cm^2

Diagram **NOT** accurately drawn

June 2007

G

F

E

D

D
A03

7 The diagram shows Rob's patio.
All the corners are right angles.
The patio is made up of square paving
stones each 50 cm by 50 cm.
Work out how many of these paving
stones are needed to tile Rob's patio.

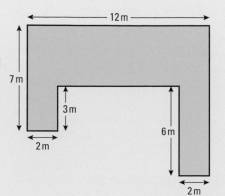

Diagram **NOT**
accurately drawn

A03

8 A room has four interior walls.

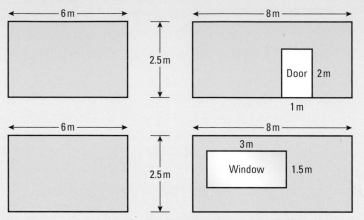

Diagram **NOT**
accurately drawn

Alesha paints the walls with emulsion paint. She does not paint the door.
A 3 litre tin of emulsion paint covers 30 m² of wall.
Work out how many 3 litre tins she needs to buy. Show all your working.

C
A02

9

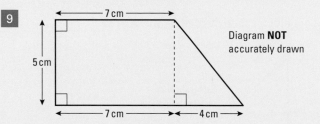

Diagram **NOT**
accurately drawn

Work out the area of the shape.

Nov 2008

A02

10 The diagram shows a rectangle inside a triangle.
The triangle has a base of 12 cm and a height of 10 cm.
The rectangle is 5 cm by 3 cm.
Work out the area of the region shown shaded in the
diagram.

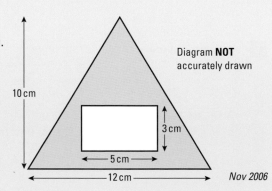

Diagram **NOT**
accurately drawn

Nov 2006

11

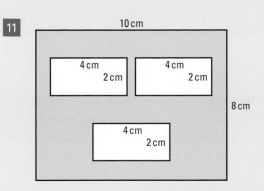

Diagram **NOT** accurately drawn

The diagram shows 3 small rectangles inside a large rectangle.
The large rectangle is 10 cm by 8 cm.
Each of the 3 small rectangles is 4 cm by 2 cm.
Work out the area of the region shown shaded in the diagram.

June 2007

12

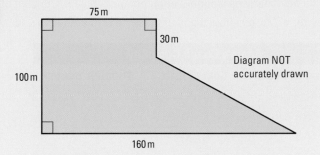

Diagram NOT accurately drawn

The diagram shows the plan of a field.
The farmer sells the field for £3 per square metre.
Work out the total amount of money the farmer should get.

March 2007

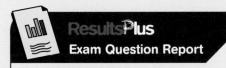

ResultsPlus
Exam Question Report

84% of students answered this question poorly.
Some candidates confused perimeter with area,
and had trouble with the area of a triangle.

The following question helps you to develop both your ability to select and apply a method (AO2) and your ability to solve problems using your skills of interpretation (AO3). Your AO3 skills are particularly required as you will need to work through several steps to solve this problem. There are also some functional elements as this is a real-life situation and there is a problem to solve.

Example

Adam runs a coach company. He has 6 small coaches, 4 medium coaches, 3 large coaches and 1 double-decker coach.

The table gives information on how many passengers each coach can seat, the cost of hiring the coach and a driver for a day, and how many of these coaches Adam owns.

Adam's Coach Company			
Coach type	Number of seats	Cost of hire	Number owned
Small	25	£100	6
Medium	38	£110	4
Large	54	£120	3
Double-decker	78	£140	1

Rachel wants to hire some coaches from Adam to take 222 people out for the day.
What is the cheapest way for Rachel to do this?

Solution

As the number of seats increases, the cost goes down proportionally. Therefore you need to use the largest coach, the double-decker, first.

1 double-decker	£140	78
3 large	£360 +	162 +
	£500	240 seats

This leaves 144 people to fit in. This could be done with three large coaches but would leave 18 empty seats.

1 double-decker	£140	78
2 large	£240	108
1 medium	£110 +	38 +
	£490	224 seats

If two large coaches were used then this would leave 36 people to fit in, so a medium coach would be needed as well

The cheapest way is £490 and there are only two spare seats.

Now try these

1 Sam is a salesman. He gets paid expenses when he drives his car on company business.
He gets paid 45p for each mile he drives.
He also gets paid a meal allowance.
Here is Sam's time and mileage sheet for one week.

Meal Allowance
Lunch £8.50
Dinner £22*

*Only paid if Sam arrives home after 8 pm

Day	Miles driven	Lunch claimed	Time arrived home
Monday	180	Yes	9 pm
Tuesday	48		5 pm
Wednesday	64	Yes	8.30 pm
Thursday	33		5 pm
Friday	75	Yes	7.30 pm

Work out Sam's total expenses for the week.

2 Lynsey took part in a sponsored swim. Her target was to raise £100 for charity. Her nan promised her that she would make up the £100 if Lynsey did not raise enough.
Here is Lynsey's sponsor form.
Lynsey swam 32 lengths in a pool of length 40 m.
Will her nan have to give her any money?
You must explain your answer.

Sponsor	Amount
Ali	£5
Rob	25p for each length
Will	30p for each length
Mum	50p for each length
Jade	2p for each metre

3 Here are the rates charged for Mr Pitkin's telephone.

Line rental	£29.36 each quarter
Daytime cost	4p for each minute
Evening and weekend	3p for each minute
To mobiles	11p for each minute
National rate	8p for each minute

Here are the details of calls made by Mr Pitkin one quarter.

Type of call	Minutes
Daytime	78
Evening	312
To mobiles	42
National rate	25

Calculate Mr Pitkin's telephone bill for that quarter.

The following question helps you to develop your ability to select and apply a method (AO2) as there is more than one possible way to raise the fees. The question also tests your ability to analyse a problem and generate a strategy to solve it (AO3).

Example

Fred's Gym charges an annual membership fee of £250, plus an additional weekly fee of £5.

The gym wish to increase the amount paid annually by each member to £600. Work out two different ways they could raise their fees.

Solution

Current amount raised
= £5 × 52 + £250 = £510

← Find the cost of membership for 1 year at the current rate.

Additional amount
= £600 − £510 = £90

← Decide how much more is needed.

Method 1
Increase annual fee to £250 + £90 = £340.
Same weekly fee of £5.

Experiment with changing the membership fee and the weekly fee to give the new cost.

Method 2
Reduce annual fee to £80 and increase weekly fee to £10.

There are many other possibilities.

Now try these

1 I have some 5p coins and some 2p coins.
If I use an even number of 5p coins I cannot make a total of 23p. Explain why not.

2 A group of 5 friends bought 10 items at the Gold Medal Chip Shop.

> **Gold Medal Chip Shop**
> Chips 95p
> Fish £2.25
> Sausage £1.80

The total bill came to £15.10.
What did they buy?

3 Gina needs exactly 17 kg of pasta for her restaurant. Pasta can be bought in 3 kg packs for £5.40 each or 5 kg packs for £8.50 each.
How many packs of each size should she buy to get the best value for money?

4 A bar of chocolate costs 50p.
The diagram shows some information on how much of this goes to different people when a bar of chocolate is sold.
What percentage of the cost of a bar of chocolate goes to the growers?

Growers	Retailers	Others
	12.5p	35.5p

5 A cereal company is considering a special deal on its 500 g packets of Tasty Flakes.

Which is the best deal?

6 In a sale all of the items in a shop are reduced by 20%. After the sale the prices are increased by 40%.
What is the overall effect on prices?
Explain your answer.

The following questions help you develop your ability to select and apply a method to get the correct solution (AO2).

Example 1

Susie rides her bike from home to school.

Susie's home is 1.8 km from school.

On her journey from home to school the front wheel on her bike rotates exactly 1000 times.

Work out the diameter of the front wheel on her bike.

Give your answer correct to 3 significant figures.

Solution 1

$1.8 \, \text{km} = 1.8 \times 1000 = 1800 \, \text{m}$

$1800 \times 100 = 180\,000 \, \text{cm}$ ← Change the units to centimetres to get a sensible answer.

$180\,000 \div 1000 = 180 \, \text{cm}$ for 1 rev ← Find the length of one revolution.

$C = \pi d$ so $180 = \pi d$
$d = 180 \div \pi = 57.2958 = 57.3 \, \text{cm}$ ← Use the formula for the circumference of a circle to find the diameter.

Example 2

$PQRS$ is a parallelogram.

PQ is parallel to SR.

PS is parallel to QR.

PQ has a length of $(2a + 3)$ centimetres.

PS has a length of $(a + 2)$ centimetres.

The perimeter of the parallelogram is 25 cm.

Find the length of PQ.

Solution 2

Since opposite sides of a parallelogram are equal, the perimeter can be written as
$$2(2a + 3) + 2(a + 2) = 25$$
$$4a + 6 + 2a + 4 = 25$$
$$6a + 10 = 25$$
$$6a = 15$$
$$a = 2.5$$
$$PQ = 2 \times 2.5 + 3 = 8 \, \text{cm}$$

Now try these

1 The three angles of a triangle are x, $2x$ and $3x$.
Find the value of x.

2 *BCEF* is a parallelogram. *FAB* is a right-angled triangle.
ECD is an isosceles triangle.
ABC is a straight line. *FED* is a straight line.
Find the value of x.

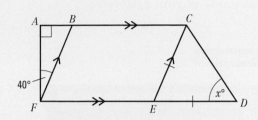

3 James rides his bike to school. His bike wheel has a diameter of 75 cm.
His bike wheel rotates 1500 times on his trip to school.
How far does James live from school?

4 The diagram shows the design for a window.

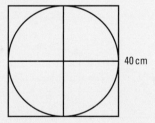

40 cm

The window is square with sides of length 40 cm.
Strips of lead are put round the edge of the window and on the window to make the pattern shown within the square.
Work out the total length of all the strips of lead.

5 The length of a rectangle is $(2x + 3)$ cm. The width of the rectangle is 5 cm.
The area of the rectangle is 35 cm².
Find the perimeter of the rectangle.

INTERPRETING AND DISPLAYING DATA

The following question helps you develop both your ability to select and apply a method (AO2) and your ability to solve problems using your skills of interpretation and proof (AO3). The AO2 skills come into parts (a) to (c) as, to find your estimates, you could either use a scatter graph with a line of best fit or put the values in order and estimate numerically. The AO3 elements come into part (d) where you are asked to comment on the reliability of your estimates.

Example The table below shows the reading speed and IQ of 10 students.

Reading speed (words per minute)	130	125	131	285	95	187	235	165	123	145
IQ	100	85	135	135	90	120	130	110	98	95

Use this information to estimate

a The reading speed of a student with an IQ of 120.

b The IQ of a student with a reading speed of 230.

c The reading speed of a student with an IQ of 80.

d Comment on the reliability of your answers.

Solution

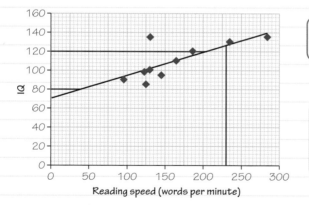

> Drawing a scatter graph will help with this question.

> Draw lines to show where you are reading off from your scatter graph. It shows the examiner how you are making your estimates.

a 205

b 126

c 40

d The scatter graph shows that there is positive correlation between IQ and reading speed. A line of best fit can be drawn. The answers to parts **a** and **b** are reasonably reliable as they lie within the range of the data. Part **c**, however, lies outside the data, so the estimate is unreliable.

* **1** The table below shows the predictions made for the highest temperature each day in Oxford.

Day	Monday	Tuesday	Wednesday	Thursday	Friday	Saturday	Sunday
Temperature °C	6	7	6	7	8	10	7

Use a suitable graph or chart to display this information.

* **2** The table below shows the amount of money a local council is investing in services.

Services	Roads & Transport	Education	Housing	Other
Amount of money (£million)	16	14	30	12

Use a suitable graph or chart to display this information for inclusion in the council's annual report.

3 The table below shows the price of a book and the number of pages it contains.

Number of Pages	100	145	150	75	140	200	90	180
Cost of book	£5	£6.50	£7	£3	£6.50	£9	£4	£9.50

Use this information to estimate

a The cost of a book with 160 pages.

b The number of pages in a book costing £5.

c The cost of a book with 50 pages.

d Comment on the reliability of your answers.

* **4** Mr Smith and Miss Khan predict the positions of 8 swimmers in a race.

Actual	1	2	3	4	5	6	7	8
Mr Smith	6	3	4	2	5	1	8	7
Miss Khan	3	1	2	4	6	7	5	8

Which teacher is closest to predicting the actual positions?

The following question helps you develop your ability to select and apply a method (AO2) and your ability to analyse and interpret problems (AO3).

Example The side of a shed is the shape of a trapezium as shown in the diagram.

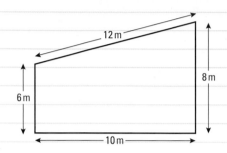

The side is to be given two coats of paint. The paint is sold in 1 litre cans costing £3 each.
1 litre of paint covers 15 square metres.
How much will it cost to paint the wall?

Solution

Using the formula
Area (trapezium) $= \frac{1}{2}(a + b)h$

$$= \frac{1}{2}(6 + 8)10$$

$$= 70 \, m^2$$

> You need to find the area of the side of the shed.
> You may choose to use the formula, or divide the shape into a rectangle and a triangle.
> The formula is given on the formulae sheet.

Dividing the shape
Area of rectangle $= 6 \times 10 = 60 \, m^2$

Area of triangle $= \frac{1}{2} \times 10 \times 2 = 10 \, m^2$

Total area $= 70 \, m^2$

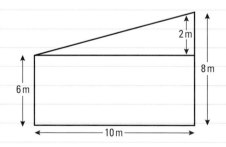

The number of tins of paint required for one coat is $70 \div 15$.

> Total area ÷ area covered by 1 tin

The number of tins needed for two coats is $140 \div 15 = 9.3$.
So 10 tins will be needed.
The cost of the paint will be £30.

1 This shape is made by joining six squares.

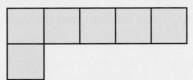

Find two shapes which have the same area but different perimeters.

2 The diagram shows a wall which is to be built with bricks.
The bricks measure 200 mm × 100 mm.
They are sold in packs of 100. One pack costs £35.
Find the cost of the bricks.

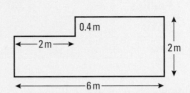

3 The diagram shows a rectangular path around a lawn. The path is 1 m wide.

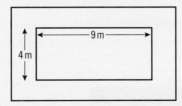

Gravel costs £124 per tonne.
1 tonne of gravel covers 15 m².
Work out the cost of covering the path with gravel.

4 Find the perimeter of three different rectangles which each have an area of 36 cm².

5 The diagram shows a bathroom wall in the shape of a trapezium. The wall is to be painted.

The paint chosen is sold in 1 litre cans costing £4 each. 1 litre covers 12 square metres.
How much will it cost to paint the wall?

AVERAGE AND RANGE

The following question helps you develop both your ability to select and apply a method (AO2) and your ability to solve problems using your skills of interpretation and proof (AO3). The AO2 skills come into part (a) as there is more than one way to compare the players so you will need to choose a method. Your AO3 skills are needed in part (c) as you will need to work through several steps to solve this problem. There are also some functional elements as average performance over a number of competitions is often used to decide places on teams in sport.

Example   The table below shows the number of runs scored by cricketers over a number of matches.

	Runs scored in 10 matches									
Jones	42	0	19	49	72	29	31	22	16	20
Phillips	68	14	64	24	18	4	5	64	32	32
Bakir	18	24	35	9	2	0	8			76
Aziz	30	31	29	28	32	34	25	35	29	30

Bakir did not play in all of the matches.

a Compare and contrast the players' scores.

b Cooke's mean score over 3 matches was 42.

    Write down 3 scores which could have given him this mean.

Smith played in the first 9 matches and scored a mean of 30 runs.

c How many runs does he need to score in the final match to give him a better average than the players in the table?

Solution   a

	Mean	Range
Jones	30	72
Phillips	32.5	64
Bakir	21.5	76
Aziz	30.3	10

> Compare the mean and range for each player.

Bakir has the lowest mean average (21.5) and he is the least consistent player with a range of 76.
Aziz is the most consistent player with a range of 10 but he never gets a really high score.
Phillips has the highest mean score (32.5) and he is more consistent than Jones and Bakir.
Jones has the second lowest mean average and he isn't very consistent.

b Cooke's average was 42. He scored a different number of runs in each match.
    Total score over 3 matches is $3 \times 42 = 126$
    Scores could be 41, 42 and 43.

c In 9 matches Smith scores $9 \times 30 = 270$ runs.
    The best average so far is Phillips' (32.5).
    In 10 matches an average of $32.5 = 325$ runs.

> Find the total number of runs he needs to improve on.

    So, to improve on this Smith needs 56 runs.

> Find the difference between Smith and Phillips. Smith needs 1 more.

* 1 Hatton's netball team scored the following goals in 8 matches.

| 41 | 33 | 35 | 38 | 36 | 36 | 35 | 38 |

Georgie's team scored the following goals in 6 matches.

| 47 | 29 | 52 | 31 | 38 | 28 |

Ann claims Hatton's team did the best. Is she correct?

Explain your answer.

A02
A03

D

2 Three meals have a mean average price of £8.97.

All of the meals have a different price.

Write down the possible price of each meal.

3 Write down two sets of numbers with the same mean but

a a different median

b a different mode

c a different range.

A02

4 A hospital outpatients target has stated that waiting times will be on average 1 hour.

The waiting times (in minutes) of the first 9 patients are given below.

| 54 | 75 | 49 | 45 | 72 | 63 | 29 | 78 | 85 |

What is the maximum time the next patient can be kept waiting if they are to meet their target?

* 5 The weights (g) of 10 lambs were recorded by Farmer Pearce.

| 1000 | 1250 | 2000 | 1875 | 1400 | 1650 | 1325 | 1125 | 1450 | 1700 |

A02

Farmer Hicken also has 10 lambs.

The range of weights for Farmer Hicken's lambs was 2.5 kg.

The mean weight of Farmer Hicken's lambs was 1.5 kg.

Test the hypothesis that Farmer Hicken's lambs were bigger than Farmer Pearce's lambs.

* 6

A03

	5 day forecast – Florida					5 day forecast – Cuba, Varadero					5 day forecast – Egypt, Red Sea				
	Monday	Tuesday	Wednesday	Thursday	Friday	Monday	Tuesday	Wednesday	Thursday	Friday	Monday	Tuesday	Wednesday	Thursday	Friday
High	27°C	27°C	27°C	27°C	27°C	25°C	26°C	26°C	27°C	30°C	27°C	25°C	25°C	26°C	27°C
Low	13°C	14°C	16°C	16°C	16°C	19°C	21°C	21°C	21°C	17°C	17°C	16°C	14°C	16°C	17°C
Humidity	0%	1%	5%	10%	20%	0%	10%	20%	20%	20%	0%	0%	0%	0%	0%

Sonja and John are planning their honeymoon for after their wedding in November.

They are trying to decide where to go.

The weather for a week in November at three different resorts is shown in the charts above.

Recommend one of these resorts, giving evidence to show it has the best weather.

The following question tests both your ability to select and apply a method in the context of choosing the best strategy to win a game (AO2) and your ability to analyse and interpret problems (AO3). The AO2 skills are used in the first part. Showing the outcome space in a table is the most efficient way to compare the probabilities but some students may just produce a list of possible outcomes. Your AO3 skills are needed in both parts of this question as you will need to give a reasoned explanation for your answer.

Example

In a game the player has a choice of throwing 1 or 2 dice.
The winning score is 6.
Joshua says 'You are more likely to win if you throw 2 dice'.

a Is Joshua correct? Explain your answer.
b Is it possible to have a winning score which is equally likely whether 1 or 2 dice are thrown? Explain your answer.

Solution

a 1 dice: numbers thrown are 1, 2, 3, 4, 5, 6; probability of a 6 is $\frac{1}{6}$.
2 dice: the outcome space is

	1	2	3	4	5	6
1	2	3	4	5	6	7
2	3	4	5	6	7	8
3	4	5	6	7	8	9
4	5	6	7	8	9	10
5	6	7	8	9	10	11
6	7	8	9	10	11	12

> Work out the probability by writing down the outcome space.

> The outcome space is useful in explaining your answer.

The probability of a total of 6 with two dice is $\frac{5}{36}$.
So, Joshua is incorrect. Throwing 1 dice is more likely to win the game.

b The likelihood of any number 1 to 6 being thrown on a single dice is $\frac{1}{6}$.
The likelihood of a number greater than 6 being thrown on a single dice is 0.
The only number with a probability of $\frac{1}{6}$ when 2 dice are thrown is 7, which is not possible on a single dice.
It is impossible to choose a winning score which is equally likely whether 1 or 2 dice are thrown.

> **Now try these**

1 A class did a survey of how they travelled to school.
The results are shown in the table below.

	Number of People
Car	9
Bus	7
Walk	10
Cycle	4

Explain why the probability of a student selected at random travelling to school by bus is not $\frac{1}{4}$.

2 Ahmed says 'I like multiple choice tests because I can get half the marks by guessing'.
a Is he correct?
b Explain your answer.

3 Three coins are tossed together.
What is the probability that they all come down the same – all heads or all tails?
Explain your answer.

4 The advert shows a special offer for lunch in a restaurant.
If the customer chooses a starter, a main and a sweet at random,
what is the probability of them choosing soup, a pie and fruit?

Meal Deal

Starter   Soup
   Orange Juice

Main   Meat Pie
   Fish Pie
   Veggie Burger

Sweet   Ice Cream
   Fruit
   Cheese

5 In a game of chance a shape is chosen at random.
The table shows the shapes and the probability that they will be chosen.

☺	✈	📖	🔔
0.2	0.4		0.1

Work out the probability that the book will be chosen.

6 A travel agent collected information from 100 people about when they took their holidays.
He found that 26 went in September, 42 preferred an August holiday and the rest went in July.
A total of 57 took their holiday in a hotel, 16 of these in September. 19 people took a self-catering holiday in August.
Work out the probability of a person selected at random taking a holiday in a hotel in August.

F
A03

A03

E
A02
A03

A02
A03

D

C
A02
A03

The Olympic Games is held every four years in different countries around the world. The Olympic committee picked London to host the 2012 games because campaigners promised to improve local communities and encourage involvement in sport.

People around the world live in different time zones. Time zones are based on Greenwich Mean Time (GMT). They state how many hours in front or behind of British time they are.

QUESTION

1. The time in Kingston is 5 hours behind the time in London. If the Jamaican sprint team leave Kingston at 2.32 pm local time and fly to London on a flight that takes 11 hours and 45 minutes, what time will they arrive in London?

Kingston   11 hours 45 minutes   London

QUESTION

2. An American visitor to the Olympics plans to travel to France after the Games. He needs to change $500 into Euros before flying to Paris. How many euros will he get?

The exchange rate is £1 = $1.10 £1 = €1.65

MODERN PENTATHLONS

Fencing – 35 bouts

Baseline 25 victories = 1000 points
Each victory over 25 gains
an extra 24 points.
Each victory below 25 loses
24 points.

Running – 3000 m

Baseline 10 minutes = 1000 points
Each second faster gains an
extra 4 points.
Each second slower loses
4 points.

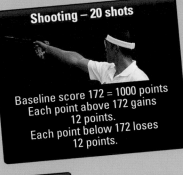

Shooting – 20 shots

Baseline score 172 = 1000 points
Each point above 172 gains
12 points.
Each point below 172 loses
12 points.

Show jumping – 15 jumps

Baseline is a clean round
(no mistakes)
Within the time limit = 1200 points.
Athletes are penalised 28 points for
each mistake.

Swimming – 200 m

Baseline 2 min 30 sec = 1000 points
Every 0.1 second faster gains
an extra point.
Every 0.1 second slower loses
a point.

QUESTION

3. Modern Pentathlons consist of five events: shooting, fencing, swimming, running and show jumping. For each event competitors are scored points above or below a baseline. Which of the three athletes below has the highest total score?

Athlete 3

Swimming 2 min 24.0 sec
Shooting 165 points
Running 10 min 4 sec
Show jumping 1 mistake
Fencing 18 victories

Athlete 2

Swimming 3 min 2.3 sec
Shooting 193 points
Running 9 min 34 sec
Show jumping 0 mistakes
Fencing 18 victories

Athlete 1

Swimming 2 min 45.7 sec
Shooting 190 points
Running 8 min 52 sec
Show jumping 3 mistakes
Fencing 27 victories

LINKS

◉ For **Question 1** you need to understand how to use time. You learnt about this in **Chapter 11**.

◉ You learnt about decimals in **Chapter 5**. You will need to be able to use decimals in calculations for **Question 2**.

◉ For **Question 3** you need to be able to add and subtract numbers from a baseline. You learnt how to do this in **Chapter 1**.

LEARNING TO DRIVE

To pass your driving test you need to know stopping distances for different speeds.

Learning to drive a car can be expensive so it is a good idea to write a budget before starting lessons.

1. When revising for the theory part of the driving test, information from a table is easier to remember than from a composite bar chart.

2. Use the information below to estimate the cost of passing the driving test. You need a theory test guide and a minimum of 30 hours of lessons. You also need to pass the theory test and the practical test.

QUESTION

QUESTION

Draw and complete a suitable table (for the different speeds) from the composite bar chart.

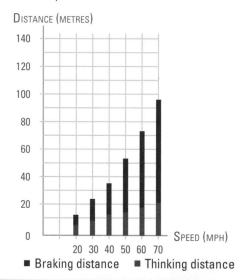

DISTANCE (METRES)

SPEED (MPH)

■ Braking distance   ■ Thinking distance

L

Learner Test Information

Theory test fees	Price
Standard fee for car and motorcycle	£31
Multiple choice questions for bus and lorry drivers	£35
Hazard perception for bus and lorry drivers	£15
Driver CPC theory test case studies for bus and lorry drivers	£30
Potential approved driving instructor	£90

Practical test type	Weekday price/ evening and weekend price
Car	£62 / £75
Tractor and other specialist vehicle	£62 / £75
Motorcycle; module 1	£15.50 / £15.50
Motorcycle; module 2	£15.50 / £15.50
Lorry and bus	£115 / £141
Driver CPC	£55 / £63
Car and trailer	£115 / £141

LINKS

- For **Question 1** you need to be able to interpret composite bar charts. You learnt how to do this in **Chapter 12**.

- You learnt how to round in **Chapter 1**. You will need to do this in **Question 2**.

- You need to be able to read two-way tables in **Question 3**. You learnt this in **Chapter 26**.

When buying a second-hand car there are lots of factors to consider. A used car guide can help you estimate the price and avoid paying too much.

Theory Test Guide
£11.99

Drive UK
£22.50 / hour
First lesson free

3. Josh sees an 0352 reg car with a mileage of 87 000 for sale privately. To get some idea of what price he might have to pay he consults a guide. There are two tables. Table 1 shows the price according to age and a set mileage, and Table 2 shows an adjustment for mileage above the set mileage.

These prices are adjusted when the mileage differs from the stated value by more than 1000 miles. The amount depends on the dealer price, the code and how many thousands of miles more or less there are on the clock.

Example: If the dealer price is £4000 and the code is D, then for each 1000 miles over the stated value the price has to be reduced by £130.

Work out an estimated price for the car that Josh has seen for sale.

QUESTION

Table 1

Table 2

AUTO TRADE **GUIDE**

For sale

Reg	New (£)	Dealer (£)	Private (£)	Trade (£)	Mileage	Code
0202	9400	2105	1875	1840	80 000	B
0252	9795	2205	1900	1845	80 000	B
0352	9895	2355	2090	2050	70 000	B
0303	9995	2470	2175	2135	70 000	B
0353	9995	2595	2435	2475	60 000	B

Dealer price	Code				
	A	B	C	D	E
Up to £999	0	0	10	20	30
£1000–£1999	0	5	35	50	55
£2000–£3999	5	10	60	80	120
£4000–£6999	10	25	95	130	210
£6999–£10 999	30	50	120	180	390
£11 000+	50	60	160	200	420

HEALTHY LIVING

FS

> **A** balanced diet is central to overall good health. To help consumers make healthy choices nutritional information is printed on most food items.

1. Samantha invites her friends round for dinner. Together they eat a bag of chicken nuggets and a pizza, which they cut into eight slices.

   Samantha has 4 slices of pizza with 5 chicken nuggets. Daisy has 3 slices of pizza with 10 nuggets and Darren has the rest of the pizza with 12 nuggets.

   The tables show some nutritional information about the pizza and the chicken nuggets.

Pizza

Typical values	Per 100 g	Per slice
Energy	1000 kJ	494 kJ
Protein	9.3 g	4.6 g
Carbohydrates	28.7 g	14.2 g
Fat	9.6 g	4.8 g
Fibre	2.3 g	1.1 g
Salt	1.0 g	0.5 g

Chicken Nuggets

Typical values	Per 100 g	Per nugget
Energy	1150 kJ	189 kJ
Protein	9.7 g	1.6 g
Carbohydrates	18.2 g	3.0 g
Fat	18.2 g	3.0 g
Fibre	1.7 g	0.3 g
Salt	1.0 g	0.2 g

QUESTION

Compare the amount of fat and salt eaten by each of the three friends.

Healthy Choice Fitness Gym

LINKS

- ◉ For **Question 1** you need to be able to read data from tables. You learnt how to do this in **Chapter 3**.

- ◉ You need to understand dual bar charts for **Question 2**. You learnt about them in **Chapter 12**.

- ◉ You learnt about interpreting pie charts in **Chapter 12**. You will use this in **Question 3**.

2. In 2007–08 there were approximately 600 000 children in Reception classes and 650 000 in Year 6.

The National Child Measurement Program details on weight are shown in the dual bar chart.

QUESTION

Compare the numbers of obese and overweight children in Reception and Year 6.

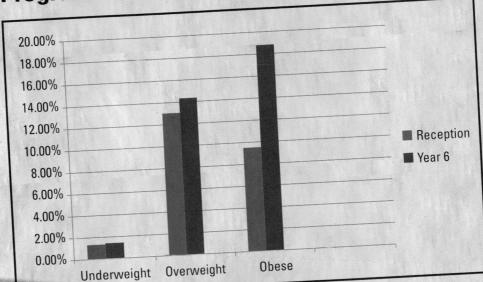

The National Child Measurement Program details on weight 2007–08

Rochelle's exercise type

Flexibility
Aerobic
Anaerobic

3. Rochelle joins the gym to improve her fitness. Her instructor gives her a pie chart showing the proportion of time she should spend on three types of exercise.

QUESTION

When Rochelle goes to the gym and spends 2 hours exercising, approximately how long should she spend on each type of exercise?

MONEY MANAGEMENT

If you are 16, 17 or 18 and you want to continue studying, you might be entitled to Education Maintenance Allowance (EMA).

QUESTION

1. Sarah's household income is £24 000 a year. This entitles her to EMA. She also has a part-time job for which she gets £5.12 an hour, and time and a half on Sundays. As well as studying, she works 12 hours during the week and 8 hours on Sunday. Provided she earns less than £6,475 per year, she pays no tax. How much money per week does Sarah get? Don't forget to include her EMA.

My weekly work schedule

	Mon	Tues	Wed	Thurs	Fri	Sat	Sun
08:00							
09:00							
10:00							
11:00							
12:00							
13:00							
14:00							
15:00							
16:00							
17:00							

YOUR ENTITLEMENT TO EMA

Your household income	How much EMA you get
up to £20 817 per year	£30 per week
£20 818–£25 521 per year	£20 a week
£25 522–£30 810 per year	£10 a week
more than £30 810 per year	no entitlement to EMA

QUESTION

2. Employees not contributing to a private pension scheme pay 11% NIC on all earnings above £95 a week. If you get paid £5.50 an hour and you work 78 hours in 3 weeks, how much will you pay in NICs?

NIC

When working in any job you need to pay National Insurance Contributions (NICs). Your payments entitle you to social security benefits, including the State Pension. The amount you pay depends on how much you earn.

INCOME TAX

One of the largest taxes for many employees is income tax. The amount you earn is broken into bands and you pay a fixed percentage for all the money in that band.

0% on earnings less than £6475
20% on earnings between £6475 and £34 800
40% on earnings above £34 800

QUESTION

3. If you earn £37 000 a year, how much income tax and national insurance will you pay?

LINKS

- For **Question 1** you need to be able to calculate with decimals. You learnt how to do this in **Chapter 5**.

- You learnt how to use percentages in calculations in **Chapter 19**. You will need to use this for **Question 2**.

- For **Question 3** you need to know how to use a calculator for complex sums. You learnt this in **Chapter 10**.

KITCHEN DESIGN

Britain has one of the highest rates of home ownership in Europe. Home owners often decorate and redesign their properties to increase their value and suit their own tastes.

QUESTION

1. The Chan family are going to replace their kitchen with a total refit. They intend using standard units which can easily be fitted. These units come in three sizes: 600 mm x 600 mm, 600 mm x 1200 mm and 600 mm x 300 mm.

The cooker, fridge-freezer and washing machine are all 600 mm x 600 mm.
The sink unit is 600 mm x 1200 mm.
Cupboard units can be any of the three sizes.

The cooker and fridge-freezer must not be next to each other or near the sink.
The Chan family's kitchen measures 3 m x 4 m.
Using a scale of 1 m = 5 cm, draw a plan of a suitable design for their kitchen.
The picture may give you some ideas.
Do not forget to show the door and windows.

LINKS

◉ For **Question 1** you will need to understand scale drawings.
   You learnt how to do this in **Chapter 7**.

◉ You learnt how to convert measurements from one unit to another in **Chapter 11**. You will need to be able to do this in **Question 2**.

◉ For **Question 3** you will need to be able to use ratio in your calculation.
   You learnt how to do this in **Chapter 24**.

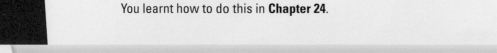

2. Before anything else is done, the entire floor is to be tiled. There are two sizes of tile available.

305 mm x 305 mm sold in packs of 6 at £6 per pack
330 mm x 330 mm sold in packs of 9 at £10 per pack

Tiles can be cut without waste, but for appearance's sake only two pieces can be cut from any one tile. Decide how this is to be done and work out a costing.

3. They are also going to tile behind the sink unit. The area of this tiling is 1500 mm x 600 mm Wall tiles come in various sizes, typically 10 cm x 10 cm and 15 cm x 15 cm. The 10 cm square tiles are sold in boxes of 25 and the 15 cm square tiles in boxes of 44.

Cost:
10 cm tiles
White £3.99 a box
Coloured £4.99 a box.
15 cm tiles
White £8.99 a box
Coloured £10.99 a box
Individual tiles of either size or colour can be bought for 50p.

Using a design that has white and coloured tiles in the ratio 3 : 2, work out how many tiles are needed and do a costing.

Floor & Wall tiles

Milano

Floor tiles 1
Size 305 x 305 mm
Pack quantity 6

White £6.00

Floor tiles 2
Size 330 x 330 mm
Pack quantity 9

White £10.00

Roma

Wall tiles
Size 10 x 10 cm
Pack quantity 25

Optional colours

White £3.99
Color £4.99

Lisbon

Wall tiles
Size 15x15cm
Pack quantity 44

Optional colours

White £8.99
Color £10.99

32

UNIVERSITY

Moving away to university is the first time many teenagers leave home. The majority of university students rent houses with small groups of friends.

QUESTION

1. Elaine and 4 friends decide to pay £1122 a month to rent a five-bedroom house near campus. Some rooms are bigger than others so the rent for each room is calculated as a proportion of the floor space of the house. How much can they each expect to pay?

Elaine's bedroom

Kitchen and living room

Ryan's bedroom

Ground floor

Bathroom

Saria's bedroom

Rashid's bedroom

Danielle's bedroom

1st floor

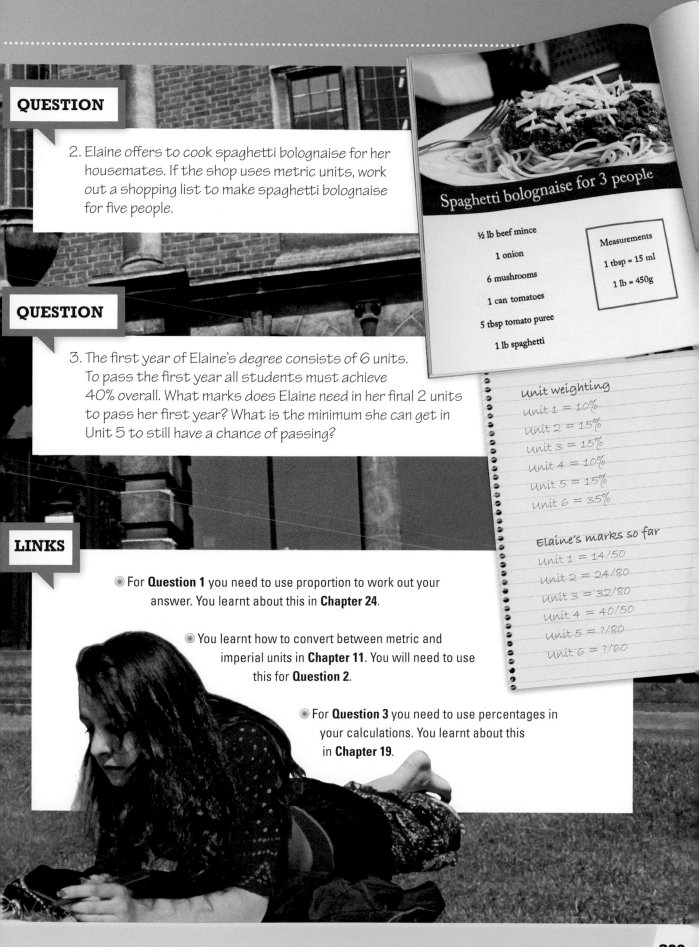

QUESTION

2. Elaine offers to cook spaghetti bolognaise for her housemates. If the shop uses metric units, work out a shopping list to make spaghetti bolognaise for five people.

Spaghetti bolognaise for 3 people

½ lb beef mince

1 onion

6 mushrooms

1 can tomatoes

5 tbsp tomato puree

1 lb spaghetti

Measurements

1 tbsp = 15 ml

1 lb = 450g

QUESTION

3. The first year of Elaine's degree consists of 6 units. To pass the first year all students must achieve 40% overall. What marks does Elaine need in her final 2 units to pass her first year? What is the minimum she can get in Unit 5 to still have a chance of passing?

Unit weighting

Unit 1 = 10%

Unit 2 = 15%

Unit 3 = 15%

Unit 4 = 10%

Unit 5 = 15%

Unit 6 = 35%

Elaine's marks so far

Unit 1 = 14/50

Unit 2 = 24/80

Unit 3 = 32/80

Unit 4 = 40/50

Unit 5 = ?/80

Unit 6 = ?/60

LINKS

For **Question 1** you need to use proportion to work out your answer. You learnt about this in **Chapter 24**.

You learnt how to convert between metric and imperial units in **Chapter 11**. You will need to use this for **Question 2**.

For **Question 3** you need to use percentages in your calculations. You learnt about this in **Chapter 19**.

15 GRAPHS 1

Marathon runners often use a graph which shows how their pace varied throughout a race. Looking at a graph is the easiest way to see their running pattern quickly and easily and help them plan the next one.

Objectives

In this chapter you will:
- write down and plot the coordinates of a point in any of the four quadrants
- find the coordinates of the midpoint of a line
- write down the equations of vertical and horizontal lines and draw them
- draw straight-line graphs of the form $y = mx + c$ and $x + y = k$, with or without using a table of values
- give the equation of any straight line drawn on a coordinate grid.

Before you start

You need to know:
- how to find points along a number line
- how to use maps and plans to find a position.

15.1 Coordinates of points in the first quadrant

◉ Objectives

◉ You can write down the coordinates of a point
in the first quadrant.

◉ You can plot the coordinates of a point in the
first quadrant.

◈ Why do this?

GPS systems use coordinates to find your position,
for example so that the emergency services can
find you.

◈ Get Ready

Tall Tree Island
On this map if you start at O and go 1 square to the right and
4 squares up you get to the Lookout.

1. Describe how you get from
   a the Lookout to the Tall Tree
   b the Tall Tree to the Beach
   c the Water Hole to the Beach.

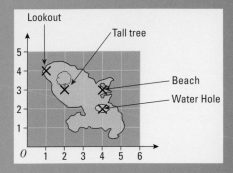

◉ Key Points

◉ The position of a point on a coordinate grid is described as two numbers.

◉ The number of units across (the **x-coordinate**) is written first, and the number of units up (the **y-coordinate**)
is written second (x, y is in alphabetical order).

◉ Given **coordinates** can be plotted on a grid.

◉ Example 1

Here is the plan of part of a zoo.
It is drawn on a coordinate grid.
Write down the names of the animals
at the following coordinates.

a (1, 2)   **b** (4, 3)
c (6, 7)   **d** (0, 5)

a lions   **b** elephants
c penguins   **d** gorillas

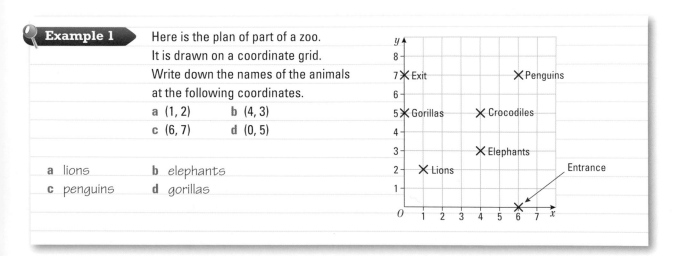

Example 2 On squared paper, draw a coordinate grid and number it to 10 across the page and up the page. Join these points up in the order given.

(2, 4)　(4, 6)　(7, 4)　(4, 2)　What shape have you drawn?

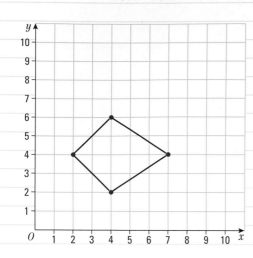

The shape is a kite.

Exercise 15A

Questions in this chapter are targeted at the grades indicated.

G

1　Here is the plan of part of a theme park. It is drawn on a coordinate grid.

　a　Write down the names of the attractions at the following coordinates.

　　i　(1, 2)　　　ii　(4, 3)　　　iii　(6, 7)

　　iv　(0, 6)　　　v　(3, 6)

　b　Write down the coordinates where you will find the:

　　i　entrance　　ii　big dipper　　iii　exit

　　iv　ghost ride　　v　swinging boats.

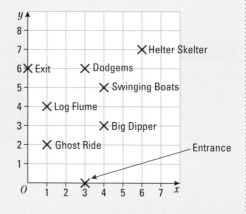

2　Here is a coordinate grid.

　a　Write down the letter of the point with the following coordinates.

　　i　(1, 2)　　　ii　(3, 4)　　　iii　(5, 7)

　　iv　(0, 4)　　　v　(0, 0)

　b　Write down the coordinates of the following points.

　　i　O　　　　ii　C　　　　iii　D

　　iv　G　　　　v　E

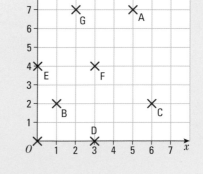

3　Here is a coordinate grid.

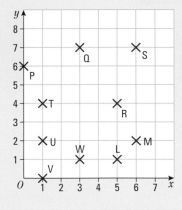

a　Write down the letter of the point with coordinates

　　i　(1, 2)　　　ii　(1, 4)　　　iii　(6, 7)
　　iv　(0, 6)　　　v　(1, 0)

b　Write down the coordinates of the points

　　i　W　　　　ii　M　　　　iii　R
　　iv　L　　　　v　Q

4　On squared paper, draw a coordinate grid and number it from 0 to 10 across the page and 0 to 10 up the page. Join these points up in the order given.

(5, 3)　(1, 3)　(2, 1)　(7, 1)　(9, 3)　(5, 3)　(5, 9)　(9, 4)　(0, 4)　(5, 9)

5　On squared paper, draw a coordinate grid and number it from 0 to 10 across the page and 0 to 10 up the page. Join each of these sets of points in the order given.

a　(1, 1)　(4, 1)　(4, 4)　(1, 4)　(1, 1)　　　　b　(6, 1)　(9, 1)　(9, 6)　(6, 6)　(6, 1)
c　(0, 6)　(5, 6)　(0, 10)　(0, 6)　　　　　　　　d　(5, 7)　(5, 10)　(8, 10)　(5, 7)

6　On squared paper, draw a coordinate grid and number it from 0 to 8 across the page and 0 to 8 up the page.
a　Plot the points P at (1, 2), Q at (7, 2) and R at (7, 5).
b　Mark the position of point S so that PQRS is a rectangle.
c　Write down the coordinates of point S.

7　On squared paper, draw a coordinate grid and number it from 0 to 8 across the page and 0 to 8 up the page.
a　Plot the points A at (1, 2), B at (6, 2) and C at (8, 5).
b　Mark the position of point D so that ABCD is a parallelogram.
c　Write down the coordinates of point D.

15.2 Coordinates of points in all four quadrants

◎ Objectives

- ◉ You can write down the coordinate of a point in any of the four quadrants.
- ◉ You can plot the coordinate of a point in any of the four quadrants.

◈ Get Ready

On squared paper, draw a coordinate grid and number it from 0 to 4 across the page and 0 to 4 up the page.
1.　Plot the points A at (1, 3), B at (3, 3), C at (3, 1) and D at (1, 1).
2.　Write down the coordinates of the point at the centre of the shape you have drawn.

Key Points

- The horizontal **axis** is called the **x-axis**.
- The vertical axis is called the **y-axis**.
- A coordinate grid is divided into four regions (**quadrants**) by the x- and y-axes.
- The point O is called the **origin** and has coordinates (0, 0).

Example 3 Write down the coordinates of the points P, Q, R and S.

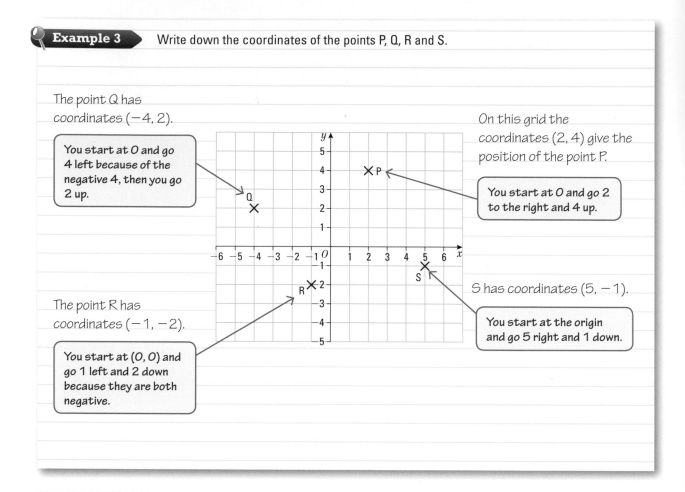

The point Q has coordinates (−4, 2).

You start at O and go 4 left because of the negative 4, then you go 2 up.

On this grid the coordinates (2, 4) give the position of the point P.

You start at O and go 2 to the right and 4 up.

The point R has coordinates (−1, −2).

You start at (0, 0) and go 1 left and 2 down because they are both negative.

S has coordinates (5, −1).

You start at the origin and go 5 right and 1 down.

Exercise 15B

F

1 Write down the coordinates of all the points A to L marked on the coordinate grid.

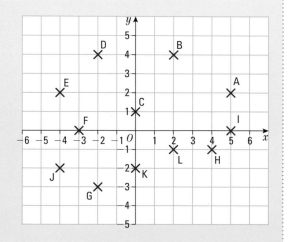

2 Draw a coordinate grid similar to the one above with the x-axis marked from -6 to $+6$ and the y-axis marked from -4 to $+5$.

Plot the following points and join them up in the order given.

$(-6, 1)$	$(-4, -2)$	$(4, -2)$	$(5, 1)$	$(-6, 1)$	$(-2, 1)$	$(0, 4)$
$(4, 4)$	$(4, 1)$	$(5, 1)$	$(5, 4)$	$(6, 4)$	$(6, 3)$	$(5, 3)$

3 On a coordinate grid make your own picture and write down the coordinates of all the points in your picture.

F

A02

15.3 Finding the midpoint of a line segment

Objectives

- You can find the coordinates of the midpoint of a line in the first quadrant.
- You can find the coordinates of the midpoint of a line in all four quadrants.

Why do this?

You might need to find the midpoint if you and a friend have agreed to meet halfway between your house and theirs.

Get Ready

The number 5 is halfway between 0 and 10.
The number 15 is halfway between 10 and 20.
To find the halfway point add the two numbers and divide by 2.

1. Find the number halfway between:

a 20 and 30   **b** 30 and 40   **c** 4 and 6   **d** 5 and 7
e -2 and 4   **f** -2 and 6   **g** 4 and -2 **h** -2 and 5.

Key Points

- The **midpoint** of a line is halfway along the line.
- To find the midpoint you add the x-coordinates and divide by 2 and add the y-coordinates and divide by 2.

Example 4 Work out the coordinates of the midpoint of the line segment PQ where P is (2, 3) and Q is (7, 11).

x-coordinate $2 + 7 = 9$ ← Add the x-coordinates and divide by 2.
$9 \div 2 = 4\frac{1}{2}$

y-coordinate $3 + 11 = 14$ ← Add the y-coordinates and divide by 2.
$14 \div 2 = 7$

Midpoint is $(4\frac{1}{2}, 7)$.

Example 5 Work out the midpoint of the line AB.

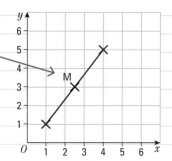

M is the midpoint of the line AB.

You can find this by measuring or by looking at the coordinates.
A has coordinates (1, 1)
B has coordinates (4, 5)

x-coordinate $1 + 4 = 5$ $5 \div 2 = 2.5$ ← Add the x-coordinates and divide by 2.

y-coordinate $1 + 5 = 6$ $6 \div 2 = 3$ ← Add the y-coordinates and divide by 2.

M has coordinates $(2.5, 3)$ or $(2\frac{1}{2}, 3)$.

Exercise 15C

1 Work out the coordinates of the midpoint of each of the line segments shown on the grid.

  a OA   **b** BC   **c** DE
  d FG   **e** HJ

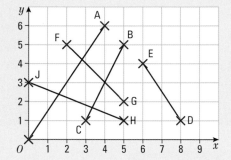

2 Work out the coordinates of the midpoint of each of these line segments.

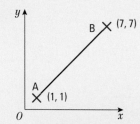

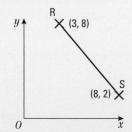

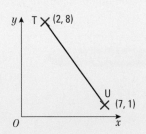

3 Work out the coordinates of the midpoint of each of these line segments.

  a AB when A is (1, 1) and B is (9, 9)   **b** PQ when P is (2, 4) and Q is (6, 9)
  c ST when S is (5, 8) and T is (2, 1)   **d** CD when C is (1, 7) and D is (7, 2)
  e UV when U is (2, 3) and V is (6, 8)   **f** GH when G is (2, 6) and H is (7, 3)

Example 6   Find the midpoint of RS.

M is the midpoint of the line RS.

You can find this by measuring or by looking at the coordinates.
R has coordinates $(-3, 1)$
S has coordinates $(3, 4)$

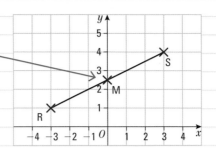

x-coordinate $-3 + 3 = 0$ $0 \div 2 = 0$ ← Add the x-coordinates and divide by 2.

y-coordinate $1 + 4 = 5$ $5 \div 2 = 2.5$ ← Add the y-coordinates and divide by 2.

M has coordinates $(0, 2.5)$ or $(0, 2\frac{1}{2})$.

Exercise 15D

1   Work out the coordinates of the midpoint of
each of the line segments shown on the grid.

a OA	**b** BC	**c** DE
d FG	**e** HJ	**f** KL
g MN	**h** PQ	**i** ST
j UV		

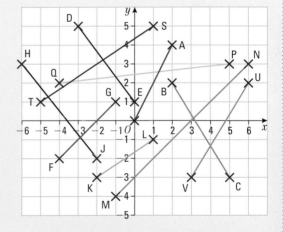

2   Work out the coordinates of the midpoint of each of these line segments.

a    **b** 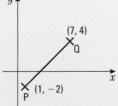   **c** 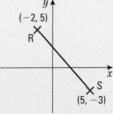   **d**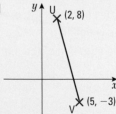

3   Work out the coordinates of the midpoint of each of these line segments.

a   AB when A is $(-1, -1)$ and B is $(9, 9)$ **b**   PQ when P is $(2, -4)$ and Q is $(-6, 9)$

c   ST when S is $(5, -8)$ and T is $(-2, 1)$ **d**   CD when C is $(1, 7)$ and D is $(-7, 2)$

e   UV when U is $(-2, 3)$ and V is $(6, -8)$ **f**   GH when G is $(-2, -6)$ and H is $(7, 3)$

15.4 Drawing and naming horizontal and vertical lines

⊙ Objectives

- You can write down the equations of vertical and horizontal lines.
- You can draw the vertical and horizontal lines with equations of the form $x = n$ and $y = m$.

? Why do this?

This would allow you to plot a route as you would be able to describe exactly where you were, where you were going and which route you would take to get there.

⬆ Get Ready

1. Make a copy of the coordinate grid and plot the following sets of points: (2, −3), (2, −2), (2, −1), (2, 0), (2, 1), (2, 2), (2, 3), (2, 4). Join the points up with a straight line.
2. On the same grid, plot these points: (−3, 3), (−2, 3), (−1, 3), (0, 3), (1, 3), (2, 3), (3, 3), (4, 3). Join the points up with a straight line.
3. Write down the coordinates of the point where the two lines meet.

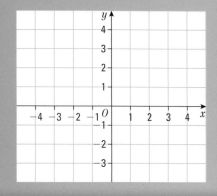

🔍 Key Points

- A vertical line on the grid has the form $x = n$. For example, on this graph all the x-coordinates are 2 so the line is called $x = 2$. The equation of the line is $x = 2$.

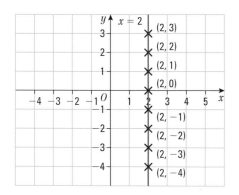

- A horizontal line on the grid has the form $y = m$. In this graph all the y-coordinates are −2 so the line is called $y = -2$. The equation of the line is $y = -2$.

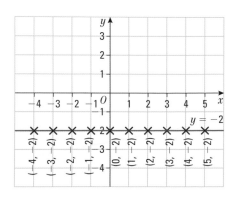

Exercise 15E

1 Write down the equations of the lines marked **a** to **d** in this diagram.

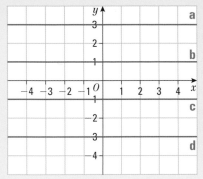

Examiner's Tip

The x-axis has equation $y = 0$.
The y-axis has equation $x = 0$.

2 Write down the equations of the lines labelled **a** to **d** in this diagram.

3 Draw a coordinate grid with x- and y-axes labelled from -5 to 5. On the grid draw and label the graphs of
  a $x = 4$ **b** $x = -2$ **c** $x = -4$ **d** $x = 1$

4 Draw a coordinate grid with axes labelled from -5 to 5. On the grid draw and label the graphs of
  a $y = 4$ **b** $y = -2$ **c** $y = -4$ **d** $y = 1$

5 Draw a coordinate grid with axes labelled from -5 to 5. On the grid draw and label the graphs of
  a $y = 3$ **b** $x = -1$
  c Write down the coordinates of the point where the two lines cross.

15.5 Drawing slanting lines

Objectives

- You can use a table of values to draw straight-line graphs with positive gradients.
- You can use a table of values to draw straight-line graphs with negative gradients.

Why do this?

In science, you might be asked to plot the results of your experiments on a graph. The graph might be a straight line.

Get Ready

Draw these lines.

1. $y = 2$ **2.** $x = -4$ **3.** $y = \frac{1}{2}$ **4.** $y = -3$
                                                                $x = 3$

Key Points

◉ Lines that slant upwards   have a positive **gradient**.

◉ Lines that slant downwards   have a negative gradient.

◉ To draw a straight-line graph with a given equation:
  ◉ make a table of values, selecting some values for x
  ◉ substitute the values of x into the equation
  ◉ plot the points from the table of values on the grid
  ◉ draw in the line.

Example 7   Draw the graph of $y = 2x - 1$.

1. Make a table of values, selecting some values for x.

x	−1	0	1	2	3
$y = 2x - 1$					

2. Substitute the values of x into $y = 2x - 1$.

x	−1	0	1	2	3
$y = 2x - 1$	−3	−1	1	3	5

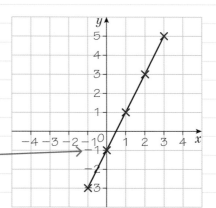

3. Plot the points on the grid.

4. Draw in the line.

ResultsPlus
Watch Out!

Don't forget to draw in the line.

Exercise 15F

D

1 **a** Copy and complete the tables of values for the straight-line graphs below.
  b On a coordinate grid with the x-axis drawn from -3 to $+3$ and y-axis drawn from -10 to $+10$, draw the graphs of $y = x - 1$, $y = 2x - 4$, $y = 3x + 1$, $y = x + 4$ and $y = 4x + 1$.

i

x	−3	−2	−1	0	1	2	3
$y = x - 1$		−3		−1		1	

ii

x	−3	−2	−1	0	1	2	3
$y = 2x - 4$		−8		−4		0	

iii

x	−3	−2	−1	0	1	2	3
$y = 3x + 1$	−8			1			10

iv

x	−3	−2	−1	0	1	2	3
$y = x + 4$		2		4			7

v

x	−2	−1	0	1	2
$y = 4x + 1$	−7				9

2 Draw the graphs of these straight lines on a coordinate grid with axes drawn from −3 to +3 on the x-axis and −10 to +10 on the y-axis.

   a $y = 2x + 1$ **b** $y = 2x + 3$ **c** $y = 2x - 3$ **d** $y = 2x + 2$

3 Draw the graphs of these straight lines on a coordinate grid with axes drawn from −3 to +3 on the x-axis and −6 to +6 on the y-axis.

   a $y = x + 1$ **b** $y = x + 3$ **c** $y = x - 3$ **d** $y = x + 2$

4 Draw the graphs of these straight lines on a coordinate grid with axes drawn from −4 to +4 on the x-axis and −10 to +10 on the y-axis.

   a $y = 2x - 1$ **b** $y = 3x + 2$ **c** $y = 2x - 2$ **d** $y = 3x - 2$

5 Draw the graphs of these straight lines on a coordinate grid with axes drawn from −4 to +4 on the x-axis and −6 to +6 on the y-axis.

   a $y = \frac{1}{2}x + 1$ **b** $y = \frac{1}{2}x + 3$ **c** $y = \frac{1}{2}x - 3$ **d** $y = \frac{1}{2}x - 2$

Example 8   Draw the graph of $y = -2x - 1$.

1. Make a table of values, selecting some values for x.

x	−3	−2	−1	0	1
$y = -2x - 1$					

2. Substitute the values of x into $y = -2x - 1$.

x	−3	−2	−1	0	1
$y = -2x - 1$	5	3	1	−1	−3

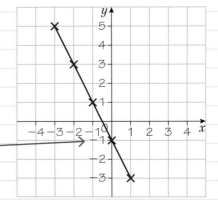

3. Plot the points on the grid.

4. Draw in the line.

Exercise 15G

D

1 **a** Copy and complete the tables of values for the straight-line graphs below.
   b Draw the graphs of these straight lines on a coordinate grid with the x-axis drawn from -3 to $+3$ and y-axis drawn from -10 to $+10$.

i

x	-3	-2	-1	0	1	2	3
$y = -x - 1$		1		-1		-3	

ii

x	-3	-2	-1	0	1	2	3
$y = -2x - 4$		0		-4		-8	

iii

x	-3	-2	-1	0	1	2	3
$y = -3x + 1$	10			1			-8

iv

x	-3	-2	-1	0	1	2	3
$y = -x + 4$		6		4			1

v

x	-2	-1	0	1	2
$y = -4x + 1$	9				-7

2 Draw the graphs of these straight lines on a coordinate grid with axes drawn from -3 to $+3$ on the x-axis and -10 to $+10$ on the y-axis.

   a $y = -2x + 1$ **b** $y = -2x + 3$

   c $y = -2x - 3$ **d** $y = -2x + 2$

3 Draw the graphs of these straight lines on a coordinate grid with axes drawn from -3 to $+3$ on the x-axis and -6 to $+6$ on the y-axis.

   a $y = -x + 1$ **b** $y = -x + 3$

   c $y = -x - 3$ **d** $y = -x + 2$

4 Draw the graphs of these straight lines on a coordinate grid with axes drawn from -4 to $+4$ on the x-axis and -10 to $+10$ on the y-axis.

   a $y = -2x - 1$ **b** $y = -3x + 1$

   c $y = -2x - 2$ **d** $y = -3x - 2$

5 Draw the graphs of these straight lines on a coordinate grid with axes drawn from -4 to $+4$ on the x-axis and -6 to $+6$ on the y-axis.

   a $y = -\frac{1}{2}x + 1$ **b** $y = -\frac{1}{2}x + 3$

   c $y = -\frac{1}{2}x - 3$ **d** $y = -\frac{1}{2}x - 2$

15.6 Drawing straight-line graphs without a table of values

◎ Objectives

- ◉ You can draw straight-line graphs of the form $y = mx + c$ using the intercept and gradient when the gradient is positive.
- ◉ You can draw straight-line graphs of the form $y = mx + c$ using the intercept and gradient when the gradient is negative.
- ◉ You can draw straight-line graphs by using intercepts on the x- and y-axes for graphs of the type $x + y = k$.

⟨?⟩ Why do this?

This could be a good way of predicting the results of an experiment before you complete it.

⟨↑⟩ Get Ready

Use a table of values to draw these straight lines.

1. $y = x + 4$ **2.** $y = x + 5$ **3.** $y = 2x + 4$

What do you notice about the lines in questions **1** and **3**?

🔍 Key Points

- ◉ The equation of a straight-line graph can be written in the form $y = mx + c$. The number on its own (c) tells you where the straight line crosses the y-axis.

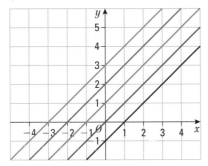

Here are five straight-line graphs.
$$y = x + 3$$
$$y = x + 2$$
$$y = x + 1$$
$$y = x + 0$$
$$y = x - 1$$

- ◉ The number (m) in front of the x tells you the gradient (steepness) of the line. If the number is positive, for each square you move to the right you move up by the number in front of the x. If the number is negative, for each square you move to the right you move down by the number in front of the x.

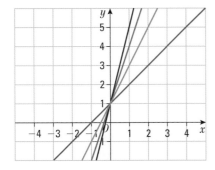

Here are four straight-line graphs.
$$y = x + 1$$
$$y = 2x + 1$$
$$y = 3x + 1$$
$$y = 4x + 1$$

- ◉ To draw a straight line from the equation $y = mx + c$:
  - ◉ mark the point (c) where the line will cross the y-axis
  - ◉ find out the gradient – how many squares you go up (or down) each square you move to the right – from the number (m) in front of the x
  - ◉ join up the points with a straight line.
- ◉ When the line is in the form $x + y = c$ the c tells you where the graph cuts the x-axis and the y-axis.

Example 9 On the grid, draw the graph of $y = 2x + 1$.

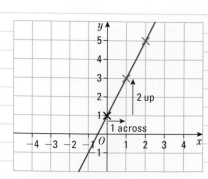

Step 1 Mark the point **X** where the line will cross the y-axis.

Step 2 For each square you move to the right you then go up 2 because there is a 2 in front of the x.

Step 3 Join up the points with a straight line.

Exercise 15H

For each question, draw the graphs of all the straight lines on the same coordinate grid with the x-axis drawn from -3 to $+3$ and y-axis drawn from -10 to $+10$. What do you notice about the set of graphs for each question?

1 **a** $y = x + 1$ **b** $y = x + 2$ **c** $y = x + 3$
   d $y = x - 1$ **e** $y = x - 2$

2 **a** $y = 2x + 1$ **b** $y = 2x + 2$ **c** $y = 2x + 3$
   d $y = 2x - 1$ **e** $y = 2x - 2$

3 **a** $y = 3x + 1$ **b** $y = 3x + 2$ **c** $y = 3x + 3$
   d $y = 3x - 1$ **e** $y = 3x - 2$

4 **a** $y = 4x + 1$ **b** $y = 4x + 2$ **c** $y = 4x + 3$
   d $y = 4x - 1$ **e** $y = 4x - 2$

5 **a** $y = \frac{1}{2}x + 1$ **b** $y = \frac{1}{2}x + 2$ **c** $y = \frac{1}{2}x + 3$
   d $y = \frac{1}{2}x - 1$ **e** $y = \frac{1}{2}x - 2$

Example 10 On the grid, draw the graph of $y = -2x + 1$.

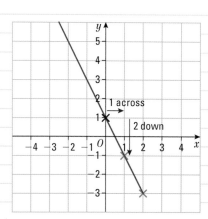

Step 1 Mark the point **X** where the line will cross the y-axis.

Step 2 For each square you move to the right you go down 2 because there is a -2 in front of the x.

Step 3 Join up the points with a straight line.

Exercise 15I

For each question, draw the graphs of all the straight lines on the same coordinate grid with the x-axis drawn from -3 to $+3$ and y-axis drawn from -10 to $+10$. What do you notice about the set of graphs for each question?

1
  a $y = -x + 1$ b $y = -x + 2$ c $y = -x + 3$
  d $y = -x - 1$ e $y = -x - 2$

2
  a $y = -2x + 1$ b $y = -2x + 2$ c $y = -2x + 3$
  d $y = -2x - 1$ e $y = -2x - 2$

3
  a $y = -3x + 1$ b $y = -3x + 2$ c $y = -3x + 3$
  d $y = -3x - 1$ e $y = -3x - 2$

4
  a $y = -4x + 1$ b $y = -4x + 2$ c $y = -4x + 3$
  d $y = -4x - 1$ e $y = -4x - 2$

5
  a $y = -\frac{1}{2}x + 1$ b $y = -\frac{1}{2}x + 2$ c $y = -\frac{1}{2}x + 3$
  d $y = -\frac{1}{2}x - 1$ e $y = -\frac{1}{2}x - 2$

D

Example 11   On the grid, draw the graph of $x + y = 4$.

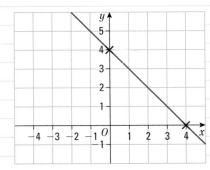

Step 1 Mark the value of the number on the x-axis and the y-axis.

Step 2 Join these points up with a straight line.

Exercise 15J

For each question, draw the graphs of all the straight lines on the same coordinate grid with the x-axis drawn from -3 to $+8$ and y-axis drawn from -3 to $+8$.

1
  a $x + y = 3$ b $x + y = 5$ c $x + y = 2$

2
  a $x + y = -2$ b $x + y = -3$ c $x + y = 1$

3
  a $x + y = -1$ b $x + y = 2.5$ c $x + y = 0$

D

15.7 Naming straight-line graphs

◎ Objective

○ You can give the equation of any straight line drawn on a coordinate grid.

⬥ Why do this?

If you have plotted a graph from values you have found in an experiment, you might need to find the relationship between them. This is what you do when you give the equation of a straight line.

⬆ Get Ready

Draw these graphs.

1. $y = -4x + 2$ **2.** $y = 2x + 5$ **3.** $x + y = 6$

🔍 Key Point

◉ When you have a straight line for which you need to find the equation, you need to work out the gradient (the m in the equation) and look to see what the **intercept** on the y-axis is (the c in the equation). You then put these values in the equation $y = mx + c$ to find the equation of the line.

🔍 Example 12

Write down the equations of these straight lines.

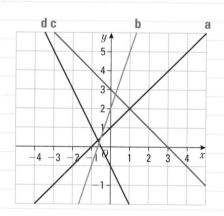

Line **a** cuts the y-axis at 1.
Gradient (slope) is 1 (it goes up 1 for every 1 across).
Equation is $y = x + 1$.

Line **b** cuts the y-axis at 2.
Gradient (slope) is 3 (it goes up 3 for every 1 across).
Equation is $y = 3x + 2$.

Line **c** cuts the x-axis at 3.
Cuts the y-axis at 3.
Equation is $x + y = 3$.

Line **d** cuts the y-axis at -1.
Gradient (slope) is -2 (it goes down 2 for every 1 across).
Equation is $y = -2x - 1$.

Exercise 15K

1 Write down the equations of these straight lines.

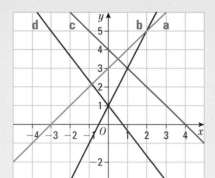

2 Write down the equations of these straight lines.

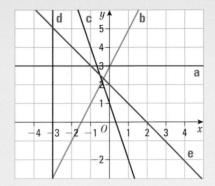

3 Write down the equations of these straight lines.

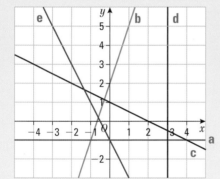

4 Write down the equations of these straight lines.

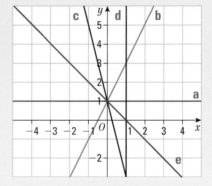

Chapter review

- The position of a point on a coordinate grid is described as two numbers.
- The number of units across (the x-**coordinate**) is written first, and the number of units up (the y-**coordinate**) is written second.
- Given **coordinates** can be plotted on a grid.
- The horizontal **axis** is called the x-**axis**.
- The vertical axis is called the y-**axis**.
- A coordinate grid is divided into four regions (**quadrants**) by the x- and y-axes.
- The point O is called the **origin** and has coordinates $(0, 0)$.
- The **midpoint** of a line is halfway along the line.
- To find the midpoint you add the x-coordinates and divide by 2 and add the y-coordinates and divide by 2.

- Horizontal lines are $y =$ lines.
- Vertical lines are $x =$ lines.

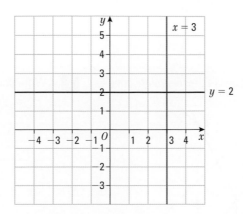

- Lines that slant upwards / have a positive **gradient**.

- Lines that slant downwards \ have a negative gradient.

- Lines that have an equation such as $y = 2x + 3$ have a gradient of 2 and cut the y-axis at (0, 3).
- To draw a straight-line graph with a given equation:
    - make a table of values, selecting some values for x
    - substitute the values of x into the equation
    - plot the points from the table of values on the grid
    - draw in the line.
- When you have a straight line for which you need to find the equation, you need to work out the gradient (the m in the equation) and look to see what the **intercept** on the y-axis is (the c in the equation). You then put these values in the equation $y = mx + c$ to find the equation of the line.

Review exercise

F

1
  a Write down the coordinates of the point A.
  b Write down the coordinates of the point B.

  N is the point $(-3, 2)$.
  c On a copy of the grid, mark the point N with a cross ($\times$).
     Label it N.

  M is another point.
  The x-coordinate of M is the same as the x-coordinate of N.
  The y-coordinate of M is the same as the y-coordinate of B.
  d Write down the coordinates of the point M.

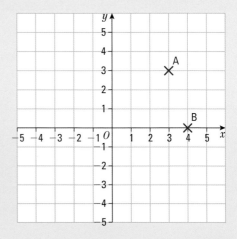

Nov 2007

2

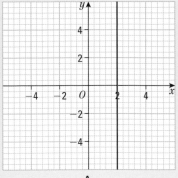

A

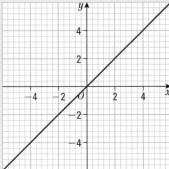

B

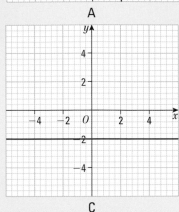

C

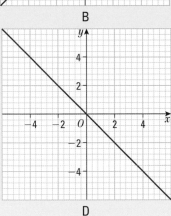

D

The diagrams show four graphs.

Here are four equations.

i $y = x$ **ii** $y = -2$ **iii** $y = -x$ **iv** $x = 2$

Match the letter of each graph with its equation.

3   **a** Copy and complete the table of values for $y + 2x = 6$

x	−1	0	1	2	3
y		6	4		

  b Draw a graph of $y + 2x = 6$ for values of x from −1 to 3.

  c Use your graph to find an estimate for the value of y when $x = 2.4$.

  d Use your graph to find an estimate for the value of x when $y = 6.6$.

4   On the coordinate grid draw the graph of $y = 2x - 3$.

Use values of x from −2 to +2.

ResultsPlus
Exam Question Report

89% of students scored poorly on this question.
The most common incorrect answer was to
plot just one point, generally (2, −3), sometimes
joining this point to the axes.

E

D

D A02

5 Draw the graph of $y = 2x - 3$ for values of x from -1 to 3. *May 2009*

A02

6 **a** Draw a graph of $2y + x = 8$ for values of x from -2 to 4.
b On the same axes draw the graph of $y = x$ for values of x from -2 to 4.
c P is the point on the graph of $2y + x = 8$ for which the value of x is the same as the value of y.
Estimate the value of x.

7 **a** Write down the coordinates of the points
    i P   **ii** Q   **iii** R
b **i** Copy the graph and join the points
      to form triangle PQR.
    ii Write down the mathematical name of
      triangle PQR.
c Write down the coordinates of the
    midpoint of
    i the side QR
    ii the side PQ
   iii the side PR.
d PQRS is a rectangle. Write down the
    coordinates of the point S.

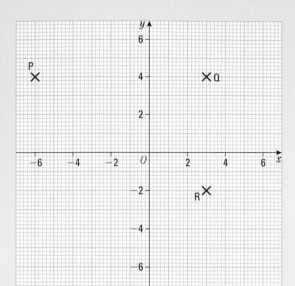

C

8 Here is a grid of centimetre squares.
a Write down the coordinates of the points
    i B   **ii** C.
ABCD is a quadrilateral.
The area of ABCD is 24 cm^2.
b Write down the coordinates of D.
E is another point on the grid.
The coordinates of the midpoint of AE are $(-1, 2)$.

A03

c Find the coordinates of E.

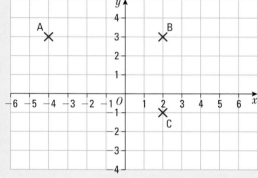

Specimen Paper 2009

9 Write down the equation of each of the lines shown in the grid.

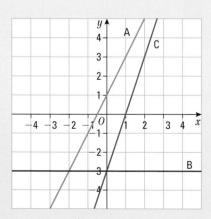

The manager of a supermarket is asked to summarise the amount, in pounds, spent by shoppers at his supermarket. He wishes to give the average amount and an indication of how the actual amounts vary from this. What figures should he give and how can he calculate them?

Objectives

In this chapter you will:
- find the mean, mode and median of a set of data and discuss the advantages and disadvantages of these three measures of average
- find the range of a set of data
- compare distributions
- use stem and leaf diagrams and frequency tables to find the mode, median and range of a set of data
- find the modal class and the interval containing the median of grouped data
- estimate the mean of grouped data.

Before you start

You need to:
- understand frequency tables
- understand grouped data and grouped frequency tables.

16.1 Finding the mode, the median and the mean

◎ Objective

◉ You can find the mode, the median and the mean of a set of data.

? Why do this?

This will help you to understand detailed weather reports where they talk about average rainfall and hours of sunlight.

◈ Get Ready

1. Write these numbers in order of size, smallest first.
   a 325   284   336   296   302
   b 0.6   0.59   0.55   0.625   0.61

2. The frequency table shows the number of times balls of different colours are picked from a bag.

Colour	Red	Yellow	Blue	Green
Frequency	6	10	8	14

   a How many times was a ball picked out?
   b Which ball was picked most?
   c How many more times was a blue ball picked than a red one?

🔍 Key Points

◉ We often describe a numerical data set by giving a single value that is representative of all the values in the set. We call this value an **average**. For example, we might say that 'The average height of members of a certain basketball team is 190 cm.'
◉ Three different averages are commonly used: the **mode**, the **median** and the **mean**.
◉ We may also be interested in how values are spread out. A common measure that is used is the **range**.

Finding the mode

🔍 Key Point

◉ The mode of a set of data is the value that occurs most frequently.

🔍 Example 1

Find the mode of each of the following sets of data.

a 3   4   6   7   8   7   5
b 13   27   26   32   42   27   8   10   13
c 7   6   9   3   5   8   2
d Black   Green   Blue   Black   Red

a Mode = 7   ← 7 occurs twice, the other numbers appear only once.

b Mode = 13 and 27 ← Both 13 and 27 occur twice. There are two modes.

c There is no mode. ← All numbers occur with the same frequency.

d The mode is black. ← Black appears twice. Each other colour only appears once.

average   mode   median   mean   range

Exercise 16A

Questions in this chapter are targeted at the grades indicated.

F

1. Find the mode for each of the following data sets.
   a 2   7   8   3   2   9   4   2
   b 10   12   14   10   15   12   11   18
   c 3   6   8   7   4   9   1   2
   d dog   cat   rabbit   dog   fish   mouse   dog

2. Briony did a spelling test every week for ten weeks. Her scores are given below.
   11   16   18   16   12   18   17   16   12   16
   Find the mode for these data.

3. Bo made a list of the colours of all the cars in a showroom. His list is shown below.
   blue   silver   white   red   red   black   silver   silver   green   silver   black
   Write down the mode for these data.

4. The hourly rates of pay for eight workers are listed below.
   £8.20   £8.50   £7.20   £8.80   £9.20   £7.20   £8.20   £9.90
   Write down the mode for these data.

Finding the median

Key Points

◉ The median is the middle value when the data are ordered from the smallest to the largest.

◉ If there are two middle values in a set of data, the median is halfway between them.

Example 2   Eleven people applied for a job in an office. Their typing speeds, in words per minute, were
36, 42, 58, 47, 43, 42, 38, 48, 52, 46 and 46.
Find the median typing speed.

> Arrange the numbers in order, smallest to largest.

36   38   42   42   43   **46**   46   47   48   52   58
The median is 46.

> The middle value is 46. There are 5 speeds lower than this and 5 speeds higher.

Example 3   The midday temperatures, in degrees Celsius, at a seaside town were
18, 21, 18, 16, 25, 18, 17, 28, 22 and 19.
Find the median temperature.

> Put the numbers in order first.

16   17   18   18   **18**   **19**   21   22   25   28
The median is 18.5°C

> There are two middle values: 18°C and 19°C.

> Halfway between 18 and 19 is 18.5.

Exercise 16B

1. Find the median for the following data sets.
   a 2   3   3   4   5   6   7   8   8
   b 3   5   6   6   8   8   9   10
   c 5   5   6   7   7   8
   d 4   2   8   6   7   9   4   3   2
   e 2   5   6   3   7   8   1   9

E

2 The minimum night-time temperatures (in °C), in February, in seven towns are recorded below.

1.0   1.2   1.4   1.2   1.3   1.2   1.5

Find the median temperature.

3 The CO_2 emissions (in thousands of tonnes) for mopeds in Scotland over eight years are shown below.

38   38   39   41   44   40   41   39

Find the median CO_2 emissions.

A03

4 The annual earnings of six managers in a factory are shown below.

£20 500   £25 700   £30 900   £27 500   £22 000   £26 500

a Work out the median earnings.

b The factory wishes to employ another manager. Recommend a salary range for them to use in an advert.

Calculating the mean

Key Point

⦿ The mean of a set of data is the sum of the values divided by the total number of observations.

$$\text{mean} = \frac{\text{sum of the values}}{\text{number of values}}$$

⦿ A short way of writing the mean uses the Greek letter sigma Σ to represent the sum of a set of values.

⦿ The mean can be worked out using the statistical functions on a scientific calculator. The method for inputting the data will vary between different scientific calculators, and the instruction leaflet will explain how to enter and process the statistical data.

Example 4   Find the mean of the numbers 2, 4, 6, 6, 8, 9, 8 and 5.

$2 + 4 + 6 + 6 + 8 + 9 + 8 + 5 = 48$ ← Total up the values.

$\text{mean} = \frac{48}{8} = 6$ ← Divide by the number of values to get the mean.

Example 5   The weekly consumptions of electricity, in units, during an 8-week period were 340, 350, 340, 355, 340, 345, 340 and 450 units.

a Find the mode.   b Find the median.   c Find the mean.

a The mode is 340 units ← 340 occurs four times, each of the other values occurs once.

b 340   340   340   **340**   **345**   350   355   450

There are two middle values: 340 and 345.

The median is 342.5 units.

Halfway between is 342.5.

c The total of all values = 340 + 340 + 340 + 340 + 345 + 350 + 355 + 450 = 2860 ← Find the total sum.

$\text{mean} = \frac{2860}{8} = 357.5$ units ← Divide by the number values.

Watch Out!

Some students use the term average – make sure you specify mean, mode or median.

⚙ Exercise 16C

1 Find the mean of the following data sets.

a 3   5   5   6   8   9

b 2   5   4   7   5   3   8   6

c 7   5   8   9   9   7

2 The number of school meals bought by 10 children in one week were

2   4   3   5   5   4   2   5   5   5

a Work out the mean number of meals.

The school has 250 children.

b Work out an estimate for the number of school meals they should provide each day.

3 The heights, in centimetres, of 11 members of a cricket team were

172   174   190   190   185   186   182   189   190   185   192

a Find the mode.

b Work out the median height.

c Work out the mean height.

4 The numbers of emails received by a sample of six people during one week were

44   72   107   155   214   197

a Find the mode.

b Work out the median number of emails.

c Work out the mean number of emails.

E

A03

16.2 Knowing the advantages and disadvantages of the three types of average

◎ Objective

○ You can discuss the advantages and disadvantages of different measures of average.

⟐ Why do this?

You need to be able to determine which type of average to use. For example, if you needed to find out the most popular football team, you would use the mode.

◈ Get Ready

1. For the following data set find

82   54   65   36   48   48   79

a The mode

b The median

c The mean

Key Points

● The three types of averages are mode, median and mean.

Each of the averages is useful in different situations.

The table shows which measure of average to use.

Measure	Advantages	Disadvantages
MODE Use the mode when the data is non-numeric or when asked to choose the most popular item.	Extreme values (outliers) do not affect the mode. Can be used with categorical data.	There may be more than one mode. There may not be a mode, particularly if the data set is small.
MEDIAN Use the median to describe the middle of a set of data that does have an extreme value.	Not influenced by extreme values.	Not as popular as mean. Actual value may not exist.
MEAN Use the mean to describe the middle of a set of data that *does not* have an extreme value.	Most popular measure. Can be used for further calculations. Uses all the data.	Affected by extreme values.

Example 6

Shuabur recorded the number of spam emails he received every day for seven days.
Here are his results:

3   7   18   21   24   24   29

a Find the mode, median and mean of these numbers.

b Comment on the results.

a Mode = 24 ← There are two 24 values.

Median = 21 ← There are three values greater than 21 and three lower.

Total = 3 + 7 + 18 + 21 + 24 + 24 + 29 = 126 ← Sum the values and divide by the number of values.

Mean = $\frac{126}{7}$ = 18

b The mode of 24 has only one higher value.
This is because the number in the sample is small.
The mean has only two values that are less;
the two low values 3 and 7 have affected it.
The median is unaffected by the two low values.
The median is the best average to represent these data.

Comment on how well the averages represent the numbers as a whole, and give reasons why they might not be representative.

Exercise 16D

1. Jenny decides that she will use the mean to represent the following prices of jumpers.

   £4   £5   £3   £7   £5   £42

   Write down a reason why the mean is not a good average to use.

   A02 A03

2. The prices of cars in a showroom were

   £5750   £5750   £7750   £7550   £7950   £10 750   £11 550   £14 700

   The garage manager puts up a poster saying 'AVERAGE PRICE £5750'.

   a Write down the name of the average he is using.

   b Is this a fair average to use? Give a reason for your answer.

   A02 A03

3. Andy records the number of pieces of junk mail he receives each day for a week.

   2   8   2   7   6   17   0

   a Find the mode.

   b Work out the median.

   c Work out the mean.

   d Comment on the values of these three averages.

* 4. The following are the times, in minutes, it takes some people to travel to work.

   10   10   13   16   17   13   40

   Which average is the best to represent these data?

   Give a reason for your choice.

16.3 Finding the range

Objectives

- You can calculate the range of a set of data.
- You can compare data sets using a measure of average and a measure of range.

Why do this?

If you were trying to negotiate a better rate of pay you might want to find the range of wages that your friends who had similar jobs earned, so that you could make sure you were getting a fair amount.

Get Ready

1. Write down the highest and lowest values in the following data sets.

   a 2   7   4   3   5   13

   b 22   25   26   19   24   29

   c 150   161   152   130   145   156

Key Points

- The range of a set of data is the difference between the highest value and the lowest value. The range tells you how spread out the data are.

  range = highest value − lowest value

- The average and the range together give a description of the frequency distribution of the data.

- To compare the distributions of sets of data you need to give a measure of average and a measure of **spread**.

Example 7

Here are the times, in minutes, that Fiona took going to school in the morning and returning after school in the afternoon.

Going to school: 20, 18, 21, 17, 25, 28, 22, 23
Going home: 14, 17, 23, 28, 24, 12, 24, 32

a Work out the range for each type of journey.
b Which journey time was the most consistent? Give a reason for your answer.

a Range going to school = 28 − 17 = 11 minutes. ← | Take the smallest value from the largest. |
Range going home = 32 − 12 = 20 minutes.

b The time taken to go to school was the most consistent.
The range was smaller. ← | Don't forget to give a reason. |

Example 8

Samples were taken from two machines that filled jars of coffee. The weights of coffee, in grams, were

Machine 1: 187, 192, 195, 198, 200, 200, 203, 205, 210, 210
Machine 2: 193, 194, 196, 199, 200, 200, 202, 204, 205, 207

a Find the mean and range of the data for each machine.
b Comment on these results.

a Machine 1

$$\text{Mean} = \frac{187 + 192 + 195 + 198 + 200 + 200 + 203 + 205 + 210 + 210}{10} = 200\,g$$

Range = 210 − 187 = 23 g

| The mean = $\dfrac{\text{total of values}}{\text{total frequency}}$ |

| Take the smallest value from the largest value. |

Machine 2

$$\text{Mean} = \frac{193 + 194 + 196 + 199 + 200 + 200 + 202 + 204 + 205 + 207}{10} = 200\,g$$

Range = 207 − 193 = 14 g

b Both machines filled the jars with the same mean amount of coffee.
Machine 2 had a smaller range. The weights of coffee were less spread out which shows the machine was more consistent.

Exercise 16E

1 Find the range for each of the following sets of data.

a   2   5   7   8   9   20   21

b   3   7   3   5   7   14   5   13

c   112   115   118   117   118   113

2 The scores in a game were

12   14   11   17   23   25   22

Petra says the range is 10. Is she correct?

You must explain your answer.

3 The table gives the exam marks in economics and psychology for a group of students taught by two different teachers.

Student	A	B	C	D	E	F	G	H	I
Economics	72	70	63	87	83	56	88	44	65
Psychology	55	65	57	68	70	55	59	60	62

a   Work out the range for each subject.

b   The headteacher wants to know which subject had the most consistent marks and also whether students are being stretched. Give a reason for your answer.

4 Samples were taken from two machines filling bottles of mineral water.

The amounts of water (ml) in the bottles were

Machine 1:   30   29   30   30   29   32

Machine 2:   30   29   29   34   30   28

a   Find the range for each machine.

b   Find the mean for each machine.

c   Comment on your answers to parts a and b.

*5 The heights of two teams of footballers (cm) were

Max Rangers: 170   172   180   190   184   179   176   183   186   190   170

Red United: 179   190   187   170   180   182   163   188   181   190   179

Calculate the mean and range for each team and compare and contrast the frequency distributions of the heights of the two teams.

16.4 Using stem and leaf diagrams to find averages and range

◎ Objectives

◉ You can use a stem and leaf diagram to find the mode and median of a set of data.

◉ You can use a stem and leaf diagram to find the range.

⍰ Why do this?

A stem and leaf diagram is a good way to put data into order.

◈ Get Ready

Write these numbers in order of size, from smallest to largest.

1. 35   42   26   58   **2.** 152   151   154   153   **3.** 0.5   0.2   0.1   0.4   **4.** 0.1   0.11   0.12   0.9

Key Point

● A **stem and leaf diagram** makes it easy to find the mode, the median and the range of a set of data.

Example 9   Here are the times, in minutes, taken by 12 people to complete a crossword puzzle.

35, 48, 42, 35, 38, 56, 34, 28, 52, 18, 43, 27

a   Write these data as a stem and leaf diagram.

b   Write down the mode of these data.

c   Find the median of these data.

d   Work out the range of these data.

> First write down the numbers whose tens digit is 1.
> Then write down those numbers whose tens digits are 2, 3, 4 and finally 5.
> The digit that each number begins with is called the stem.
> The following digit is called the leaf.

a   18 28 27 35 35 38 34 48 42 43 56 52

Stem	Leaf
1	8
2	8 7
3	5 5 8 4
4	8 2 3
5	6 2

> Under stem write the numbers 1 to 5. Opposite each stem write the leaves. Don't worry about the order. This gives you an unordered stem and leaf diagram.

Key: 1|8 stands for 18 minutes

Stem	Leaf
1	8
2	7 8
3	4 5 5 8
4	2 3 8
5	2 6

> Order the leaves to give an ordered stem and leaf diagram.

b   The mode is 35 minutes.

> 35 appears twice, the other values only once.

c   The median is $\dfrac{35 + 38}{2} = 36.5$ minutes.

> You can find the middle numbers by counting in from each end.

d   The range is $56 - 18 = 38$ minutes.

> The largest and smallest values are the first leaf and the last leaf. The range is the difference between them.

Exercise 16F

D

1   Here is an unordered stem and leaf diagram.

0	1 6 3 4 2
1	2 6 2 4 2
2	4 2 5 7 3 9
3	3 0 2 5 8 4 6
4	7 3 6 2

Key:
1|2 stands for 12

Draw this as an ordered stem and leaf diagram.

D

2   Keith has a job that involves driving to different shops each day.
Here are the distances, in kilometres, that he drove during April.

| 8 | 10 | 21 | 17 | 9 | 31 | 22 | 6 | 9 | 15 |
| 17 | 22 | 17 | 14 | 39 | 25 | 26 | 18 | 12 | 27 |

Draw a stem and leaf diagram to represent these data.

3   Here is a stem and leaf diagram showing the numbers of cars sold by a garage group over each of a
number of weeks.

```
0 │ 9
1 │ 2  6  7  8                          Key:
2 │ 1  3  5  5  5  6  6  6  6           2│1 stands for 21 cars
3 │ 2  2  2  4  6  7
4 │ 0
```

  a   Write down the number of weeks represented in this diagram.

  b   Write down the mode for these data.

  c   Find the median number of cars.

  d   Work out the range of these data.

4   Here is a list of the number of minutes patients had to wait to see a dentist during one day at a
dental surgery.   A02

| 10 | 5 | 23 | 8 | 14 | 16 | 3 | 2 | 12 | 24 | 22 | 7 |
| 15 | 18 | 23 | 30 | 23 | 16 | 16 | 20 | 3 | 5 | 2 | 18 |

Draw a suitable diagram that the dentist could use to

  a   display these data

  b   show a measure of spread

  c   show a measure of average.

16.5 Using frequency tables to find averages for discrete data

◉ Objective

○ You can use a frequency table to find averages.

◈ Why do this?

If your dance club was debating whether to move
location it could work out the change to average
journey time as this might affect membership
levels.

⊕ Get Ready

1. When a die was thrown 30 times the number 6 came up 5 times.

  a   What was the frequency of the number 6?

  b   What was the sum of all the 6s added together?

Key Points

● The mode is the number that has the highest frequency.

● The median is the number that is the middle value or halfway between the middle values if there are two of them.

● The mean is the sum of the values divided by the number of values (the number of values is the total frequency).

● For discrete data in a frequency table:

mean $= \dfrac{\sum f \times x}{\sum f}$ where f is the frequency, x is the variable and $\sum$ means 'the sum of'.

Example 10 The table shows information about the number of children in a sample of families.

Number of children in family	1	2	3	4	5
Frequency	3	11	9	5	6

a Write down the mode of these data.

b Find the median of these data.

c Work out the mean of these data.

a The mode is 2 children. ← 2 has the highest frequency, which is 11.

b

Number of children in family x	Frequency f	Frequency × number of children $f \times x$
1	3	3
2	11	22
3	9	27
4	5	20
5	6	30
Total	34	102

There are 11 families with 2 children in each so the number of children is $11 \times 2 = 22$ children.

The total number of families is 34.

The total number of children is the sum of all the $f \times x$ values.

The total frequency is 34 so the median will be between the 17th and 18th values. ← There will be 16 values either side.

There are 3 families with 1 child in them.

3 lies between the 14th and 23rd values.

There are $3 + 11 = 14$ families with 2 or 1 children.

There are $3 + 11 + 9 = 23$ families with 3, 2 or 1 children.

The 17th and 18th values must both be 3.

The median is 3.

c The mean is $\dfrac{102}{34} = 3$ children. ← Mean $= \dfrac{\text{Total number of children}}{\text{Total number of families}} = \dfrac{\sum f \times x}{\sum f}$

Exercise 16G

A03 **D**

1 A council wanted to provide extra parking on an estate.
They asked a sample of households how many cars they had.
The results are shown in the frequency table.

Number of cars	Frequency
0	0
1	10
2	7
3	6
4	2

  a Write down the mode of these data.

  b Find the median number of cars.

  c Work out the mean number of cars.

  d There are 130 households on the estate and each household currently has one parking space.
Recommend the number of extra parking spaces the council should provide. Explain your answer.

2 A sample of a tomato crop was taken and each tomato was weighed.
The weights to the nearest 5 g are shown in the frequency table.

Weight of tomatoes	Frequency
55	2
60	5
65	10
70	6
75	2

  a Write down the mode of these data.

  b Find the median weight of the tomatoes.

  c Work out the mean weight of the tomatoes.

3 In an experiment with peas the number of peas per pod was recorded.
The results are shown in the frequency table.

Number of peas per pod	Frequency
1	0
2	0
3	3
4	7
5	11
6	12
7	15
8	12
9	10

  a Write down the mode of these data.

  b Find the median number of peas per pod.

  c Work out the mean number of peas per pod.

16.6 Working with grouped data

Objectives

- You can find the modal class for grouped data.
- You can find a class interval containing the median of grouped data.

Why do this?

If you are collecting continuous data, such as times taken to swim 100 m, you need to have class intervals in your frequency table.

Get Ready

1. Write whether each of the following types of data is continuous or discrete.
   a Number of items in a shopping basket
   b Size of feet
   c Waist size

2. Write down whether each statement is true or false.
   a $10 > 12$ b $0.1 < 0.2$ c $0.17 > 0.6$

Key Points

- If you do not know the exact data in each class interval, you cannot give an exact value for the mode or the median and can only estimate the mean value.

- The class interval with the highest frequency is called the **modal class**.

- You can only write down the class interval in which the median falls.

Example 11 The frequency table gives information about the number of letters, l, in a sample of people's surnames.

Class interval	Frequency
3 to 5	1
6 to 8	3
9 to 11	5
12 to 14	4
15 to 17	2

This is discrete data. No whole number appears in two classes.

There will be $3 + 1 = 4$ values less than or equal to 8.

a Find the modal class.
b Find the class into which the median falls.

Look for the class with the highest frequency.

a The modal class is 9 to 11.
b There are 15 names in total so the median will be the 8th value.
   There are $3 + 1 = 4$ names that are less than 9 letters long
   and $5 + 3 + 1 = 9$ names that are less than 12 letters long,
   so the median is in the class interval 9–11.

Example 12 ▶ The frequency table gives information about the lengths, l in mm, of leaves from a certain plant.

Class interval	Frequency f
$20 \leqslant l < 25$	6
$25 \leqslant l < 30$	8
$30 \leqslant l < 35$	13
$35 \leqslant l < 40$	14
$40 \leqslant l < 45$	9

This is continuous data.

There will be $6 + 8 = 14$ values less than 30.

a Find the modal class.

b Find the class into which the median falls.

a The modal class is $35 \leqslant l < 40$.

b There are 50 leaves so the median will be halfway between the 25th and 26th values.
There are 14 values less than 30 and 27 less than 35.
The median falls in the class $30 \leqslant l < 35$.

Exercise 16H

1 A group of students were asked how many times they visited a library in a term.
The results are shown in the frequency table.

Class interval	0 to 2	3 to 5	6 to 8	9 to 11
Frequency	0	6	10	6

a Write down the modal class.

b Find the class into which the median falls.

2 A group of students did a mental arithmetic test.
The results are shown in the frequency table.

Class interval	1 to 5	6 to 10	11 to 15	16 to 20
Frequency	1	9	15	5

a Write down the modal class.

b Find the class into which the median falls.

3 The frequency table gives the diameter, d in mm, of 48 balls of lead used in a quality control investigation.

a Write down the modal class.

b Find the class into which the median falls.

Class interval	Frequency
$0.7 \leqslant d < 0.9$	2
$0.9 \leqslant d < 1.1$	4
$1.1 \leqslant d < 1.3$	16
$1.3 \leqslant d < 1.5$	12
$1.5 \leqslant d < 1.7$	14

E

16.7 Estimating the mean of grouped data

Objective

- You can estimate the mean of grouped data.

Why do this?

If data is collected and arranged in a grouped frequency table, you will not have exact data values, so you won't be able to calculate the mean exactly. Using the middle values of the class intervals provides good estimates to work with.

Get Ready

1. Which number is halfway between:

   a 6 and 8   **b** 56 and 64   **c** 75 and 76   **d** 0.75 and 0.85   **e** 100 000 and 150 000?

Key Point

- An estimate for the mean of grouped data can be found by using the midpoint of the class interval and the formula $\dfrac{\sum fx}{\sum f}$ where f is the frequency and x is the class midpoint.

Example 13   Work out an estimate for the mean length of the people's surnames given in Example 11.

Class interval	Frequency f	Class midpoint x	$f \times x$
3 to 5	1	4	4
6 to 8	3	7	21
9 to 11	5	10	50
12 to 14	4	13	52
15 to 17	2	16	32
Totals	15		159

The middle value of the class 3–5 is 4.

The middle value of the class 6–8 is 7. The three people in the class 6–8 might not all have surnames 7 letters long. This is why it is an **estimated** mean.

Estimated mean $= \dfrac{\sum f \times x}{\sum f} = \dfrac{159}{15} = 10.6$

You can now use the formula.

Example 14   Work out an estimate for the mean length of the leaves given in Example 12.

Class interval	Frequency f	Class midpoint x	$f \times x$
$20 \leqslant l < 25$	6	22.5	135
$25 \leqslant l < 30$	8	27.5	220
$30 \leqslant l < 35$	13	32.5	422.5
$35 \leqslant l < 40$	14	37.5	525
$40 \leqslant l < 45$	9	42.5	382.5
Totals	50		1685

ResultsPlus
Examiner's Tip

Remember to use the class midpoint when estimating the average.

Estimated mean $= \dfrac{1685}{50} = 33.7$ mm.

Exercise 16I

1 In a healthy eating investigation the canteen supervisor at Conville College recorded the numbers of packets of crisps bought per month by a sample of students.
The results are shown in the frequency table.

Find an estimate for the mean number of packets bought.

Class interval	Frequency f
1 to 3	1
4 to 6	9
7 to 9	15
10 to 12	5

A02 C

2 A store is worried about the reliability of its lift. It records the number of times it breaks down each week over a period of 28 weeks.
The results are shown in the frequency table.
Find an estimate for the mean number of breakdowns.

Class interval	Frequency f
0 to 1	20
2 to 3	3
4 to 5	4
6 to 7	1

A02

3 Emma recorded the length of time (t), in minutes, each of her business phone calls took over a period of one month.
The results are shown in the frequency table.

Find an estimate for the mean time for her business phone calls.

Class interval	Frequency
$0 \leqslant t < 5$	9
$5 \leqslant t < 10$	10
$10 \leqslant t < 15$	9
$15 \leqslant t < 20$	7
$20 \leqslant t < 25$	5

A02

Chapter review

◉ The **mode** of a set of data is the value that occurs most frequently.

◉ The **median** is the middle value when the data are ordered from the smallest to the largest.

◉ If there are two middle values in a set of data the median is halfway between them.

◉ The **mean** of a set of data is the sum of the values divided by the total number of observations.

$$\text{mean} = \frac{\text{sum of the values}}{\text{number of values}}$$

◉ **Range** = highest value − lowest value

◉ To compare the distributions of sets of data you need to give a measure of **average** and a measure of **spread**.

◉ A **stem and leaf diagram** makes it easy to find the mode, median and range of a set of data.

◉ For discrete data in a frequency table:

$$\text{mean} = \frac{\sum f \times x}{\sum f}$$ where f is the frequency, x is the variable and $\sum$ means 'the sum of'.

◉ For grouped data:

  ◉ the class interval with the highest frequency is called the **modal class**

  ◉ you can only write down the class interval in which the median falls

  ◉ an estimate for the mean of grouped data can be found by using the midpoint of the class interval.

Review exercise

F

1 Mary threw a dice 24 times.
Here are the 24 scores.

3 5 3 4 1 2 4 5 6 2 3 4
3 1 4 3 2 3 5 5 3 4 2 1

a Complete the frequency table.　　**b** Write down the mode.

Score	Tally	Frequency
1		
2		
3		
4		
5		
6		

ResultsPlus
Exam Question Report

Most candidates did very well on part **a** of this question as they knew to write tallies and then show the frequencies.

March 2007

E

2 The weekly incomes of five people are shown in the table.

a Work out the mean income per week.

b Work out the median income per week.

c Work out the range of weekly incomes.

d How much more did Mrs Chown earn than Mrs Basingi?

Name	Income
Mr Rahman	£420
Mrs Basingi	£365
Mr Clarke	£400
Mr Abson	£280
Mrs Chown	£430

3 Here are ten numbers.

7 6 8 4 5 9 7 3 6 7

a Work out the range.　　　　　　**b** Work out the mean.　　　*Nov 2008*

4 Jason collected some information about the heights of 19 plants.
This information is shown in the stem and leaf diagram.

```
1 | 1  2  3  4
2 | 3  3  5  9  9           Key:
3 | 0  2  2  6  6  7        4|8 means 48 mm
4 | 1  1  4  8
```

Find the median.　　　　　　　　　　　　　　　　　　　*Nov 2008*

5 Peter rolled a 6-sided dice ten times.
Here are his scores.

3 2 4 6 3 3 4 2 5 4

a Work out the median of his scores.

b Work out the mean of his scores.

c Work out the range of his scores.　　　　　　　　　*June 2007*

6 Here are the weights, in kg, of 8 people.

63　65　65　70　72　86　90　97

a Write down the mode of the 8 weights.　　**b** Work out the range of the weights.　　*June 2007*

7 Five positive numbers have a mode of 5, a median of 5 and a mean of 4.

Write down as many possible combinations of five numbers that give these statistics as you can.

8 The stem and leaf diagram shows the ages, in years, of all the workers in a small factory.

2	0	2	2	5	7				
3	3	3	4	4	4	5			
4	5	6	6	6	6	8	9	9	9
5	2	4	4	6	7	9			
6	0	2	5						

Key:
4|5 stands for 45 years

a Work out the number of workers.

b Write down the mode of these data.

c Find the median of these data.

d Work out the range of these data.

9 A group of girls went to a college dance. They each bought a new dress.

The costs of the dresses were

£22 £22 £22 £28 £32 £36 £40 £40 £45 £180

a Write down the mode of these data.

b Find the median price.

c Work out the mean price.

d Which of the three averages worked out in parts **a**, **b** and **c** best describes the price the girls paid? Give a reason for your answer.

10 Samples of apples were taken from two trees. One was an eating apple tree and the other was a cooking apple tree.

The weights, in grams, of the apples were

Eating apple	135	135	140	138	142	150	132
Cooking apple	140	136	150	160	138	162	150

a Find the means and ranges for these data.

b Compare the frequency distributions of the weights of the two types of apple.

11 Zoe recorded the weights, in kilograms, of 15 people.
Here are her results.

87 51 46 77 74 58 68 78
48 63 52 64 79 60 66

a Complete the ordered stem and leaf diagram to show these results.

b Write down the number of people with a weight of more than 70 kg.

c Work out the range of the weights.

4	
5	
6	
7	
8	

March 2009

D

12 Zach has 10 CDs.
The table gives some information about
the number of tracks on each CD.

a Write down the mode.

b Work out the mean.

Number of tracks	Frequency	
11	1	
12	3	
13	0	
14	2	
15	4	

June 2009

ResultsPlus
Exam Question Report

81% of students answered part **b** of this question
poorly.

13 Here are the ages, in years, of 15 teachers.

35   52   42   27   36
23   31   41   50   34
44   28   45   45   53

Draw an ordered stem and leaf diagram to show this information. You must include a key.   *May 2008 adapted*

14 Explain why the sentence 'The majority of spiders in this country have more than the average number of legs' can be true. You need to state which average is being used.

15 Ali found out the number of rooms in each of 40 houses in a town.
He used the information to complete the frequency table.

Number of rooms	Frequency	
4	4	
5	7	
6	10	
7	12	
8	5	
9	2	

Ali said that the mode is 9. Ali is wrong.

a Explain why.

b Calculate the mean number of rooms.   *Nov 2007*

A03

16 A group of university students did a maths test. The table shows their scores.

Males	42	22	65	42	70	50	45
Females	25	90	55	26	95	50	87

Using your understanding of averages and range, compare the males' and the females' scores.

17 One teacher is responsible for the distribution of milk at break time in each school.
There are p classes and each class is sent $x + 2$ bottles of milk, where x is the number of children present in a class.
The teacher uses $\Sigma x + 2p$ to work out how many bottles are needed.

a Is $\Sigma x + 2p$ an expression, an equation or a formula?

There are 80 children in Banjo school.
The numbers attending Banjo school one day are shown in the table.

Class	1	2	3	4
Numbers attending	20	18	18	20

b Work out the number of bottles of milk required for that day.

c The dairy send Banjo school 84 bottles of milk per day.
Discuss whether or not you think this is a suitable number.

* 18 Class 5A take six maths tests every year, each one out of 100. Meena has a mean score of 64 marks per test for the first five tests of the year. Her parents have promised her a bicycle if she can achieve a mean score of 70. What mark would she have to get in the sixth test to achieve this mean score?

19 A small factory pays salaries to 8 workers, a manager and an owner.
The salaries they earn are shown in the table.

	Salary
Workers	£10 000
Manager	£40 000
Owner	£180 000

Depending on the average you use, the average wage of people in the factory is vastly different.
If you were negotiating for a higher salary for the workers, which average would you use?
If you were negotiating for the management to keep the salaries low, which average would you use?
Explain your answers.

20 The hourly wages, in pounds, of the employees in a factory were recorded.
The results are shown in the frequency table.

Hourly wage £s	7 to 9	10 to 12	13 to 15	16 to 18	19 to 21
Frequency	5	20	20	10	5

a Write down the modal class of these data.

b Find the class interval that contains the median hourly wage.

c Estimate the mean hourly wage.

C

21 Oliver measured the heights (h), in cm, of the leek plants in his garden.
Here are his results.

Class interval	Frequency
$25 \leqslant h < 27$	5
$27 \leqslant h < 29$	10
$29 \leqslant h < 31$	13
$31 \leqslant h < 33$	15
$33 \leqslant h < 35$	7

Exam Question Report

91% of students scored poorly on this question because they did not use the midpoint of the range to find the mean of grouped data.

a Write down the modal class.

b Find the class into which the median height falls.

c Work out an estimate for the mean height of the leeks.

22 Josh asked 30 students how many minutes they each took to get to school.
The table shows some information about his results.
Work out an estimate for the mean number of minutes taken by the 30 students.

Time (t minutes)	Frequency
$0 < t \leqslant 10$	6
$10 < t \leqslant 20$	11
$20 < t \leqslant 30$	8
$30 < t \leqslant 40$	5

Nov 2008

23 Vanessa made 80 phone calls last month.
The table gives information about the length of the calls.

Length of call (t minutes)	Frequency	
$0 < t \leqslant 10$	20	
$10 < t \leqslant 20$	32	
$20 < t \leqslant 30$	14	
$30 < t \leqslant 40$	9	
$40 < t \leqslant 50$	5	

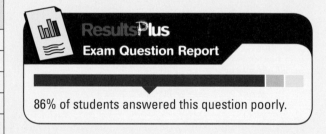

Exam Question Report

86% of students answered this question poorly.

Work out an estimate for the mean length of the calls.

March 2008

24 80 people work in Joe's factory.
The table shows some information about the annual pay of these 80 workers.

Annual pay (£x)	Number of workers
$10\,000 < x \leqslant 14\,000$	32
$14\,000 < x \leqslant 16\,000$	24
$16\,000 < x \leqslant 18\,000$	16
$18\,000 < x \leqslant 20\,000$	6
$20\,000 < x \leqslant 40\,000$	2

a Write down the modal class interval.

b Find the class interval that contains the median.

June 2007

In zorbing or sphereing a participant inside an inner capsule in a large, transparent ball rolls along the ground or downhill. The record for the fastest speed in a sphere is a rather scary 52 km per hour.

⊚ Objective

In this chapter you will:
- ⊚ learn how to work out the area and circumference of circles and part circles.

⬨ Before you start

You should be able to identify and name parts of a circle, including:
- ⊚ circumference
- ⊚ radius
- ⊚ diameter
- ⊚ arc
- ⊚ chord
- ⊚ tangent
- ⊚ segment
- ⊚ sector.

17.1 Circumference of a circle

◉ Objective

○ You can remember and use the formula to find the circumference of a circle.

⍰ Why do this?

Bicycle designers might need to know how far a wheel goes on each revolution.

◈ Get Ready

1. Name each part of the circle, centre O, drawn in red.

   a   b   c

2. Copy and complete the following statements by putting a number on the dotted line.

   a The diameter = the radius × ......

   b The radius = the diameter × ......

🔍 Key Points

◉ Draw some circles and measure the circumference and diameter of each one.
  You should find that the circumference of any circle is just over 3 times the diameter.
  This value cannot be worked out exactly but it is about 3.142.
  We use the Greek letter π (pi) for this value. You can find it on your calculator.

◉ For all circles: $\dfrac{\text{circumference}}{\text{diameter}} = \dfrac{C}{d} = \pi$

◉ This gives us a formula to calculate the circumference of a circle:
  $C = \pi \times d$ or $C = \pi d$.

◉ As the diameter is twice the radius, we can also use this formula:
  $C = 2 \times \pi \times r$ or $C = 2\pi r$.

ResultsPlus
Examiner's Tip

You will need to learn these formulae for your exam.

🔍 Example 1

A circle has a radius of 5.7 cm.
Work out the circumference of the circle.
Give your answer to 1 decimal place.

5.7 cm

$C = 2\pi r$

$= 2 \times \pi \times 5.7$ ← You need to write down all the numbers you are putting into your calculator.

$= 35.81415$ ← Always show your unrounded answer before you round it.

$= 35.8\,\text{cm (to 1 d.p.)}$ ← This number rounds to 35.8.

Example 2

Find the circumference of a circle with diameter 13 cm.
Give your answer to 1 decimal place.

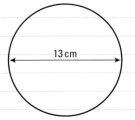

13 cm

$C = \pi d$ ← | Use this formula as we are given the diameter.

$= \pi \times 13$ ← | Use the π key or use $\pi = 3.142$.

$= 40.8407$ ← | Write down at least six digits from your calculator.

$= 40.8$ cm (to 1 d.p.) ← | Give the units with your answer.

Results Plus
Examiner's Tip

Do not just write down the formula. Make sure you write down the numbers you are putting into the formula.

Exercise 17A

Questions in this chapter are targeted at the grades indicated.

C

1 Work out the circumferences of circles with the following radii.
   a 6.3 m   b 9.5 cm   c 4.2 cm   d 12.5 mm   e 29.4 cm

2 Work out the circumferences of circles with the following diameters.
   a 6.9 cm   b 10.1 mm   c 5.3 cm   d 9.7 cm   e 5 m

3 A circular pond has a radius of 1.8 metres. Work out the circumference of the pond.

4 A circular plate has a radius of 16 cm. Work out the circumference of the plate.

5 A penny-farthing bicycle has a large wheel and a small wheel.
   The large wheel has a diameter of 1.43 metres and the small wheel has a radius of 0.15 metres.

   a Work out the circumference of the large wheel.
   b Work out the circumference of the small wheel.
   c Work out how many times the small wheel has to turn when the large wheel turns once.
   Give your answers to 2 decimal places.

Example 3   The circumference of a circle is 78.4 cm. Work out its radius.
Give your answer to 2 decimal places.

$C = 2\pi r$

$r = \dfrac{C}{2\pi}$ If you divide both sides by 2π, you can use this formula.

$r = \dfrac{78.4}{2\pi} = 12.477747$ On the calculator you need to press: $78.4 \div 2 \times \pi$.

$= 12.48$ cm (to 2 d.p.)   Note that the number is rounded to 2 d.p.

In the same way, if you are asked to find the diameter, given the circumference, you can use the formula

$d = \dfrac{C}{\pi}$

Exercise 17B

Give your answers to 1 decimal place in each of the following questions.

1.   Work out the diameters of circles with the following circumferences.
   a   45.6 m   b   20 cm   c   58.1 cm   d   37.2 mm   e   100 cm

2.   Work out the radii of circles with the following circumferences.
   a   30.4 cm   b   71.8 mm   c   64 cm   d   93.2 cm   e   49.5 m

3.   A trundle wheel is used to measure a garden path.
   The circumference of the trundle wheel is 188 cm.

Work out the diameter of the trundle wheel.

17.2 Area of a circle

⦿ Objective

⦿ You can remember and use the formula to find the area of a circle.

⦹ Why do this?

Being able to work out the area of a circle could be helpful if you want to use a square or rectangular cake tin instead of a circular one.

⬥ Get Ready

Use your calculator to work out the following calculations.
Give your answers correct to 2 decimal places.

1. $\pi 5^2$ **2.** $3^2\pi$ **3.** $(7\pi)^2$ **4.** $(9\pi)^2$

⬥ Key Points

◉ Take a circle and cut it into lots of equal sectors.

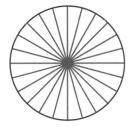

Circumference $= C$
radius $= r$

◉ Rearrange these sectors as shown in the diagram.

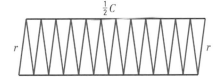

◉ When there are lots of sectors this gets closer to a rectangle.

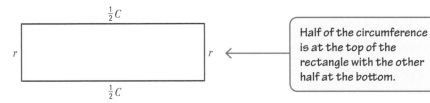

Half of the circumference is at the top of the rectangle with the other half at the bottom.

◉ The area of the circle $=$ the area of the rectangle

$$= \tfrac{1}{2}C \times r$$

Area of a rectangle $=$ base $\times$ height

$$= \tfrac{1}{2} \times 2\pi r \times r$$

$$= \pi r^2$$

$A = \pi r^2$

Area $= \pi \times$ radius $\times$ radius

Results Plus

Examiner's Tip

You will need to learn this formula for your exam.

Example 4 A circle has a radius of 8.2 cm.
Work out the area of the circle.
Give your answer to 3 significant figures.

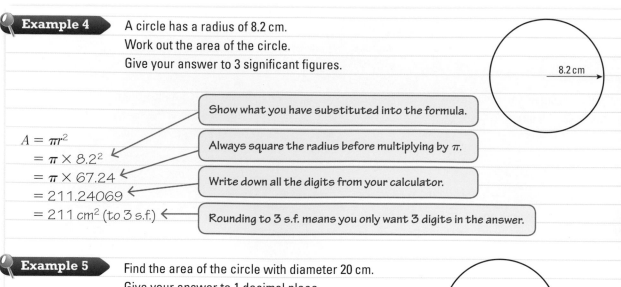

Show what you have substituted into the formula.

Always square the radius before multiplying by π.

Write down all the digits from your calculator.

Rounding to 3 s.f. means you only want 3 digits in the answer.

$A = \pi r^2$
$\quad = \pi \times 8.2^2$
$\quad = \pi \times 67.24$
$\quad = 211.24069$
$\quad = 211 \text{ cm}^2 \text{ (to 3 s.f.)}$

Example 5 Find the area of the circle with diameter 20 cm.
Give your answer to 1 decimal place.

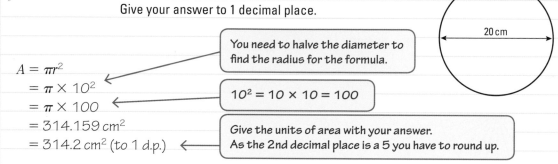

You need to halve the diameter to find the radius for the formula.

$10^2 = 10 \times 10 = 100$

Give the units of area with your answer.
As the 2nd decimal place is a 5 you have to round up.

$A = \pi r^2$
$\quad = \pi \times 10^2$
$\quad = \pi \times 100$
$\quad = 314.159 \text{ cm}^2$
$\quad = 314.2 \text{ cm}^2 \text{ (to 1 d.p.)}$

Exercise 17C

Give your answers to 3 significant figures in each of the following questions.

1 Work out the areas of circles with the following radii.
   a 5.1 m   **b** 3 cm   **c** 8.7 cm   **d** 15.2 mm   **e** 9.4 cm

2 Work out the areas of circles with the following diameters.
   a 30 cm   **b** 24.4 mm   **c** 7.4 cm   **d** 12.3 cm   **e** 8 m

3 A goat is tied to a post in the middle of a field covered in grass.
He is tied so that he can eat the grass within 3.8 m of the post.
Work out the area of grass from which he cannot eat.

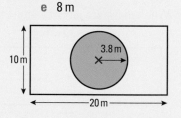

4 Rosie makes some jam. She covers the tops of the jars with
circular pieces of material of diameter 8.6 cm.
Work out the area of material covering one jar.

5 The diagram shows a square of side 9 cm inside a circle of radius 11 cm.
   a Work out the area of the circle.
   b Work out the area of the square.
   c Work out the area of the shaded part.

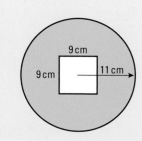

6 The diagram shows a rectangular card measuring 32 cm by 16 cm. Eight circles of radius 2 cm are cut out so that the card can hold eight pots of yoghurt.
Work out the area of the card that is left.

A03 C

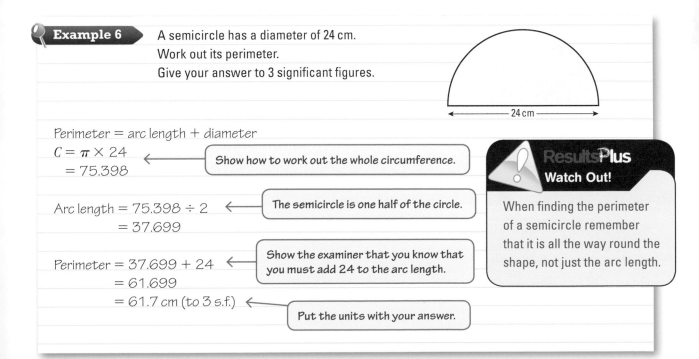

17.3 Area and perimeter of half and quarter circles

◎ Objective

● You can find the perimeter and area of a half circle and a quarter circle.

② Why do this?

Many shapes come in half circles and quarter circles such as flower beds and you may need to calculate the area to know how much soil to get.

◈ Get Ready

Use your calculator to work out the following, giving your answer to 2 decimal places.

1. $\dfrac{7^2 \pi}{2}$ 2. $\dfrac{(2\pi)(31)}{4}$ 3. $\dfrac{(5\pi)^2}{2}$ 4. $\dfrac{21\pi}{4}$

◉ Key Points

● The perimeter of a **semicircle** is the diameter + half the circumference.
● The perimeter of a quarter circle is the diameter + one quarter the circumference.
● The area of a semicircle is half the area of the circle.

Example 6   A semicircle has a diameter of 24 cm.
Work out its perimeter.
Give your answer to 3 significant figures.

⟵ 24 cm ⟶

Perimeter = arc length + diameter
$C = \pi \times 24$ ⟵ Show how to work out the whole circumference.
  $= 75.398$

Arc length = 75.398 ÷ 2   ⟵ The semicircle is one half of the circle.
       $= 37.699$

Perimeter = 37.699 + 24   ⟵ Show the examiner that you know that you must add 24 to the arc length.
     $= 61.699$
     $= 61.7$ cm (to 3 s.f.) ⟵ Put the units with your answer.

Results Plus
Watch Out!

When finding the perimeter of a semicircle remember that it is all the way round the shape, not just the arc length.

Exercise 17D

Give your answers to 3 significant figures in each of the following questions.

1   Calculate the perimeter and the area of each sector.

a
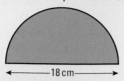
←——18 cm——→

b
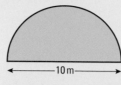
←——10 m——→

c
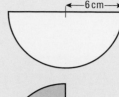
←—6 cm—→

d

←—4 cm—→

e

←—6.2 cm—→

f
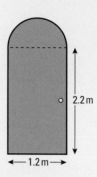
←—8.1 m—→

2   A door is in the shape of a rectangle with a semicircle on top.
The width of the door is 1.2 m.
The height of the rectangular part of the door is 2.2 m.
a   Calculate the area of the door.
The door is to be covered with brown leather. The leather costs
£22.49 per square metre and comes in 2 m widths.
b   What is the cost of covering the door?

2.2 m

←1.2 m→

3   The diagram shows a triangle inside a quarter of a circle.
a   Work out the area of the shaded segment.
b   Work out the perimeter of the whole shape.

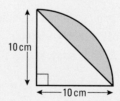

10 cm

←—10 cm—→

Chapter review

● To find the circumference (C) of a circle when given the radius (r) or diameter (d), use the formulae
$C = \pi d$
$C = 2\pi r$

● To find the diameter (or radius) of a circle when given the circumference, use the formula
$d = \dfrac{C}{\pi}$

● To find the area (A) of a circle when given the radius (or diameter), use the formula $A = \pi r^2$.

Review exercise

1 A circle has a radius of 6 cm.
A square has a side of length 12 cm.
Work out the difference between the area of the
circle and the area of the square.
Give your answer correct to one decimal place.

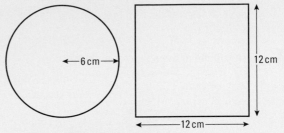

Diagram **NOT** accurately drawn

Nov 2008

2 The diagram shows two small circles inside a large circle.
The large circle has a radius of 8 cm.
Each of the two small circles has a diameter of 4 cm.

a Write down the radius of each of the small circles.

b Work out the area of the region shown shaded in the diagram.
Give your answer correct to one decimal place.

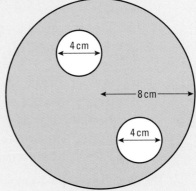

Diagram **NOT** accurately drawn

Nov 2008

3 The radius of this circle is 8 cm.
Work out the circumference of the circle.
Give your answer correct to 2 decimal places.

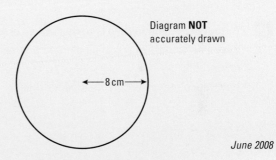

Diagram **NOT**
accurately drawn

June 2008

4 The diagram shows a semicircle.
The radius of the semicircle is 10 cm.
Calculate the area of the semicircle.
Give your answer correct to 3 significant figures.
State the units of your answer.

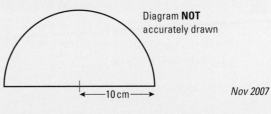

Diagram **NOT**
accurately drawn

Nov 2007

5 The diagram shows a circle.
The radius of the circle is 10 cm.
Calculate the area of the circle.
Give your answer correct to 3 significant figures.
State the units with your answer.

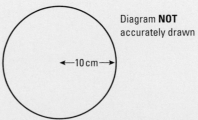

Diagram **NOT**
accurately drawn

Nov 2007

C

6 The diameter of a wheel on Harry's bicycle is 0.65 m.
Calculate the circumference of the wheel.
Give your answer correct to 2 decimal places.

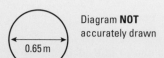

Diagram **NOT** accurately drawn

0.65 m

June 2007

7 Here is a tile in the shape of a semicircle.

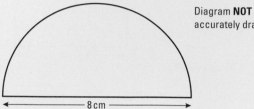

Diagram **NOT** accurately drawn

8 cm

The diameter of the semicircle is 8 cm.
Work out the perimeter of the tile.
Give your answer correct to 2 decimal places.

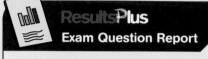

ResultsPlus
Exam Question Report

89% of students answered this question poorly. Many students incorrectly used the formula for area of a circle (πr^2).

June 2009

8 The top of a table is a circle.
The radius of the top of the table is 50 cm.
a Work out the area of the top of the table.

The base of the table is a circle.
The diameter of the base of the table is 40 cm.
b Work out the circumference of the base of the table.

June 2007

A02 A03

9 A ring-shaped flowerbed is to be created around a circular lawn of radius 2.55 m.
Roses costing £4.20 are to be planted approximately every 50 cm around the flowerbed.
How much money will be needed for roses?

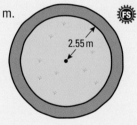

2.55 m

A02 A03

10 The diagram shows a garden that includes a lawn, a vegetable patch, a circular pond and a flowerbed. All measurements are shown in metres.
The lawn is going to be re-laid with with turf costing £4.60 per square metre.
How much will this cost?

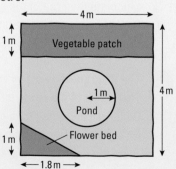

4 m

1 m

Vegetable patch

4 m

1 m

Pond

Flower bed

1 m

1.8 m

18 CONSTRUCTIONS AND LOCI

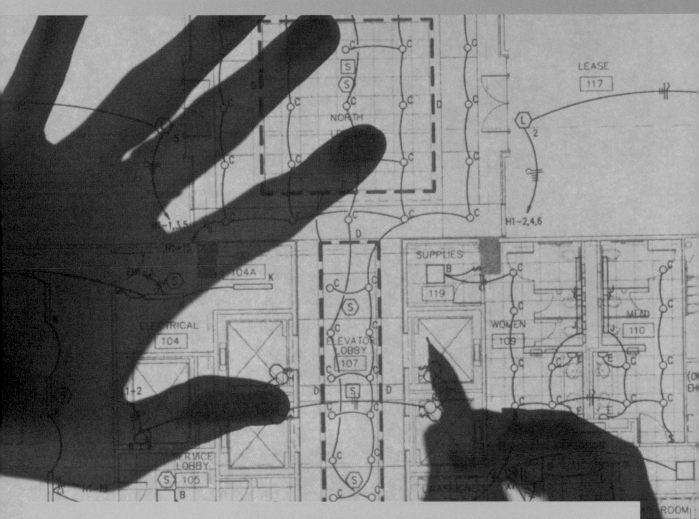

Architects do scale drawings of projects they are working on for both planning and presentation purposes. Originally these were done on paper using ink and copies had to be made by hand. Later they were done on tracing paper so that copying was easier. Today, computer-generated drawings have largely taken over. The first program that could create these was released in the 1960s and now for many of the top architecture firms these have been replaced by architectural animation.

◉ Objectives

In this chapter you will:
- learn how to draw the five basic constructions using a ruler and compass only
- learn how to accurately construct the locus of a set of points and draw regions associated with loci.

◈ Before you start

You need to:
- know that 'bisect' means 'cut in half'
- be able to measure and draw lines accurately
- have a pencil, eraser, ruler, protractor and compasses.

18.1 Constructions

◉ Objectives

○ You can bisect angles and lines by using compasses.
○ You can produce certain angles and diagrams using compasses.

⟐ Why do this?

Architects and engineers need to be able to accurately measure angles when designing and constructing buildings.

⬥ Get Ready

1. Draw a line of length 5 cm.
2. Draw a circle of radius 5 cm.

⟐ Key Points

◉ Constructing an angle of 60°.

A ——————— B

Start with a line.

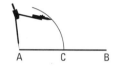
A   C   B

Open your compasses. Put the point at A and draw an arc that cuts the line. Label the point C.

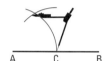
A   C   B

Keeping your compasses the same width, put the point at C. Draw an arc to cut the first one.

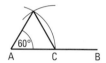
A 60° C   B

Join up to get a 60° angle. This is an equilateral triangle.

◉ Bisecting an angle.

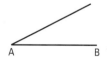
A   B

Start with an angle.

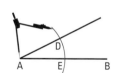
A   E   B

Put the point of the compasses at A and draw an arc that cuts both lines. Label the points D and E.

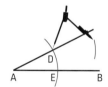
A   E   B

Put the point at D and draw an arc between the two sides of the angle. Without adjusting your compasses, place the point at E and draw an arc to cut the first one. Label the point where they cross F.

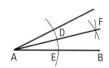
A   E   B

Draw a straight line from F to A to bisect the angle.

◉ To construct angles of sizes other than 60°, use the idea of starting with 60°, or a multiple of 60°, and bisecting the angle. For example, to construct an angle of 30°, first construct an angle of 60° and then bisect it.

◉ Bisecting a line.

A   B

Start with a line.

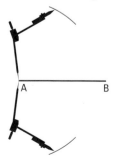
A   B

Open your compasses to more than half the line length. Put the point at A and draw an arc above and below the line.

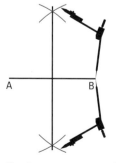
A   B

Keeping your compasses the same, put the point at B and draw arcs above and below the line to cross the other arcs.

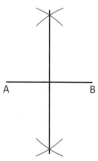
A   B

Join the crosses to make a line. This line will bisect AB.

● Constructing a perpendicular line from a point on a given line.

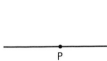
Start with a line and a point P on that line.

Put your compass on point P and draw arcs on the line either side of point P.

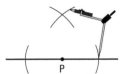
From each of the arcs drawn, use compasses to draw two intersecting arcs above the line.

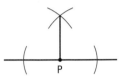
Join the intersecting arcs to point P. This line is perpendicular to the given line at P.

● Constructing a perpendicular line from a point to a given line.

Start with a line and a point P above that line.

Put your compass on point P and draw arcs on the line either side of point P.

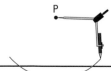
From each of the arcs drawn, use compasses to draw two intersecting arcs below the line.

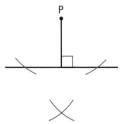
Use the intersecting arcs to draw a line from point P to the line below. This line is perpendicular to the given line.

Example 1   Construct a regular hexagon inside a circle.

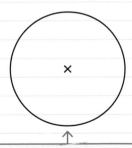

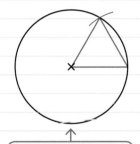

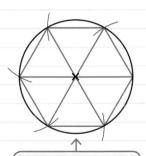

↑
Start with a circle. A regular hexagon has angles at the centre which are each $\frac{360}{6} = 60°$.

↑
Draw an angle at the centre which is 60°. Draw in a triangle.

↑
Draw in all the other 60° angles, drawing in the triangles.

Exercise 18A

Questions in this chapter are targeted at the grades indicated.

For each of the following questions use one or more of the five constructions.

1   Construct each of the following angles.
   a 60°   b 90°   c 30°   d 45°   e 15°

2   Draw lines of the lengths shown. Then bisect each of the lines.
   a 8 cm   b 6 cm   c 7 cm   d 9 cm

3   Construct equilateral triangles with sides of the following lengths.
   a 5 cm   b 8 cm   c 7 cm   d 6 cm

C

C

4 Construct squares with sides of the following lengths.

    a 6 cm   **b** 4 cm   **c** 8 cm   **d** 5 cm

5 Draw a triangle. Bisect each of its sides.

Extend each of the bisectors so they meet at a point X. Put the point of your compass at point X. Draw a circle through each of the vertices of the triangle.

6 Draw pairs of parallel lines that have a distance between them of:

    a 3 cm   **b** 5 cm   **c** 3.5 cm   **d** 4.5 cm

7 Draw a 5 cm line. Mark a point P on your line. Construct a perpendicular line at point P.

8 Draw a 5 cm line. Mark a point P at least 4 cm above your line. Construct a perpendicular line from point P down to the line.

9 Construct an equilateral triangle of side 6 cm. Construct a perpendicular line from a vertex to the opposite side of the triangle.

10 Draw a regular hexagon in a circle of radius 4 cm.

11 Draw a regular octagon in a circle of radius 4 cm.

12 Draw a regular pentagon in a circle of radius 4 cm.

18.2 Loci

Objectives

- You can interpret a locus as a set of points.
- You can draw a locus that obeys a rule.
- You can use constructions to draw loci.

Why do this?

Seismologists can use loci to work out the area that an earthquake is likely to affect and damage.

Get Ready

1. Draw accurately an equilateral triangle of side 5 cm.

Key Points

- The locus is a set of points that obey a given rule. Loci is the plural of locus.

- The **locus** of points which are the same distance (**equidistant**) from a single point is a circle.

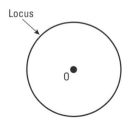

- The locus of points that are the same distance from two points is a line that is the **perpendicular bisector** of the line joining the two points.

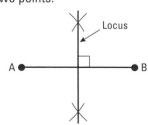

- The locus of points the same distance from two lines is the **bisector** of the angle between the lines.

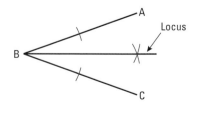

Example 2

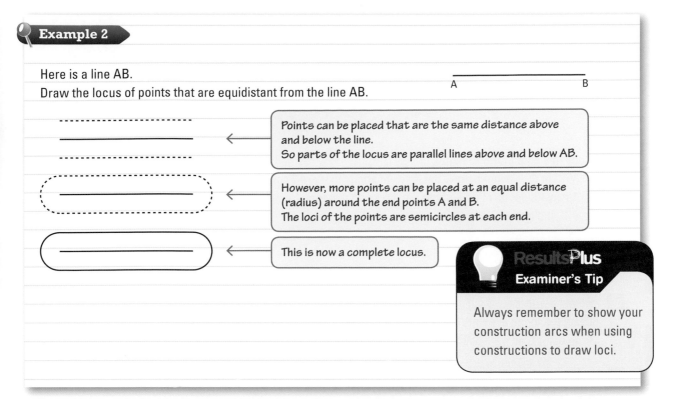

Here is a line AB.
Draw the locus of points that are equidistant from the line AB.

A _____ B

> Points can be placed that are the same distance above and below the line.
> So parts of the locus are parallel lines above and below AB.

> However, more points can be placed at an equal distance (radius) around the end points A and B.
> The loci of the points are semicircles at each end.

> This is now a complete locus.

Results Plus
Examiner's Tip

Always remember to show your construction arcs when using constructions to draw loci.

Exercise 18B

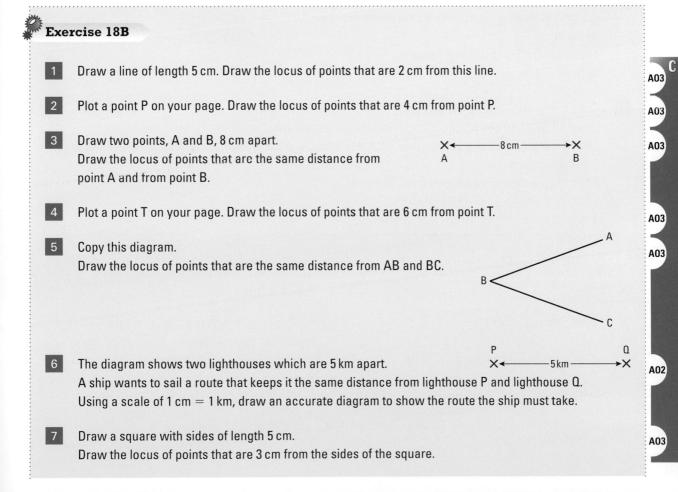

1   Draw a line of length 5 cm. Draw the locus of points that are 2 cm from this line.
A03

2   Plot a point P on your page. Draw the locus of points that are 4 cm from point P.
A03

3   Draw two points, A and B, 8 cm apart.
    Draw the locus of points that are the same distance from
    point A and from point B.
    X←——8 cm——→X
    A   B
A03

4   Plot a point T on your page. Draw the locus of points that are 6 cm from point T.
A03

5   Copy this diagram.
    Draw the locus of points that are the same distance from AB and BC.
A03

6   The diagram shows two lighthouses which are 5 km apart.
    P   Q
    X←——5 km——→X
    A ship wants to sail a route that keeps it the same distance from lighthouse P and lighthouse Q.
    Using a scale of 1 cm = 1 km, draw an accurate diagram to show the route the ship must take.
A02

7   Draw a square with sides of length 5 cm.
    Draw the locus of points that are 3 cm from the sides of the square.
A03

C

8 This rectangle has a width of 3 cm and a length of 5 cm.

   a Within the rectangle, draw the locus of points that are 1 cm from DC.

   b Draw the locus of points that are the same distance from AD and from DC.

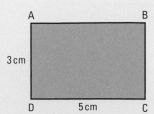

9 Copy this equilateral triangle.

   a Within the triangle, draw the locus of points that are 2 cm from point A.

   b Draw the locus of points that are the same distance from AB and from BC.

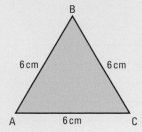

10 Points S and T are 6 cm apart.
Draw the locus of points that are the same distance from point S as from point T.

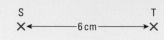

11 This triangle represents a field.
The scale of the plan is 1 cm = 2 m.
The farmer places an electric wire 1 m from each side of the field to stop the cows getting near to the perimeter of the field.
Copy the plan and draw on it where the electric wire should be placed.

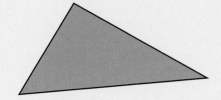

12 Copy the triangle from question 11.
Draw the locus of points that are equidistant from two adjacent sides.
Then draw the locus of points that are equidistant from another two adjacent sides, and then the third set of adjacent sides.
Mark clearly the point where all three loci meet.

18.3 Regions

⊙ Objectives

- You can interpret a locus as a region.
- You can draw a region that obeys a rule.
- You can use constructions to draw regions.

❓ Why do this?

A school may only take students from a certain catchment area. Understanding regions would enable you to work out if you lived in the necessary area.

◈ Get Ready

1. Plot a point P on your page. Draw the locus of points that are 5 cm from point P.
What shape have you drawn? What do all the points within the shape have in common?

Key Points

- Sometimes the locus is a **region** of space.
- The locus of points that are no more than a given distance from a single point is the area within a circle.

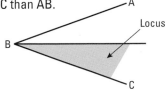

Locus

- The diagram shows the locus of points that are closer to BC than AB.

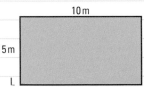

Locus

Example 3

This diagram is a plan of a yard. A light at corner L can illuminate the yard up to a distance of 3 m from the light. Indicate this region on the plan.

10 m

5 m

L

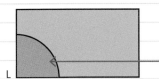

The region is part of a circle of radius 3 m from corner L. The region inside this quarter circle should be shaded.

L

The shaded region represents that part of the yard that can be illuminated from corner L.

ResultsPlus
Watch Out!

Don't forget to shade the region.

Exercise 18C

1 Copy the point A.
Shade the region that is less than 3 cm from point A.

✕ A

2 Copy the rectangle ABCD.
a Shade the region of points that are less than 2 cm from B.
b Shade the region of points that are less than 1 cm from BC.

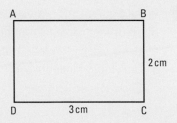

3 Copy the rectangle PQRS.
a Shade the region of points that are closer to PR than to PS.
b Shade the region of points that are less than 1 cm from SR.
c Shade the region of points that are both closer to PR than to PS, **and** are less than 1 cm from SR.

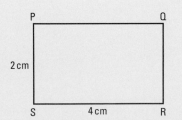

C

A03

4 The diagram represents a garden.

Plants must be planted in the garden so they are more than 1 metre from the edges of the garden.

a Copy the diagram and shade the region in which the plants must **not** be planted.

b A sprinkler is placed at C. It can spread water up to 2 m from the sprinkler. On your diagram, shade the region that can be watered by the sprinkler.

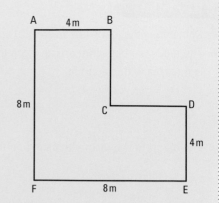

A03

5 This rectangle is a plan of a swimming pool. The half of the pool nearest to side AB is reserved for school lessons.

Copy the rectangle and shade the region of the pool reserved for school lessons.

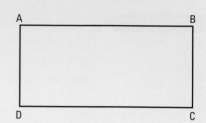

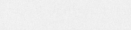

A03

6 The diagram represents three radio masts.

Signals from each radio mast can be received up to 20 km away.

Copy the diagram and shade the region in which signals from all three masts can be received.

Scale: 1 cm = 10 km.

A03

7 In the diagram, AB represents the coast, with a lighthouse at point L.

Ships cannot come nearer than 2 km from the coast.

a Use a scale of 1 cm = 1 km to represent this on a plan.

b On a foggy night the light can only be seen up to 3 km from the lighthouse. Show this on your plan.

8 Copy this triangle.

a Show the region of points that are less than 2 cm from A.

b Show the region of points that are less than 2 cm from BC.

c Shade the region of points that are nearer to AB than AC.

d Is there a region in which there are some
points that satisfy all three of these
conditions?
Indicate this clearly on
your diagram.

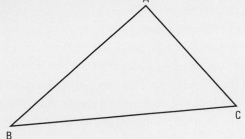

9 Copy this rectangle.

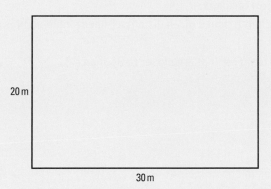

The rectangle represents a plan of a compound which has dimensions of 20 m by 30 m, and a scale of
1 cm = 5 m. Lights are placed on the walls all around the compound. The lights can illuminate a region
no more than 5 m from the compound wall, both inside and outside the compound. Shade the region
illuminated on your plan.

10

The above is a plan of a hall. Draw an accurate plan, using a scale of 1 cm = 2 m.

Then draw the following regions on the plan.

a The locus of points that are nearer to BC than AB.

b The locus of points that are within 3 m of point H or of point C.

c The locus of points that are more than 2 m from FE.

Indicate clearly any points that are within all of the regions indicated.

Chapter review

- You have learnt the following constructions:
  - constructing an angle of 60°
  - bisecting an angle
  - bisecting a line
  - constructing a perpendicular line from a point on a given line
  - constructing a perpendicular line from a point to a given line.
- The locus is a set of points that obey a given rule. Loci is the plural of locus.
- The **locus** of points which are the same distance (**equidistant**) from a single point is a circle.
- The locus of points that are the same distance from two points is a line that is the **perpendicular bisector** of the line joining the two points.
- The locus of points the same distance from two lines is the **bisector** of the angle between the lines.
- Sometimes the locus is a **region** of space.
- The locus of points that are no more than a given distance from a single point is the area within a circle.
- The diagram shows the locus of points that are closer to BC than AB.

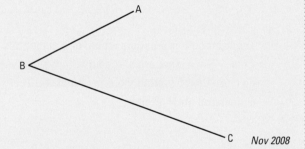

Review exercise

1 Use ruler and compasses to construct the bisector of angle ABC.
You must show all your construction lines.

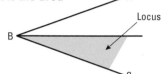

Nov 2008

2 Draw the locus of all points which are equidistant from the points A and B.

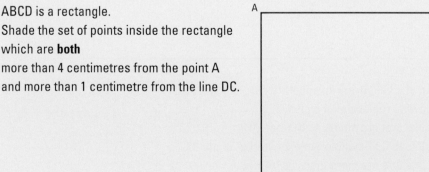

A✕   ✕B

Nov 2008

3 Use ruler and compasses to construct an equilateral triangle with sides of length 6 centimetres.
You must show all your construction lines.   *June 2008*

4 ABCD is a rectangle.
Shade the set of points inside the rectangle which are **both**
more than 4 centimetres from the point A
and more than 1 centimetre from the line DC.

5   ABC is a triangle. Make an accurate copy of ABC.
Shade the region inside the triangle which is **both**
less than 4 centimetres from the point B
and closer to the line AC than the line BC.

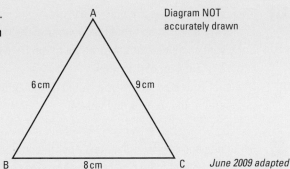

Diagram NOT
accurately drawn

A

6 cm   9 cm

B   8 cm   C

June 2009 adapted

6   There are two sprinklers in the garden, C and D.

C   D
✕←—5 m—→✕

They are 5 m apart. Draw a plan of this situation.
The sprinklers can water the garden for up to a distance of 3 m in all directions.
Show on your plan the part of the garden that is getting double the water.

A03

7   This diagram shows a quadrangle.
Copy the plan shown, using a scale of 1 cm = 5 m.
  a A path goes from D so that it is the same distance
     from AD as from CD.
     Draw this path on your plan.
  b A flower bed is dug into the quadrangle at C so that
     the plants are no further than 5 m from C.
     Draw the flower bed on your plan.
  c A fence is put up that is exactly 5 m from side BC. Draw this fence on your plan.

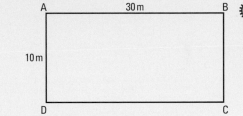

A   30 m   B

10 m

D   C

A03

8   A goat is tied to one wall of a shed.
The shed has dimensions 8 m by 4 m.
Draw a diagram to show the area of grass that can be eaten
by the goat when the rope has a length of **a** 2 m and **b** 4 m.

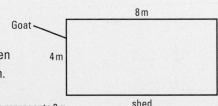

8 m

Goat

4 m

shed

Scale: 1 cm represents 2 m

A03

9   A gardener plants two shrubs 60 cm apart. This is the minimum distance that can be allowed
between the shrubs that he is planting. Draw a diagram to show the gardener the area in which a third
shrub cannot be planted.

A03

10   The diagram shows a plan of a compound.
The compound has a length of 32 m and a width of 20 m,
and is surrounded by a fence.
Draw a scale plan of the compound.
  a A security guard walks around the outside of the compound
     at a distance of 3 m from the fence.
     Draw the path taken by the security guard on your plan.
  b Another guard walks around the inside of the compound at a distance of 2 m from the fence. On your
     plan show the path taken by this guard.

32 m

20 m

A03

C

A03

11 This diagram shows the plan of a room.
Lights are going to be fitted to the walls of the room.
Each light can illuminate an area up to 2 m from where the light is fixed.
Draw a scale plan of the room.
Show on your plan where you would fix the lights so that the entire
perimeter of the room is illuminated.

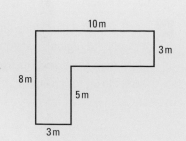

A03

12 Mr Jones and Mr Parry live next door to each other. Their houses are 24 m apart. A cable needs
to be run between the two houses. Mr Jones and Mr Parry want the cable to be laid as far from their
houses as possible. Draw a plan to show how the cable could be laid so they are both happy.

A03

13 As a bicycle moves along a flat road, draw the locus of
a the yellow dot, a point on the centre of the wheel
b the green dot, a point on the circumference of the wheel.

19 PERCENTAGES

Relative humidity is a measure of how much water is held by air at a given temperature as a percentage of the maximum amount of water that air could hold. Under a rainforest canopy, on the forest floor, relative humidity is an amazing 95%. The continent with the lowest humidity is Antarctica where humidity is just 0.03%. This is because water vapour freezes in the extreme low temperatures so very little remains in the air.

⊚ Objectives

In this chapter you will:
- convert between percentages, fractions and decimals
- put lists of percentages, fractions and decimals in order of size
- work out a percentage of a quantity and express one number as a percentage of another.

⟐ Before you start

You need to be able to:
- simplify a fraction
- order fractions
- multiply fractions and decimals
- convert between fractions and decimals.

19.1 Converting between percentages, fractions and decimals and ordering them

⊙ Objectives

- You can convert between percentages, fractions and decimals.
- You can write a list of percentages, fractions and decimals in order of size.

⊘ Why do this?

You need to be able to convert between percentages, fractions and decimals to make sure that you are getting the best discounts in a shop.

⬦ Get Ready

1. What fraction of the shape is shaded?

2. Write $\frac{40}{100}$ in its simplest form.

3. Write $\frac{27}{100}$ as a decimal.

🔑 Key Points

- Per cent means 'out of 100'.
- 40 per cent means 40 out of 100 or $\frac{40}{100}$.
- 40 per cent is written as 40%.

 This large square is divided into 100 small squares.
40 of the 100 small squares are shaded.
40% of the large square is shaded.

- You can write percentages as decimals or fractions.
- You can write decimals or fractions as percentages.
- You can put a list with fractions, decimals and percentages in order of size by changing them to the same type of number.

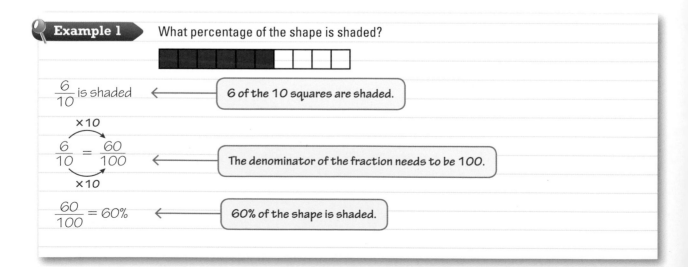

🔍 Example 1

What percentage of the shape is shaded?

$\frac{6}{10}$ is shaded ⟵ 6 of the 10 squares are shaded.

$\overset{\times 10}{\frac{6}{10} = \frac{60}{100}}$ ⟵ The denominator of the fraction needs to be 100.

$\frac{60}{100} = 60\%$ ⟵ 60% of the shape is shaded.

Exercise 19A

Questions in this chapter are targeted at the grades indicated.

G

1 What percentage of each shape is shaded?

a 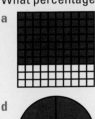 b 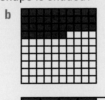 c

d e f

2 For each shape in question **1** write down the percentage of the shape that is not shaded.

3 a Copy this shape and shade 70% of it.

b Copy this shape and shade 40% of it.

4 60% of the houses in a street have a satellite dish.
What percentage of the houses do not have a satellite dish?

5 Louise has some flowers. 25% of the flowers are yellow. What percentage of the flowers are not yellow?

6 73% of bingo players are female. What percentage of bingo players are male?

7 On a wall the tiles are white or green or yellow. 15% of the tiles are green. 20% of the tiles are yellow.
What percentage of the tiles are white?

Example 2 Write 37% as a decimal.

$37\% = \dfrac{37}{100}$ ← Write the percentage as a fraction.

$\dfrac{37}{100} = 37 \div 100 = 0.37$ ← Change the fraction into a decimal.

so $37\% = 0.37$

Exercise 19B

G

1 Write these percentages as decimals.

a 50% b 45% c 62% d 95%
e 29% f 30% g 3% h 7%

2 Write 125% as a decimal.

3 A shop reduced its prices by 12.5%. Write 12.5% as a decimal.

4 A savings account has an interest rate of 3.2%. Write 3.2% as a decimal.

F

Example 3 Write 65% as a fraction in its simplest form.

$65\% = \dfrac{65}{100}$ ← Write the percentage as a fraction.

$\overset{\div 5}{\dfrac{65}{100}} = \dfrac{13}{20}$ ← Simplify the fraction by dividing both 65 and 100 by a common factor (see Section 1.10).
$\underset{\div 5}{}$

so $65\% = \dfrac{13}{20}$ ← $\dfrac{13}{20}$ cannot be simplified.

Exercise 19C

1 Write these percentages as fractions in their simplest form.
   a 60% b 75% c 35% d 90%
   e 5% f 80% g 84% h 32%

2 64% of the spectators at a football match were male.
   Write down the fraction of the spectators that were male. Give your fraction in its simplest form.

3 24% of students cycled to school. What fraction of students cycled to school?
   Give your fraction in its simplest form.

4 A jacket is made from 55% silk and 45% linen. The shopkeeper wants to put on the label the fraction of silk the jacket is made of. Write 55% as a fraction in its simplest form.

5 Write these percentages as fractions in their simplest form.
   a 12.5% b 2.5% c 37.5% d $17\frac{1}{2}\%$

Example 4 Write these numbers in order of size. Start with the smallest number.
   $\dfrac{3}{8}$ 0.4 35%

$\dfrac{3}{8} = 0.375$ ← Change $\dfrac{3}{8}$ into a decimal.

0.4 ← 0.4 is already a decimal.

$35\% = 0.35$ ← Change 35% into a decimal.

0.35 0.375 0.4 ← Write the decimals in order of size (see Section 5.2).

35% $\dfrac{3}{8}$ 0.4 ← Write each number in its original form.

ResultsPlus Examiner's Tip
Remember to show your working out so that an examiner can follow your reasoning.

Exercise 19D

1 **a** Write 23% as a decimal.

    b Write $\frac{1}{4}$ as a decimal.

    c Which is bigger, 23% or $\frac{1}{4}$?

2 **a** Write 74% as a decimal.

    b Write $\frac{7}{10}$ as a decimal.

    c Which is bigger, 74% or $\frac{7}{10}$?

3 Write each list in order of size, starting with the smallest number.

    a $\frac{1}{2}$ 48%   0.45   **b** 55% $\frac{6}{10}$ 0.53

    c 0.7 $\frac{3}{4}$ 68%   **d** 27% $\frac{3}{10}$ 0.2

4 Which is bigger, 15% or $\frac{7}{40}$?

5 Write each list in order of size, starting with the smallest number.

    a 0.4 $\frac{1}{2}$ 45%   30% $\frac{1}{3}$

    b 0.12 $\frac{1}{20}$ 15% $\frac{1}{5}$ 10%

    c $\frac{2}{3}$ 68%   0.63 $\frac{13}{20}$ 0.6

    d 70% $\frac{27}{40}$ 0.65 $\frac{3}{5}$ 62%

G

F

Mixed exercise 19E

1 What percentage of the shape is shaded?

2 15% of the cars in a car park are red. What percentage of the cars are not red?

3 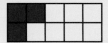

    a What fraction of the shape is shaded?

    b What percentage of the shape is shaded?

    c Copy the shape and shade in more squares so that 80% of the shape is shaded.

4 Write 70% as a decimal.

5 Write 37% as a fraction.

6 Write 28% as fraction. Give your answer in its simplest form.

G

F

7 Write these numbers in order of size. Start with the smallest number.

$\frac{3}{8}$ 35%   0.3 $\frac{1}{3}$ 0.25

E A02 A03

8 Two students did a test. Sam gained $\frac{2}{5}$ of the total marks. Ryan gained 45% of the total marks. Who did better? Explain your answer.

19.2 Finding percentages of quantities

Objectives

○ You can work out a percentage of a quantity.

○ You can use percentages in real-life problems.

Why do this?

This is useful if you need to work out your exam marks as a percentage of the total to see if you have reached the 75% pass mark.

Get Ready

1. Work out $32 \div 4$.   **2.** Work out $80 \div 5$.   **3.** Work out $45 \div 10$.

Key Points

◉ You should know these percentages and their fraction and decimal equivalents.

Percentage	1%	10%	25%	50%	75%
Decimal	0.01	0.1	0.25	0.5	0.75
Fraction	$\frac{1}{100}$	$\frac{1}{10}$	$\frac{1}{4}$	$\frac{1}{2}$	$\frac{3}{4}$

◉ If a percentage can be written as a simple fraction it is easy to work out a percentage of a quantity without using a calculator. For example:

 ◉ to work out 50% of a quantity you work out $\frac{1}{2}$ of it

 ◉ to work out 25% of a quantity you work out $\frac{1}{4}$ of it.

◉ To work out a percentage of a quantity using fractions you should:

 ◉ write the percentage as a fraction, and then

 ◉ multiply the fraction by the quantity.

Example 5

Work out 30% of 50. Do not use a calculator.

Method 1

$10\% = \frac{1}{10}$ ← 10% is equivalent to $\frac{1}{10}$.

$\frac{1}{10}$ of $50 = 50 \div 10 = 5$ ← To find $\frac{1}{10}$ of 50, divide 50 by 10.

10% of 50 = 5

so 30% of $50 = 3 \times 5$ ← 30% is 3 lots of 10%.

     = 15

Method 2

$\frac{30}{100} \times 50$ ← Replace of with times.

30% is $\frac{30}{100}$.

30×0.5 ← Divide by bottom $\frac{50}{100} = 0.5$.

= 15 ← Multiply.

Exercise 19F

1 Work out
  a 50% of £24   b 50% of 80 kg   c 25% of 32 m   d 50% of 76p
  e 25% of £200   f 75% of 12 cm   g 75% of $40   h 25% of £92

2 Work out:
  a 10% of £60   b 10% of 70 km   c 20% of 70 km   d 30% of £120
  e 20% of £80   f 15% of 20 kg   g 70% of 300 ml   h 35% of £40

3 Simon's salary last year was £35 400. He saved 10% of his salary. Simon wants to buy a car costing £3650. Has he saved enough?

4 A packet of breakfast cereal contains 750 g of cereal plus '20% extra free'.
Work out how much extra cereal the packet contains.

5 Jamal earns £28 000 in one year. He gets £1000 tax free. On the remainder he pays income tax at 20%. Work out how much income tax he pays in that year.

6 Hannah earns £6.80 per hour. She is given a pay rise of 5%. Work out how much extra she gets per hour.

7 The price of a new sofa is £480. Leah pays a deposit of 15% of the price.
Work out the deposit she pays.

8 The normal cost of a suit is £120. In a sale the cost of the suit is reduced by 35%.
Work out how much the cost of the suit is reduced by in the sale.

9 Rahma pays income tax. She pays 20% on the first £37 400 of her income and 40% on income over £37 400. Rahma's income last year was £59 400. Work out how much income tax she paid last year.

Example 6   The normal price of a television is £375.
Harry is given a discount of 24%.
Work out the discount that Harry is given.

To work out 24% of 375 it is quicker to use a calculator.

$24\% = \frac{24}{100}$ ←   [Change the percentage to a fraction.]

$\frac{24}{100} \times 375 = 90$ ←   [Key in 2 4 ÷ 1 0 0 × 3 7 5 =]

Harry is given a discount of £90.

Exercise 19G

1 Work out
  a 12% of £40   b 86% of 45 kg   c 54% of £370   d 37% of 640 km
  e 48% of 330 ml   f 23% of $90   g 8% of £170   h 92% of 1500 m

D

2 There are 250 boats in a harbour. 46% of the boats are yachts.
How many yachts are there in the harbour?

3 Alan invested £1200 in a savings account. At the end of the year he received 4% interest.
Work out how much interest he received.

4 There are 225 students in Year 10.
24% of these students study history.
How many of the students study history?

5 Moira's salary is £48 000. Her employer agrees to increase her salary in line with inflation. The rate of inflation this year is 3%. Work out the amount her salary has increased by.

6 In a restaurant a service charge of 12.5% is added to the cost of the meal.
Work out the service charge when the cost of the meal is £60.

7 VAT is charged at the rate of $17\frac{1}{2}$%. Work out how much VAT will be charged on:
a   a ladder costing £84
b   a garage bill of £130.

8 The rate of simple interest is 3% per year. Work out the simple interest paid on £500 in one year.

A02
A03

9 The cash price of a washing machine is £670.
A Credit Plan requires a deposit of 5% of the cash price and 24 monthly payments of £28.
Which is the cheapest way to buy the washing machine. Explain your answer.

C
A02

10 A 100 g tub of margarine has the following nutrional content.

fat	38 g
sodium	1.3 g
carbohydrate	2.8 g
protein	0.1 g

a   What percentage of the margarine is fat?
b   How many grams of fat would there be in a 250 g tub?

A03

11 Roger bought 50 pineapples at 80p each.
He sold all the pineapples.
On each of the first 36 pineapples he made a 35% profit.
On each of the remaining pineapples he made a 40% loss.
Work out the overall profit or loss that Roger made.

19.3 Using percentages

◎ Objectives

○ You can increase and decrease a quantity by a given percentage.
○ You can use a multiplier to work out a percentage increase or percentage decrease.

⊘ Why do this?

Banks and building societies use percentages for interest rates.

⊕ Get Ready

1. Work out 25% of 60.   **2.** Work out 30% of 90.   **3.** Work out 37% of 40.

Key Points

⦿ To increase a quantity by a percentage, work out the increase and add this to the original quantity.

⦿ To decrease a quantity by a percentage, work out the decrease and subtract this from the original quantity.

⦿ An alternative method is to work out the multiplier for an increase or decrease and then multiply the original amount by the multiplier to find the new amount. (See Example 7, method 2.)

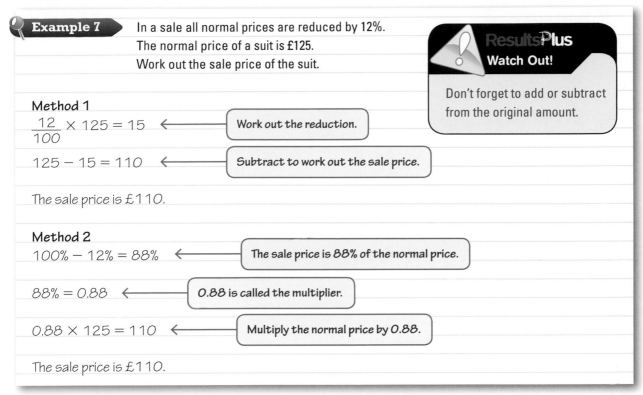

Example 7

In a sale all normal prices are reduced by 12%.
The normal price of a suit is £125.
Work out the sale price of the suit.

ResultsPlus
Watch Out!

Don't forget to add or subtract from the original amount.

A02

Method 1

$\dfrac{12}{100} \times 125 = 15$ ← Work out the reduction.

$125 - 15 = 110$ ← Subtract to work out the sale price.

The sale price is £110.

Method 2

$100\% - 12\% = 88\%$ ← The sale price is 88% of the normal price.

$88\% = 0.88$ ← 0.88 is called the multiplier.

$0.88 \times 125 = 110$ ← Multiply the normal price by 0.88.

The sale price is £110.

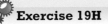

Exercise 19H

1. A packet contains 500 g of cereal plus 20% extra cereal. Work out the weight of cereal in the packet.

E

2. Karen's salary is £26 500. Her salary is increased by 3%. Work out her new salary.

3. The price of rail fares increased by 11%. Before the increase the price of a ticket was £87. Work out the price of the ticket after the increase.

4. A travel company reduced the prices of its holidays by 12%. What is the new price of a holiday which was originally priced at £695?

5. A car tyre costs £48 plus VAT at $17\frac{1}{2}\%$. Work out the total cost of the car tyre.

6. VAT at $17\frac{1}{2}\%$ is added to a telephone bill of £76. Work out the total bill.

7. Katie invests £3600. The interest rate is 3.5% per year. How much will Katie have in her account at the end of one year?

8. Vicky invests £1500 at 4% per year simple interest. Work out the value of her investment after one year.

D

9 Raja bought a car for £8000. In one year the value of the car depreciated by 10%.
Work out the value of the car one year after he bought it.

10 The normal price of a pack of croissants is £1.96.
The normal price is reduced by 25%.
Work out the price after the reduction.

11 A store reduced all normal prices by 15% in a two-day sale.
Work out the sale price of:
a a drill with a normal price of £70
b a lawnmower with a normal price of £180
c a tin of paint with a normal price of £14.

A03

12 In a super-sale a shop reduces its sale prices by a further 10%.

> **SALE**
> $\frac{1}{2}$ **off**
> normal prices
>
> **PLUS**
> an extra
> 10% off
> sale prices

In the super-sale, Steve buys a camera with a normal price of £240.
How much does he pay?

A03

13 Riverside Garage has a loyalty scheme for customers who buy their cars from the garage.
The scheme gives customers a discount on the cost of labour and the cost of parts.
The percentage discount depends on the age of the car.

Age of vehicle (years)	Labour discount	Parts discount
4	10%	5%
5	12.5%	5%
6	15%	5%
7	17.5%	10%
8	20%	10%
9	25%	10%
10 or older	30%	10%

Alan bought a new car from Riverside Garage in September 2005.
Today, Riverside Garage carried out some repairs on the car.

Copy and complete the bill for the repairs.

Item	Cost before discount	% discount	Cost after discount
Labour	£120	..............%	£..............
Parts	£84	..............%	£..............
Riverside Garage		Total before VAT	£..............
		VAT at $17\frac{1}{2}$%	£..............
		Total with VAT	£..............

19.4 Writing one quantity as a percentage of another

◎ Objective

◉ You can write one quantity as a percentage of another.

⊘ Why do this?

Pay rises, profits and losses are often expressed as percentages.

⬥ Get Ready

1. Express 7 as a fraction of 10.

2. Express 48 as a fraction of 72.

3. $\frac{3}{10} = \frac{?}{100}$

🌐 Key Points

◉ To write one quantity as a percentage of another quantity:
   ◉ write the first quantity as a fraction of the second quantity
   ◉ convert the fraction to a percentage.

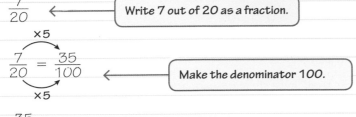

Example 8 Write 7 out of 20 as a percentage. Do not use a calculator.

$\frac{7}{20}$ ← Write 7 out of 20 as a fraction.

$\frac{7}{20} \overset{\times 5}{\underset{\times 5}{=}} \frac{35}{100}$ ← Make the denominator 100.

$\frac{35}{100} = 35\%$

⚙ Exercise 19I

1. **a** Write 7 out of 10 as a percentage.
   b Write £23 out of £50 as a percentage.
   c Write 13 kg out of 20 kg as a percentage.
   d Write 12p out of £1 as a percentage.
   e Write 120 m*l* out of 200 m*l* as a percentage.

2. There are 25 trees in a park. Of these trees, 18 are oak trees.
   What percentage of the trees are oak trees?

3. Tom planted 400 flower seeds. Of these seeds, 360 germinated.
   What percentage of the seeds germinated?

4. A glass contains 500 m*l* of drink. 400 m*l* of the drink is water. What percentage of the drink is water?

5. Chloe scored 24 out of 60 in a test. Write 24 out of 60 as a percentage.

6. 50 people in a club voted to elect a secretary.
   There were three candidates.
   Abi got 40% of the votes, Laura got 12 votes and Faisal got the remaining votes.
   What percentage of the votes did Faisal get?

G

A02

Example 9   A magazine has 72 pages. Forty-five of the pages have advertisements on them.
What percentage of the pages have advertisements on them?

$\dfrac{45}{72}$ ← Write 45 out of 72 as a fraction.

$\dfrac{45}{72} = 0.625$ ← Change the fraction to a decimal using a calculator.

$0.625 \times 100 = 62.5$ ← Multiply the decimal by 100.

62.5% of the pages have advertisements.

Exercise 19J

1 a Write 525 g as a percentage of 750 g.
   b Write £8.40 as a percentage of £120.
   c Write 126 ml as a percentage of 350 ml.
   d Write 312 km as a percentage of 480 km.
   e Write 90p as a percentage of £2.50.

2 A football team played 45 matches. The team won 18 of these matches.
What percentage of the matches did the team win?

3 There are 32 students in a class. On Friday, six of these students were absent.
What percentage of the students were absent on Friday?

4 There are 1650 students in a school. 297 of the students are in Year 11.
What percentage of the students are in Year 11?

5 120 g of cheese contains 18.6 g of carbohydrates and 5.4 g of protein.
What percentage of the cheese is:
   a carbohydrates   b protein?

6 A film audience consists of 108 males and 132 females. What percentage of the audience is female?

7 Sam is mixing sand and gravel.
A mixture of 16 bucketfuls is 25% sand. He wants to make a mixture of 50% sand.
How many bucketfuls of sand must Sam add to make the mixture 50% sand?

Chapter review

● Per cent means 'out of 100'.
● You can write percentages as decimals or fractions.
● You can write decimals or fractions as percentages.
● You can put a list with fractions, decimals and percentages in order of size by changing them to the same type of number.

● You should know these percentages and their fraction and decimal equivalents.

Percentage	1%	10%	25%	50%	75%
Decimal	0.01	0.1	0.25	0.5	0.75
Fraction	$\frac{1}{100}$	$\frac{1}{10}$	$\frac{1}{4}$	$\frac{1}{2}$	$\frac{3}{4}$

● If a percentage can be written as a simple fraction it is easy to work out a percentage of a quantity without using a calculator.

● To work out a percentage of a quantity using a written method you should:
  ● write the percentage as a fraction, and then
  ● multiply the fraction by the quantity.

● To increase a quantity by a percentage, work out the increase and add this to the original quantity.

● To decrease a quantity by a percentage, work out the decrease and subtract this from the original quantity.

● An alternative method is to work out the multiplier for an increase or decrease and then multiply the original amount by the multiplier to find the new amount.

● To write one quantity as a percentage of another quantity:
  ● write the first quantity as a fraction of the second quantity
  ● change the fraction to a percentage.

Review exercise

1 In a survey, 42 out of 60 students said they would prefer to go to a theme park.
   Write 42 out of 60 as a percentage.

2 a Write 10% as a decimal.
   b Write 4% as a decimal.
   c Write 26% as a fraction.
      Give your answer in its simplest form.   *June 2007*

3 The weight of a coin is 25% nickel and 75% copper.
   a Write 25% as a decimal.
   b Write 25% as a fraction.
      Give your answer in its simplest form.

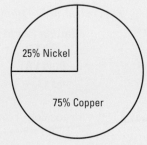

25% Nickel

75% Copper

4 a Write 60% as a fraction. Give your fraction in its simplest form.
   b 55% of the students in a school are female.
      What percentage of students are male?   *Specimen 2006*

5 a Work out 50% of £60.
   b Work out 25% of 20 metres.   *May 2009*

6 180 of the 600 counters in a bag are red.
   Work out 180 as a percentage of 600.   *May 2009*

7 Write 9 out of 12 as a percentage.   *Nov 2008*

G

F

8   **a**   Write 92% as a decimal.
       b   Write 3% as a fraction.
       c   Work out 5% of 400 grams.   *May 2008*

9   A hotel has 56 guests.
       35 of the guests are male.
       a   Work out 35 out of 56 as a percentage.

       40% of the 35 male guests wear glasses.
       b   Write the number of male guests who wear glasses as a fraction of the 56 guests.
          Give your answer in its simplest form.   *Nov 2007*

E

10   The weight of a coin is 8 grams. 25% of the weight is nickel and 75% of the weight is copper. A bag
       contains 10 coins. Work out the weight of
       a   copper
       b   nickel.   *June 2006*

D

11   The weight of some biscuits is 125 g. 18% of the weight is fat. Work out the weight of the fat.

12   A furniture store reduced its prices in a sale. Work out the sale price of:
       a   a bookcase with 50% off the normal price of £498
       b   a bed with 40% off the normal price of £595
       c   a fitted kitchen with 45% off the normal price of £12 000.

13   The population of a town is 54 000. In ten years' time the population is expected to have increased by
       12%. What is the population expected to be in ten years' time?

A02
A03

14   In a maths test, Majda scored 27 out of 40. In the same test Emma scored 70%.
       Who got the better score? Explain your answer.

A03

* **15**   Jack wants to buy a new shed. There are three shops that sell the shed he wants.

      Sheds For U   **Garden World**   **Ed's Sheds**
        25% off   **£210** $\frac{1}{3}$ **off**
    normal price of   **plus VAT**   **usual price of**
        £320   **at** $17\frac{1}{2}$**%**   **£345**

       Jack wants to pay as little as possible. From which of these three shops should Jack buy his shed?

16   A concert ticket costs £45 plus a booking charge of 15%.
       Work out the total cost of a concert ticket.

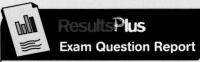

Results **Plus**
Exam Question Report

84% of students answered this question poorly.
Many candidates incorrectly added £15 to £45,
giving an answer of £60.

                                                             June 2007

17 The normal price of a cat basket is £20.
In a sale, the manager reduces the price of the cat basket by 15%.
Work out the price of the cat basket in the sale.

Nov 2008

D

18 The normal price of a computer game is £21.30. The normal price is reduced by 20%.
What is its reduced price?

C

19 Students went to London or to York on a school trip.
The table shows the some information about the children.

a What percentage of all the girls went to York?

b What percentage of all those who went to London were boys?

c What percentage of all the children went to London?

	London	York
Boys	27	13
Girls	18	22

A02

*20 Rachael is a sales manager.
Last year, Rachael had a 10% pay rise. This year, she had a 5% pay rise.
Ziggy says, 'Rachael has had a 15% pay rise over the two years.'
Is Ziggy correct? Explain your answer.

A03

21

	Number of girls	Number of boys
Year 10	108	132
Year 11	90	110

A02

The table gives information about Year 10 and Year 11 at Mathstown School.

a Work out the percentage of students in Year 10 who are girls.

Mathstown School had an end-of-term party.

40% of the students in Year 10 and 70% of the students in Year 11 went to the party.

b Work out the percentage of all students in Years 10 and 11 who went to the party.

Nov 2004

20 THREE-DIMENSIONAL SHAPES

The photo shows a work of art by the artists Christo and Jeanne-Claude in which they wrapped the Pont Neuf Bridge in Paris in 40,876 m² (454,178 sq ft) of silky golden fabric. In order to wrap the bridge, they needed to work out the surface area so they could calculate the amount of fabric required.

◎ Objectives

In this chapter you will:
- ◉ learn how to recognise and draw 3D shapes
- ◉ learn how to draw and interpret plans and elevations
- ◉ find the volumes of shapes made from cuboids
- ◉ find the volumes and surface areas of prisms
- ◉ understand the effect of enlargement for perimeter, area and volume of shapes and solids
- ◉ convert between area and volume measures.

◈ Before you start

You need to be:
- ◉ familiar with two-dimensional shapes.

20.1 Recognising three-dimensional shapes

Objectives

- You can recognise 3D shapes.
- You can count the vertices, edges and faces of 3D shapes.

Why do this?

A designer or architect needs to be able to describe the shape they want to build. The correct names for 3D shapes make it easier to explain what the product will look like.

Get Ready

1. Can you remember the names of these two-dimensional shapes?

Key Points

- Here are some **three-dimensional** shapes.

Cube

Cylinder

Cuboid

Sphere

Cone

> ### ResultsPlus
> ### Examiner's Tip
>
> You need to know the names of all of these shapes.

- **Pyramids** have a base, which can be any shape, and sloping triangular sides.

Triangular pyramid called a tetrahedron

Rectangular-based pyramid

- A **prism** has two parallel faces and a number of rectangular sides joining them.

Triangular prism

Pentagonal prism

Octagonal prism

- The flat surfaces of a 3D shape are called **faces**.
- The lines where two faces meet are called **edges**.
- The point (corner) at which edges meet is called a **vertex**. The plural of vertex is vertices.

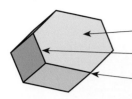

Face (two-dimensional surface).

Edge (where two faces meet).

Vertex (corner where three or more edges meet).

Example 1

a Name this shape.
b How many faces does it have?
c How many edges does it have?
d How many vertices does it have?

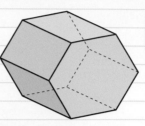

Two parallel hexagons
Rectangular sides

a Hexagonal prism
b 8 faces
c 18 edges
d 12 vertices

The parallel faces are called the cross-section of the prism. The cross-section of this prism is a hexagon.

Exercise 20A

Questions in this chapter are targeted at the grades indicated.

G

1 Name this shape.

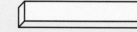

2 This 3D shape has a circular base.
Name the 3D shape.

3 A prism has a base which has five sides. What type of prism is it?

4 This pyramid has a special name. Name this pyramid.

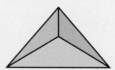

5 Look at the picture. Copy and complete the table with as many 3D shapes as you can find.

	Shape	Object
1	sphere	football
2		
3		
4		
5		
6		
7		
8		
9		
10		
11		
12		

F

Exercise 20B

1 Copy and complete the table below.

	Shape		Faces	Edges	Vertices
A	Cube				
B	Pentagonal prism				
C	Triangular prism				
D	Square-based pyramid				
E	Cuboid				
F	Tetrahedron				
G	Octagonal prism				

2 What is the shape of the cross-section of this prism?

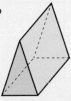

3 Draw a sketch of a prism with a pentagonal cross-section.

4 A pyramid has six triangular faces.
  a What is the shape of its other face?   **b** What type of pyramid is it?

A03 E

20.2 Drawing 3D shapes

⊙ Objectives

○ You can draw 3D shapes using isometric paper.
○ You can draw nets and recognise solids from their nets.

◈ Why do this?

A designer must make an accurate 3D drawing of the container they wish to make and draw an accurate 2D plan that can be made from a flat piece of card or other material.

◈ Get Ready

Draw these shapes.

1. A cube   **2.** A triangular prism   **3.** A square-based pyramid

Key Points

- **Isometric paper** will help you to make scale drawings of three-dimensional objects.
- Isometric paper must be the right way up – **vertical** lines down the page and no **horizontal** lines.
- A **net** is a pattern of flat (2D) shapes that can be folded to make a hollow **solid** shape.

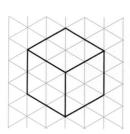

This cube has sides of 2 cm.

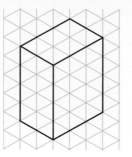

This cuboid has height 4 cm, length 3 cm and width 2 cm.

This prism has a triangular face.

Shapes can be joined together.

This cuboid can be made from a piece of card shaped like this.

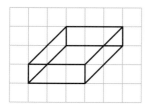

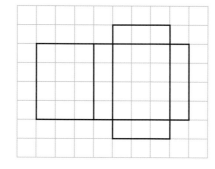

> **Results Plus**
> **Examiner's Tip**
>
> Draw the shape at an angle as if you are looking from the bottom-right corner.

There are other possible nets to make the same cuboid.

Exercise 20C

E

1 On isometric paper draw a cube of side 3 cm.

2 Use isometric paper to draw a cuboid with height 2 cm, width 4 cm and length 3 cm.

3 The diagram shows a shape made up of three cubes.
On isometric paper draw a different shape made up of the same three cubes.

isometric paper   vertical   horizontal   net   solid

E

4 Use isometric paper to make full-sized drawings of these prisms.

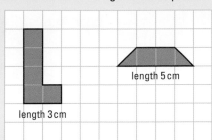

length 5 cm

length 3 cm

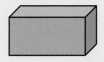 **Exercise 20D**

E

1 The diagrams below show some solids.

1   **2**   **3**   **4**

Three nets are shown below (not drawn to scale).

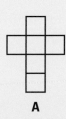   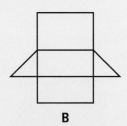   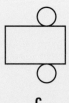

A   **B**   **C**

  a Write which net can be folded to make each solid.

  b Draw a net for the missing solid.

2 Sketch the following solids and their nets.

  a hexagonal prism   **b** tetrahedron

3 Charlie has some sheets of metal of length 2 m and width 1.5 m.
He wants to make containers to collect different types of rubbish.
What shape should he use for the container for

  a the collection of waste paper from an office

  b the collection of waste engine oil from a garage?

In each case, make a sketch of the net for the container.
Explain why you have chosen a container of this shape.

 A02 A03

20.3 Plans and elevations

Objective

● You can draw and interpret plans and elevations.

Why do this?

Architects, engineers and draughtsmen must be able to represent 3D objects with accurate 2D drawings.

Get Ready

1. A, B and C have been drawn from a different viewpoint. For each, write down the letter of the congruent 3D shape.

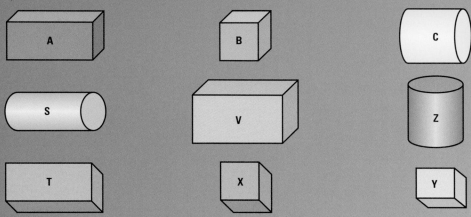

Key Points

● **Plans** and **elevations** show the 2D view of a 3D object drawn from different angles.

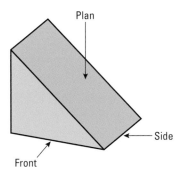

● The view from above is called the plan.

● The view from the front is called the **front elevation**.

● The view from the side is called the **side elevation**.

Example 2

Use squared paper to help you to draw the plan, front elevation and side elevation of this triangular prism.

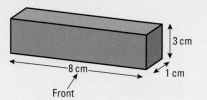

Plan

Side elevation

Front elevation

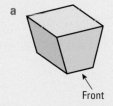

Plan

Dotted lines help to show how the plan and elevations fit together.

Front elevation

Side elevation

Exercise 20E

1 Use squared paper to make an accurate drawing of the plan, front elevation and side elevation for this cuboid.

3 cm

8 cm

1 cm

Front

2 Sketch the plan, front elevation and side elevation for these shapes.

a

Front

b

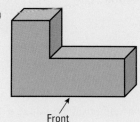

Front

c

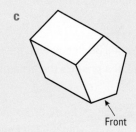

Front

3 The diagram shows the plan and front elevation of an object.

a Sketch the side elevation.

b Draw a 3D sketch of the shape.

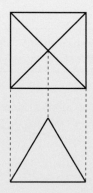

D

A03

A03

20.4 Volumes

◈ Get Ready

1. What is the area of this rectangle? Each small square has side 1 cm.

2. What is the area of this shape? Each small square has side 1 cm.

3. Use the formula
   area = length × width
   to find the area of this shape.

3 cm

9 cm

4. Find the area of these two-dimensional shapes.

a
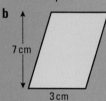
4 cm
5 cm

b
7 cm
3 cm

c
3 cm
4 cm
2 cm

d
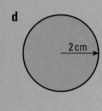
2 cm

🔍 Key Points

- The **volume** of a 3D shape is the amount of space it takes up. The diagram shows a cube of side 1 cm. Its volume is 1 cm³.

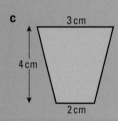

- The volume of a 3D shape with measurements in centimetres is the number of centimetre cubes it contains.
- The volume of a prism is the area of the **cross-section** × its length.

Example 3

The diagram shows a cuboid made from centimetre cubes. Find the volume of the cuboid.

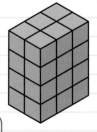

Remember some of the cubes are hidden. Try building the shape using multilink.

Multiply the length by the width (3 × 2) to find the number of cubes (6) in each layer.

There are 6 centimetre cubes in each layer.

Multiply this by the height (4) to give the number of cubes altogether (24).

There are 4 layers.

The volume of the cuboid is 6 × 4 = 24 cm³.

So the volume of the cuboid = length × width × height. When the lengths are measured in metres, volume is measured in m³.

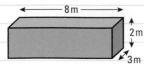

Example 4   Work out the volume of the cuboid shown below.

Volume of the cuboid = length × width × height
Volume = 8 × 3 × 2 m³ = 48 m³

ResultsPlus
Examiner's Tip

Remember, you need to specify the units in all of your answers.

Exercise 20F

1   The diagrams below show prisms that have been made from centimetre cubes.
Find the volume of each prism.

a
b
c

2   Work out the volumes of these cuboids.

a    3 cm   1 cm   1 cm

b 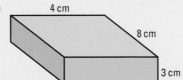   4 cm   8 cm   3 cm

3   Work out the volume of the following cuboids.
a   length 7 m, width 8 m and height 4 m
b   length 17 mm, width 12 mm and height 3 mm

4   Work out the volume of a cube of side 7 cm.

5   A cuboid has volume 150 cm³. Its width is 3 cm and its height is 10 cm.
Find the length of the cuboid.

6   A cereal packet was measured and found to have height 35 cm, width 20 cm and depth 10 cm.
Give the dimensions of a box that will hold 24 cereal packets.

A03

Example 5 Work out the volume of this prism.

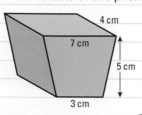

ResultsPlus
Examiner's Tip

If you can't remember how to use the formula, you can split the trapezium into a rectangle and two triangles.

The cross-section is a trapezium. ← | Decide on the shape of the cross-section. |

area of a trapezium $= \frac{1}{2}(a + b) \times h$

$= \frac{1}{2} \times (3 + 7) \times 5$ ← | Put in the lengths you know. |

$= 25 \, cm^2$ ← | Don't forget the units. |

Volume $= 25 \times 4$
$= 100 \, cm^3$

Exercise 20G

C

1 The diagram shows a prism.
The cross-section of this prism is a hexagon with an area of $5 \, m^2$.
If the length of the prism is 3 m, what is the volume of the prism?

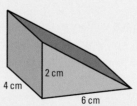

2 The diagram shows the cross-section of a prism of length 8 cm.
Work out the volume of the prism.

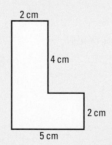

3 The diagram shows a prism with a right-angled triangle as its cross-section.
Work out the volume of the prism.

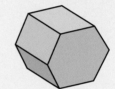

A02
A03
4 A metal bar of length 30 cm has a square cross-section of side 5 cm.
The metal is to be recast into triangular rods of length 10 cm. The cross-section of the rods is a right-angled triangle as shown.
  a Work out the volume of the metal bar.
  b Work out the volume of each rod.
  c Work out the maximum number of rods that can be made from the bar.
  d Work out how much metal is left over.

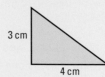

Example 6 Work out the volume of this cylinder.

A cylinder is a prism with a circular cross-section.

Volume of a cylinder = Area of the circular end × height

$$= \pi \times r^2 \times h$$
$$= \pi \times 5 \times 5 \times 2 = 50\pi$$
$$= 157.0796327$$
$$= 157\ cm^3 \text{ (to 3 significant figures)}$$

Put in the values you know for the radius and the height.

Use the π button on your calculator.

Exercise 20H

In this exercise give your answers to 3 significant figures. Do not forget to give the units.

1 Find the volume of these cylinders.

a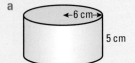
b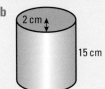
c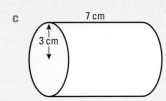

2 A hexagonal prism has a length of 6 cm. If the area of the cross-section is 9 cm², calculate its volume.

3 A triangular prism has length 12 cm. The triangular face has base 8 cm and height 9 cm. Calculate the volume of the prism.

4 A water tank on a farm is in the shape of a cylinder. It has a radius of 2 m and a height of 2.5 m and it is full of water.
The farmer uses water from the tank to fill troughs for horses. The trough is shown below.

A02
A03

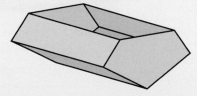

The cross-section of the trough is a pentagon of area 0.8 m².
Each trough is 1.2 m long.
Work out:
a how many troughs the farmer can fill from the water tank
b how much water is left over.

C

20.5 Surface area

- ○ You can find the surface area of a prism.
- ○ You can find the surface area of a cylinder.

? Why do this?

The surface area is the amount of space on an object available for design, information or advertising.

⬦ Get Ready

1. Find the area of this two dimensional shape.

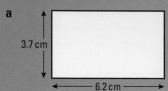

a
3.7 cm
6.2 cm

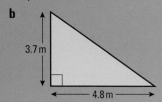

b
3.7 m
4.8 m

c
4.3 cm

🌐 Key Points

- ◉ The **surface area** of a prism is the area of the net that can be used to build the shape.
- ◉ Area is measured in square units — mm^2, cm^2, m^2 and km^2 are common.
- ◉ Each of the six faces of a cube is a square, so the surface area of a cube is 6 times the area of one face.
- ◉ The area of each square face of this cube is $2 \times 2 = 4\,cm^2$.
  So the surface area of the cube is $6 \times 4 = 24\,cm^2$.
- ◉ A cylinder is a prism with a circular cross section.

2 cm
2 cm
2 cm

🔍 Example 7 Work out the surface area of this cuboid.

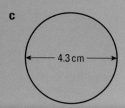

4 cm
9 cm
3 cm

9 cm

A   3 cm

B   4 cm

Sketch the net of the cuboid and label with the lengths you know. Label the shapes you will find the area of with a letter.

C   3 cm

4 cm E   D   F   4 cm

3 cm   9 cm   3 cm

Area of rectangle A	$= 3 \times 9 = 27\,cm^2$	
Area of rectangle B	$= 4 \times 9 = 36\,cm^2$	
Area of rectangle C	$= 3 \times 9 = 27\,cm^2$	
Area of rectangle D	$= 4 \times 9 = 36\,cm^2$	
Area of rectangle E	$= 3 \times 4 = 12\,cm^2$	
Area of rectangle F	$= 3 \times 4 = 12\,cm^2$	
Total surface area	$= 150\,cm^2$	

Results Plus
Examiner's Tip

Show the examiner evidence that you can find the area of each shape. Marks will be awarded for this step!

Results Plus
Examiner's Tip

Remember units of area are cm^2.

Exercise 20I

1   Find the surface area of this cuboid.

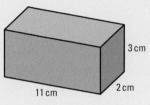

3 cm
2 cm
11 cm

E

2   The diagram shows a piece of cheese.
Work out the surface area of the cheese.

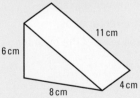

11 cm
6 cm
4 cm
8 cm

A02

3   A cereal packet in the shape of a cuboid has height 35 cm, width 20 cm and depth 20 cm.
Work out the surface area of the cereal packet.

A02

4   The diagram shows a plastic part from a child's toy.
Work out the surface area of the part.

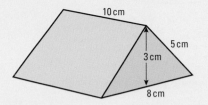

10 cm
5 cm
3 cm
8 cm

A02   D

5   A cube has a surface area of 24 cm².
Work out the length of the side of the cube.

A02
A03   C

6   A block of icing is sold in the shape of a prism.
The cross section of the prism is in the shape of a square.
The length of the prism is twice the length of the side of the square.
The surface area of the prism is 90 cm².
Work out the length of the side of the square.

A03

Example 8   Work out the surface area of this cylinder.

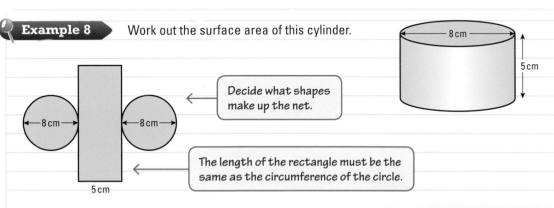

8 cm
5 cm

Decide what shapes
make up the net.

8 cm
8 cm
5 cm

The length of the rectangle must be the
same as the circumference of the circle.

Circumference $= \pi d = 8\pi = 25.13$ cm
Area circle $= \pi r^2 = \pi \times 4 \times 4 = 50.27$ cm²
Area rectangle $= 5 \times 25.13 = 125.65$ cm²
Total surface area of a cylinder $= 2 \times$ circle $+$ rectangle
$= 2 \times 50.27 + 125.65$
$= 226.19$ cm²

Results Plus
Examiner's Tip

Showing your method ensures you
get all the marks!

C

Exercise 20J

1   The diagram shows a sketch of the net of a cylinder.

   a Work out the length of the rectangle which makes up the
      covered surface.
   b Work out the area of the curved surface.
   c Work out the area of the circular ends.
   d Work out the total surface area of the cylinder.

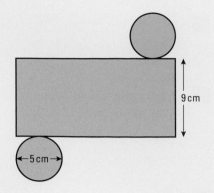

2   Find the surface area of these cylinders.

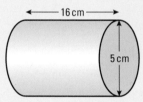

3   The diagram shows a can of food.

A label giving information about the product completely covers the curved surface area.
Work out the area of the label.

20.6 Perimeter, area and volume

⊙ **Objectives**

● You understand the effect of enlargement for perimeter, area and
  volume of shapes and solids.
● You understand that enlargement does not have the same effect on area
  and volume.
● You can use simple examples of the relationship between enlargement
  and area and volume of simple shapes and solids.

⟐ **Why do this?**

When using scale models it is
important to be able to find the
surface area and volume of the
full-sized object.

⟐ **Get Ready**

Find the area of these shapes.

1.
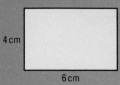

2.
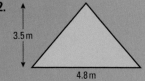

3.

Key Points

● The diagram shows squares of side 1 cm, 2 cm, 3 cm and 4 cm.
   The squares are all similar.

Square A has side 1 cm   perimeter 4 cm   It takes 1 cm^2 to fill the square.

Square B has side 2 cm   perimeter 8 cm   It takes 4 cm^2 to fill the square.

Square C has side 3 cm   perimeter 12 cm   It takes 9 cm^2 to fill the square.

Square D has side 4 cm   perimeter 16 cm   It takes 16 cm^2 to fill the square.

● When a shape is enlarged each length is multiplied by the scale factor.
   Enlarged perimeter = original perimeter × scale factor

● The area is length × length. Both lengths must be multiplied by the scale factor.
   Enlarged area   = (length × scale factor) × (length × scale factor)
                 = length × length × scale factor × scale factor
                 = original area × scale factor^2

● The diagram shows a cube of side 1 cm,
   a cube of side 2 cm and a cube of side 3 cm.

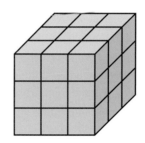

● The volume of the cube of side 1 cm is $1 × 1 × 1 = 1$ cm^3.
● The volume of the cube of side 2 cm is $2 × 2 × 2 = 8$ cm^3.
● The volume of the cube of side 3 cm is $3 × 3 × 3 = 27$ cm^3.
● Enlarged volume = (length × scale factor) × (length × scale factor) × (length × scale factor)
                      = length × length × length × scale factor × scale factor × scale factor
                      = original volume × scale factor^3

Example 9   The diagram shows rectangles A and B made up of cm squares.

Rectangle A has a perimeter of 10 cm and an area of 6 cm^2.

Rectangle A is enlarged by a scale factor of 3 to become rectangle B.

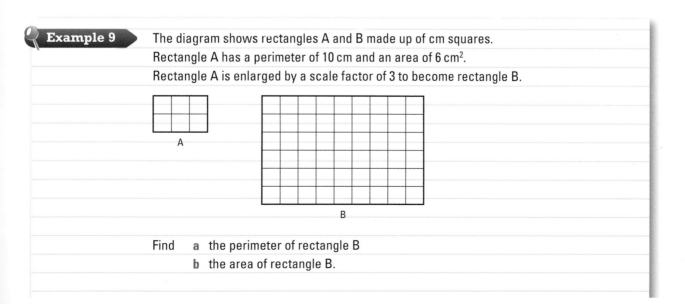

Find   **a**   the perimeter of rectangle B.
        b   the area of rectangle B.

a Perimeter of A = 10 cm
   Perimeter of B = 3 × 10 = 30 cm ⟵

> Multiply the perimeter by the scale factor.

> Multiply the area by the square of the scale factor.

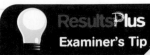

ResultsPlus
Examiner's Tip

Show the examiner evidence that you know the scale factor is squared when you are looking for the area. This will be worth a mark!

b Area of A = 6 cm²
   Area of B = 6 × 3² = 6 × 9 = 54 cm²

Exercise 20K

E

1 The diagram shows a rectangle A.
   a Find the perimeter of rectangle A.
   b Find the area of rectangle A.
   c On squared paper draw an enlargement P of rectangle A with scale factor 2.
   d Find the perimeter of rectangle P.
   e Find the area of rectangle P.
   f On squared paper draw an enlargement R of rectangle A with scale factor 3.
   g Find the perimeter of rectangle R.
   h Find the area of rectangle R.
   i Work out the value of   i $\dfrac{\text{Perimeter P}}{\text{Perimeter A}}$ ii $\dfrac{\text{Perimeter R}}{\text{Perimeter A}}$
      Write down anything you notice about these results.
   j Work out the value of   i $\dfrac{\text{Area P}}{\text{Area A}}$ ii $\dfrac{\text{Area R}}{\text{Area A}}$
      Write down anything you notice about the results.

C A03

   k S is an enlargement of rectangle A with scale factor 8.
      i What is the perimeter of rectangle S?
      ii What is the area of rectangle S?

2 A rectangle with length 4 cm and width 5 cm is enlarged with scale factor 4.
   a Find the new length and width.
   b Find the perimeter of the enlarged rectangle.
   c Find the area of the enlarged rectangle.

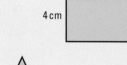

5 cm
4 cm

3 The triangle shown with area 20 cm² is enlarged with scale factor 3.
   Find the area of the new triangle.

8 cm

4 The diagram shows a scale drawing of a flowerbed.
   The real flowerbed is 50 times bigger than the drawing.
   a The perimeter of the drawing is 1.2 m.
      Find the perimeter of the real flowerbed.
   b The area of the drawing is 1 m².
      Find the area of the real flowerbed.

5   A triangle with base 4 cm and area 6 cm² is enlarged.
    The enlarged triangle has base 12 cm.
    Work out the area of the enlarged triangle.

6   A photograph with length 4 cm and width 6 cm is enlarged.
    The enlarged rectangle has length 16 cm.
    Work out the perimeter and area of the enlargement.

A02

Example 10   The diagram shows a cuboid P with length 10 cm and volume 40 cm³.
                 The cuboid is enlarged with scale factor 4.
                 Find:
                 a the length of the enlarged cuboid
                 b the volume of the enlarged cuboid.

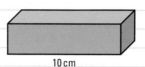

10 cm

a Length = 4 × 10 cm = 40 cm.
b Volume = 4 × 4 × 4 × 40 = 2560 cm³.

Exercise 20L

1   **a** Find the scale factor of the enlargement.
    b Find the volume of cuboid B.

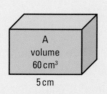

A
volume
60 cm³
5 cm

B

10 cm

C

2   The small cylinder has volume 20 cm³.
    The large cylinder is an enlargement of the small cylinder.
    Find the volume of the large cylinder.

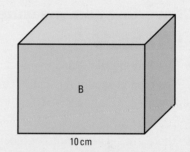

5 cm

15 cm

A03

3   Two similar jugs have heights 4 cm and 7 cm respectively.
    The smaller jug holds 50 cm³.
    How much does the larger jug hold?

4   Two cylindrical tins have heights 5 cm and 9 cm.
    The large tin is an enlargement of the small tin.
    The volume of the small tin is 375 cm³.
    Find the volume of the large tin.

5 A cereal company is designing a large box to hold individual packets of cereal.

The individual packets are cuboids with height 10 cm, width 8 cm and depth 4 cm.

The large box must be an enlargement of an individual packet.

Work out the height, width and depth of a large box which will hold 25 packets.

20.7 Converting units of measure

◎ Objectives

○ You can convert between area measures.

○ You can convert between volume measures.

○ You can convert between units of volume and units of capacity.

❓ Why do this?

When objects are enlarged it is often necessary to use different units. Measurements on a map may be in millimetres or centimetres, whilst in real life they are in miles or kilometres.

◈ Get Ready

1. Convert the following measurements to the units given.

   a 250 cm to m   **b** 350 mm to cm   **c** 6.3 m to cm

   d 350 m to km   **e** 1.5 km to m   **f** 3.6 cm to mm

Converting area measures

🔑 Key Points

◉ The two squares A and B are congruent. They are exactly the same size and shape.

The area of square A is 1 m × 1 m = 1 m^2.

The area of square B is 100 cm × 100 cm = 10 000 cm^2.

$$1 m^2 = 10\,000 cm^2$$

◉ There are similar results for other units.

Length	Area
1 cm = 10 mm	1 cm^2 = 10 × 10 = 100 mm^2
1 m = 100 cm	1 m^2 = 100 × 100 = 10 000 cm^2
1 km = 1000 m	1 km^2 = 1000 × 1000 = 1 000 000 m^2

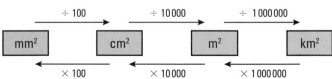

◉ When changing from a large unit to a smaller one, multiply.

◉ When changing from a small unit to a larger one, divide.

🔍 Example 11 ▸ Change 3.6 m^2 to cm^2.

$3.6 m^2 = 3.6 × 10\,000 = 36\,000 cm^2$. ←

> Metres are bigger than cm so the answer must be bigger. This is a clue to multiply.

Example 12 Change 4 375 000 m² to km².

> Metres are smaller than km so the answer must be smaller. This is a clue to divide.

$$4\,375\,000 \text{ m}^2 = 4\,375\,000 \div 1\,000\,000 = 4.375 \text{ km}^2.$$

Exercise 20M

1 Change to cm².
   a 3 m²
   b 4.5 m²
   c 300 mm²
   d 34 mm²

2 Change to m².
   a 6 km²
   b 0.4 km²
   c 20 000 cm²
   d 3450 cm²

3 The area of a pane of glass is 14 500 cm².
   Write down the area of the pane of glass in m².

4 A plan shows the area of a square flowerbed as 225 cm².
   The real flowerbed is also a square but its sides are 10 times bigger.
   Find the area of the real flowerbed, giving your answer in m².

5 A door has height 2 m and width 70 cm.
   a Write down the area of the door in m².
   b Write down the area of the door in cm².

6 Dave is tiling the wall in a bathroom. The wall is 3 m long and is to be tiled to a height of 1.5 m.
   The tiles are square with length 15 cm.
   How many tiles are needed?

7 The diagram shows a wall with a window in it.
   Work out the shaded area.
   Give your answer in m².

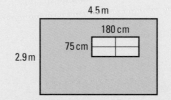

8 The perimeter of a rectangle is 1.5 m.
   The length of the longest side is 45 cm.
   Find the area of the rectangle.

Converting volume measures

Key Points

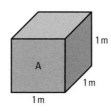
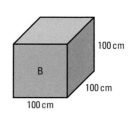

- The two cubes A and B are congruent.
  They are exactly the same size and shape.
  The volume of cube A is 1 m × 1 m × 1 m = 1 m³.
  The volume of cube B is 100 cm × 100 cm × 100 cm = 1 000 000 cm³.

$$1 \text{ m}^3 = 1\,000\,000 \text{ cm}^3$$

There are similar results for other units.

Length	Volume
1 cm = 10 mm	$1\,cm^3 = 10 \times 10 \times 10 = 1000\,mm^3$
1 m = 100 cm	$1\,m^3 = 100 \times 100 \times 100 = 1\,000\,000\,cm^3$
1 km = 1000 m	$1\,km^3 = 1000 \times 1000 \times 1000 = 1\,000\,000\,000\,m^3$

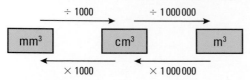

Litres are often used to measure the capacity or amount a container can hold.

$1\,litre = 1000\,cm^3$

$1\,cm^3 = 1\,ml$

Example 13   a Change $7\,m^3$ to cm^3 b Change $500\,000\,cm^3$ to m^3

a $7 \times 1\,000\,000 = 7\,000\,000\,cm^3$
b $500\,000 \div 1\,000\,000 = 0.5\,m^3$

Example 14   a Change 2.5 litres to cm^3 b Change $5000\,cm^3$ to litres

a $2.5 \times 1000 = 2500\,cm^3$
b $5000 \div 1000 = 5\,litres$

Exercise 20N

1 Change to cm^3.
   a $4\,m^3$ b $4.5\,ml$ c $400\,mm^3$ d 3 litres

2 Change to litres.
   a $400\,ml$ b $5600\,cm^3$ c $1\,m^3$ d $3500\,mm^3$

3 How many mm^3 are there in 1 litre?

4 The diagram shows a cuboid.
   Work out the volume of the cuboid in
   a cm^3
   b mm^3.

   4 cm 3 cm 12 cm

5 The petrol tank of a car holds 42 litres of fuel.
   How many cm^3 is this?

6 A bottle of medicine holds 0.5 litres.
   How many $5\,cm^3$ doses are contained in the bottle?

7 A swimming pool has length 50 m, width 9 m and depth 1.5 m.
   How much water does it hold?
   Give your answer in litres.

8 Individual bars of soap are cuboids of size 8 cm × 6 cm × 4 cm.
The bars are placed in the large cardboard container shown in the diagram.
How many bars can be packed into the container?

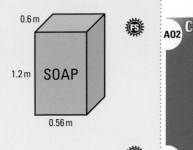

9 A path is to be resurfaced with 1 cubic metre of concrete.
The path is 20 m long and 2.4 m wide.
What is the depth of the concrete in cm?

10 A cylinder holds 26 litres of molten metal.
The metal is to be made into cubes of side 3 cm.
How many cubes can be made?

Chapter review

- Some examples of **three-dimensional** shapes are: **cube, cylinder, cuboid, sphere** and **cone**.
- **Pyramids** have a base which can be any shape, and sloping triangular sides.
- A **prism** has two parallel faces and a number of rectangular sides joining them.
- The flat surfaces of a 3D shape are called **faces**.
- The lines where two faces meet are called **edges**.
- The point (corner) at which edges meet is called a **vertex**. The plural of vertex is vertices.
- **Isometric paper** will help you to make scale drawings of three-dimensional objects.
- Isometric paper must be the right way up – **vertical** lines down the page and no **horizontal** lines.
- A **net** is a pattern of flat (2D) shapes that can be folded to make a hollow **solid** shape.
- **Plans** and **elevations** show the 2D view of a 3D object drawn from different angles.
- The view from above is called the plan.
- The view from the front is called the **front elevation**.
- The view from the side is called the **side elevation**.
- The **volume** of a 3D shape is the amount of space it takes up.
- The volume of a 3D shape with measurements in centimetres is the number of centimetre cubes it contains.
- The volume of a prism is the area of the **cross-section** × its length.
- The **surface area** of a prism is the area of the net that can be used to build the shape.
- A cylinder is a prism with a circular cross section.
- Enlarged perimeter = original perimeter × scale factor.
- Enlarged area = original area × scale factor^2.
- Enlarged volume = original volume × scale factor^3.

Length	Area
1 cm = 10 mm	$1 \text{ cm}^2 = 10 \times 10 = 100 \text{ mm}^2$
1 m = 100 cm	$1 \text{ m}^2 = 100 \times 100 = 10\,000 \text{ cm}^2$
1 km = 1000 m	$1 \text{ km}^2 = 1000 \times 1000 = 1\,000\,000 \text{ m}^2$

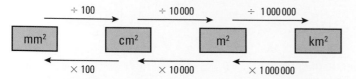

Length	Volume
1 cm = 10 mm	$1 \text{ cm}^3 = 10 \times 10 \times 10 = 1000 \text{ mm}^3$
1 m = 100 cm	$1 \text{ m}^3 = 100 \times 100 \times 100 = 1\,000\,000 \text{ cm}^3$
1 km = 1000 m	$1 \text{ km}^3 = 1000 \times 1000 \times 1000 = 1\,000\,000\,000 \text{ m}^3$

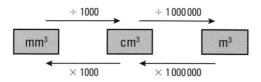

- Litres are often used to measure the capacity or amount a container can hold.
  1 litre = 1000 cm³
  1 cm³ = 1 m*l*

Review exercise

1 Here is a solid prism made from centimetre cubes.

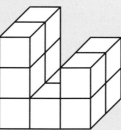

Work out the volume of the solid prism.

Nov 2008

2 Write down the mathematical name of each of these two 3D shapes.

i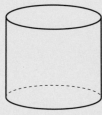
ii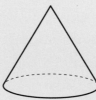

Nov 2008

3   Find the volume of this prism.

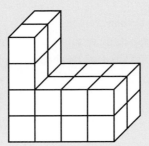

Diagram **NOT**
accurately drawn

represents 1 cm³

June 2008

4   Here is a diagram of a cuboid.
Write down the number of
i faces   ii edges   iii vertices.

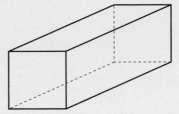

June 2008

5   Here is a diagram of a 3D prism.
Write down the number of
i faces   ii edges   iii vertices.

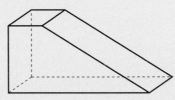

Nov 2007

6   Work out the volume of the cuboid.

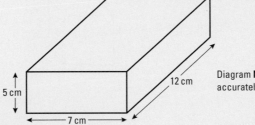

12 cm

Diagram **NOT**
accurately drawn

5 cm

7 cm

Nov 2008

7   The diagram shows a pyramid with a square
base.

Diagram **NOT**
accurately drawn

3 cm

3 cm

3 cm

3 cm

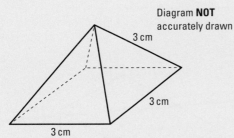

The length of each side of the base is 3 cm.
The length of each sloping edge is 3 cm.
On a copy of the grid of centimetre squares,
draw an accurate net of the pyramid.

Nov 2008

E

8 The diagram shows some nets and some solid shapes.
An arrow has been drawn from one net to its solid shape.
Draw an arrow from each of the other nets to its solid shape.

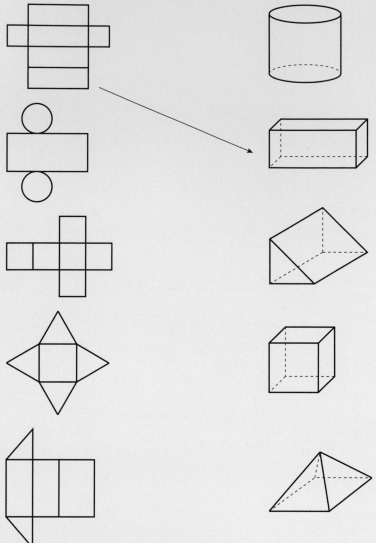

Nov 2008

9 The diagram shows a prism drawn on a centimetre isometric grid.

a On a centimetre grid, draw the front elevation of the prism from the direction marked by the arrow.

b On a centimetre grid, draw a plan of the prism.

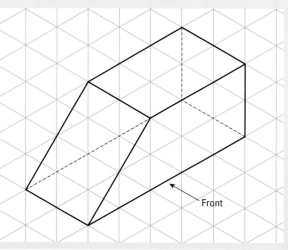

Front

Nov 2007

10 The diagram shows a solid object made of 6 identical cubes.

  a On a centimetre grid, draw the side elevation of the solid object from the direction of the arrow.

  b On a centimetre grid, draw the plan of the solid object.

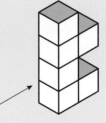

June 2007

11 The diagram shows a solid cuboid.
On a triangular isometric grid, make an accurate full-size drawing of the cuboid.

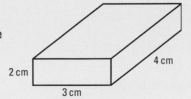

4 cm
2 cm
3 cm

June 2007

12 The diagram shows a cuboid.
The cuboid has
a volume of 300 cm³,
a length of 10 cm,
a width of 6 cm.
Work out the height of the cuboid.

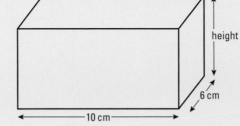

height

Diagram **NOT** accurately drawn

6 cm

10 cm

Nov 2006

13

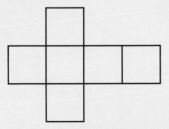

Here is a net of a 3D shape.
The diagrams show four 3D shapes.

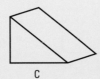

A   B   C   D

Write down the letter of the 3D shape that can be made from the net.

June 2007

14 The diagram represents a solid made from 5 identical cubes.
On the grid below, draw the view of the solid from direction A.

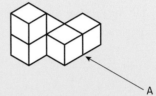

A

Nov 2008

E

D

415

D

15 Here are the front elevation, side elevation and the plan of a 3D shape.
Draw a sketch of the 3D shape.

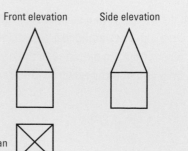

Front elevation   Side elevation

Plan

June 2008

16 Cereal boxes are packed into cartons.
A cereal box measures 4 cm by 6 cm by 10 cm.
A carton measures 20 cm by 30 cm by 60 cm.

The carton is completely filled with cereal boxes.
Work out the number of cereal boxes that will completely fill **one** carton.

Box

10 cm

6 cm

4 cm

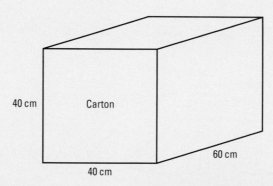

Diagram **NOT** accurately drawn

60 cm

Carton

30 cm

20 cm

Nov 2007

A02 A03

17

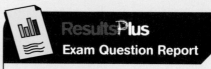

40 cm   Carton

60 cm

40 cm

Diagram **NOT** accurately drawn

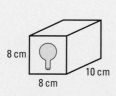

8 cm

8 cm   10 cm

A light bulb box measures 8 cm by 8 cm by 10 cm.
Light bulb boxes are packed into cartons.
A carton measures 40 cm by 40 cm by 60 cm.
Work out the number of light bulb boxes which can completely fill **one** carton.

ResultsPlus
Exam Question Report

85% of students answered this question poorly. Some students calculated that there were 5 in a row, 5 high and 6 back, but added these numbers together rather than multiplying them.

June 2007

18 Louise makes chocolates.
Each box she puts them in has
Volume = 1000 cm³
Length = 20 cm
Width = 10cm.

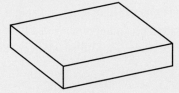

a Work out the height of a box.

Louise makes 350 chocolates.
Each box will hold 18 chocolates.

b Work out
  i how many boxes Louise can fill completely,
  ii how many chocolates will be left over.

19 A piece of plastic has been designed as part of a game.
The plan and front elevation of a prism have been
drawn on centimetre squared paper.

a On squared paper draw a side elevation of the shape.
b Draw a 3D sketch of the shape.
c Find the volume of the prism.

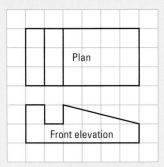

Plan

Front elevation

20 A solid cylinder has a radius of 4 cm and a height of 10 cm.
Work out the volume of the cylinder.
Give your answer correct to 3 significant figures.

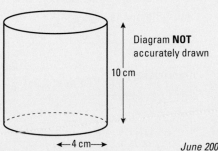

Diagram **NOT**
accurately drawn

10 cm

←4 cm→

June 2008

21 An oil drum is in the shape of a cylinder.
The lid has been removed.
Calculate the surface area of the oil
drum without the lid.

0.8 m

1.5 m

22 Here is a triangular prism.
Calculate the volume of the prism.

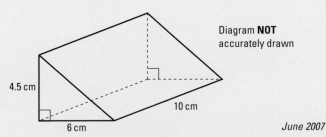

Diagram **NOT**
accurately drawn

4.5 cm

10 cm

6 cm

June 2007

23 Rainfall on a flat rectangular roof 10 m by 5.5 m flows into a cylindrical tub of diameter 3 m.
Find, in cm, the increase in depth of the water in the tub caused by a rainfall of 1.2 cm.
Give your answer correct to 2 significant figures.

24 Cylindrical cans of radius 4 cm and height 10 cm are filled from a drum containing 0.5 m³ of oil.
a Calculate the number of cans filled.
b Calculate the quantity of oil left over in cm³.

* **25** Danni is going to redecorate her bedroom.
She wants to wallpaper all four walls and fit new carpet and underlay.
The dimensions of her room are shown in the diagram.

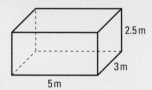

The carpet that she likes costs £18.50 per m² and the underlay £3.50 per m² fully fitted.
The wallpaper costs £25 per roll and the shop has a handy table that she can use to estimate how many rolls she needs to buy.

Room perimeter (m)	Height (m)			
	2.1 to < 2.25	2.25 to < 2.4	2.4 to < 2.55	2.55 to < 2.7
10	4	5	5	5
11	5	5	5	5
12	5	6	6	6
13	6	6	7	7
14	6	7	7	7
15	7	7	8	8
16	7	8	9	9
17	8	8	9	9
18	8	9	10	10
19	9	9	10	10
20	10	10	12	12

Calculate how much Danni will have to budget for the decoration.

26 You are planning a party for 30 children.

You buy some concentrated orange squash and some plastic cups.

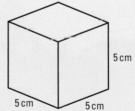

Each plastic cup will have 150 m*l* of drink in it. (150 m*l* = 150 cm³)

a Check that the plastic cup shown can hold 150 m*l* of drink. Use the formula:

$$\text{volume} = \frac{\pi \times h \times d^2}{4}$$

Each of the 30 children at the party will have a maximum of three drinks of orange squash.

Each plastic cup is to be filled with 150 m*l* of drink.

The squash needs to be diluted as shown on the bottle label.

A bottle of concentrated orange squash contains 0.8 litres of squash and costs £1.25.

b How many bottles of concentrated orange squash do you need for the party?
How much will they cost in total?

27 The triangular prism has length 10 cm and volume 90 cm³.

The small prism is a scale model for a larger prism which is a part for
a piece of machinery.

The length of the larger prism is 30 cm.

All of its lengths are enlarged by the same scale factor.

Find the volume of the larger prism.

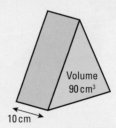

Diagram **NOT**
accurately drawn

28 **a** A solid cube has sides of length 5 cm.
Work out the total surface area of the cube.
State the units of your answer.

The volume of the cube is 125 cm³.

b Change 125 cm³ into mm³.

Nov 2009

29 Work out the total surface area of the triangular prism.

Diagram **NOT**
accurately drawn

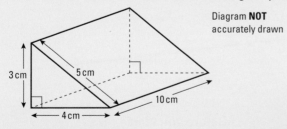

May 2008

30 The volume of this cube is 8 m³.
Change 8 m³ to cm³.

Diagram **NOT**
accurately drawn

June 2007

21 EQUATIONS AND INEQUALITIES

In algebra, letters are used to represent unknown numbers. A sky diver can work out how long it takes, t, to go from a speed of 4 metres per second to a speed of 20 metres per second. The sky diver would need to solve the equation $4 + 10t = 20$.

◉ Objectives

In this chapter you will:
- ◉ set up simple equations
- ◉ solve equations with one unknown value
- ◉ use equations to solve problems
- ◉ find approximate solutions to complicated equations using trial and improvement
- ◉ understand and use inequality signs
- ◉ show inequalities on a number line
- ◉ solve inequalities.

◉ Before you start

You need to:
- ◉ know that letters can represent numbers
- ◉ be able to collect like terms
- ◉ know simple number bonds.

21.1 Using simple equations

Objectives

- You can write simple equations using letters.
- You can set up simple equations from word problems.
- You can solve simple equations using number bonds.

Why do this?

You might use a simple equation to work out how long it would take you to get drive to your friend's house as you would know the distance and the average speed you could travel.

Get Ready

1. Write down

  a 13×6 **b** 7×19 **c** 15×7

Key Point

- In an equation, a symbol or letter represents an unknown number. For example

$5 \times \boxed{} + 2 = 17$ and

$5x + 2 = 17$ are equations.

Example 1   Write these equations using letters.

    a $\boxed{} + 9 = 15$ **b** $4 \times \boxed{} - 3 = 21$

a $x + 9 = 15$ ←   Replace $\boxed{}$ with a letter, usually x.
b $4x - 3 = 21$

Example 2   I think of a number.
I multiply the number by 3, and the answer is 21.

    a Write this problem as an equation.
    b What number did I think of?

a $3x = 21$ ←   Let x stand for the number I thought of. Write $3 \times x$ as $3x$.

b $3 \times 7 = 21$ ← $21 \div 3 = 7$
The number I thought of was 7.

Exercise 21A

Questions in this chapter are targeted at the grades indicated.

In questions 1–10, write each equation using a letter.

1 $\boxed{} - 7 = 9$

2 $\boxed{} + 3 = 11$

3 $8 \times \boxed{} = 32$

4 $9 + \boxed{} = 20$

G

G

5. $7 \times \square - 3 = 32$

6. $\square \times 3 = 21$

7. $3 \times (\square + 1) = 24$

8. $\square \times 4 + 5 = 13$

9. $(5 + \square) \times 2 = 16$

10. $7 + 5 \times \square = 27$

In questions 11–15
a express the problem as an equation
b find the number.

11. Amy thinks of a number and adds 7 to it.
The answer is 11.

12. Ben thinks of a number and multiplies it by 5.
The answer is 45.

13. Irina thinks of a number. She adds 2 to it and multiplies the result by 4.
The answer is 24.

14. David thinks of a number. He multiplies it by 2 and subtracts 5 from the result.
The answer is 13.

15. Hannah thinks of a number. She multiplies it by 6 and adds 7 to the result.
The answer is 31.

21.2 Solving equations with one operation

◎ Objective

● You can use the balancing method of solving equations for one operation.

❖ Why do this?

Equations can be used to work out unknown quantities.

⬆ Get Ready

Express each problem as an equation.
1. Chloe thinks of a number. She multiplies it by 3 and subtracts 4 from the result. The answer is 14.
2. Abdul thinks of a number. He multiplies it by 6 and adds 2. The result is 26.
3. Sarah thinks of a number. She multiplies it by 7 and subtracts 4. The result is 17.

🔍 Key Points

◉ You can solve an equation by rearranging it so that the letter is on its own on one side of the equation.
◉ To rearrange an equation you can:
   ◉ add the same number to both sides
   ◉ subtract the same number from both sides
   ◉ multiply both sides by the same number
   ◉ divide both sides by the same number.
   This is called the **balance method**.

Example 3 Solve the equation $d + 4 = 9$.

$$d + 4 = 9$$
$$d + 4 - 4 = 9 - 4 \quad \longleftarrow \boxed{\text{Subtract 4 from both sides.}}$$
$$d = 5$$

ResultsPlus
Examiner's Tip

'Solve' means find the value of the letter.

Example 4 Solve the equation $a - 7 = 4$.

$$a - 7 = 4$$
$$a - 7 + 7 = 4 + 7 \quad \longleftarrow \boxed{\text{Add 7 to both sides.}}$$
$$a = 11$$

Exercise 21B

Solve these equations.

1. $a + 6 = 7$
2. $y + 3 = 5$
3. $h + 2 = 9$
4. $p - 5 = 4$
5. $q - 3 = 7$
6. $d - 6 = 2$
7. $x + 3 = 3$
8. $t - 4 = 0$
9. $r + 7 = 10$
10. $k + 2 = 3$
11. $n + 1 = 2$
12. $x - 2 = 3$
13. $m + 7 = 12$
14. $y - 7 = 9$
15. $w + 5 = 5$
16. $q - 10 = 2$
17. $5 + p = 7$
18. $6 + t = 6$
19. $a + 19 = 31$
20. $21 + x = 21$
21. $p - 15 = 23$
22. $7 = a + 3$
23. $6 = b + 5$
24. $10 = v + 10$

Exercise 21C

Solve these equations.

1. $a + 5 = 10$
2. $p - 4 = 7$
3. $q - 3 = 5$
4. $x + 7 = 15$
5. $y + 4 = 17$
6. $s + 12 = 15$
7. $x - 7 = 15$
8. $y - 4 = 17$
9. $s - 12 = 15$
10. $a + 5 = 6$
11. $p - 5 = 6$
12. $c + 17 = 21$
13. $5 + a = 6$
14. $11 + p = 16$
15. $12 + q = 12$
16. $4 = a + 2$
17. $5 = b + 3$
18. $12 = c - 3$
19. $10 = p + 5$
20. $11 = y - 10$
21. $15 = t + 10$
22. $12 = p + 12$
23. $12 = p - 12$
24. $p + 12 = 12$

F

F

Example 5 Solve the equation $5x = 25$.

$$5x = 25$$

$$\frac{5 \times x}{5} = \frac{25}{5}$$ ← Divide both sides by 5.

$$x = \frac{25}{5} = 5$$ ← Cancel $\frac{5 \times x}{5} = x$

Exercise 21D

Find the value of the letter in these equations.

1 $3a = 6$ **2** $4p = 8$ **3** $5p = 15$

4 $6s = 18$ **5** $2k = 10$ **6** $7u = 28$

7 $2g = 14$ **8** $5l = 35$ **9** $6j = 12$

10 $8f = 32$ **11** $3r = 27$ **12** $5v = 45$

Example 6 Solve the equation $\frac{y}{3} = 2$.

$$\frac{y}{3} = 2$$

$$\frac{y \times 3}{3} = 2 \times 3$$ ← Multiply both sides by 3.

$$y = 2 \times 3$$
Cancel $\frac{y \times 3}{3}$
$$y = 6$$

Exercise 21E

Find the value of the letter in these equations.

1 $\frac{a}{2} = 5$ **2** $\frac{b}{5} = 4$ **3** $\frac{s}{4} = 3$

4 $\frac{c}{6} = 5$ **5** $\frac{t}{4} = 6$ **6** $\frac{s}{8} = 9$

7 $\frac{h}{6} = 12$ **8** $\frac{f}{4} = 7$ **9** $\frac{d}{3} = 15$

10 $\frac{a}{3} = 15$ **11** $\frac{b}{5} = 8$ **12** $\frac{r}{4} = 13$

13 $\frac{a}{12} = 5$ **14** $\frac{b}{2} = 16$ **15** $\frac{k}{3} = 16$

Mixed exercise 21F

Solve these equations.

1 $a + 4 = 5$

2 $b + 3 = 6$

3 $c + 4 = 9$

4 $p - 3 = 6$

5 $q - 2 = 2$

6 $d - 6 = 2$

7 $2p = 6$

8 $4r = 8$

9 $3s = 6$

10 $4 + r = 7$

11 $6 + e = 7$

12 $7 + p = 7$

13 $\dfrac{a}{2} = 6$

14 $\dfrac{b}{5} = 12$

15 $\dfrac{s}{5} = 6$

F

E

21.3 Solving equations with two operations

Objective

⊙ You can use the balancing method of solving equations for two operations.

Get Ready

Solve the equations.

1. $9 + q = 16$ **2.** $x - 5 = 11$ **3.** $5r = 30$ **4.** $\dfrac{p}{7} = 4$

Key Points

⊙ In an equation with two operations, deal with the $+$ or $-$ first.
⊙ Solutions to equations can be whole numbers, fractions or decimals.
⊙ Solutions to **linear equations** can be negative.

Example 7   Solve

   a $3a + 2 = 11$ **b** $4p - 3 = 13$ **c** $\dfrac{x}{5} + 2 = 6$

a $3a + 2 = 11$

   $3a + 2 - 2 = 11 - 2$ ← Take 2 from both sides.

   $3a = 9$

   $3a \div 3 = 9 \div 3$ ← Divide both sides by 3.

   $a = 3$

b $4p - 3 = 13$

$4p - 3 + 3 = 13 + 3$ ← Add 3 to both sides.

$4p = 16$

$4p \div 4 = 16 \div 4$ ← Divide both sides by 4.

$p = 4$

c $\frac{x}{5} + 2 = 6$

$\frac{x}{5} + 2 - 2 = 6 - 2$ ← Take 2 from both sides.

$\frac{x}{5} = 4$

$\frac{x}{5} \times 5 = 4 \times 5$ ← Multiply both sides by 5.

$x = 20$

Exercise 21G

Solve these equations.

1 $2a + 1 = 5$ **2** $2a - 1 = 5$ **3** $3a + 2 = 8$

4 $3a - 5 = 4$ **5** $3p + 7 = 7$ **6** $3p + 7 = 13$

7 $q + 5 = 17$ **8** $5r - 6 = 4$ **9** $6t - 12 = 18$

10 $7f - 12 = 9$ **11** $2r - 11 = 15$ **12** $10a - 5 = 5$

13 $10a + 5 = 5$ **14** $4d + 7 = 19$ **15** $5c - 2 = 18$

16 $\frac{a}{3} + 2 = 3$ **17** $\frac{z}{5} + 1 = 2$ **18** $\frac{r}{6} + 4 = 7$

19 $\frac{s}{4} + 6 = 9$ **20** $\frac{b}{3} + 7 = 13$ **21** $\frac{c}{4} - 2 = 4$

22 $\frac{f}{3} - 6 = 3$ **23** $\frac{h}{2} - 4 = -2$ **24** $\frac{x}{5} - 1 = 2$

Example 8   Solve $5m - 8 = 3$.

$5m - 8 = 3$

$5m - 8 + 8 = 3 + 8$ ← Add 8 to both sides.

$5m = 11$

$5m \div 5 = 11 \div 5$ ← Divide both side by 5.

$m = \frac{11}{5}$ or $2\frac{1}{5}$ or 2.2

ResultsPlus
Examiner's Tip

Unless the equation says give your answer in its simplest form, an improper fraction such as $\frac{11}{5}$ is ok.

Exercise 21H

Solve the equations.

1 $2a + 3 = 6$ 2 $2a - 4 = 3$ 3 $3a + 7 = 15$

4 $3a - 6 = 7$ 5 $5p + 7 = 15$ 6 $5p - 7 = 15$

7 $5e + 3 = 3$ 8 $4t + 3 = 9$ 9 $8j - 7 = 5$

10 $7c - 4 = 7$ 11 $8k + 3 = 5$ 12 $3d - 7 = 3$

13 $9u + 7 = 9$ 14 $4q - 4 = 5$ 15 $7y + 6 = 15$

E

Example 9   Solve $3k + 11 = 2$.

$$3k + 11 = 2$$
$$3k + 11 - 11 = 2 - 11$$
$$3k = -9$$
$$3k \div 3 = -9 \div 3$$
$$k = -3$$

Take 11 from both sides.

Divide both sides by 3.

Remember negative number ÷ positive number = negative number
(see Sections 1.8 and 1.9).

Exercise 21I

Solve these equations.

1 $2a + 3 = 1$ 2 $2a + 5 = 1$ 3 $2a + 9 = 1$

4 $3a + 8 = 5$ 5 $3a + 7 = 1$ 6 $5p + 12 = 2$

7 $2s + 7 = -3$ 8 $5p - 2 = -12$ 9 $4k - 5 = -9$

10 $8h + 10 = 2$ 11 $4y + 12 = -8$ 12 $3e + 47 = 20$

13 $6t - 12 = -12$ 14 $3w + 4 = 1$ 15 $2c + 15 = 11$

16 $13a + 9 = 9$

E

Mixed exercise 21J

Solve these equations.

1 $2s + 4 = 10$ 2 $5d + 3 = 18$ 3 $8m - 7 = 33$

4 $4h - 2 = 14$ 5 $4k + 7 = 43$ 6 $3y + 7 = 13$

E

E

7 $5p + 2 = 9$	**8** $4f + 4 = 17$	**9** $3s - 6 = 5$
10 $-7g - 4 = 12$	**11** $4f - 5 = 12$	**12** $5k - 12 = 6$
13 $-3s - 15 = 2$	**14** $6j - 3 = 19$	**15** $9b + 7 = 2$
16 $-2r + 12 = 5$	**17** $5t + 15 = -2$	**18** $7y - 15 = -21$
19 $3e - 5 = -6$	**20** $-4f - 7 = -2$	**21** $5g + 17 = 15$
22 $4h + 4 = 0$	**23** $-3c - 5 = 0$	**24** $8s + 9 = 4$
25 $\frac{z}{2} + 2 = 4$	**26** $\frac{x}{5} - 3 = 2$	**27** $\frac{p}{2} - 5 = -3$
28 $\frac{c}{3} + 4 = -2$	**29** $\frac{a}{8} - 1 = 5$	**30** $-\frac{e}{3} + 2 = 10$

D

21.4 Solving equations with brackets

◎ Objective

○ You can solve equations that have brackets.

◆ Get Ready

Solve these equations.

1. $3x - 4 = 23$ **2.** $\frac{b}{6} + 2 = 4$ **3.** $-4x + 4 = 6$

🔍 Key Point

◉ In an equation with brackets, expand the bracket first.

🔍 Example 10   Solve the equation $5(2x + 3) = 7$.

$$5(2x + 3) = 7$$
$$5 \times 2x + 5 \times 3 = 7 \quad \longleftarrow \quad \boxed{\text{Multiply each term inside the bracket by 5 to expand the brackets.}}$$
$$10x + 15 = 7$$
$$10x + 15 - 15 = 7 - 15 \quad \longleftarrow \quad \boxed{\text{Take 15 from both sides.}}$$
$$10x = -8$$
$$10x \div 10 = -8 \div 10 \quad \longleftarrow \quad \boxed{\text{Divide both sides by 10.}}$$
$$x = -\frac{8}{10} \text{ or } -\frac{4}{5} \text{ or } -0.8$$

Example 11 Solve the equation $\dfrac{y-7}{4} = 2$.

$\dfrac{y-7}{4} = 2$ ← In this expression $\dfrac{y-7}{4}$, the division sign acts as a bracket so $\dfrac{y-7}{4} = \frac{1}{4}(y-7)$.

$\frac{1}{4}(y-7) = 2$

$4 \times \frac{1}{4}(y-7) = 4 \times 2$ ← Multiply both sides by 4.

$y - 7 + 7 = 8 + 7$ ← Add 7 to both sides.

$y = 15$

Exercise 21K

Solve the equations.

1 $5(a-5) = 70$

2 $6(b+5) = 30$

3 $\dfrac{c}{6} = 4$

4 $3(d-5) = 15$

5 $5(e+2) = 40$

6 $\dfrac{f+4}{5} = 4$

7 $4g + 5 = 29$

8 $\dfrac{h}{3} - 5 = 2$

9 $4(m-4) = 12$

10 $9p - 1 = 2$

11 $6(q+5) = 30$

12 $5v + 3 = 7$

13 $\dfrac{x}{3} + 7 = 5$

14 $3(y-1) = 2$

15 $3c + 5 = 2$

16 $2(b-3) = 3$

17 $3(2d-5) = 27$

18 $\dfrac{n-3}{6} = 2$

19 $\dfrac{t+10}{6} = 1$

20 $\dfrac{3c+4}{3} = 2$

D

C

21.5 **Solving equations with letters on both sides**

Objective

- You can solve equations that have letters on both sides.

Why do this?

Knowing how to solve equations helps you solve other problems such as finding one weight when given another.

Get Ready

Solve the equations.

1. $\dfrac{g+2}{4} = 6$

2. $\dfrac{a}{10-2} = 3$

3. $4(b-5) = 4$

Key Point

- In an equation with a letter on both sides, use the balance method to rearrange the equation so that the letter is on one side only.

Example 12 Find the value of p in the equation $5p - 2 = 3p + 6$.

$5p - 2 = 3p + 6$
$5p - 3p - 2 = 3p - 3p + 6$ ← Take $3p$ from both sides.
$2p - 2 = 6$
$2p - 2 + 2 = 6 + 2$ ← Add 2 to both sides.
$2p = 8$
$p = 4$ ← Divide both sides by 2.

Exercise 21L

Solve the equations.

1 $2a + 9 = a + 5$ 2 $3c - 1 = c + 9$ 3 $5p - 7 = 2p + 11$

4 $8b + 9 = 3b + 14$ 5 $9q - 8 = 2q + 13$ 6 $x + 13 = 5x + 1$

7 $4d + 17 = 8d - 3$ 8 $7y = 2y + 15$ 9 $3n + 14 = 5n$

10 $5k + 1 = 2k + 1$ 11 $4u + 3 = 2u + 8$ 12 $7r - 3 = 2r + 9$

13 $6v - 7 = 3v + 7$ 14 $9t + 5 = 4t + 9$ 15 $7m - 2 = 3m + 8$

16 $3g + 4 = 9g - 1$ 17 $5b + 6 = 7b + 5$ 18 $2h + 7 = 8h - 1$

19 $3e = 7e - 18$ 20 $9f = 3f + 4$

21.6 Solving equations with negative coefficients

Objective

● You can solve equations with negative coefficients using the balance method.

Get Ready

Solve these equations.

1. $10a = 5a + 5$ **2.** $3b + 9 = 6b - 3$ **3.** $8c - 5 = 3c + 10$

Key Points

● A **coefficient** is the number in front of an unknown.
● You solve equations with negative coefficients using the balance method you have used in the previous sections.

Example 13　Solve the equations.

$$\textbf{a}\quad 7 - 3x = 19 \qquad \textbf{b}\quad 4 - 3x = 7 - 5x$$

a
$$7 - 3x = 19$$
$$7 = 19 + 3x \quad \leftarrow \boxed{\text{Add } 3x \text{ to both sides.}}$$
$$7 - 19 = 19 - 19 + 3x \quad \leftarrow \boxed{\text{Take 19 from both sides.}}$$
$$-12 = 3x$$
$$-12 \div 3 = 3x \div 3 \quad \leftarrow \boxed{\text{Divide both sides by 3.}}$$
$$-4 = x$$
$$So\ x = -4 \quad \leftarrow \boxed{\text{Rewrite so that it is in the form } x =.}$$

b
$$4 - 3x = 7 - 5x \quad \leftarrow \boxed{\text{Add } 5x \text{ to both sides.}}$$
$$4 + 2x = 7$$
$$2x = 3 \quad \leftarrow \boxed{\text{Subtract 4 from both sides.}}$$
$$x = \frac{3}{2} \text{ or } 1\frac{1}{2} \quad \leftarrow \boxed{\text{Divide both sides by 2.}}$$

Exercise 21M

Solve these equations.

1. $8 - x = 6$
2. $9 - 2x = 1$
3. $40 - 3x = 1$
4. $3x + 2 = 10 - x$
5. $4(x + 1) = 11 - 3x$
6. $9 - 2x = x$
7. $9 - 5x = 3x + 1$
8. $2 - x = x$
9. $1 - 6x = 9 - 7x$
10. $5 - 6x = 9 - 8x$
11. $3 - 4x = 8 - 9x$
12. $17 - 6x = 5 - 3x$
13. $3 - 4x = 15$
14. $7 - 6x = 7$
15. $8 - 2x = 3$
16. $5 + 2x = 8 - 3x$
17. $8 + 3x = 1 - 4x$
18. $5(4 - x) = 5 + 4x$
19. $13 - 2x = 3 - 7x$
20. $3 - 9x = 5 - 6x$

C

21.7 Using equations to solve problems

Objective

- You can set up equations from mathematical and practical situations and solve the problem.

Why do this?

You can use equations to solve problems, such as finding the distance travelled with a given starting velocity, time and acceleration.

Get Ready

1. Two of the angles in a triangle are 140° and 20°. What is the third angle?
2. There are three angles at a point. Two of the angles are 110° and 90°. What is the third angle?
3. The perimeter of a square is 40 cm. What are the side lengths?

Key Point

- You can solve problems in mathematics and other subjects by setting up equations and solving them.

Example 14

In the diagram, ABC is a straight line.
Work out the size of angle DBC.

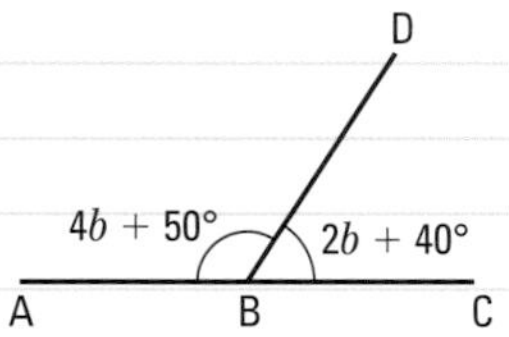

$4b + 50 + 2b + 40 = 180$ ← Write an equation in terms of b, using the sum of the angles on a straight line = 180°.

$6b + 90 = 180$ ← Collect the terms.

$6b = 180 - 90$

$6b = 90$ ← Divide both sides by 6.

$b = 15$

angle DBC $= 2b + 40$

$= 2 \times 15 + 40$ ← Substitute $b = 15$.

$= 30 + 40$

$= 70°$

Exercise 21N

E A03 **1** I think of a number. I multiply it by 7 and subtract 9.
The result is 47. Find the number.

A03 **2** I think of a number. I multiply it by 3 and subtract the result from 50.
The answer is 14. Find the number.

D A03 **3** The sizes of the angles of a triangle are $a + 30°$, $a + 40°$ and $a - 10°$.
Find the size of the largest angle.

4 The diagram shows three angles at a point.
Find the size of each angle.

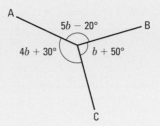

5 The lengths, in centimetres, of the sides of a triangle are $3x - 4$, $x + 5$ and $15 - 2x$.
The perimeter of the triangle is 24 cm.
Find the length of each side.

6 I think of a number. I multiply it by 7 and subtract 6 from the result.
The answer is the same as when I multiply the number by 4 and add 27 to the result.
Find the number.

7 The length of each side of a square is $2y - 5$ centimetres. The perimeter of the square is 36 cm.
Find the value of y.

8 Gwen is 39 years older than her son. She is also 4 times as old as he is.
Find Gwen's age.

9 The length of a rectangle is 3 cm greater than its width. The perimeter of the rectangle is 54 cm.
Find its length.

10 The diagram shows a rectangle.
Find the values of x and y.

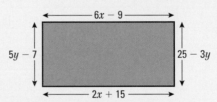

21.8 Solving equations by trial and improvement

◉ Objective

◉ You can find approximate solutions to equations using trial and improvement.

⬆ Get Ready

1. The angles in a triangle are $x°$, $(2x + 20)°$ and $(3x + 40)°$. Find x.
2. I am 30 years older than my daughter. Our combined age is 54. How old is my daughter?
3. The length of each side of an equilateral triangle is $x + 3$ cm. The perimeter is 36 cm. Find x.

🔍 Key Points

◉ There are some equations that can only be solved using the method known as '**trial and improvement**'.
◉ It is helpful to set out your work in a table (see Example 15).

Example 15 Use trial and improvement to solve the equation $x^3 + x = 16$.
Give your answer correct to 2 d.p.

> $x^3 + x = 16$ is a cubic equation. The highest power of x is 3.

Method
Estimate a value for x.
Calculate $x^3 + x$ using your estimate.
Compare your answer with 16.
If your answer is too small, choose a bigger value for x.
If your answer is too big, choose a smaller value for x.
Keep repeating this process until you find a value for x correct to 2 d.p. which makes $x^3 + x$ as close as possible to 16.

x	$x^3 + x$	Bigger or smaller than 16?
2	10	Too small
3	30	Too big
2.5	18.125	Too big
2.4	16.224	Too big
2.3	14.467	Too small
2.35	15.327 875	Too small
2.36	15.504 256	Too small
2.37	15.682 053	Too small
2.38	15.861 272	Too small
2.39	16.041 919	Too big
2.385	15.951 416 63	Too small

> Try $x = 2.385$ to find out whether the solution is closer to 2.38 or 2.39.

The solution lies between 2.385 and 2.39.
So $x = 2.39$ to 2 d.p.

Exercise 21O

C

1. Use a trial and improvement method to solve $x^3 + x = 8$, giving your answer correct to 2 d.p.

2. Use trial and improvement to solve $\dfrac{x^2 + x}{5} = 17$, giving your answer correct to 2 d.p.

3. Use a trial and improvement method to solve $x^3 + 4x = 100$, giving your answer correct to 1 d.p.

4. Use trial and improvement to solve $\dfrac{x^3}{2 + x} = 50$, giving your answer correct to 1 d.p.

5. Use trial and improvement to solve $\dfrac{x^3 + 1}{x} = 10$, giving your answer correct to 2 d.p.

6. Use trial and improvement to solve $2x^3 + 2x = 50$, giving your answer correct to 2 d.p.

7. The equation $x^3 - 4x = 24$ has a solution between 3 and 4.
   Use a trial and improvement method to find this solution. Give your answer correct to 1 d.p.
   You must show all your working.

21.9 **Introducing inequalities**

⊙ Objectives

⊙ You know and understand inequality signs.
⊙ You can put in the correct inequality sign to make a statement true.

⟳ Why do this?

Inequalities can be used to compare quantities.

The number of apples is greater than the number of pears.

⬦ Get Ready

Put these numbers in order. Start with the smallest.

1. −15   −19   24   5   1.34

🔑 Key Points

◉ > means **greater than**.
◉ ⩾ means **greater than or equal to**.
◉ < means **less than**.
◉ ⩽ means **less than or equal to**.

Example 16 Put the correct inequality sign between each pair of numbers to make a true statement.

a 7, 8   **b** 9, 6

a 7 < 8 ⟵ | Use < as 7 is less than 8. |

b 9 > 6 ⟵ | Use > as 9 is greater than 6. |

Examiner's Tip

Sometimes in the examination the question will ask for integers. These are the same as positive and negative whole numbers.

Example 17 Write down the values of x that are whole numbers and satisfy these inequalities.

a $3 < x < 8$ **b** $-3 \leqslant x < 2$

a $3 < x < 8$ ⟵ | This means x is greater than 3 but less than 8. These whole numbers would satisfy this statement. |
   4, 5, 6, 7

b $-3 \leqslant x < 2$ ⟵ | This means x is greater than or equal to −3 and less than 2. |
   −3, −2, −1, 0, 1 ⟵ | Write the numbers in order of size. |

Exercise 21P

D

1 Put the correct sign ($<$ or $>$ or $=$) between each pair of numbers to make a true statement.

a 4, 6   b 5, 2   c 12, 8   d 6, 6

e 15, 8   f 3, 24   g 10, 3   h 0, 0.1

i 6, 0.7   j 4.5, 4.5   k 0.2, 0.5   l 4.8, 4.79

2 Write down whether each statement is true or false. If it is false, write down the pair of numbers with the correct sign.

a $6 > 4$ b $2 > 6$ c $6 > 6$ d $6 > 8$

e $6 < 5$ f $8 = 14$ g $7 < 6.99$ h $6 > 6.01$

i $7 < 0$ j $4 < 4$ k $6 = 4$ l $6 > 0.84$

C

3 Write down the values of x that are whole numbers and satisfy these inequalities.

a $4 < x < 6$ b $3 < x < 8$

c $0 \leqslant x < 4$ d $3 < x < 6$

e $1 < x \leqslant 4$ f $2 < x < 6$

g $4 \leqslant x < 7$ h $-2 \leqslant x < 4$

i $-1 < x < 5$ j $-2 < x \leqslant 6$

k $-3 \leqslant x < 3$ l $-4 \leqslant x \leqslant 2$

m $0 < x < 5$ n $-1 < x \leqslant 4$

o $-5 \leqslant x < 0$ p $-3 \leqslant x \leqslant 3$

21.10 Representing inequalities on a number line

◎ Objective

◦ You can show inequalities on a number line.

⬙ Get Ready

Write the values of x that are whole numbers and satisfy these inequalities.

1. $-3 \leqslant x \leqslant 4$ **2.** $-5 \leqslant x < 2$ **3.** $-6 \leqslant x \leqslant 7$

🔍 Key Points

◉ You can show **inequalities** on a number line.

◉ An empty circle shows that the value is not included and a filled circle shows that the value is included (see Example 19).

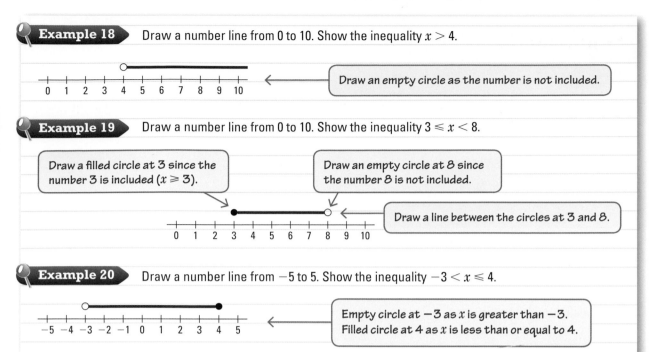

Example 18 Draw a number line from 0 to 10. Show the inequality $x > 4$.

Draw an empty circle as the number is not included.

Example 19 Draw a number line from 0 to 10. Show the inequality $3 \leqslant x < 8$.

Draw a filled circle at 3 since the number 3 is included ($x \geqslant 3$).

Draw an empty circle at 8 since the number 8 is not included.

Draw a line between the circles at 3 and 8.

Example 20 Draw a number line from −5 to 5. Show the inequality $-3 < x \leqslant 4$.

Empty circle at −3 as x is greater than −3.
Filled circle at 4 as x is less than or equal to 4.

Exercise 21Q

1 Draw six number lines from 0 to 10. Show these inequalities.

    a $x > 6$ **b** $x > 5$ **c** $x < 4$ **d** $x > 8$ **e** $x < 6$ **f** $x > 9$

2 Draw ten number lines from 0 to 10. Show these inequalities.

    a $3 < x < 7$ **b** $5 < x < 8$ **c** $5 \leqslant x < 8$ **d** $7 < x \leqslant 9$

    e $4 \leqslant x \leqslant 6$ **f** $2 < x \leqslant 8$ **g** $3 \leqslant x < 5$ **h** $4 < x < 7$

    i $5 \leqslant x < 6$ **j** $2 < x \leqslant 5$

3 Draw ten number lines from −5 to 5. Show these inequalities.

    a $-3 \leqslant x < 4$ **b** $-2 < x < 5$ **c** $-1 < x \leqslant 3$ **d** $-4 \leqslant x \leqslant 0$

    e $0 < x < 4$ **f** $-3 < x \leqslant 2$ **g** $-4 \leqslant x < 1$ **h** $0 \leqslant x \leqslant 3$

    i $-5 \leqslant x < 2$ **j** $-2 \leqslant x < 1$

4 Write down the inequalities represented on these number lines.

    a 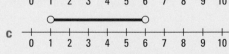   **b**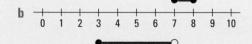

    c 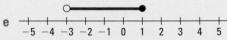   **d**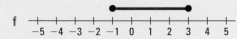

    e 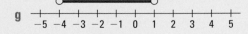   **f**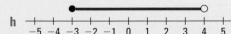

    g   **h**

21.11 Solving inequalities

Objective

○ You can solve linear inequalities.

Why do this?

Businesses use inequalities to help them maximise profits or minimise costs.

Get Ready

Show these on a number line.

1. $-6 < x < 2$ **2.** $-5 \leqslant x < 3$ **3.** $1 < x \leqslant 8$

Key Point

◉ You can solve inequalities in the same way as linear equations except you must not:

　◉ multiply both sides by the same negative quantity

　◉ divide both sides by the same negative quantity.

Example 21　　a　Solve the inequality $2x - 1 < 4$.

　　　　　　　　b　Show the solution on a number line.

a　$2x - 1 < 4$　　　　　　　Add 1 to both sides.

　　　$2x < 5$

　　　$x < 2\frac{1}{2}$　　　　　　　Divide both sides by 2.

b
```
├──┼──┼──┼──┼──┼──┼──┼──○
-4 -3 -2 -1  0  1  2  3  4
```

Example 22　　a　Solve the inequality $2x + 3 \leqslant 5x + 7$.

　　　　　　　　b　Write down the smallest integer that satisfies this inequality.

a　$2x + 3 \leqslant 5x + 7$　　　　　Subtract $2x$ from both sides.

　　　$3 \leqslant 3x + 7$

　　　$-4 \leqslant 3x$　　　　　　　Subtract 7 from both sides.

　　　$x \geqslant -1\frac{1}{3}$

Divide both sides by 3.

$-4 \leqslant 3x$ is the same as $3x \geqslant -4$

b　The smallest integer that satisfies this inequality is -1.

Example 23　　Find all the integers that satisfy the inequality $-9 \leqslant 3x < 5$. Draw a diagram to help.

$-9 \leqslant 3x < 5$　　　　Divide each term in the inequality by 3.

$-3 \leqslant x < \frac{5}{3}$

The integer solutions are $-3, -2, -1, 0, 1$

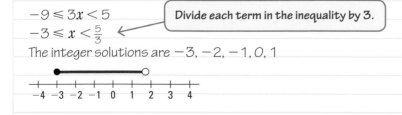

```
├──●──┼──┼──┼──○──┼──┼──┤
-4 -3 -2 -1  0  1  2  3  4
```

Exercise 21R

In questions 1–12, solve the inequality.

1 $x + 3 < 7$ **2** $x - 1 \geqslant 5$ **3** $2x \leqslant 12$

4 $\dfrac{x}{3} > 2$ **5** $x - 4 < 5$ **6** $5x > 20$

7 $x + 9 \geqslant 9$ **8** $3x - 7 \leqslant 8$ **9** $4x + 3 \geqslant 15$

10 $5x - 7 < 3$ **11** $7x - 2 > 3x + 10$ **12** $4x - 3 \leqslant 9x - 8$

In questions 13–18, solve the inequality and show the solution on a number line.

13 $4x > 11$ **14** $6x \leqslant 3$ **15** $3x + 7 \geqslant 1$

16 $8x - 3 > 7$ **17** $2x + 5 < 2$ **18** $7x - 5 \leqslant 3x - 2$

In questions 19–27, find all the integers that satisfy the inequality.

19 $4 \leqslant 2x \leqslant 8$ **20** $-9 \leqslant 3x < 6$ **21** $-15 < 5x \leqslant 5$

22 $0 \leqslant 6x < 24$ **23** $-16 < 4x \leqslant 0$ **24** $2 \leqslant 3x < 7$

25 $-7 < 5x \leqslant 15$ **26** $-5 < 2x < 5$ **27** $-10 < 3x < 0$

In questions 28–54, solve the inequality.

28 $8x < 20$ **29** $4x \geqslant 3$ **30** $5x > -15$

31 $3x \geqslant -8$ **32** $\dfrac{x}{4} > -2$ **33** $21 < 6x$

34 $4x - 9 \geqslant 2$ **35** $6x + 7 \leqslant 3$ **36** $8x - 1 > 6$

37 $9 < 7x + 2$ **38** $5x + 3 \geqslant 2x + 9$ **39** $7x + 2 \leqslant 3x - 2$

40 $8x - 1 > 5x - 6$ **41** $9x - 7 < 5x + 3$ **42** $2x + 9 \geqslant 7x - 6$

43 $2(x - 3) \geqslant 8$ **44** $5(x + 2) > 10$ **45** $3(x + 1) < x + 9$

46 $7 - x \leqslant 1$ **47** $8 - 3x > 2$ **48** $2 - 5x < 6$

49 $7 - 2x \geqslant 3x + 2$ **50** $4(x - 3) \leqslant 3 - x$ **51** $10 - 3x > 2x - 1$

52 $6 - 5x \leqslant 2 - 3x$ **53** $3 - 5x \geqslant 4 - 7x$ **54** $11 - 2x < 2 - 5x$

55 Solve the inequality $7x + 5 > 4x - 9$.
Write down the smallest integer that satisfies it.

56 Solve the inequality $3x + 4 \leqslant 1 - 2x$.
Write down the largest integer that satisfies it.

Chapter review

- In an equation, a symbol or letter represents an unknown number.
- You can solve an equation by rearranging it so that the letter is on its own on one side of the equation.
- To rearrange an equation you can:
  - add the same number to both sides
  - subtract the same number from both sides
  - multiply both sides by the same number
  - divide both sides by the same number.

  This is called the **balance method**.
- In an equation with two operations, deal with the $+$ or $-$ first.
- Solutions to equations can be whole numbers, fractions or decimals.
- Solutions to **linear equations** can be negative.
- In an equation with brackets, expand the bracket first.
- In an equation with a letter on both sides, use the balance method to rearrange the equation so that the letter is on one side only.
- A **coefficient** is the number in front of an unknown.
- You solve equations with negative coefficients using the balance method.
- You can solve problems in mathematics and other subjects by setting up equations and solving them.
- There are some equations that can only be solved using the method known as '**trial and improvement**'.
- $>$ means **greater than**.
- $\geqslant$ means **greater than or equal to**.
- $<$ means **less than**.
- $\leqslant$ means **less than or equal to**.
- You can show **inequalities** on a number line.
- An empty circle shows that the value is not included and a filled circle shows that the value is included.
- You can solve inequalities in the same way as linear equations except you must not:
  - multiply both sides by the same negative quantity
  - divide both sides by the same negative quantity.

Review exercise

1 Solve these equations.

a $a + 7 = 12$	**b** $c - 4 = 6$	**c** $3p = 21$
d $\dfrac{d}{4} = 3$	**e** $5x + 4 = 19$	**f** $6b - 7 = 17$
g $a + 7 = 3$	**h** $5b = -30$	**i** $c - 2 = -3$
j $2e = 11$	**k** $3h + 7 = 1$	**l** $4m + 5 = 2$
m $6p + 19 = 2$	**n** $6q + 7 = 3$	

2 Solve these equations.

a $2r + 7 = r + 10$	**b** $3x - 2 = x + 8$	**c** $5c + 4 = 2c + 19$	**d** $3b + 4 = b + 5$
e $5d - 2 = 2d + 3$	**f** $7y - 9 = 2y - 5$	**g** $3t + 8 = 6t + 1$	**h** $2w = 8w - 15$
i $7u - 6 = 4u - 15$	**j** $5w + 8 = 3w - 5$	**k** $3y - 5 = 7y + 5$	

E

D

3 **a** Solve $4x - 1 = 7$ **b** Solve $5(2y + 3) = 20$

4 **a** Solve $4x + 3 = 19$ **b** Solve $4y + 1 = 2y + 8$ **c** Simplify $2(t + 5) + 13$

5 Solve $4(x + 3) = 6$

6 **a** Solve $2(x - 2) = 10$ **b** Solve $4(y + 1) = 10$

7 Solve these equations.
   a $3(a + 5) = a + 21$ **b** $5(b - 4) = 2b + 1$ **c** $7c - 2 = 3(c + 6)$
   d $6(d - 2) = 5(d - 1)$ **e** $8(e - 1) = 5(e + 2)$ **f** $9(f - 2) = 2(f + 3)$
   g $4(2m + 1) = 3(5m - 1)$ **h** $2(3t + 4) = 5(2t - 1)$ **i** $8 - a = 5$
   j $13 - 3b = 1$ **k** $5 - 4d = 3$ **l** $5 - 3g = 1$
   m $4 - 3p = 18$

8 The perimeter of a rectangle is 120 cm.
   The length of a rectangle is 4 times its width.
   Find the length of the rectangle.

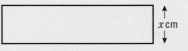

9 In the diagram, all measurements are in centimetres.
   ABC is an isosceles triangle.
   $AB = 2x$
   $AC = 2x$
   $BC = 10$
   a Find an expression, in terms of x, for the perimeter of
   the triangle. Simplify your own expression.
   b The perimeter of the triangle is 34 cm.
   Find the value of x.

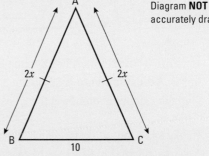

Diagram **NOT**
accurately drawn

10 Use the method of trial and improvement to find solutions to these equations.
   a $x^3 + 3x - 20 = 0$ (to 2 d.p.) **b** $x^3 - 20x - 3 = 0$ (to 2 d.p.)
   c $x^3 - 4x^2 - 5 = 0$ (to 2 d.p.) **d** $x^3 - 2x^2 = 25$ (to 2 d.p.)
   e $x^3 + 5x = 26$ (to 1 d.p.) **f** $x^3 - 2 = 2x$ (to 2 d.p.)

11 The equation $x^3 + 10x = 21$ has a solution between 1 and 2. Use a trial and improvement method
   to find this solution. Give your answer correct to 1 d.p. You must show ALL your working.

12 Tariq and Yousef have been asked to find the solution, correct to 1 decimal place, of the equation
   $x^3 + 2x = 56$

   a Work out the value of $x^3 + 2x$ when $x = 3.65$.

   Tariq says 3.6 is the solution.
   Yousef says 3.7 is the solution.

   b Use your answer to part **a** to decide whether Tariq or Yousef is correct.
   You must give a reason.

13 In this quadrilateral, the sizes of the angles, in degrees, are

$x + 10$ $2x$ $2x$ 50

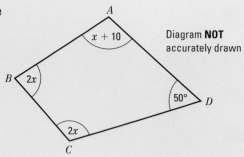

Diagram **NOT** accurately drawn

   a Use this information to write down an equation in terms of x.

   b Work out the value of x.

June 2008

14 Uzma has £x. Hajra has £20 more than Uzma. Mabintou has twice as much as Hajra.
The total amount of money they have is £132.
Find how much money they each have.

15 Jake has 3 sticks, A, B and C. Stick B is 5 cm longer than stick A. Stick C is 4 times the length of stick B.
Stick C is also 3 times the sum of the lengths of A and B.
Find the lengths of the 3 sticks.

16

A

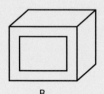

B

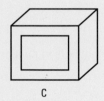

C

Here are 3 boxes. Box A has £x. Box B has £4 more than box A. Box C has one third of the money in box B. Altogether there is £24 in the 3 boxes.
Find the amount of money in the 3 boxes.

17 Becky has 4 more CDs than Emil. Justin has twice as many CDs as Becky.
The total number of CDs they have altogether is 32.

   a Form an equation.

   b Work out how many CDs Justin has.

18 **a** The equation $x^3 + 4x^2 = 100$ has a solution between 3 and 4.
Find this solution. Give your answer correct to one decimal place.
You must show ALL your working.

   b The diagram shows a cuboid.
The base of the cuboid is a square.
The height of the cuboid is 4 cm more than the width.
The volume of the cuboid is 100 cm³.
Write down the height of the cuboid, correct to 1 decimal place.

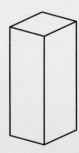

Nov 2006 adapted

19 A garden room is 3 metres longer than it is wide.
Given that its area is 14 m², use trial and improvement to find an estimate for its perimeter.

20 Show each inequality on a number line.

    a $x > 1$ **b** $x \leqslant 3$ **c** $x \leqslant 0$

    d $-2 \leqslant x < 1$ **e** $-1 < x \leqslant 3$ **f** $1 \leqslant x < 4$

21 Write down the inequalities represented on these number lines.

a 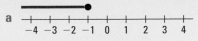   **b**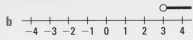

c    **d**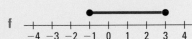

e    **f**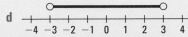

22 For each of these inequalities, list the integers that satisfy it.

    a $-3 \leqslant x < 1$ **b** $0 < x \leqslant 3$ **c** $-2 \leqslant x \leqslant 4$ **d** $-4 < x < -1$

23 Solve each inequality.

    a $x - 6 > 4$ **b** $6x \leqslant 30$ **c** $2x - 5 < 4$

    d $5x + 11 \leqslant 1$ **e** $8x + 9 \geqslant 4x + 3$ **f** $7x - 1 < 4x - 1$

    g $3x - 1 < 5x$ **h** $2(x - 3) < 7$ **i** $4 - x \leqslant x + 8$

24 Solve each inequality and show the solution on a number line.

    a $2x < 5$ **b** $4x \geqslant -2$ **c** $3x - 4 > 1$

    d $6x + 7 \leqslant 1$ **e** $9x - 5 < 4x + 5$ **f** $6x + 7 < 8x + 7$

25 **a** Solve the inequality $6x < 7 + 4x$.

    b Expand and simplify $(y + 3)(y + 4)$.

26 **a** Solve $5 - 3x = 2(x + 1)$.

    b $-3 < y \leqslant 3$ y is an integer.

      Write down all possible values of y.

27 **a** Solve $7x > 21$.

    b Solve $5y + 1 \geqslant 3y + 13$.

28

Diagram **NOT** accurately drawn

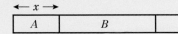

Here are 3 rods.

The length of rod A is x cm.

Rod B is 4 cm longer than rod A.

The length of rod C is twice the length of rod B.

The total length of all 3 rods is L cm.

    a Show that $L = 4x + 12$.

The total length of all 3 rods must be less than 50 cm.

    b Write down the inequality that must be satisfied.

    c Work out the range of possible values of x.

Specimen Paper 2009

A02

29 Tom has three parcels.

The weight of the first parcel is x kg.

The weight of the second parcel is 2 kg more than the first.

The weight of the third parcel is 3 kg more than the second.

The total weight of the parcels must be less than or equal to 20 kg.

Work out the largest possible weight of the heaviest parcel.

30

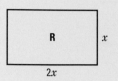 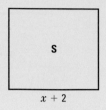

The perimeter of the rectangle R is less than the perimeter of the square S.

Write down the range of values of x.

The graph of a quadratic function is a curve called a parabola. A javelin would follow the path of a perfect parabola if air resistance, wind spin and rotation didn't affect it. Javelin throwing was part of the Ancient Greek Olympics. In one of their events competitors had to throw the javelin from the back of a galloping horse.

◎ Objectives

In this chapter you will:
- draw and interpret a range of linear and non-linear graphs
- draw and use conversion graphs
- draw and interpret distance–time graphs, including those that include sections which are curved
- use a table of values to draw quadratic graphs
- use graphs of quadratic functions to solve quadratic equations.

◆ Before you start

You need to know:
- how to plot a coordinate
- how to substitute a number for x in a linear equation.

22.1 Interpreting and drawing the graphs you meet in everyday life

◉ Objectives

- ◉ You can interpret and draw straight-line graphs through the origin, e.g. to find the cost of items.
- ◉ You can interpret and draw straight-line graphs not through the origin, e.g. to find the total bill given a basic cost plus cost per unit in gas bills or mobile phone bills.
- ◉ You can interpret and draw graphs of water filling different-shaped containers.
- ◉ You can interpret the shape of graphs, e.g. filling and emptying a bath.

❓ Why do this?

You will often see graphs in newspapers and you need to be able to interpret the information on them.

⬦ Get Ready

1. Cereal bars cost 20p each.
   Copy and complete this table for the cost of buying cereal bars.

Number of cereal bars	1	2	3	4	5	6	7	8	9	10
Cost in pence	20									200

🔑 Key Points

- ◉ A straight-line graph that goes through the origin means that the more items you buy, the more it will cost you. The cost is related to the number of items you buy. This is the type of graph you get with a Pay as you go mobile phone. If you don't use the phone there is no cost.

Pay as you go

The more minutes you use the more it costs.

If you don't use the phone there is no cost.

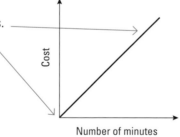

- ◉ A straight-line graph that does not cross the vertical axis at the origin means there is a basic charge and then the cost is related to the number of items you buy. This is the type of graph you get with a mobile phone on a contract. If you don't use the phone you still have to pay the monthly cost.

Contract

The more minutes you use the more it costs.

If you don't use the phone you still have to pay the monthly cost.

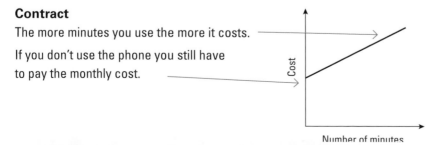

- Containers with straight vertical sides fill at a **constant** rate.

- Containers with sides that bulge out fill quickly at first. The rate slows down until the widest point of the bulge, then the rate speeds up again.

- Containers with sides that curve in fill slowly at first. The rate speeds up until the narrowest point of the curve, then the rate slows down again.

- Containers that are thin fill faster than containers that are fatter.

- The thinner the container, the steeper the graph.
  You can see the difference in the graphs of these containers.
  They have the same height.

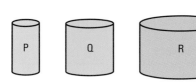 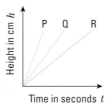

- Containers with slanting sides have curved line graphs. They start steep and get less steep as the diameter gets bigger.

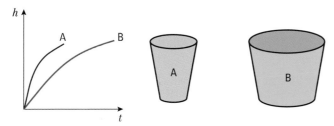

- Containers with sides that bulge out have a curved line graph and they start steep, get less steep and then get steep again.

- The graph for the top of the flask is a straight line.

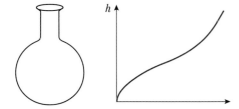

Example 1

Sam sells packets of crisps at 30p each.
He makes a table of values to help him remember what to charge people when they buy different numbers of packets of crisps.
Plot a graph to show this information.

> This pattern goes up in 30s.

Number of packets	1	2	3	4	5	6	7	8
Cost in pence	30	60	90	120	150	180	210	240

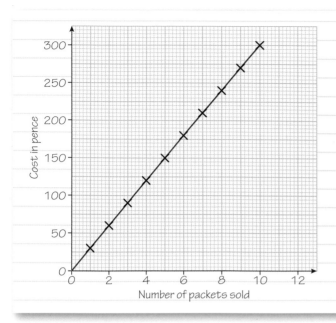

You plot the number of packets against the cost in pence.

When the points are plotted and joined you get a straight line. This is why it is a linear relationship.

You can extend the plotted points to 9 and 10 packets and so read off the cost of these numbers of packets.

 Example 2

Sharon is a taxi driver. She charges £2 before the journey starts and then £1.50 for every 2 minutes of the journey.

Use the information to make up a table of values for taxi rides up to 20 minutes long.

Plot a graph to show this information.

Journey time in minutes	0	2	4	6	8	10	12	14	16	18	20
Cost in pounds		2	3.50	5	6.50	8	9.50	11	12.50	14	15.50	17

ResultsPlus

Examiner's Tip

Make sure you understand the scale of your graph before you draw it or read values from it.

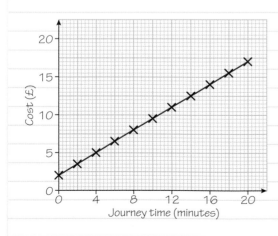

On a coordinate grid plot the points but be careful how you read the scales.

Each little square on the horizontal axis is 0.4 minutes or 48 seconds.

Each little square on the vertical axis is 50p.

Notice the graph starts part way up the vertical axis and not at the origin.

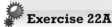 **Exercise 22A**

1 The table shows the cost of potatoes per kg.

Weight in kg	1	2	3	4	5
Cost in pence	30	60	90	120	150

a Draw a graph for this table.

b Work out how much 2.5 kg of potatoes would cost.

c Extend the graph to work out the cost of 6 kg of potatoes.

2 The table shows the cost of ice lollies.

Number of ice lollies	1	2	3	4	5
Cost in pence	25	50	75	100	125

a Draw a graph for the cost of ice lollies from the table.

b Extend the graph and then use it to work out the cost of
   i 8 ice lollies   ii 6 ice lollies.

3 The table shows the number of litres of petrol left in a car's petrol tank on a journey.

Travelling time in hours	1	2	3	4	5	6	7	8
Number of litres left	55	50	45	40	35	30	25	20

a Draw a graph from the information given in the table.

b How many litres were in the tank at the start of the journey (after 0 hours)?

c How many litres were in the tank after $5\frac{1}{2}$ hours?

4 A car uses 2 litres of petrol for every 5 km it travels.
   a Copy and complete the table showing how much petrol the car uses.

Distance travelled in km	0	5	10	15	20	25
Petrol used in litres	0	2	4			

b Draw a graph from the information in your table.

c Work out how much petrol is used to travel 4 km.

d Work out how many kilometres have been travelled by the time 15 litres of petrol have been used.

5 The water in a reservoir is 144 m deep. During a dry period the water level falls by 4 m each week.
   a Copy and complete this table showing the expected depth of water in the reservoir.

Week	0	1	2	3	4	5	6	7	8
Expected depth of water in m	144	140							

b Draw a graph from the information in your table.

c How deep would you expect the reservoir to be after 10 weeks?

If the water level falls to 96 m, the water company will divert water from another reservoir.

d After how long will the water company divert water?

F

6 Sally has a Pay as you go mobile phone. She pays 40p for each minute she uses her phone.

 a Copy and complete this table of values for the cost of using Sally's phone.

Minutes used	0	5	10	15	20	25	30	35	40	45	50
Cost in pounds	0		4								20

 b Plot the points in the table on a coordinate grid and draw a graph to show the cost of using Sally's phone.

 c Use your graph to find the cost of using her phone for 32 minutes.

 d One month Sally paid £8.40 to use her phone. For how many minutes did Sally use her phone that month?

7 Bob has a contract phone. He pays £15 each month and then 10p for each minute he uses his phone.

 a Copy and complete this table of values for the cost of using Bob's phone.

Minutes used	0	5	10	15	20	25	30	35	40	45	50
Cost in pounds	15		16								20

 b Plot the points in the table on a coordinate grid and draw a graph to show the cost of using Bob's phone.

 c Use your graph to find the cost of Bob using his phone for 32 minutes.

 d One month Bob paid £17 to use his phone. For how many minutes did Bob use his phone that month?

8 Kieran buys his gas from a company that charges 50p for each unit of gas he uses.

 a Copy and complete this table of values for the cost of gas used by Kieran.

Units used	0	10	20	30	40	50	60	70	80	90	100
Cost in pounds	0		10								50

 b Plot the points in the table on a coordinate grid and draw a graph to show the cost of using gas.

 c Use your graph to find the cost of using 32 units of gas.

 d One month Kieran paid £45 for gas. How many units of gas did Kieran use that month?

E

9 Jamie buys his electricity from a company that charges £20 each month and then 25p for each unit of electricity he uses.

 a Copy and complete this table of values for the cost of using electricity for Jamie.

Units used	0	10	20	30	40	50	60	70	80	90	100
Cost in pounds	20		25								45

 b Plot the points in the table on a coordinate grid and draw a graph to show the cost of using electricity.

 c Use your graph to find the cost of using 32 units of electricity.

 d One month Jamie paid £38 for electricity. How many units of electricity did Jamie use that month?

10 The graph shows the cost of using a mobile phone for one month on three different tariffs.

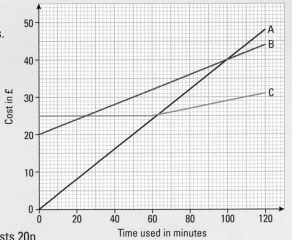

The three tariffs are

Tariff 1   Rental £20   every minute costs 20p
Tariff 2   Pay as you go   every minute costs 40p
Tariff 3   Rental £25   first 60 minutes free then each minute costs 10p

a Match each tariff with the letter of its graph.

Fiona uses her mobile phone for about 60 minutes each month.

* b Explain which tariff would be the cheapest for her to use.
   You must give the reasons for your answer.

* 11 Jodie wants to buy a new phone. She has a choice of one of these tariffs.

Tariff	Monthly payment	Cost of calls per minute
A	£0	35p
B	£10	20p
C	£20	10p
D	£25	5p

Jodie uses her phone for about 50 minutes each month.
Which tariff should Jodie choose?

Example 3   Match these containers to their graphs when they are filled with water at a constant rate.

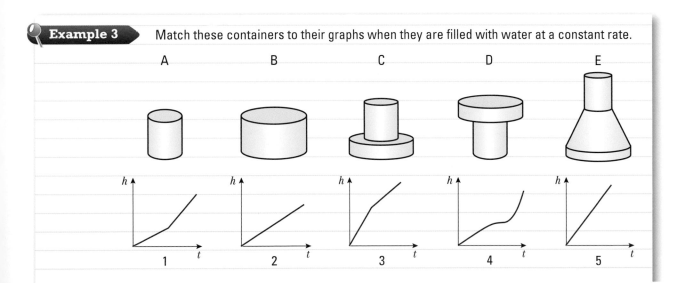

A and 5 go together since A has straight sides and is thin.

B and 2 go together since B has straight sides and is wider than A.

C and 1 go together since C is wide with straight sides and then narrow with straight sides.

D and 3 go together since D is narrow with straight sides and then wide with straight sides.

E and 4 go together since E has three sections which can be identified on the graph.

Exercise 22B

In questions 1 to 4, liquid is poured at a constant rate into the containers. The height of the liquid in the container h in cm is plotted against the passage of time t in seconds.

1 Match these containers with their graphs.

A   B   C   D   E

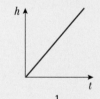 1 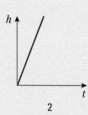 2 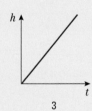 3   4 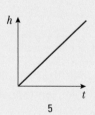 5

2 Match these containers with their graphs.

A   B  C   D   E

1 2 3 4 5

3   Match these containers with their graphs.

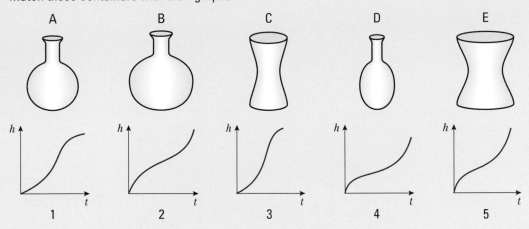

4   Match these containers with their graphs.

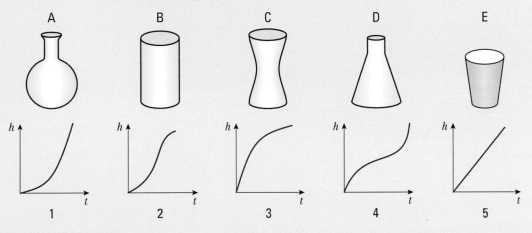

5   Liquid is poured into each of these containers at a constant rate.
    For each set of containers draw, on the same graph, the height of the liquid h against the time t in seconds.

    a

    b

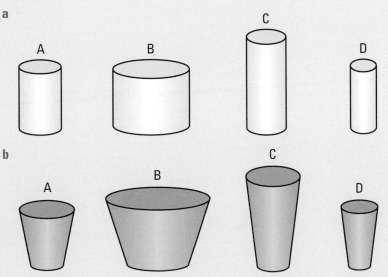

C

c

A

B

C

D

d

A

B

C

D

e

A

B

C

D

* 6   Here is a graph that shows the height of water in a bath.
      For each part of the graph, describe what may have
      happened.

* 7   Here is a graph that shows the height, in cm, of the water in a bath.
      Explain, giving the heights and the times, what happened at each stage of the process.

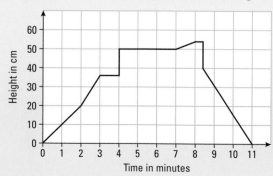

* 8   Here is a graph that shows the height, in metres, of a hot air balloon during a flight.

Describe the flight of the hot air balloon.

22.2 Drawing and interpreting conversion graphs

◎ Objectives

- ◉ You can use a conversion graph to change one unit to another.
- ◉ You can draw a conversion graph to change one unit to another.

◈ Why do this?

A conversion graph can help you convert money from one currency to another when you go on holiday.

◈ Get Ready

One euro is worth 80 pence. This can be written as €1 = 80p or £0.80.

1. Write in pence or pounds   **a** €5   **b** €10   **c** €20   **d** €100

2. Write in euros   **a** £1.60   **b** £8   **c** £24   **d** £16

◈ Key Point

- ◉ **Conversion graphs** are used to change measurements in one unit to measurements in a different unit. They can also be used to change between money systems in different countries.

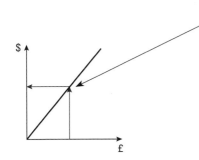

To change £ to $ you read up from the £ to the line and then read across to the $.

To change $ to £ you read across from the $ to the line and then read down to the £.

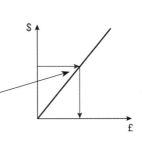

Exercise 22C

F

1   This graph can be used to change between pounds (£) and Hong Kong dollars.

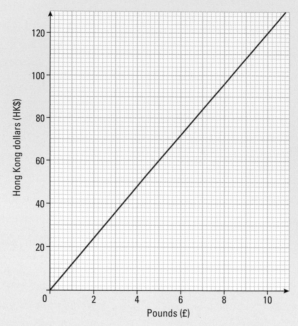

a   Use the graph to change these amounts to Hong Kong dollars (HK$).
   i £10   ii £5   iii £8   iv £100   v £200
b   Use the graph to change these amounts to pounds.
   i HK$60   ii HK$30   iii HK$90   iv HK$600   v HK$1200

2   Copy the table and use the temperature conversion graph to complete it.

°C	5	20		28			35	80		40
°F			80		50	100			200	

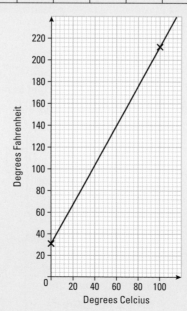

3 a Draw a conversion graph from pounds to kilograms. Use the fact that 0 pounds is 0 kilograms and 50 kg is approximately 110 pounds.
   On your graph draw axes for kilograms and pounds using scales of 1 cm = 10 pounds and 2 cm = 10 kg. Plot the points (0, 0) and (50, 110) and join them with a straight line.

   b Copy and complete this table using your conversion chart to help you.

Kilograms	0			45	30	15			35	50
Pounds	0	10	20				50	14		110

4 Copy this table and then use the information in the table to draw a conversion graph from inches into centimetres. Use your graph to help you fill in the missing values.

Inches	0	1	2				9	8		12
Centimetres	0			10	15	20			25	30

5 Copy this table and then use the information in the table to draw a conversion graph from miles into kilometres. Use your graph to help you fill in the missing values.

Miles	0	5		40		30			24	50
Kilometres	0		16		36		72	20		80

6 Copy this table and then use the information in the table to draw a conversion graph from acres into hectares. Use your graph to help you fill in the missing values.

Hectares	0			12	15	17			3	20
Acres	0	20	30				24	45		50

22.3 Drawing and interpreting distance–time graphs

◉ Objective

- ◉ You can interpret straight-line distance–time graphs.
- ◉ You can draw straight-line distance–time graphs.
- ◉ You can interpret distance–time graphs where the line is curved.

◈ Why do this?

Graphs help us to understand information more easily.

◈ Get Ready

1. Lauren travelled 90 miles in 3 hours. What was her average speed?
2. Anna travelled at 50 miles per hour for 2 hours. How far did she travel?
3. Idris travelled 100 miles at 50 miles per hour. How long did it take him?

🟣 Key Points

- ◉ On **distance–time graphs**:
  - ◉ time is always on the horizontal axis
  - ◉ distance is always on the vertical axis
  - ◉ a slanting line means movement is taking place
  - ◉ a horizontal line means no movement is taking place, the object is stationary.
- ◉ To work out speed you divide the distance travelled by the time taken.

Example 4

Mary travels to work by bus.

She walks the first 750 metres in 10 minutes, waits at the bus stop for 5 minutes, then travels the remaining 3000 metres by bus. She arrives at the work bus stop 21 minutes after she set off from home.

    a   Draw a distance–time graph of her journey.

    b   Work out the average speed of the bus in kilometres per hour.

a

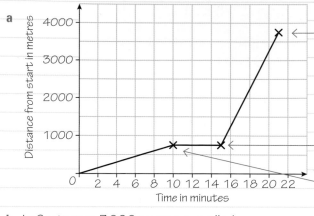

Plot the final point at (21 minutes, 750 + 3000 metres).

Plot the point (15 minutes, 750 metres).

Plot the point (10 minutes, 750 metres).

b   In 6 minutes, 3000 metres travelled

In 60 minutes, $3000 \times 10 = 30\,000$ m

$30\,000 \div 1000 = 30$ km per hour

Join the points.
Waiting at the bus stop means that the line is horizontal for 5 minutes.

Example 5

This is a graph showing the journey made by an ambulance.

On the graph from 0 to A the ambulance travels 10 km in 10 minutes. From A to B the ambulance travels 20 km in 10 minutes. From B to C the ambulance does not go anywhere for 5 minutes. The 30 km journey back to base takes 15 minutes.

Work out the speed of the ambulance for each part of the journey.

0 to A

10 km in 10 minutes

In 60 minutes 60 km could be travelled

Speed = 60 km per hour

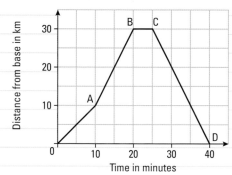

A to B

20 km in 10 minutes

In 60 minutes 120 km could be travelled

Speed = 120 km per hour

B to C

0 km in 5 minutes

Speed = 0 km per hour

C to D

30 km in 15 minutes

In 60 minutes 120 km could be travelled

Speed = 120 km per hour

Example 6

The distance fallen by a stone when it is dropped from a cliff is shown on this graph.

a What distance did the stone fall in 2 seconds?

b How long did the stone take to fall 32 metres?

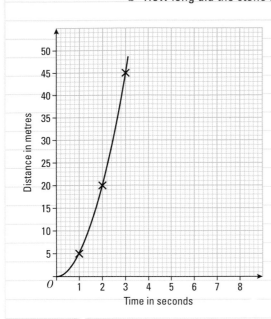

a The stone fell 20 metres in 2 seconds.

> Draw a line up from 2 seconds to meet the curve. Then draw across to the distance axis. It cuts at 20 metres.

b The stone took 2.5 seconds to fall 32 metres.

> Draw a line across from 32 metres to the curve. Then draw down to the time axis. It cuts it at about 2.5 seconds.

Exercise 22D

1 Jane walked to the shops, did some shopping then walked home again.

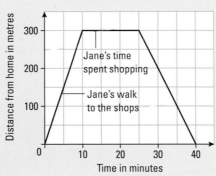

a How many minutes did it take Jane to walk to the shops?

b How far away were the shops?

c How many minutes did Jane spend shopping?

d How many minutes did it take Jane to walk home?

e Work out the speed at which Jane walked to the shops.
First give your answer in metres per minute, then change it to km per hour.

f Work out the speed at which Jane walked back from the shops.
First give your answer in metres per minute, then change it to km per hour.

C

A02

2 Here is a graph of David's journey by car to see his aunt.

   a Write a story of the journey, explaining what happened during each part of it.

   b Work out David's speed during each part of the journey. Give your answers in km per hour.

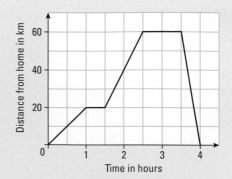

A02

3 This graph shows Tom's and Sarah's journeys. Tom sets off from London at 08:00 and travels to a town 90 km away to meet his girlfriend Sarah. He stops for a rest on the way. Once he gets to Sarah's he turns around and drives straight home because he discovers that she set off for London some time ago to see him.

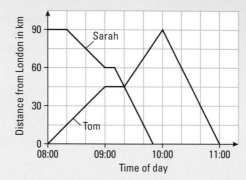

   a Describe Tom's journey in detail, explaining after what distance he stopped on the way and for how long.

   b Describe Sarah's journey in detail, explaining after what distance she stopped on the way and for how long.

   c At what time did Sarah and Tom pass each other and what distance were they from London when they passed each other?

A02

* **4** Imran has a bath. The graph shows the depth of the bath water.

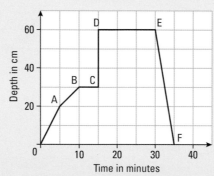

He starts at 0 by turning the hot and cold water taps on.

Between 0 and point A on the graph the depth of water goes up 20 cm in 5 minutes.

Explain what happens between points A and B, B and C, C and D, D and E, and E and F on the graph and how long each part of the process lasts.

* **5** Daniel walked to the post box near his house to post a letter. It took him 4 minutes to walk to the post box, which was 400 m away. Daniel chatted to a friend for 2 minutes and he walked home in 3 minutes. Use graph paper to draw a distance–time graph for this journey.

* **6** Kirsty took a trip in a hot air balloon. The balloon rose 400 metres in the air in one hour and stayed at this height for two and a half hours. The balloon then came back to the ground in half an hour. The hot air balloon company wants to display the journey as a distance–time graph. Draw a distance–time graph for this balloon flight.

7 Annabel travels to school. She walks 250 metres to the bus stop in 4 minutes, waits at the bus stop for about 5 minutes and then travels the remaining 1000 metres by bus. She arrives at the bus stop outside the school 15 minutes after she sets off from home.
   a   Draw a distance–time graph of the journey.
   b   Work out the speed of the bus, first in metres per minute, then in km per hour.

* **8** Mae went shopping by car. She drove the 10 miles to the shops in 30 minutes. She stayed at the shops for 30 minutes and then started to drive home. The car then broke down after 5 minutes when she had travelled 4 miles from the shops. It took 10 minutes to repair the car and another 5 minutes to get home. Draw a distance–time graph for Mae's journey.

9 Use the graph in Example 6 to find
   a   the distance fallen by the stone in
      i   1.5 seconds   ii   3 seconds
   b   the time taken for the stone to fall
      i   40 metres   ii   25 metres.

10 Karen skis down a mountain. The graph shows her run.

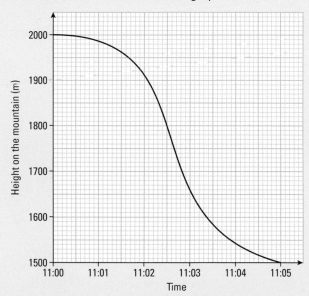

   a   From the graph, write down the height Karen was at after
      i   1 minute   ii   2 minutes 30 seconds   iii   4 minutes 15 seconds.
   b   Use the graph to write down the time at which Karen was at the following heights.
      i   1900 m   ii   1750 m   iii   1625 m

C

11 The speed of a ball when it is dropped is shown in the following table of values.

Distance in metres	0	5	10	15	20	25
Speed in metres per second	0	10	14	17	20	22

a Draw a graph using the information given in the table.

b Use the graph to work out the speed when the distance fallen is 12 metres.

c Use the graph to work out the distance fallen when the speed is 18 metres per second.

22.4 Drawing quadratic graphs

Objectives

- You can use a table of values to draw graphs of the form $y = ax^2 \pm b$.
- You can use a table of values to draw graphs of the form $y = ax^2 \pm bx \pm c$.
- You know the effect of putting a negative sign in front of the x^2.

Why do this?

You would be able to plot a graph of the trajectory of your ball if you were playing basketball.

Get Ready

Remember BIDMAS? (See Section 9.4.)

When you substitute numbers into these expressions you do the indices first and then the other processes.

e.g. Find the value of $3x^2$ when $x = 5$ $\qquad 3 \times 5^2 = 3 \times 25 = 75$

1. When $x = 3$, find the value of

a x^2 b $5x^2$ c $x^2 + 2x$

d $4x^2 + 3x$ e $x^2 - 5x$ f $2x^2 - 10x$

Key Points

- To draw a quadratic graph (e.g. $y = x^2$):
  - make a table of values, selecting some values for x

x	-3	-2	-1	0	1	2	3
$y = x^2$							

  - substitute the values of x into $y = x^2$

x	-3	-2	-1	0	1	2	3
$y = x^2$	9	4	1	0	1	4	9

  - plot the points on a grid
  - draw in the line.
- For quadratic graphs such as $y = ax^2 + b$:
  - the number (b) that is on its own moves the graph up or down
  - the number (a) that is in front of the x^2 brings the graph closer to the y-axis
  - if there is a minus sign in front of the x^2 then the graph turns upside down.

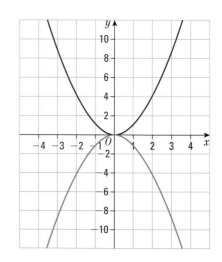

Example 7 On a coordinate grid with values of x from -4 to $+4$, draw the graphs of

i $y = x^2$ ii $y = x^2 + 1$ iii $y = x^2 - 1$ iv $y = x^2 - 2$

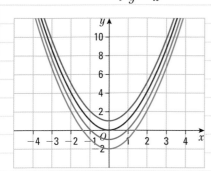

1. Make a table of values, selecting some values for x.

x	-3	-2	-1	0	1	2	3
$y = x^2$							

2. Substitute the values of x into $y = x^2$.

x	-3	-2	-1	0	1	2	3
$y = x^2$	9	4	1	0	1	4	9

x	-3	-2	-1	0	1	2	3
$y = x^2 + 1$	10	5	2	1	2	5	10

3. Plot the points on the grid.
4. Draw in the line.
5. Repeat for all the other lines.

x	-3	-2	-1	0	1	2	3
$y = x^2 - 1$	8	3	0	-1	0	3	8

x	-3	-2	-1	0	1	2	3
$y = x^2 - 2$	7	2	-1	-2	-1	2	7

Example 8 On a coordinate grid with values of x from -4 to $+4$, draw the graphs of

i $y = x^2$ ii $y = 2x^2$ iii $y = 3x^2$

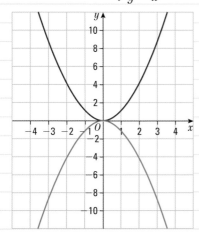

x	-3	-2	-1	0	1	2	3
$y = x^2$	9	4	1	0	1	4	9

x	-2	-1	0	1	2
$y = 2x^2$	8	2	0	2	8

x	-2	-1	0	1	2
$y = 3x^2$	12	3	0	3	12

Example 9 On a coordinate grid with values of x from -4 to $+4$, draw the graphs of

i $y = x^2$ ii $y = -x^2$

x	-3	-2	-1	0	1	2	3
$y = x^2$	9	4	1	0	1	4	9
$y = -x^2$	-9	-4	-1	0	-1	-4	-9

Exercise 22E

C

1 **a** Copy and complete the tables of values for these quadratic graphs.

 b Draw the graphs on a coordinate grid with x-axis drawn from -3 to $+3$ and y-axis drawn from -12 to $+12$.

i

x	-3	-2	-1	0	1	2	3
$y = x^2 + 2$		6		2		6	

ii

x	-3	-2	-1	0	1	2	3
$y = -x^2 - 2$		-6		-2		-6	

iii

x	-3	-2	-1	0	1	2	3
$y = -x^2 + 1$	-8			1			-8

iv

x	-3	-2	-1	0	1	2	3
$y = -x^2 + 4$		0		4			-5

v

x	-3	-2	-1	0	1	2	3
$y = x^2 + 3$		7				7	

2 **a** Copy and complete the tables of values for these quadratic graphs.

 b Draw the graphs on a coordinate grid with x-axis drawn from -3 to $+3$ and y-axis drawn from -20 to $+20$.

i

x	-3	-2	-1	0	1	2	3
$y = 2x^2 + 1$		9		1		9	

ii

x	-3	-2	-1	0	1	2	3
$y = -2x^2 + 1$		-7		1		-7	

iii

x	-2	-1	0	1	2
$y = -3x^2 + 1$	-11		1		

iv

x	-3	-2	-1	0	1	2	3
$y = 2x^2 - 1$		7		-1			

v

x	-3	-2	-1	0	1	2	3
$y = -2x^2 - 1$		-9			-1		

3 Draw these quadratic graphs on a coordinate grid with x-axis drawn from -3 to $+3$ and y-axis drawn from -15 to $+15$.

 a $y = x^2$ **b** $y = x^2 + 3$ **c** $y = x^2 - 3$ **d** $y = -x^2 + 3$ **e** $y = -x^2 - 3$

4 Draw these quadratic graphs on a coordinate grid with x-axis drawn from -3 to $+3$ and y-axis drawn from -20 to $+20$.

 a $y = x^2$ **b** $y = -x^2$ **c** $y = 2x^2$ **d** $y = -2x^2$ **e** $y = -(x + 1)^2$

5 Draw these quadratic graphs on a coordinate grid with x-axis drawn from -3 to $+3$ and y-axis drawn from -30 to $+30$.

 a $y = x^2$ **b** $y = 3x^2$ **c** $y = -3x^2 - 3$ **d** $y = -3x^2$ **e** $y = 3x^2 + 3$

Example 10 On a coordinate grid with values of x from -4 to $+4$, draw the graph of $y = x^2 + 2x - 5$.

1. Make a table of values, selecting some values for x.

x	-3	-2	-1	0	1	2	3
x^2							
$2x$							
-5							
y							

2. Substitute the values of x into $y = x^2 + 2x - 5$.

x	-3	-2	-1	0	1	2	3
x^2	9	4	1	0	1	4	9
$2x$	-6	-4	-2	0	2	4	6
-5	-5	-5	-5	-5	-5	-5	-5
y	-2	-5	-6	-5	-2	3	10

Square x
Double x
Always -5

3. Plot the points on the grid.
4. Draw in the curved line.

ResultsPlus
Watch Out!

Don't forget the rules of BIDMAS when filling in the table of values.

Exercise 22F

1 **a** Copy and complete the tables of values for these quadratic graphs.
   b Draw the graphs on a coordinate grid with x-axis drawn from -3 to $+3$ and y-axis drawn from -20 to $+20$.

i

x	-3	-2	-1	0	1	2	3
x^2	$+9$			0			
$2x$	-6			0			
$+1$	$+1$			$+1$			
$y = x^2 + 2x + 1$	4			1			

ii

x	-3	-2	-1	0	1	2	3
x^2		$+4$		0			
$3x$		-6		0			
$+2$		$+2$		$+2$			
$y = x^2 + 3x + 2$		0		2			

iii

x	-3	-2	-1	0	1	2	3
x^2	$+9$			0			
$2x$	-6			0			
-5	-5			-5			
$y = x^2 + 2x - 5$	-2			-5			

C

iv

x	-3	-2	-1	0	1	2	3
x^2	$+9$				$+1$		
$-2x$	$+6$				-2		
$+3$	$+3$				$+3$		
$y = x^2 - 2x + 3$	18				2		

v

x	-3	-2	-1	0	1	2	3
x^2	$+9$			0			
$-2x$	$+6$			0			
-3	-3			-3			
$y = x^2 - 2x - 3$	12			-3			

2 Draw these quadratic graphs on a coordinate grid with x-axis drawn from -3 to $+3$ and y-axis drawn from -15 to $+15$.

 a $y = x^2$ b $y = x^2 + 3x$ c $y = x^2 - 3x$ d $y = -x^2 + 3x$ e $y = -x^2 - 3x$

3 Draw these quadratic graphs on a coordinate grid with x-axis drawn from -3 to $+3$ and y-axis drawn from -20 to $+20$.

 a $y = x^2 + 2x + 3$ b $y = -x^2 + 2x + 3$ c $y = 2x^2 + x - 1$

 d $y = -2x^2 + x + 1$ e $y = (x - 1)^2$ f $y = (x + 1)(x - 1)$

4 Draw these quadratic graphs on a coordinate grid with x-axis drawn from -3 to $+3$ and y-axis drawn from -30 to $+30$.

 a $y = x^2 - 3x + 2$ b $y = 3x^2 - 4x$ c $y = -3x^2 + 4x$

 d $y = -3x^2 + x$ e $y = 2x^2 - 3x$ f $(x + 2)(x - 1)$

22.5 Using graphs of quadratic functions to solve equations

◎ Objective

● You can use graphs of quadratic functions to solve quadratic equations.

◈ Get Ready

Draw the graphs for these quadratic equations.

1. $y = -x^2 + 2x$ **2.** $y = 4x^2 - 2x$

◈ Key Point

◉ You can solve a **quadratic equation** by drawing the graph and then finding where the graph crosses the x-axis, where $y = 0$.

Example 11 Solve the equations **a** $x^2 - x - 3 = 0$ **b** $x^2 - x - 3 = 2$

Draw the graph of $y = x^2 - x - 3$.
Step 1: Make up a table of values.
Step 2: Plot the points.
Step 3: Draw the curve.
Step 4: Find where the curve cuts the line $y = 0$ (the x-axis).

x	-3	-2	-1	0	1	2	3
x^2	$+9$	$+4$	$+1$	0	$+1$	$+4$	$+9$
$-x$	$+3$	$+2$	$+1$	0	-1	-2	-3
-3	-3	-3	-3	-3	-3	-3	-3
$y = x^2 - x - 3$	9	3	-1	-3	-3	-1	$+3$

a $x = -1.3$ and $x = 2.3$

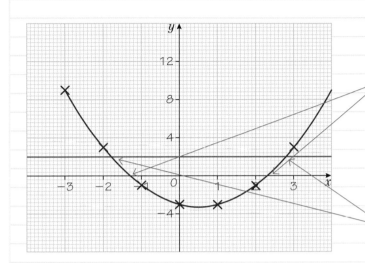

The curve meets the x-axis when $x = -1.3$ and 2.3.
These are the answers, or the solutions, to the equation.

b $x = -1.8$ and $x = 2.75$

If the equation was $x^2 - x - 3 = 2$ then you would need to read off at $y = 2$.
In this case the solutions would be $x = -1.8$ and 2.75.

Exercise 22G

1 **a** Draw the graph of $y = 2x^2 - 3x - 2$ for values of x from -2 to $+4$.
   b Use your graph to solve the equations.
     i $2x^2 - 3x - 2 = 0$ **ii** $2x^2 - 3x - 2 = 10$

2 **a** Draw the graph of $y = x^2 - 3x - 2$ for values of x from -2 to $+5$.
   b Use your graph to solve the equations.
     i $x^2 - 3x - 2 = 0$ **ii** $x^2 - 3x - 2 = 5$

3 **a** Draw the graph of $y = 2x^2 - 3x$ for values of x from -2 to $+3$.
   b Use your graph to solve the equations.
     i $2x^2 - 3x = 0$ **ii** $2x^2 - 3x = 4$

4 **a** Draw the graph of $y = x^2 - 2x$ for values of x from -2 to $+4$.
   b Use your graph to solve the equations.
     i $x^2 - 2x = 0$ **ii** $x^2 - 2x = 3$

5 **a** Draw the graph of $y = x^2 - 4x + 3$ for values of x from -1 to $+5$.
   b Use your graph to solve the equations.
     i $x^2 - 4x + 3 = 0$ **ii** $x^2 - 4x + 3 = 2$

C

Chapter review

- A straight-line graph that goes through the origin means that the more items you buy, the more it will cost you. The cost is related to the number of items you buy.
- A straight-line graph that does not cross the vertical axis at the origin means there is a basic charge and then the cost is related to the number of items you buy.
- The thinner the container the steeper the graph.
- Containers with slanting sides have curved line graphs. They start steep and get less steep as the diameter gets bigger.
- Containers with sides that bulge out have a curved line graph and they start steep, get less steep and then get steep again.
- **Conversion graphs** are used to change measurements in one unit to measurements in a different unit. They can also be used to change between money systems in different countries.
- On **distance–time graphs**:
  - time is always on the horizontal axis
  - distance is always on the vertical axis
  - a slanting line means movement is taking place
  - a horizontal line means no movement is taking place, the object is stationary.
- To work out speed you divide the distance travelled by the time taken.
- To draw a quadratic graph:
  - make a table of values, selecting some values for x
  - substitute the values of x into y
  - plot the points on a grid
  - draw in the line.
- The graphs of quadratic expressions in the form $ax^2 + bx + c$
  - have a $\cup$ shape if a is positive
  - have a $\cap$ shape if a is negative
  - cut the y-axis at $(0, c)$.
- You can solve a **quadratic equation** by drawing the graph and then finding where the graph crosses the x-axis, where $y = 0$.

Review exercise

D **A03**

1 Here are six temperature/time graphs.
For each graph describe how the temperature changes with time.

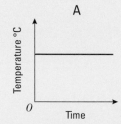

A

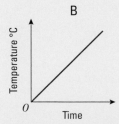

B

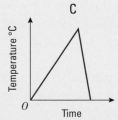

C

D

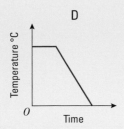

E

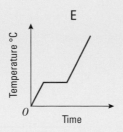

F

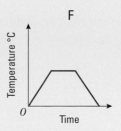

Nov 2008

2 Dave drives a truck. He uses this rule to work out how much to charge for using his truck.

Total charge (£) = number of miles travelled × 2 + 10

a Draw a graph to show how much Dave charges for distances from 0 to 50 miles.

Nick also owns a truck. He charges £2.50 for every mile travelled.

b When is it cheaper to use Nick's truck?

AO2
AO3

3 The formula $F = 2C + 30$ can be used to estimate F given the value of C, where F is the temperature in Fahrenheit and C is the temperature in Celsius.

Copy and complete the table and use it to draw the graph of F against C for values of C from 0 to 100.

C	0	20	40	60	80	100
F			110			

4 The graph shows the cost, C, of a hiring a sander for d days from Hire It.

Find:

a a formula linking C with d

b the cost of hiring the sander for 10 days.

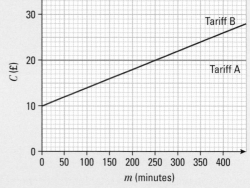

C

AO3

5 Josh is presented with a graph showing him a choice of two different mobile phone tariffs. The graph shows the cost, C, against the number of minutes, m, spent on his mobile in a particular month.

a Find the formulae of C against m for both tariffs.

b Explain in words how the tariffs are calculated.

c Advise Josh which scheme he should choose.

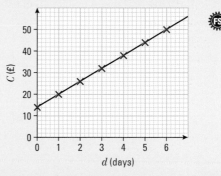

AO3

*6 Abbie has the option of joining two health clubs:

Hermes has a joining fee of £100 plus a fee of £5 per session.

Atlantis has a joining fee of £200 with a fee of £3 per session.

Which health club should she choose?

You must show all calculations and fully explain your solution.

AO2
AO3

C

7 **a** Copy and complete the table of values for
$y = x^2 - 4x - 2$.

x	-1	0	1	2	3	4	5
y		-2	-5			-2	3

b Copy the grid and draw the graph of
$y = x^2 - 4x - 2$.

c Use your graph to find estimates of the
solutions of $x^2 - 4x - 2 = 0$.

Nov 2008

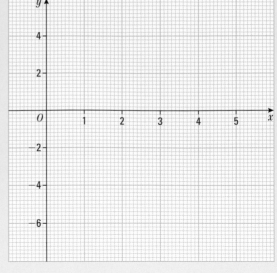

8 The graph of $y = x^2 - 2x$ has been drawn
on the grid.
Copy the graph and use it to find estimates
of the solutions of

a $x^2 - 2x = 0$

b $x^2 - 2x = 2$

c $x^2 - 2x = -1$

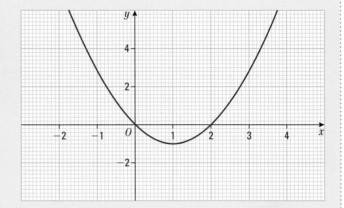

A03

9 The diagram shows a square with sides of length x cm.
Part of the square is shaded as shown in the diagram.
Given that the shaded area is 3cm^2, find an estimate for the value of x.

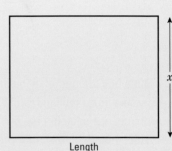

A02
A03

10 The diagram shows a rectangle.
All the measurements are in cm.
The width is x and the length is 3 cm more than the width.
The area of the rectangle is 20 cm^2.

a Draw a suitable graph.

b Find an estimate for the value of x.

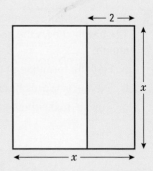

Length

23 TRANSFORMATIONS

Many fairground rides use rotation, and the G-forces created, to thrill their riders. Can you tell the order of rotational symmetry for this ride? Where do you think the centre of rotation is?

Objectives

In this chapter you will:
- learn how to distinguish between the four main transformations
- learn how to carry out and describe single and multiple transformations.

Before you start

You need to:
- be able to spot patterns in shapes in real-life situations, for example in wallpaper
- have a pencil, eraser, ruler, squared paper and tracing paper for some transformations.

23.1 Introduction

◎ Objectives

- You can recall the names of the four main transformations.
- You can identify simple transformations.

❓ Why do this?

Transformations are useful for producing designs for tiles, mosaics, wallpaper and rugs.

⬆ Get Ready

1. Pick out any letters that you think might be symmetrical in some way.

A B C D E F G H I J K L M N O P Q R S T U V W X Y Z

🌐 Key Point

- There are four main **transformations**.

Reflection	Rotation	Translation	Enlargement
 |  |  |  |

⚙ Exercise 23A

Questions in this chapter are targeted at the grades indicated.

In each of the following cases, identify which of the four transformations is being shown.

E

1

2

3

4

5

6

E

7

8

9

10

11

12

23.2 **Translations**

◎ Objectives

- ○ You can recognise a translation.
- ○ You can carry out a translation.
- ○ You can describe a translation.

❓ Why do this?

In sport, any movement can be described as a translation, from a set piece in football to a gymnastics routine.

⬦ Get Ready

1. You are standing at point A on the grid.
   You are facing to the right.
   You want to move along the lines of the grid to get to point B.
   Describe a number of moves you could make to get to point B.

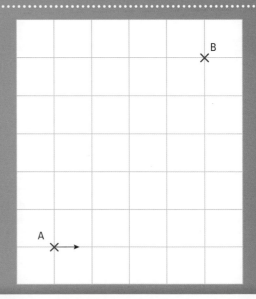

🔍 **Key Points**

◉ A **translation** is a sliding movement made from one or more moves.
In a simple translation you need to describe the distance and direction of each move.

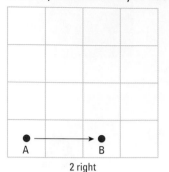

2 right

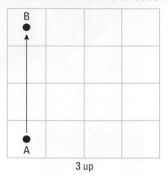

3 up

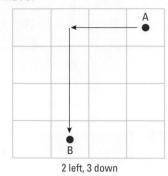

2 left, 3 down

◉ You can describe a translation by using a **column vector**, e.g. $\binom{3}{2}$.

The top number describes the movement to the right, the bottom number the movement up.

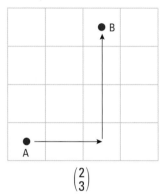

$\binom{2}{3}$

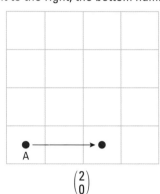

$\binom{2}{0}$

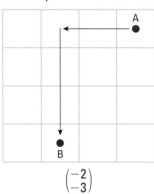

$\binom{-2}{-3}$

Notice the negative signs mean the opposite direction: left instead of right, down instead of up.

◉ In a translation:
- ◉ the lengths of the sides of the shape do not change
- ◉ the angles of the shape do not change
- ◉ the shape does not turn.

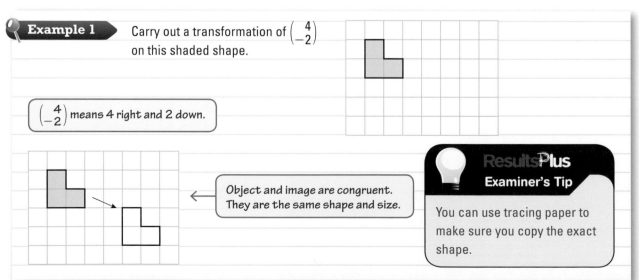

🔍 **Example 1**   Carry out a transformation of $\binom{4}{-2}$ on this shaded shape.

$\binom{4}{-2}$ means 4 right and 2 down.

Object and image are congruent. They are the same shape and size.

ResultsPlus
Examiner's Tip

You can use tracing paper to make sure you copy the exact shape.

Example 2 Describe, as a column vector, the transformation that moves triangle A to triangle B.

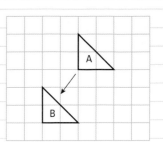

The transformation is a translation.

The movement is 2 left and 3 down.

Watch Out!

Don't forget the negative signs when writing column vectors.

The description using a column vector is

'A translation of $\begin{pmatrix} -2 \\ -3 \end{pmatrix}$.'

Exercise 23B

C

1 Copy each shape and carry out the translation described.

a $\begin{pmatrix} 2 \\ 3 \end{pmatrix}$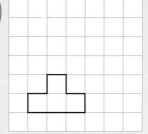

b $\begin{pmatrix} 3 \\ 0 \end{pmatrix}$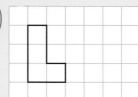

c $\begin{pmatrix} 1 \\ 2 \end{pmatrix}$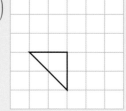

d $\begin{pmatrix} 3 \\ -1 \end{pmatrix}$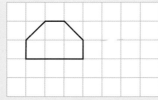

e $\begin{pmatrix} 0 \\ -2 \end{pmatrix}$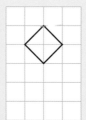

f $\begin{pmatrix} -2 \\ -4 \end{pmatrix}$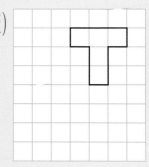

g $\begin{pmatrix} -2 \\ 4 \end{pmatrix}$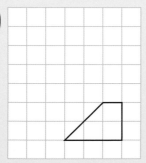

h $\begin{pmatrix} -1 \\ -3 \end{pmatrix}$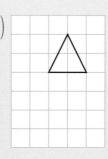

C

2 Carry out these translations on the shaded shape.

a $\begin{pmatrix} 2 \\ 3 \end{pmatrix}$ Label the shape A.

b $\begin{pmatrix} 3 \\ 2 \end{pmatrix}$ Label the shape B.

c $\begin{pmatrix} -1 \\ 2 \end{pmatrix}$ Label the shape C.

d $\begin{pmatrix} 4 \\ -2 \end{pmatrix}$ Label the shape D.

e $\begin{pmatrix} -3 \\ 0 \end{pmatrix}$ Label the shape E.

f $\begin{pmatrix} 3 \\ -2 \end{pmatrix}$ Label the shape F.

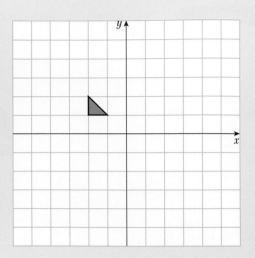

3 Describe, as a column vector, each transformation given below.

a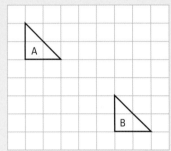

b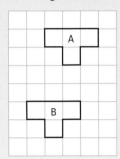

c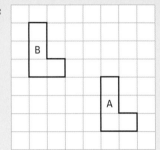

d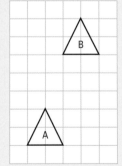

e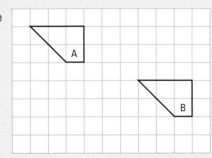

f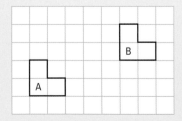

g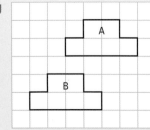

h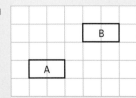

C

4 Describe, using a column vector, the transformation that moves the following shapes.

a A to B

b C to B

c B to D

d D to B

e B to A

f A to C

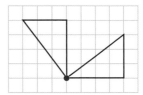

23.3 Rotations

◎ Objectives

● You can recognise a rotation.
● You can carry out a rotation.
● You can describe a rotation.

❓ Why do this?

Engineers need to check the path of a fairground ride as it rotates to make sure it does not crash into anything.

⬥ Get Ready

1. Write down the order of rotation of each of these shapes.

a    b    c   d

🔍 Key Points

● A **rotation** can be described as a fraction of a turn, or as an angle of turn.

● The direction of rotation can be **clockwise** ↻ or **anticlockwise** ↺.

● The point about which the shape is turned is called the **centre of rotation**.

● It is useful to use tracing paper to assist in rotating a shape.

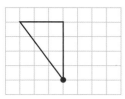 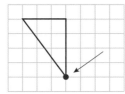 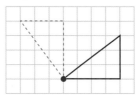

| Trace the shape onto tracing paper. | Hold the tracing paper down at the point of rotation, and rotate the tracing paper as requested. | Trace the shape in the new position. | Rotation complete. |

● The centre of rotation can sometimes be given as coordinates on a coordinate grid.

● In a rotation:

  ◦ the lengths of the sides of the shapes do not change

  ◦ the angles of the shape do not change

  ◦ the shape turns.

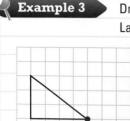

Example 3

Draw the image of the triangle after it has been rotated 90° clockwise about point O.
Label the image B.

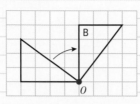

> Object and image are congruent.
> They are the same shape and size.

Exercise 23C

1 Copy each diagram. Draw the image of each shape after the rotation requested, using the point shown as centre of rotation.

a

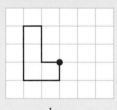

$\frac{1}{4}$ turn
clockwise

b

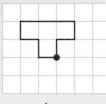

$\frac{1}{2}$ turn

c

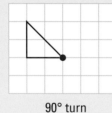

90° turn
anticlockwise

d

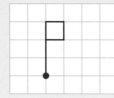

90° turn
clockwise

2 Copy each diagram. Draw the image of each shape after the rotation requested, using the point shown as centre of rotation.

a

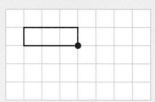

90° turn
anticlockwise

b

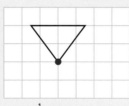

$\frac{1}{4}$ turn
clockwise

c

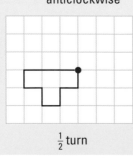

$\frac{1}{2}$ turn

d

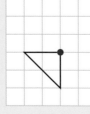

180° turn

3 Copy each diagram. Draw separate images for each shape after a rotation of 90° anticlockwise about each of the centres marked.

a

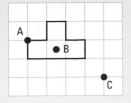

b

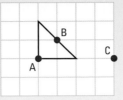

c

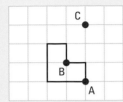

d

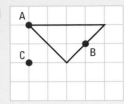

4 Copy each diagram. Draw separate images for each shape after the rotation requested, using the given point as centre of rotation.

a A rotation of A clockwise 90° about (1, 0)

b A rotation of B 90° clockwise about (−1, 2)

c A rotation of C 180° about (2, −2)

d A rotation of A anticlockwise 90° about (1, −1)

e A rotation of B 180° about (0, 3)

f A rotation of C clockwise 90° about (2, 2)

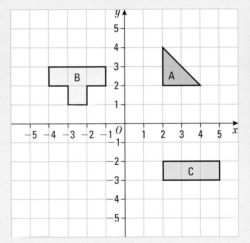

Example 4 Describe fully the transformation that maps shape A onto shape B.

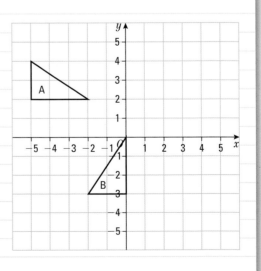

This is a rotation.

From A to B the direction is anticlockwise, and it is a 90° turn.

Using tracing paper or drawing lines (as shown) you can identify the centre of rotation as (0, 2), so the complete description is 'A rotation, 90° anticlockwise, centre (0, 2).'

> It is important to state that your transformation is a rotation.

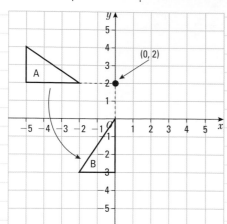

> Marks are given for stating three facts:
> (1) transformation is a rotation
> (2) direction and angle
> (3) centre of rotation.

Exercise 23D

D

1 Describe fully the rotation that maps shape A onto shape B.

a

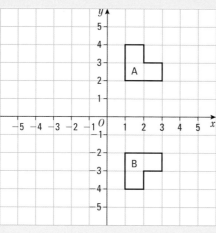

b

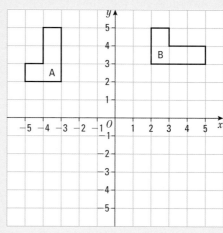

c

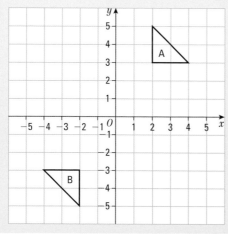

d

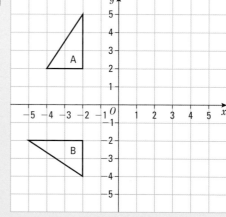

2　Describe fully the transformation that maps shape A onto shape B.

a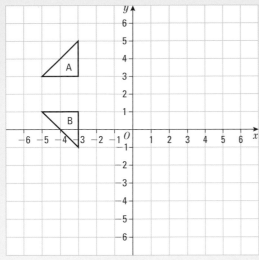

b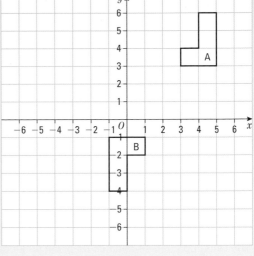

c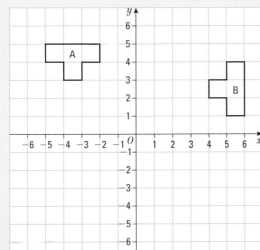

d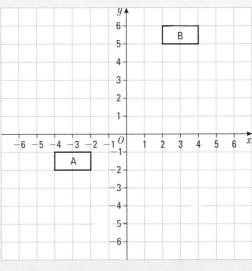

23.4 **Reflections**

◈ **Get Ready**

1. Below is a set of four shapes. Write down the number of lines of symmetry of each shape.

a 　　b 　　c 　　d

Key Points

- Images of a shape that are formed by reflecting a given shape about a **line of reflection** (or mirror line) are called **reflections** of the shape.
- In a reflection:
  - the lengths of the sides of the shape do not change
  - the angles of the shape do not change.
- It is useful to use tracing paper to assist in reflecting a shape.

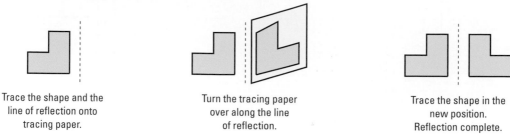

| Trace the shape and the line of reflection onto tracing paper. | Turn the tracing paper over along the line of reflection. | Trace the shape in the new position. Reflection complete. |

- The reflection can sometimes be given on a coordinate grid.
- The line of reflection can be given as the equation of a line.

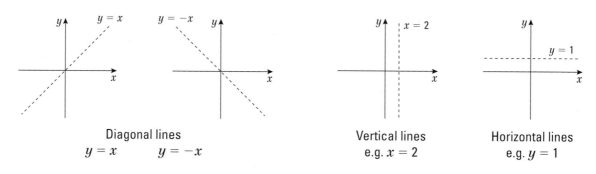

| Diagonal lines $y = x$ $y = -x$ | Vertical lines e.g. $x = 2$ | Horizontal lines e.g. $y = 1$ |

Example 5   Draw the image of this shape after it has been reflected in the mirror line.

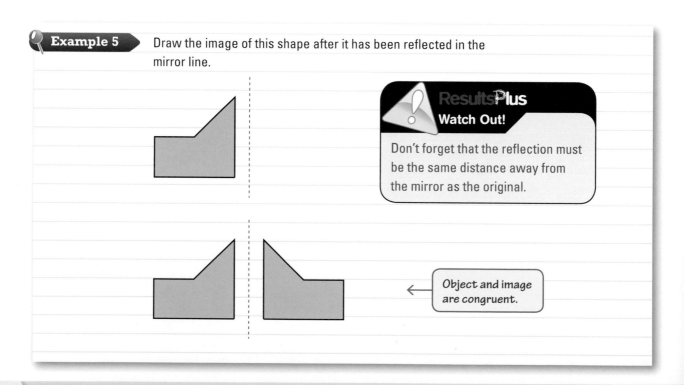

ResultsPlus
Watch Out!

Don't forget that the reflection must be the same distance away from the mirror as the original.

Object and image are congruent.

Exercise 23E

1 In each of these diagrams the dotted line is a line of reflection.
Copy each diagram and draw the reflection of the shape in the line.

a 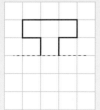   b 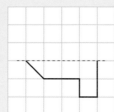   c

d 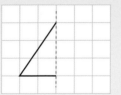   e 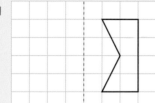   f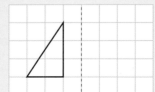

g 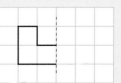   h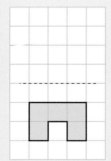

2 In each of these diagrams the dotted line is a line of reflection.
Copy each diagram and draw the reflection of the shape in the line.

a 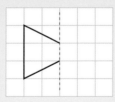   b 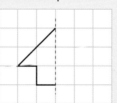   c

d 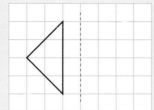   e 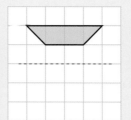   f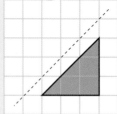

g 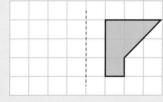   h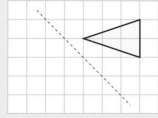

Example 6   Describe fully the transformation that maps shape A onto shape B.

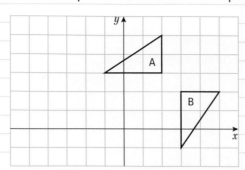

Marks are given for stating two facts:
(1) transformation is a reflection
(2) the line in which the shape has been reflected.

This is a reflection.

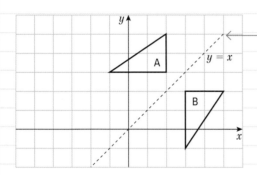

First draw the line of symmetry.
This is the line $y = x$.

The complete description of the transformation is:
'A reflection in the line $y = x$.'

Exercise 23F

1   Describe fully the reflection that maps shape A onto shape B.

a
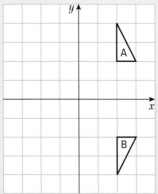

b
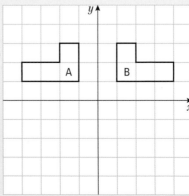

D

c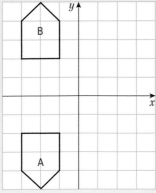

d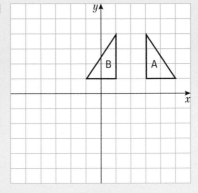

e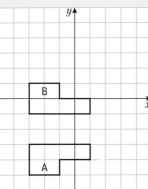

f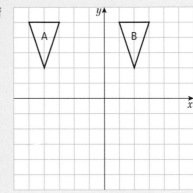

2 Describe fully the reflection that maps

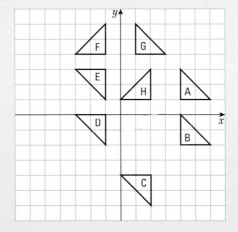

a F onto G

b D onto F

c A onto H

d A onto G

e F onto E

f B onto C

g C onto H.

3 Describe fully the reflection that maps shape A onto shape B.

a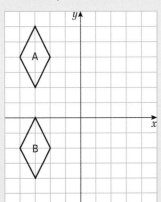

b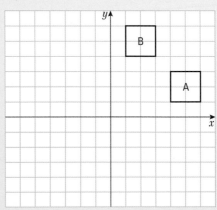

C

c
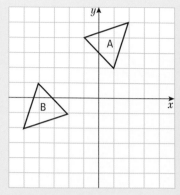

d
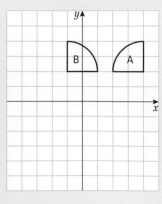

e
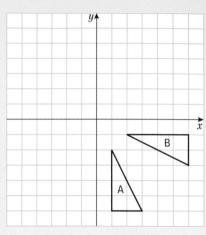

f
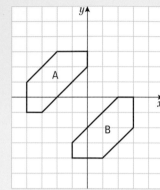

23.5 Enlargement

◉ Objectives

- ◉ You can recognise an enlargement.
- ◉ You can carry out an enlargement.
- ◉ You can describe an enlargement.

◈ Why do this?

Once an original for something exists, it can be enlarged. This can be seen in the case of Russian dolls, where each enlargement contains a smaller version of itself.

⬆ Get Ready

1. Draw this picture of a robot onto a copy of the larger grid, making the picture twice as big.

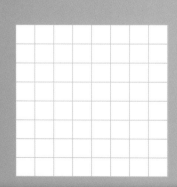

Key Points

● An **enlargement** changes the size of an object. It changes the length of its sides, but does not change its shape.

● The **scale factor** of the enlargement is the value that lengths in the original object are multiplied by to get the lengths in the image.
  For example, a scale factor of $1\frac{1}{2}$ means all the lengths are $1\frac{1}{2}$ times what they were in the original shape.

● Shapes can be enlarged from a point called the **centre of enlargement**.

● To find or use the centre of enlargement, draw in additional lines from the centre of enlargement to the vertices of the shape or shapes.

Example 7 Draw an enlargement of this shape, scale factor 2.

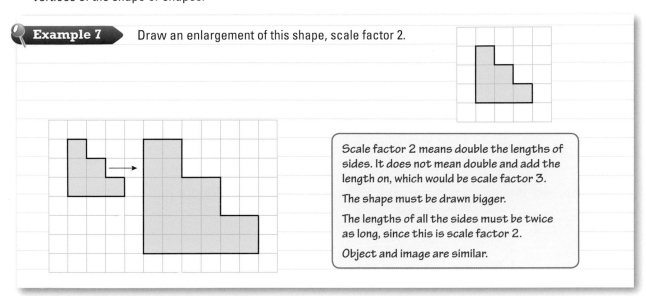

Scale factor 2 means double the lengths of sides. It does not mean double and add the length on, which would be scale factor 3.

The shape must be drawn bigger.

The lengths of all the sides must be twice as long, since this is scale factor 2.

Object and image are similar.

Exercise 23G

Copy the diagrams and enlarge each of the following shapes by the stated scale factor (sf).

1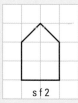
s f 2

2
s f 3

3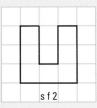
s f 2

4
s f 3

5
s f 1$\frac{1}{2}$

6
s f 2$\frac{1}{2}$

7
s f 2

8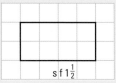
s f 3

9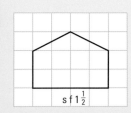
s f 1$\frac{1}{2}$

D

Example 8
Draw an enlargement of the triangle, scale factor 2, using point A as the centre of enlargement.

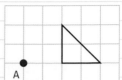

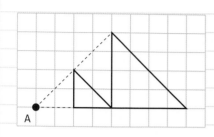

It is useful to draw in additional lines from the centre of enlargement, which will connect the vertices of the triangles. The lengths in the enlargement are twice the lengths of the original shape. The vertices are twice the distance from the centre of enlargement.

Exercise 23H

D

1 Copy each diagram onto squared paper. Enlarge each of these shapes by the stated scale factor (sf), from the given point of enlargement.

a
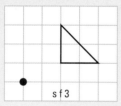
s f 3

b
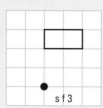
s f 3

c
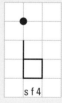
s f 2

d
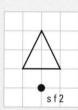
s f 2

e
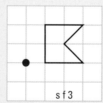
s f 3

f

s f 4

2 Copy each diagram onto squared paper. For each diagram draw two images, one from each of the points of enlargement given.

a
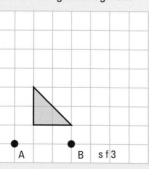
A   B   s f 3

b
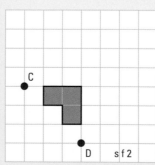
D   s f 2

c
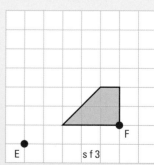
E   s f 3

3 Copy each diagram onto squared paper. For each diagram draw two images, one from each of the points of enlargement given.

a

A (0, 0)
B (3, −1)

s f 2

b

C (0, 0)
D (−2, 1)

s f 3

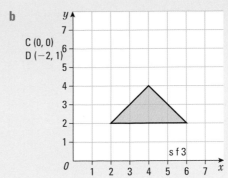

c

E (0, 0)
F (−1, 1)

s f 2

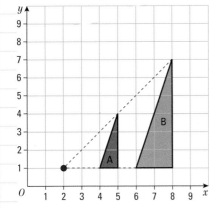

Example 9

Describe fully the transformation that maps shape A onto shape B.

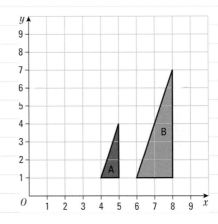

This is an enlargement.

By comparing the lengths of the sides, you can tell that the sides on shape B are twice the lengths of the sides on shape A.

By joining up the vertices (see the dotted lines) you can extend these lines so they meet at the centre of enlargement (2, 1).
So the transformation is
'An enlargement of scale factor 2, centre (2, 1).'

Marks are given for stating three facts:
(1) transformation is an enlargement
(2) the scale factor
(3) the centre of enlargement.

D

C

1   Describe fully the transformation that maps shape A onto shape B.

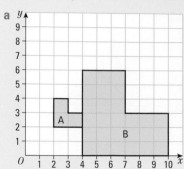
a

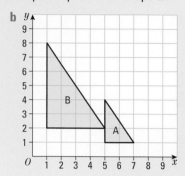
b

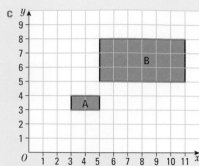
c

2   Describe fully the transformation that maps shape A onto shape B.

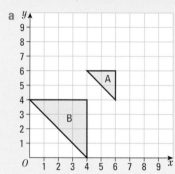
a

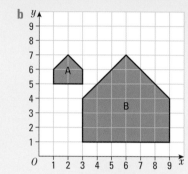
b

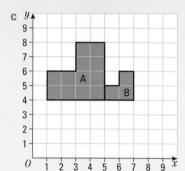
c

23.6 Combinations of transformations

⊙ Objectives

- You can carry out all four common transformations.
- You can describe all four transformations.

❓ Why do this?

Graphic designers and fabric printers would use a combination of transformations to produce designs or material for clothes.

⬆ Get Ready

1. Copy this shape.
   Rotate this shape clockwise 90° and draw the image.
   Repeat this twice more.

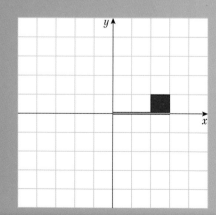

Key Point

◉ It is sometimes possible to find a single transformation that has the same effect as a combination of two transformations.

Example 10 Shape A is transformed by a reflection in the y-axis to image B. Image B is reflected in the x-axis to image C.

What single transformation maps shape A onto shape C?

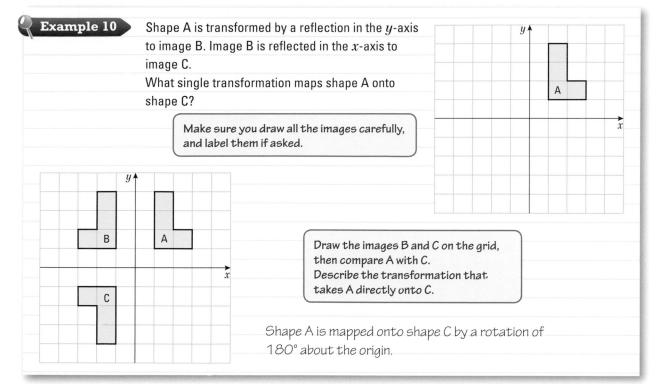

> Make sure you draw all the images carefully, and label them if asked.

> Draw the images B and C on the grid, then compare A with C.
> Describe the transformation that takes A directly onto C.

Shape A is mapped onto shape C by a rotation of 180° about the origin.

Exercise 23J

1   Copy the diagram.
  a Reflect the shape in the x-axis.
  b Reflect the image in the line $y = 2$.
  c Describe the single transformation that is equivalent to **a** followed by **b**.

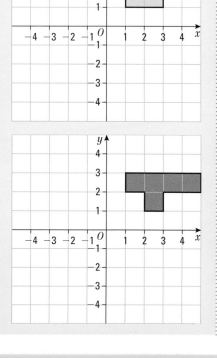

2   Copy the diagram.
  a Reflect the shape in the line $y = x$.
  b Rotate the image 90° anticlockwise about the origin.
  c Describe the single transformation that is equivalent to **a** followed by **b**.

C

3 Copy the diagram.
  a Reflect the shape A in the y-axis.
    Label the image B.
  b Reflect the image B in the x-axis.
    Label this image C.
  c Describe the single transformation that maps A onto C.

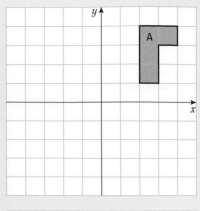

4 Copy the diagram.
  a Rotate the shape A 90° clockwise, centre (3, 3).
    Label the image B.
  b Rotate the image B 180°, centre (6, 3).
    Label this image C.
  c Describe the single transformation that maps A onto C.

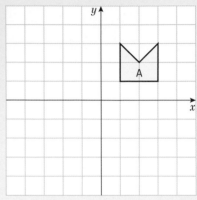

5 Triangle P has been rotated 180° about the point (1, 1) to give triangle Q.
  a Rotate triangle Q 180° about the point (3, −1).
    Label the triangle R.
  b Describe the single transformation that takes triangle P to triangle R.

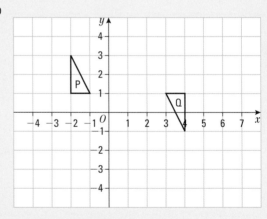

Chapter review

● There are four main **transformations**: translation, rotation, reflection and enlargement.
● A **translation** is a sliding movement made from one or more moves. In a simple translation you need to describe the distance and direction of each move.
● Another way of describing translations is to use a **column vector**, e.g. $\binom{3}{2}$. The top number describes the movement to the right or left, the bottom number the movement up or down.
● In a translation: the lengths of the sides of the shape do not change; the angles of the shape do not change; the shape does not turn.
● A **rotation** can be described as a fraction of a turn, or as an angle of turn.
● The direction of rotation can be **clockwise** or **anticlockwise**.

- The point about which the shape is turned is called the **centre of rotation**.
- It is useful to use tracing paper to assist in rotating a shape.
- The centre of rotation can sometimes be given as coordinates on a coordinate grid.
- In a rotation: the lengths of the sides of the shape do not change; the angles of the shape do not change; the shape turns.
- Images of a shape which are formed by reflecting a given shape about a **line of reflection** (or mirror line) are called **reflections** of the shape.
- In a reflection: the lengths of the sides of the shape do not change; the angles of the shape do not change.
- It is useful to use tracing paper to assist in reflecting a shape.
- The reflection can sometimes be given on a coordinate grid.
- The line of reflection can be given as the equation of a line.
- An **enlargement** changes the size of an object. It changes the lengths of it sides but does not change its shape.
- The **scale factor** of the enlargement is the value that lengths in the original object are multiplied by to get the lengths in the image.
- Shapes can be enlarged from a point called the **centre of enlargement**.
- To find or use the centre of enlargement, draw in additional lines from the centre of enlargement to the vertices of the shape or shapes.

Review exercise

1 **a** Reflect shape **A** in the y-axis.

   b Describe fully the **single** transformation which takes shape **A** to shape **B**.

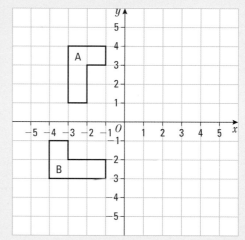

Nov 2008

2 On a copy of the grid, enlarge the shape with a scale factor of 2.

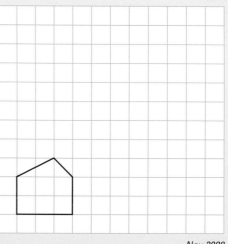

Nov 2008

E

E

3 Reflect the shaded shape in the mirror line.

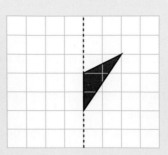

March 2007

4 On a copy of the grid, draw an enlargement of the shaded shape with a scale factor of 3.

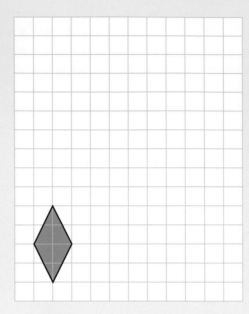

Nov 2006

5 On a copy of the grid, draw an enlargement, scale factor 2, of the shaded shape.

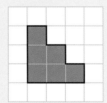

June 2009

D

6 **a** Reflect Shape **A** in the y-axis. Label your new shape **B**.

b Translate Shape **A** by the vector $\begin{pmatrix} 3 \\ -2 \end{pmatrix}$

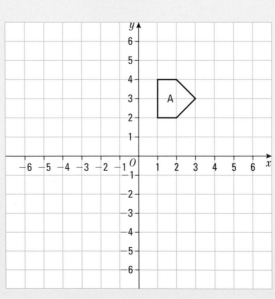

Nov 2007 adapted

7 Triangle **T** has been drawn on the grid.
  a On a copy of the grid, reflect triangle **T** in the y-axis. Label the new triangle **A**.
  b On a copy of the grid, rotate triangle **T** by a half turn, centre O. Label the new triangle **B**.
  c Describe fully the single transformation which maps triangle **T** onto triangle **C**.

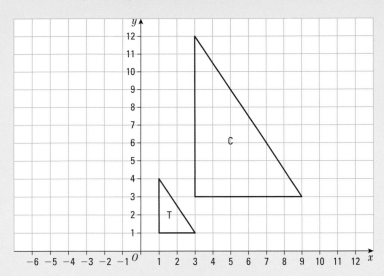

June 2007 adapted

8 On a copy of the grid, rotate the triangle a half turn about the point O.

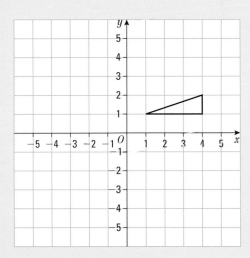

Nov 2006

9 Describe fully the single transformation that maps triangle **A** onto triangle **B**.

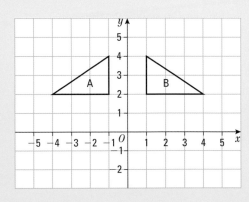

June 2009

C

10   **a** Rotate triangle **P** 180° about the point (–1, 1).
Label the new triangle **A**.

b Translate triangle **P** by the vector $\begin{pmatrix} 6 \\ -1 \end{pmatrix}$.
Label the new triangle **B**.

c Reflect triangle **Q** in the line $y = x$.
Label the new triangle **C**.

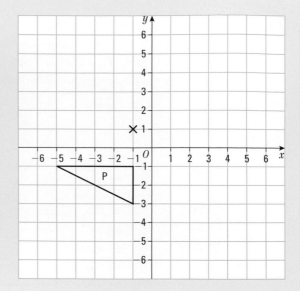

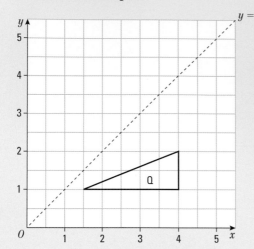

June 2008

11 Describe fully the single transformation
that will map shape **P** onto shape **Q**.

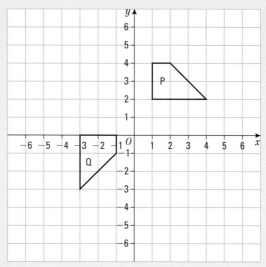

Nov 2007

12   **a** Describe fully the single transformation that
maps triangle **A** onto triangle **B**.

b On the grid, rotate triangle **A** 90° anticlockwise
about the point (–1, 1).
Label your new triangle **C**.

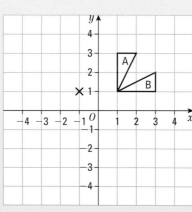

Nov 2006

C

13   **a**   On a copy of the grid, reflect triangle **P** in the line $x = 2$.

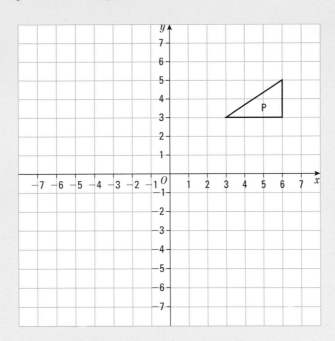

    b   Describe fully the **single** transformation that takes triangle **Q** to triangle **R**.

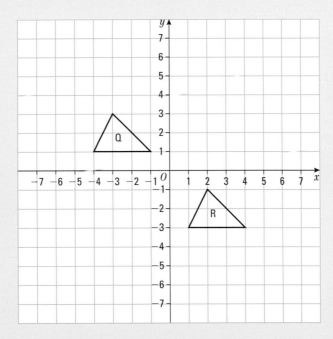

Nov 2006

C

14 Triangle **A** is reflected in the x-axis
to give triangle **B**.
Triangle **B** is reflected in the line
$x = 1$ to give triangle **C**.
Describe the **single** transformation
that takes triangle **A** to triangle **C**.

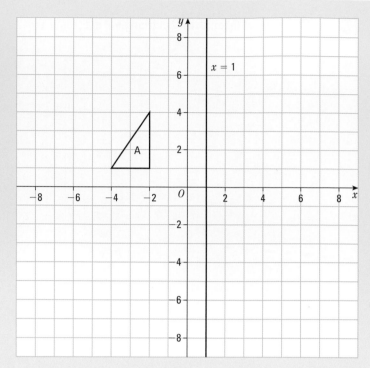

June 2008

24 RATIO AND PROPORTION

To understand bicycle gears, you need to work with ratios. Gears work by changing the distance that you move forward each time you pedal. The lowest gear of a bicycle might use a front cog with 22 teeth and a back cog with 33 teeth. That's a ratio of 0.67 : 1. So for each pedal stroke, the wheels do 0.67 of a turn. The highest gear might have a front chain wheel with 44 teeth and a back chain wheel with 11 teeth, giving a ratio of 4 : 1. So the wheel turns four times for every pedal stroke.

◎ Objectives

In this chapter you will:
- write a ratio in various forms
- use a ratio to write down a fraction
- use equivalent ratios to find unknown quantities
- solve problems using scales of maps and scale drawings and problems involving proportion
- divide a quantity in a given ratio and use the unitary method to find quantities.

◇ Before you start

You need to be able to:
- multiply and divide by an integer
- simplify a fraction
- find equivalent fractions
- convert between metric units.

24.1 Introducing ratio

◎ Objectives

- You can write down a ratio.
- You can use a ratio to write down a fraction.
- You can write a ratio in its simplest form.
- You can write a ratio in the form $1 : n$.

? Why do this?

Scales on maps and drawings are often written as ratios. You can work out the distance between two places on a map by using the map's scale.

⬆ Get Ready

1. What fraction of this rectangle is blue?

2. Write $\frac{9}{12}$ in its simplest form.

3. Write $\frac{80}{100}$ in its simplest form.

Key Points

- **Ratios** are used to compare quantities.

  The ratio of green triangles to white triangles is 3 : 2.

  The ratio of white triangles to green triangles is 2 : 3.

- Ratios can be simplified like fractions.

  To simplify a ratio you divide each of its numbers by a common factor. Common factors appeared in Section 1.10.

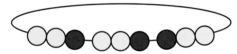

  This necklace has 6 yellow beads and 3 red beads.

  The ratio of yellow beads to red beads is 6 : 3.

  Both 6 and 3 can be divided by 3 to give the ratio 2 : 1.

  This means that for every 2 yellow beads there is 1 red bead.

- When a ratio cannot be simplified it is in its simplest form.

- It is sometimes useful to write ratios in the form $1 : n$.

  The number n is written as a whole number or a decimal.

  When one of the numbers in a ratio is 1, the ratio is in **unitary** form.

Example 1

Write down the ratio of yellow counters to blue counters.

The ratio is 5 : 4.

⬅ 5 comes first as 'yellow counters' is written before 'blue counters'.

**Exercise 24A**

Questions in this chapter are targeted at the grades indicated.

E

1 For each pattern of tiles, write down the ratio of the number of white tiles to the number of red tiles.

a

b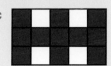

c

2 Adam is 16 years old. Sarah is 13 years old.

a What is the ratio of Adam's age to Sarah's age?

b What is the ratio of Sarah's age to Adam's age?

3

a What is the ratio of the number of circles to the number of squares?

b What is the ratio of the number of circles to the number of triangles?

c What is the ratio of the number of triangles to the number of squares?

d What is the ratio of the number of circles to the number of squares to the number of triangles?

Example 2 A box contains blue cubes and red cubes in the ratio 3 : 2.
What fraction of these cubes are blue?

For every 3 blue cubes there are 2 red cubes.

$3 + 2 = 5$

$\frac{3}{5}$ of the cubes are blue.

Exercise 24B

E

1 In a fitness centre, the ratio of the number of men to the number of women is 1 : 2.
What fraction of these people are men?

2 A vase contains red roses and white roses in the ratio 2 : 3.

a What fraction of these roses are red?

b What fraction of these roses are white?

3 A box of chocolates contains milk chocolates and dark chocolates in the ratio 5 : 3.

a What fraction of these chocolates are milk chocolates?

b What fraction of these chocolates are dark chocolates?

D

4 A box contains blue pens, red pens and black pens in the ratio 7 : 1 : 2.
    a What fraction of these pens are red?
    b What fraction of these pens are blue?

5 On a farm, $\frac{1}{2}$ of the horses are female.
    What is the ratio of female horses to male horses on this farm?

6 On the farm, $\frac{1}{3}$ of the pigs are female.
    What is the ratio of female pigs to male pigs on this farm?

Example 3   Write the ratio 15 : 20 in its simplest form.

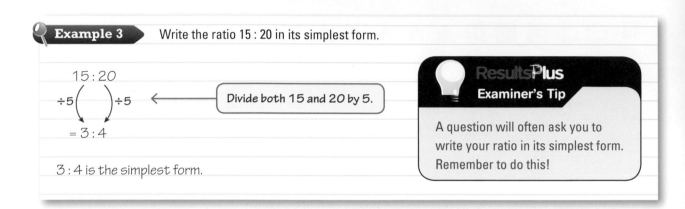

$$15 : 20$$
$$\div 5 \left(\right) \div 5 \qquad \longleftarrow \quad \boxed{\text{Divide both 15 and 20 by 5.}}$$
$$= 3 : 4$$

3 : 4 is the simplest form.

ResultsPlus
Examiner's Tip

A question will often ask you to write your ratio in its simplest form. Remember to do this!

Exercise 24C

E

1 Write these ratios in their simplest form.
    a 5 : 10   **b** 4 : 12   **c** 21 : 15   **d** 30 : 45
    e 80 : 60   **f** 24 : 72   **g** 160 : 300   **h** 600 : 2400

2 In a fruit bowl there are 8 oranges and 12 apples.
    Write down the ratio of the number of oranges to the number of apples.
    Give your ratio in its simplest form.

3 There are 200 boys and 150 girls in a school.
    Write down the ratio of the number of boys to the number of girls.
    Give your ratio in its simplest form.

4 In a bird aviary there are 20 finches, 36 canaries and 24 budgies.
    Write down the ratio of the number of finches to the number of canaries to the number of budgies.
    Give your ratio in its simplest form.

5 In a class there are 28 students. Twelve of the students are boys.
    Write down the ratio of the number of boys to the number of girls.
    Give your ratio in its simplest form.

Example 4   Write the ratio 40 minutes : 3 hours in its simplest form.

40 minutes : 3 hours

= 40 minutes : 180 minutes   ←   Make the units the same.

= 40 : 180   ←   Divide both numbers by 10

= 4 : 18   ←   Divide both numbers by 2

= 2 : 9   ←   This is the simplest form.

Exercise 24D

E

1.   Write these ratios in their simplest form.
   a   20 minutes : 1 hour   b   40p : £1   c   30 cm : 1 m   d   1 day : 6 hours

2.   Write these ratios in their simplest form.
   a   600 g : 2 kg   b   4 cm : 6 mm   c   2 hours : 45 minutes   d   75 cm : 3 m

3.   A glass contains 450 ml of water and a bottle contains 1 l of water.
   Write down the ratio of the amount of water in the glass to the amount of water in the bottle.
   Give your ratio in its simplest form.

4.   A small bag of sugar weighs 375 g and a large bag of sugar weighs 2 kg.
   Write down the ratio of the weight of the small bag to the weight of the large bag.
   Give your ratio in its simplest form.

5.   A table has a length of 1.2 m and a width of 40 cm.
   a   Find, in its simplest form, the ratio of the length of the table to the width of the table.
   b   Find, in its simplest form, the ratio of the width of the table to the length of the table.

Example 5   Write the ratio 6 : 9 in the form 1 : n.

6 : 9

÷6   (   )   ÷6   ←   Divide by 6 to make the first number into 1.

= 1 : 1.5

Exercise 24E

D

1 Write these ratios in the form $1 : n$.
   a 2 : 5   b 5 : 12   c 10 : 3   d 200 : 250

2 The length of a model car is 20 cm. The length of the real car is 360 cm.
Write down the ratio of the length of the model car to the length of the real car.
Give your answer in the form $1 : n$.

3 In a school there are 120 teachers and 1740 students. The headteacher wants to display on his website the ratio of the number of teachers to the number of students in the form $1 : n$. Work out this ratio.

4 Write these ratios in the form $1 : n$.
   a 20p : £1   b 50 g : 1 kg   c 4 cm : 8 mm   d 30 minutes : 3 hours

Mixed exercise 24F

E

1 Write, in its simplest form, the ratio of the number of red squares to the number of white squares.

2 A necklace has yellow beads and green beads only.
The ratio of the number of yellow beads to the number of green beads is 4 : 5.
   a What fraction of the beads are yellow?
   b What fraction of the beads are green?

3

Ingredient	Weight in grams
Protein	6
Carbohydrate	16
Fat	4

Write, in its simplest form, the ratio of:
   a the weight of carbohydrate to the weight of protein
   b the weight of fat to the weight of carbohydrate
   c the weight of protein to the weight of carbohydrate to the weight of fat.

4 Write these ratios in their simplest form.
   a 2 hours : 40 min   b 800 m : 2 km

5 A recipe uses 150 g of margarine and 1 kg of fruit.
Write down the ratio of the weight of margarine to the weight of fruit.
Give your ratio in its simplest form.

D

6 Write the ratio 24 : 36
   a in its simplest form   b in the form $1 : n$.

7 On a bus there are 12 adults and 27 children.
Write the ratio of the number of adults to the number of children in the form $1 : n$.

8 An art gallery issues this guidance to schools planning to visit the gallery.

D

A03

> The recommended adult/pupil ratio is:
> For Years 1 to 3, a minimum of 1 adult to every 5 pupils
> For Years 4 to 9, a minimum of 1 adult to 10 pupils
> For Years 10 onwards, a minimum of 1 adult to 15 pupils

A primary school is planning a visit to the art gallery.
The table shows information about the pupils going on the visit.

Year	Number of pupils
2	7
3	13
4	12
5	14

Work out the minimum number of adults that need to go on the visit.

24.2 Solving ratio problems

⊙ Objectives

● You can use equivalent ratios to find unknown quantities.
● You can solve problems using scales of maps and scale drawings.

⊘ Why do this?

You could use ratios if you were trying to mix a specific colour of paint for decorating your room that the shop did not stock.

◈ Get Ready

1. $\frac{1}{5} = \frac{?}{20}$
2. Which of these fractions are equivalent to $\frac{1}{4}$? $\quad \frac{2}{6} \quad \frac{2}{8} \quad \frac{3}{9} \quad \frac{3}{12}$
3. Change 3.65 centimetres to metres.

🔍 Key Points

● The ratio 15 : 20 simplifies to the ratio 3 : 4.
15 : 20 and 3 : 4 are **equivalent ratios**.
● If you know the ratio of two quantities and you know one of the quantities, you can use equivalent ratios to find the other quantity.
● Maps have a scale to tell us how a distance on the map relates to the real distance.
A scale of 1 : 25 000, for example, means that 1 cm on the map represents a real distance of 25 000 cm.
● Someone making a scale model or producing a scale drawing will also need to use a scale.

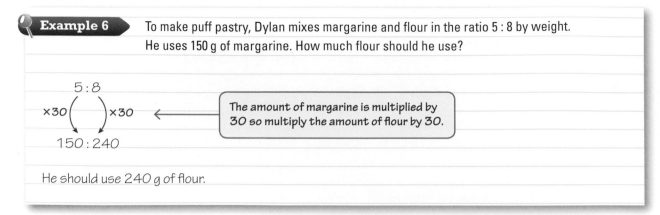

Example 6 To make puff pastry, Dylan mixes margarine and flour in the ratio 5 : 8 by weight.
He uses 150 g of margarine. How much flour should he use?

5 : 8

×30 ( ) ×30 ← The amount of margarine is multiplied by 30 so multiply the amount of flour by 30.

150 : 240

He should use 240 g of flour.

Exercise 24G

1 James makes mortar by mixing cement and sand in the ratio 1 : 5.
He uses 4 buckets of cement. Work out how many buckets of sand he uses.

2 Margaret makes porridge by mixing oats and water in the ratio 1 : 2.
Work out the number of cups of water she uses for:

   a 2 cups of oats   **b** 3 cups of oats   **c** 10 cups of oats.

3 An alloy contains iron and aluminium in the ratio 4 to 1 by weight.

   a If there is 24 kg of iron, work out the weight of aluminium.

   b If there is 15 kg of aluminium, work out the weight of iron.

4 In a recipe for making pastry the ratio of the weight of flour to the weight of margarine is 3 : 2.

   a Work out the weight of margarine needed for:

      i 60 g of flour   **ii** 300 g of flour   **iii** 450 g of flour.

   b Work out the weight of flour needed for:

      i 60 g of margarine   **ii** 100 g of margarine   **iii** 250 g of margarine.

5 Sidra is making a fruit drink.
She mixes orange juice, pineapple juice and syrup in the ratio 3 : 4 : 1.

   a If she uses 600 ml of orange juice, how much syrup will she need?

   b If she uses 1 l of pineapple juice, how much orange juice will she need?

6 In the 2008 Olympic Games, the ratio of the number of bronze medals won by Great Britain to the number of bronze medals won by Japan was 3 : 2. Great Britain won 5 more bronze medals than Japan. How many bronze medals did Great Britain win?

7 Paul makes green paint by mixing 2 parts of yellow paint with 3 parts of blue paint.
Paul has 500 ml of yellow paint and 1 litre of blue paint.
What is the maximum amount of green paint that Paul can make?

D

A03

Example 7

The scale of a map is 1 : 20 000.
Work out the real distance, in kilometres, that 9 cm on the map represents.

1 : 20 000 ← | 1 cm on the map represents a real distance of 20 000 cm. |

9 × 20 000 = 180 000 ←

| Multiply the length on the map by 20 000. |

The real distance is 180 000 cm.
180 000 ÷ 100 = 1800 m ← | Change 180 000 cm to kilometres. |
1800 ÷ 1000 = 1.8 km

9 cm on the map represents a real distance of 1.8 km.

Exercise 24H

1 Alex uses a scale of 1 : 50 to draw a plan of his bedroom.
On the plan the length of the bedroom is 8 cm.
Work out the real length of the bedroom.

2 Shannon makes a scale model of a house. She uses a scale of 1 : 12.
The height of the model house is 60 cm.
Work out the height of the real house.

3 A model of a ship is made using a scale of 1 : 600.
The length of the model ship is 40 cm.
Work out the length of the real ship.

4 The length of a car is 3 metres.
Asif makes a model of the car. He uses a scale of 1 : 12.
Work out the length, in centimetres, of the model car.

5 A company makes model cars using a scale of 1 : 18.
  a Work out the length of the real car if the length of a model car is 20 cm.
  b Work out the length of a model car if the length of the real car is 4.68 m.

6 The scale of a map is 1 : 100 000.
On the map the distance between two towns is 6 cm.
Work out the real distance between the two towns.

7 The scale of a map is 1 : 50 000.
On the map the length of a railway tunnel is 3.5 cm.
Work out the real length of the railway tunnel.

8 The scale of a map is 1 : 200 000.
The real distance between two towns is 24 km.
Work out the distance between the towns on the map.

D

24.3 Sharing in a given ratio

◉ Objective

○ You can divide a quantity in a given ratio.

? Why do this?

Builders use ratios to make sure they have the exact quantities of ingredients needed for mixing concrete.

◆ Get Ready

1. In a necklace the ratio of blue beads to yellow beads is 3 : 4.
   What fraction of these beads are yellow?
2. Write the ratio 16 : 24 in its simplest form.
3. Write the ratio 12 : 180 in the form 1 : n.

Key Point

◉ Sometimes we want to divide a quantity in a certain ratio.
   Suppose Sam and Hannah buy a box of chocolates costing £3.00.
   If Sam paid £2.00 and Hannah paid £1.00 they might decide to share the chocolates in the ratio 2 : 1.

🔍 Example 8

Heidi and Kirsty share £75 in the ratio 2 : 3.
Work out how much money each girl receives.

$2 + 3 = 5$ ← Work out the total number of shares.

$75 ÷ 5 = 15$ ← Work out the size of one share.

$15 × 2 = 30$ ← Heidi receives 2 shares.

$15 × 3 = 45$ ← Kirsty receives 3 shares.

Heidi gets £30 and Kirsty gets £45.

ResultsPlus
Watch Out!

Some students divide the total amount by the numbers in the ratio. Make sure you work out the number of shares first.

⚙ Exercise 24I

D

1 Share £80 in the ratio 1 : 4.

2 Share £24 in the ratio 3 : 5.

3 Share £45 in the ratio 2 : 3.

4 Alex and Ben share 40 sweets in the ratio 3 : 1.
  Work out how many sweets each receives.

5   At school the technician is going to make some brass. Brass is made from copper and nickel
    in the ratio 17 : 3. **D**
    Work out how much copper and how much nickel he will need to make 800 g of brass.

6   The ratio of boys to girls in a class is 4 : 5.
    There are 27 students in the class. Work out the number of girls in the class.

7   Share £40 in the ratio 1 : 3 : 4. **C**

8   Rebecca is making shortbread. She uses flour, sugar and butter in the ratio 3 : 1 : 2.
    Work out how much of each ingredient she needs to make 900 g of shortbread.

9   Three boys shared £30 in the ratio 5 : 3 : 2.
    William received the smallest amount. Work out how much William received.

10 Masud made some compost. He mixed soil, manure and leaf mould in the ratio 3 : 2 : 1.
    Masud made 120 litres of compost. Work out how much manure he used.

Mixed exercise 24J

1   A model of a helicopter is made using a scale of 1 : 72. **D**
    The length of the model helicopter is 20 cm.
    Work out the length, in metres, of the real helicopter.

2   Tom is making cakes. In a recipe the ratio of the weight of margarine to the weight of caster
    sugar is 3 : 1.
    Work out the weight of caster sugar Tom needs if the recipe uses 300 g of margarine.

3   The ratio of Martin's height to Tom's height is 7 to 8.
    Martin's height is 140 cm. What is Tom's height?

4   In a school choir the ratio of the number of boys to the number of girls is 2 : 5.
    There is a total of 35 boys and girls in the choir.
    Work out the number of girls in the choir.

5   The scale of a map is 1 : 300 000.
    On the map the distance between two towns is 4.5 cm.
    Work out the real distance, in kilometres, between the towns.

6   Ashley and Farjad share £42 in the ratio 3 : 4.
    Work out how much more money Farjad receives than Ashley.

7   An alloy contains copper, manganese and nickel in the ratio 14 : 5 : 1 by weight. **C**
    The weight of the copper is 70 kg.
    Work out the weight of the manganese and the weight of the nickel.

8   Matt needs 3 lengths of wood for a shelf display in the ratio 5 : 2 : 3.
    The plank of wood the shelves are cut from is 1200 cm long.
    Work out the length of each piece of wood for the 3 shelves.

24.4 Solving ratio and proportion problems using the unitary method

Objectives

- You can use the unitary method to find quantities.
- You can solve problems involving proportion.

Why do this?

Chefs use proportion to make sure they use the correct quantity of each ingredient.

Get Ready

1. If two cups of tea cost £2.70, what is the cost of one cup of tea?
2. If five pens cost 85p, what is the cost of one pen?
3. If four pizzas cost £7.80, what is the cost of one pizza?

Key Points

- If two quantities increase or decrease at the same rate they are in **direct proportion**.
- If 1 pound buys 2 dollars, then 2 pounds will buy 4 dollars.
  When the number of pounds is doubled, the number of dollars is also doubled.
  The number of dollars is **proportional** to the number of pounds.

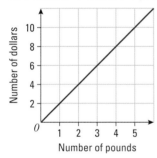

As the number of dollars is proportional to the number of pounds, the graph is a straight line passing through the origin.

- When two quantities are in direct proportion the ratio between them says the same.
- There are many examples of proportion in everyday life.
  The cost of a bag of tomatoes is directly proportional to the weight of the tomatoes.
  The cost of buying some petrol is proportional to the number of litres of petrol you put in your tank.
- You can use the unitary method to solve problems involving direct proportion.
  In the unitary method you always find the value of one item (or unit) first.
  For example, if three snack bars cost 96p, you can work out the cost of one snack bar.
  Once you know that one snack bar costs 32p you can then work out the cost of any number of snack bars.

Example 9

Seven bricks weigh 21 kg.
Work out the weight of 10 of these bricks.

$21 \div 7 = 3$ ⟵ First work out the weight of one brick.

$3 \times 10 = 30$ ⟵ Multiply the weight of one brick by 10.

10 bricks weigh 30 kg.

Exercise 24K

1. Five pens cost 75p. Work out the cost of eight of these pens.

2. Three identical stamps cost 96p. Work out the cost of seven of these stamps.

3. Eight cinema tickets cost £46. What would 10 of these tickets cost?

4. Fifteen identical pipes laid end to end make a length of 120 metres.
   What length will 12 of the pipes make if they are laid end to end?

5. The cost of 5 metres of ribbon is £6.25. Daniel wants to buy 3 metres of ribbon.
   Work out the cost of 3 metres of ribbon.

6. Three 2.5-litre tins of paint cost £44.94. Work out the cost of five of the 2.5-litre tins of paint.

7. 100 g of cheese contains 14 g of carbohydrates.
   Work out the weight of carbohydrates in 125 g of cheese.

8. 300 sheets of paper have a total thickness of 2.7 cm.
   Charlotte's paper feeder for her printer holds 4.5 cm of paper.
   How many sheets will she need to completely fill the paper feeder?

Example 10 This is a list of ingredients needed to make fruit crumble for four people.

| 350 g fruit | 100 g flour | 50 g margarine | 50 g sugar |

Work out the amount of margarine needed to make fruit crumble for 12 people.

$12 \div 4 = 3$ ⟵ 3 times as much of each ingredient is needed.

$\div 4$ then $\times 12$ is the same as $\times 3$.

$50 \times 3 = 150$ ⟵ Multiply the amount of margarine by 3.

150 g of margarine is needed.

Exercise 24L

1. This is a list of ingredients needed to make 50 cheese straws.

   100 g flour
   50 g margarine
   75 g cheese

   Connor wants to make 100 cheese straws. Work out the amount of each ingredient Connor needs.

2. This is a recipe for making Quiche Lorraine for four people.

   100 g pastry
   100 g bacon
   75 g cheese
   2 eggs
   150 ml milk

   Rhys is making Quiche Lorraine for 6 people.
   Work out the amount of each ingredient he needs.

D

3 This is a list of ingredients needed to make 20 almond biscuits.

175 g flour
75 g caster sugar
50 g ground almonds
150 g margarine

a Work out the amount of flour needed to make 40 almond biscuits.

b Work out the amount of margarine needed to make 10 almond biscuits.

c Work out the amount of ground almonds needed to make 30 almond biscuits.

Example 11

a Martin went to Spain. He changed £400 into euros.
The exchange rate was £1 = 1.08 euros. How many euros did he receive?

b When Martin came home he changed 63 euros into pounds.
The new exchange rate was £1 = 1.05 euros. How many pounds did he receive?

a 400 × 1.08 = 432 ⟵ Multiply the number of pounds by 1.08.
Martin received 432 euros.

b 63 ÷ 1.05 = 60 ⟵ Divide the number of euros by 1.05.
Martin received £60.

Exercise 24M

D

1 Hannah went on holiday to France. She changed £200 into euros.
The exchange rate was £1 = 1.08 euros. Work out how many euros Hannah received.

2 Matas changed £600 into Russian roubles. The exchange rate was £1 = 44.95 roubles.
Work out how many roubles Matas received.

3 Suha is going to the USA. The exchange rate is £1 = $1.42
a Change £300 into dollars.
b Change $355 into pounds.

4 Paolo bought a railway ticket for €84 in Italy. The exchange rate was £1 = €1.12.
Work out the cost of the ticket in pounds.

5 Danny paid 74 francs for a meal in Switzerland. The exchange rate was £1 = 1.85 francs.
Work out the cost of the meal in pounds.

Mixed exercise 24N

1 The weight of 10 identical coins is 250 g. Work out the weight of 12 of these coins.

2 Matthew was paid £75 for 12 hours' work in a shop.
At the same rate, how much would he be paid for 7 hours' work?

3 The cost of three rolls of wallpaper is £25.50. Mandeep needs five rolls of wallpaper to wallpaper her dining room. Work out the total cost.

4 £1 = 1.18 euros
   a Change £400 into euros.   **b** Change 236 euros into pounds.

5 This is a recipe for making sponge pudding for six people.
   100 g margarine
   100 g caster sugar
   2 eggs
   225 g flour
   30 ml milk
   Work out the amount of each ingredient needed to make sponge pudding for 15 people.

6 Aliyah came back from a holiday in Australia. She changed $164 into pounds.
The exchange rate was £1 = $2.05. Work out how many pounds she received.

7 This is a recipe for making fruit crumble for four people.
   350 g fruit
   50 g margarine
   100 g flour
   50 g sugar
   Hazel has only 175 g of flour.
   She has plenty of each of the other ingredients.
   Work out how many people she can make fruit crumble for.

8 Calum is on holiday in Switzerland.
He buys a pair of sunglasses.
He can pay either 80 francs or 55 euros.
Is it cheaper for Calum to pay 80 francs or to pay 55 euros?
Explain your answer.

Exchange rates
£1 = 1.12 euros
£1 = 1.68 francs

Chapter review

● **Ratios** are used to compare quantities.
● Ratios can be simplified like fractions.
   To simplify a ratio you divide each of its numbers by a common factor.
● When a ratio cannot be simplified it is in its simplest form.

- It is sometimes useful to write ratios in the form $1 : n$.
  The number n is written as a whole number or a decimal.
  When one of the numbers in a ratio is 1, the ratio is in **unitary** form.
- The ratio $15 : 20$ simplifies to the ratio $3 : 4$.
  $15 : 20$ and $3 : 4$ are **equivalent ratios**.
- If you know the ratio of two quantities and you know one of the quantities, you can use equivalent ratios to find the other quantity.
- Maps have a scale to tell us how a distance on the map relates to the real distance.
- Someone making a scale model or producing a scale drawing will also need to use a scale.
- Sometimes we want to divide a quantity in a certain ratio.
- If two quantities increase or decrease at the same rate they are in **direct proportion**.
- When two quantities are in direct proportion the ratio between them stays the same.
- You can use the unitary method to solve problems involving direct proportion.
  In the unitary method you always find the value of one item (or unit) first.

Review exercise

1 There are some sweets in a bag.
8 of the sweets are toffees. 12 of the sweets are mints.
Write down the ratio of the number of toffees to the number of mints.
Give your ratio in its simplest form.
June 2009

2 A coin is made from copper and nickel.
84% of its weight is copper.
16% of its weight is nickel.
Find the ratio of the weight of copper to the weight of nickel.
Give your answer in its simplest form.
June 2008

3 The distance from Ailing to Beeford is 2 km. The distance from Ceetown to Deeton is 800 metres.
Write as a ratio
Distance from Ailing to Beeford : Distance from Ceetown to Deeton.
Give your answer in its simplest form.

4 There are some oranges and apples in a box.
The total number of oranges and apples is 54.
The ratio of the number of oranges to the number of apples is $1 : 5$.
Work out the number of apples in the box.
June 2009

5 A garage sells British cars and foreign cars.
The ratio of the number of British cars sold to the number of foreign cars sold is $2 : 7$.
The garage sells 45 cars in one week.
Work out the number of British cars the garage sold that week.
June 2008

6 Alice builds a model of a house. She uses a scale of $1 : 20$.
The height of the real house is 10 metres.
  a Work out the height of the model.
The width of the model is 80 cm.
  b Work out the width of the real house.

7 There are 600 counters in a bag.

90 of the 600 are yellow. 180 of the 600 are red.

The rest of the counters in the bag are blue or green.

There are twice as many blue counters as green counters.

Work out the number of green counters in the bag.

May 2009

8 Here is a list of ingredients for making fudge for 6 people.

Fudge

Ingredients for 6 people

600 g of sugar

12 g of butter

480 g of condensed milk

90 ml of milk

86% of students did very well on this type of question.

Work out how much of each ingredient is needed to make fudge for 9 people.

Nov 2006

9 Ron went to Spain.

He changed £200 into euros (€).

The exchange rate was £1 = €1.40.

a How many euros did he get?

When he came home he changed €10.64 back into pounds.

The exchange rate was now £1 = €1.33.

b How many pounds did he receive?

June 2006

10 Bob lays 200 bricks in one hour.

He always works at the same speed.

He starts work at 9 am.

Bob takes 15 minutes for morning break and 30 minutes for lunch break.

Bob has to lay 960 bricks.

Work out the time at which he will finish laying bricks.

June 2006 adapted

11 The exchange rate between pounds (£) and euros (€) is £1 = €1.08 in London and €1 = 88p in Paris.

Will has £1200 to change into euros. Should he do it in London or Paris?

12 Mr Brown makes some compost.

He mixes soil, manure and leaf mould in the ratio 3 : 1 : 1.

Mr Brown makes 75 litres of compost.

How many litres of soil does he use?

Nov 2006

25 LINE DIAGRAMS AND SCATTER GRAPHS

Life expectancy over time is one variable often represented using a line graph. The line for life expectancy in the UK shows a continual increase from 1980 to the present day. In 1980, a man could expect to live to an age of about 71 years whilst the average life expectancy for a woman was 77. By 2009, the life expectancy for both sexes had gone up considerably, with average life expectancy for a baby girl at 81.5 years and for a baby boy at 77.2 years.

◉ Objectives

In this chapter you will:
- ◉ learn to produce and interpret line graphs and scatter graphs
- ◉ see if there is any linear association between two variables
- ◉ be able to draw lines of best fit
- ◉ be able to distinguish between positive, negative and zero correlation
- ◉ use a line of best fit to predict values of a variable.

◑ Before you start

You need to know:
- ◉ how to plot and read points on a graph.

25.1 Drawing and using line graphs

Objectives

- You can draw line graphs.
- You can estimate values from a line graph.

Why do this?

If you collect data using an experiment you can often see how one thing changes as you change another thing. For example, you might record the amounts of carbon monoxide in a busy street at different times of day. You can use a line graph to represent these data.

Get Ready

1. The graph shows two plotted points.
   a Write down the coordinates of the two points.
   b A third point has coordinates (2, 2.5); add this point to a copy of the graph and draw a straight line through the points.

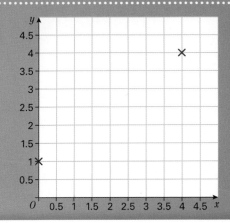

Key Points

- If you measure 10 people's heights, you are making a single observation of each member in a sample size of 10. However, if you measure their weights as well as their heights then you are making 10 pairs of observations.
- Pairs of observations can be plotted on a line graph.

Example 1

The table below gives information about the levels of carbon monoxide in a busy street.

Time of day	04:00	08:00	12:00	16:00	00:00	04:00
Carbon monoxide level (parts per million)	1	2	14	18	9	1

a Draw a line graph for these data.
b When was the amount of carbon monoxide at its highest?
c When was the amount of carbon monoxide at its lowest?

a

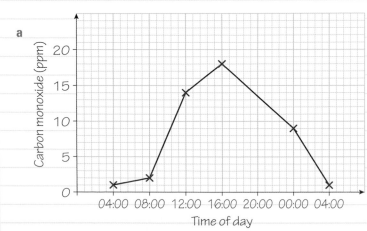

Plot the points on the graph.
Join the points with straight lines.

b 16:00 hours ← Find when the highest value occurs.

c 04:00 hours ← Find when the lowest value occurs.

Example 2 The line graph shows the rate at which water flows in a river measured on the same day for each month of a year.

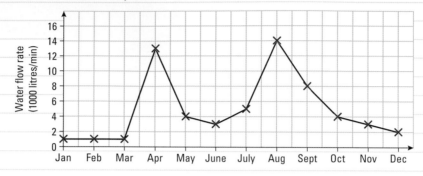

a Which month had the highest water flow rate?

b When was it lowest?

a August. ← This will be the highest point.

b January, February and March.

↑ It remains the same for these three months.

In both of the above examples time was the variable along the x-axis. This is often the case with line graphs.
It is clear that there is no pattern to these two graphs.

Exercise 25A

Questions in this chapter are targeted at the grades indicated.

E

1 Abbie has a Pay as you go mobile phone. She pays 40p for each minute she uses her phone.
She displays the cost of using the phone on the graph.

a How much does it cost Abbie to use her phone for 32 minutes?

b One month Abbie spent £8.40 on using her phone.
For how many minutes did Abbie use her phone that month?

E

2 This graph converts temperatures between Fahrenheit and Celsius.

a Use the graph to convert these temperatures to degrees Fahrenheit.
   i 20°C   **ii** 100°C   **iii** 36°C

b Use the graph to convert these temperatures to degrees Celsius.
   i 140°F   **ii** 60°F   **iii** 88°F

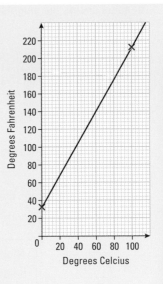

* **3** A man sells cornets from an ice cream van.
He kept a tally of the number of cornets sold during
the seven hours he worked one day.
During the day he sold a total of 500 cornets.
The table shows how many cornets he had sold, in total,
by the end of each hour.

Hours	0	1	2	3	4	5	6	7
Cornets sold	0	10	30	150	260	300	320	500

a Draw a line graph for these data.

b Write down the hour in which he sold the most cornets.

25.2 Drawing and using scatter graphs

Objective

- You can use a scatter graph to see if there is any relationship between pairs of variables.

Why do this?

Do people who are tall have larger feet?
To investigate such problems a scatter graph can be drawn.

Get Ready

What are the values of points A to D?

1.

2.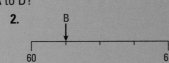

3.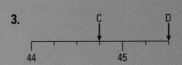

Key Point

- A **scatter graph** shows whether there is any relationship between two variables.

For example, Christopher was 168 cm tall and his foot length was 25.1 cm.
The grid shows how he plotted this point on a graph.

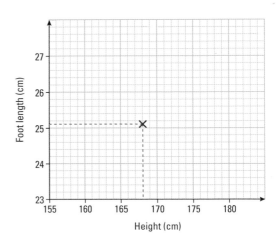

The table shows the heights and foot lengths of seven other boys.

Height (cm)	159	166	167	170	170	171	175
Foot length (cm)	24.0	24.8	24.5	25.4	25	25.7	26.0

Each of these pairs of values is plotted on the graph in the same way.
The resulting graph is called a scatter diagram or scatter graph.

The scatter graph shows the heights and foot lengths, in cm, of all eight boys.

> There are two variables: height and foot length.

There seems to be a relationship between the boys' heights and foot lengths: the greater the height, the longer the foot length.

> The pairs of values are plotted on the graph in the usual way. This cross shows Height 171, Foot length 25.7.

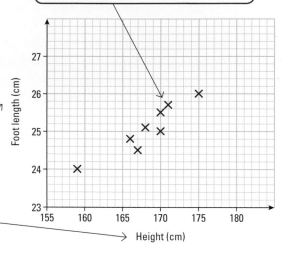

Example 3

The table below shows, for a range of cars, the engine size, in litres, and the average distance (miles) they can travel on 1 gallon of petrol.

Engine size (l)	1.4	1.6	1.8	2.0	2.5
Miles per gallon	42.8	42.2	40.3	39.8	30.4

a Draw a scatter graph for these data.
b Comment on the relationship between engine size and the average miles per gallon.

a

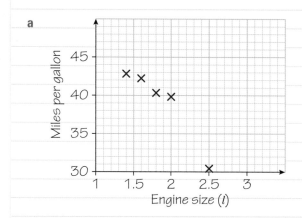

b There seems to be a relationship:
the bigger the engine the fewer miles they can travel on one gallon of petrol.

> Look to see if there is a pattern. State what the pattern is.

Exercise 25B

1   A health clinic recorded the pulse rates, in beats per minute, and the breathing rates, in breaths per minute, of 12 people. The scatter graph shows this information.

Describe the relationship between breaths per minute and pulse rate.

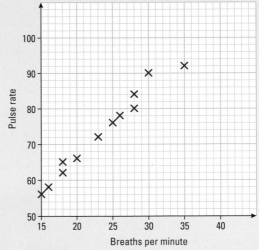

2   The heights and weights of 10 children in Year 7 are shown in the table.

| Height (cm) | 140 | 150 | 145 | 150 | 170 | 180 | 160 | 155 | 160 | 165 |
| Weight (kg) | 36 | 35 | 40 | 42 | 62 | 75 | 50 | 50 | 55 | 60 |

a   Using the scales shown on the diagram, draw a scatter graph for these data.

b   Describe the relationship between height and weight.

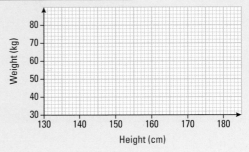

3   The data table shows the weight and top speeds of nine cars.

| Weight (kg) | 800 | 900 | 1000 | 1100 | 1200 | 1400 | 1500 | 1600 | 1700 |
| Top speed (mph) | 90 | 90 | 100 | 105 | 110 | 115 | 120 | 120 | 125 |

a   Using the scales shown on the diagram, draw a scatter graph for these data.

b   Describe the relationship between weight and top speed.

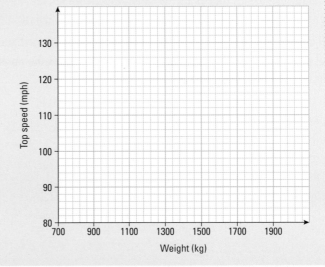

A03

D

* 4 The data show the number of pedestrian crossings and the number of pedestrian accidents in each of 12 areas over a period of six months.

Number of crossings	10	8	15	6	12	11	9	16	18	12	5	9
Number of accidents	7	10	4	10	4	8	6	3	2	3	16	7

Draw a scatter graph for these data.
Describe the relationship between the number of pedestrian crossings and the number of pedestrian accidents.

25.3 Recognising correlation

◎ Objective

● You can distinguish between positive, negative and zero correlation.

❓ Why do this?

To get your driving license you have to take a theory test and a driving test. Is there a relationship between these two? If there is, the two are correlated. It is important to be able to determine what sort of correlation there is.

◆ Get Ready

1. The number of visitors per month for a 5-month period to an outdoor water park are as follows:
   7675   5536   2462   1021   500.
   Can you suggest which months these are?

🔑 Key Points

● On the two scatter graphs in Section 25.2, it appears that there is a pattern: as one variable changes so does the other variable. We say they are correlated. A relationship between pairs of variables is called a **correlation**.

● If one variable increases as the other one increases the correlation is said to be positive. For example, in the case of height and foot length, the foot length increases as the height increases, so the correlation is positive.

● If one variable decreases as the other increases the correlation is said to be negative. For example, in the case of engine size and the number of miles per gallon, as the engine size increases the number of miles per gallon decreases, so the correlation is negative.

● If there is no relationship between the variables then there is no correlation and the correlation is said to be zero. These three possibilities are shown on the right.

Positive correlation

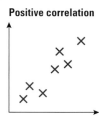

As one value increases the other one increases.

Negative correlation

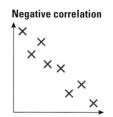

As one value increases the other decreases.

No correlation

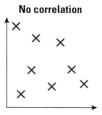

The points are random and widely spaced.

Example 4

A factory owner thought that older men performed a task at a quicker rate than young men. The scatter graph shows the ages of 12 men and the time it took them to do the task.

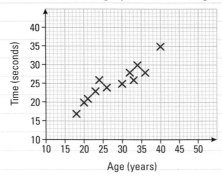

a Describe the correlation.

b Describe the relationship between age and time taken to do the task.

c Was the factory owner right?

> As one variable goes up so does the other one.

a Positive correlation.

b The greater the age the longer it took to do the task.

c The factory owner was wrong.

Exercise 25C

1 The scatter graph shows the marks achieved by a group of 10 students in a physics exam and in a woodwork exam.

Describe the correlation shown by this scatter graph.

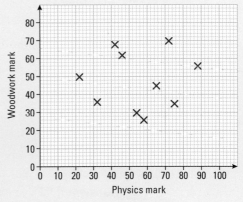

2 A garage sells second-hand cars. The scatter graph shows the ages and costs of 10 luxury cars of the same model.

a Describe the correlation.

b Describe the relationship between age and cost.

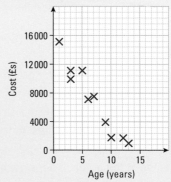

D

A03

D A03

3 Copy and complete the following table. Tick the type of correlation for each set of data.

Variables	Positive correlation	Negative correlation	No correlation
Height and Weight of people			
Intelligence and Weight of people			
Size of garden and Number of birds			
Age and Running speed of adults			
Height and Shoe size of people			
Age of cars and Engine size			
Arm length and Leg length of people			

C A03

* **4** Jacob sells bottled water from a market stall. He likes hot days because he claims that he will sell more bottles. For 10 days he records the number of bottles of water that he sells. He also gets the maximum temperature for each day from a website.
The data he collects are shown in the table.

Maximum temperature (°C)	19	27	18	24	30	22	23	16	25	27
Number of bottles of water sold	50	80	50	64	90	60	65	45	70	75

a Draw a scatter graph for these data.
b Describe the correlation and decide whether his claim is correct, with reasons.

25.4 Lines of best fit

Objective

● You can draw lines of best fit by eye.

Why do this?

A line of best fit acts as a model for the relationship. A line of best fit smoothes out the irregularities that are due to other things.

Get Ready

1. Look again at the scatter graph in question 2 on p. 523 about the correlation between the age of a garage's cars and their cost. How much would you expect an 8-year-old car to cost?

Key Points

● If the points on a scatter graph lie approximately in a straight line, the correlation is said to be **linear**. The word 'linear' means in a straight line.
● If the points are roughly in a straight line you can draw a **line of best fit** through them. You can work out the gradient of the line of best fit (see Section 15.7).
● A line of best fit is a straight line that passes as near as possible to the various points so as to best represent the trend of the graph. It does not have to pass through any of the points, but it might pass through some of them.

There should be roughly the same number of points either side of the line and the line drawn should best represent the trend of the points.

If lines of best fit were added to the scatter graphs of height and foot length in Section 25.2 and miles per gallon and engine size in Example 3, they would look like this.

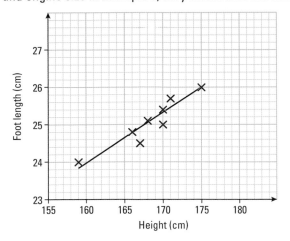 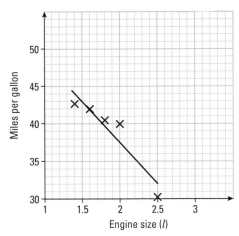

Example 5

The heights, in cm, and the weights, in kg, of 10 children are recorded.
The table below shows information about the results.

Height (cm)	38	40	42	43	45	49	49	50	51	53
Weight (kg)	145	148	147	151	152	155	157	158	160	164

a Draw a scatter graph of these data.
b Describe the correlation between the weight and height of these children.
c Draw a line of best fit on your scatter graph.

a

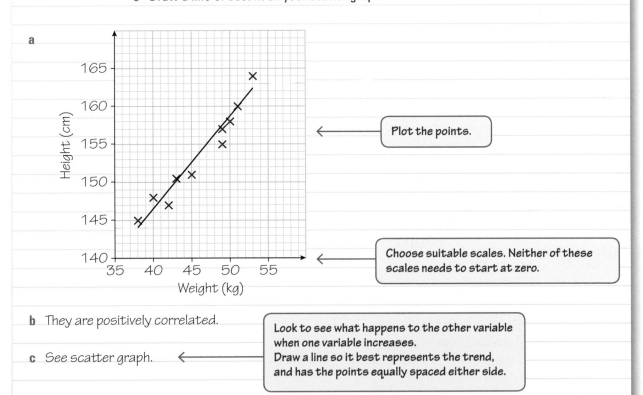

Plot the points.

Choose suitable scales. Neither of these scales needs to start at zero.

b They are positively correlated.

Look to see what happens to the other variable when one variable increases.
Draw a line so it best represents the trend, and has the points equally spaced either side.

c See scatter graph.

Exercise 25D

D

1 The scatter graph shows the engine sizes and the distance travelled on a litre of petrol for 10 different cars.

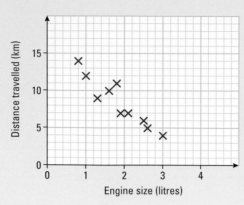

a Describe the correlation.

b Copy the scatter diagram.

c Draw a line of best fit on your diagram.

A03

d Describe the relationship between engine size and distance travelled.

C

2 A company has a hairdressing shop in each of 10 towns. The director of the company plans to expand into other towns. To make sensible decisions he has to look at profits and town sizes for the shops he already has. This information is shown in the table.

Town size (1000s)	11	15	16	20	26	30	35	42	45	50
Annual profits (£1000s)	42	46	45	48	55	50	54	58	65	70

a Copy the scatter graph and plot the remaining points.

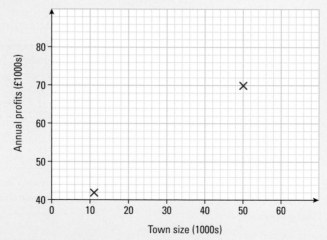

b Describe the correlation.

c Draw a line of best fit on your diagram.

A03

d Describe the relationship between town size and profits and make a recommendation for the size of town he should next expand to.

3 An NHS Trust has seven hospitals. Some data on the average number of operations and the number of operating theatres for each hospital are shown in the table.

Number of operating theatres	2	4	5	5	6	7	8
Average number of operations per week	60	80	90	85	100	130	150

a Copy the diagram and complete the scatter graph for these data.

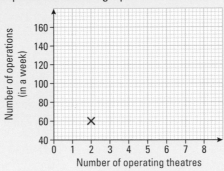

b Describe the correlation.

c Draw a line of best fit on your diagram.

d Describe the relationship between the number of operating theatres and the average number of operations per week.

A03

25.5 Using lines of best fit to make predictions

⊙ Objective

● You can use a line of best fit to predict a value of one of a pair of variables given a value for the other variable.

⊙ Why do this?

An ice cream man could use a line of best fit to estimate the number of sales when the temperature increases.

⊙ Get Ready

1. Look again at the Get Ready for Section 25.1.
What do you think would be the coordinates of the point with $x = 5$?

🔑 Key Points

● If a value of one of the variables is known you can estimate the corresponding value of the other by using the line of best fit.

For example, to estimate the height of a child weighing 47 kg, using the graph from Example 5, you draw a vertical line at 47 kg until it meets the line of best fit. You then draw a horizontal line from there and read off where it comes on the vertical scale. In this case, you read off 155 cm.

● Using the line of best fit to obtain answers that are outside the range of the plotted points may give unreliable answers.

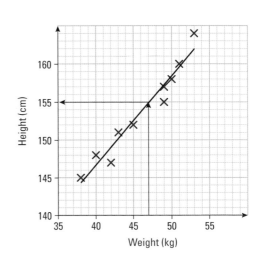

Example 6 The scatter graph gives information about the reaction times, in milliseconds, of adult women.

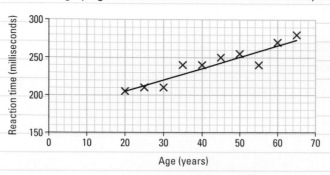

a Estimate the reaction time of a woman of 56.

b Estimate the age of a woman whose reaction time is 250 milliseconds.

c Explain why you would not estimate outside the range of readings with this scatter graph.

a 260 milliseconds.

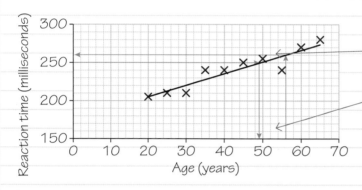

For part **a**, draw a line from 56 up to the line of best fit. From where it meets the line draw a horizontal line across to the vertical axis and read off the required value.

For part **b**, draw a horizontal line from 250 on the vertical axis across to the line of best fit. From where it meets the line draw a vertical line down to the horizontal axis and read off the required value.

b 49 years.

c If you extend the line to age 0 you find a newborn child has a reaction time of 170 milliseconds, which is obviously silly. If you extend in the other direction a person with a reaction time of 420 would be about 120 years old.

ResultsPlus
Examiner's Tip

Always draw in the lines – you may get method marks for this even if they are in the wrong place.

Exercise 25E

C
A03

1 A factory makes model cars. The scatter diagram shows some information about the numbers of models made and the cost of making them.

a Describe the correlation.

b The manager wants to estimate:

i the cost of making 50 000 models

ii the number you can make for £45 000.

Use the graph to make these estimates for the manager.

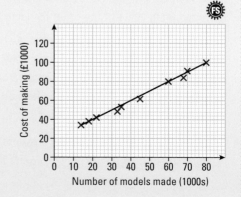

2 The scatter graph shows the midday temperature and the number of units of electricity used by a house on each of 10 days.

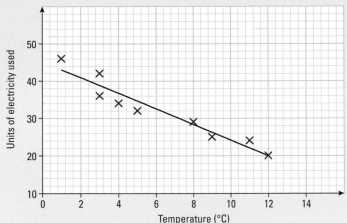

a Describe the correlation.

b Use the graph and the line of best fit to find an estimate for:

   i the number of units of electricity used when the midday temperature was 7°C

   ii the midday temperature when 35 units of electricity were used.

3 The following scatter graph shows the ages and blood pressures of a group of 10 men.

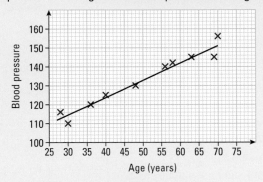

a Estimate the blood pressure of a man aged 65.

b Predict the age of a man whose blood pressure is 135.

c Describe the relationship between age and blood pressure.

Chapter review

◉ Pairs of observations can be plotted on a line graph.

◉ A scatter graph shows whether there is any relationship between two variables.

◉ A relationship between pairs of variables is called a **correlation**.

◉ If one variable increases as the other increases the correlation is said to be positive.

◉ If one variable decreases as the other increases the correlation is said to be negative.

◉ If there is no relationship between the variables then there is **no correlation** and the correlation is said to be zero.

◉ A **line of best fit** is a straight line that passes as near as possible to the various points so as to best represent the trend of the graph.

◉ If a value of one of the variables is known you can estimate the corresponding value of the other by using the line of best fit.

⚙ Review exercise

D

1 The scatter graph shows some information about 10 students.
It shows the arm length and the height of each student.

  a What type of correlation does this scatter graph show?

  b Draw a line of best fit on a copy of the scatter graph.

Another student has an arm length of 75 cm.

  c Use your line of best fit to estimate the height of this student.

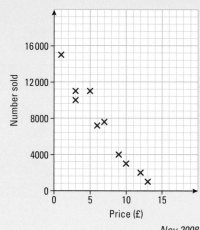

March 2009

2 A superstore sells the Clicapic digital camera.
The price of the camera changes each week.
Each week the manager records the price of the camera and the number of cameras sold that week.
The scatter graph shows this information.

  a Describe the relationship between the price of the camera and the number of cameras sold.

  b Draw a line of best fit on the scatter graph.

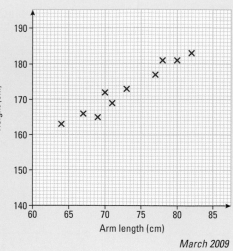

Nov 2008

3 Some students revised for a mathematics exam.
They used an internet revision site.
The scatter graph shows the amount of time seven students spent on the internet revision site and the marks the students got in the mathematics exam.

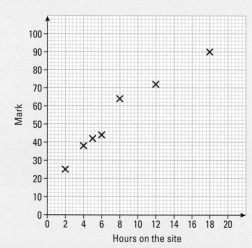

Here is the information for 3 more students.

Hours on the site	7	10	16
Mark	50	56	78

  a Plot this information on a copy of the scatter graph.

  b What type of correlation does this scatter graph show?

  c Draw a line of best fit on your scatter graph.

Nov 2008

4 The scatter graph shows some information about the age, in years, of apprentices and the time, in minutes, it takes them to learn a certain skill.

A line of best fit is drawn on the graph.

a Work out an estimate for the gradient of the line of best fit.

b Use the line of best fit to estimate how long it would take a 16.5 year old to learn the skill.

c Describe the correlation.

d What conclusions can you draw about the time it takes apprentices to learn skills?

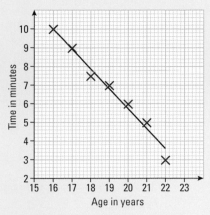

* **5** The table gives information about the marks gained by 10 students in a French exam and in a German exam. The exam was marked out of 50.

Student	A	B	C	D	E	F	G	H	I	J
French	10	10	18	25	28	33	39	42	43	46
German	11	14	21	26	35	32	42	42	45	50

a Draw a scatter diagram for these data.

b Draw in a line of best fit.

c Work out the gradient of the line of best fit.

d Work out the proportion of students that got less than 26 in at least one of the exams.

6 The scatter graph shows information for some weather stations.

It shows the height of each weather station above sea level (m) and the mean July midday temperature (°C) for that weather station.

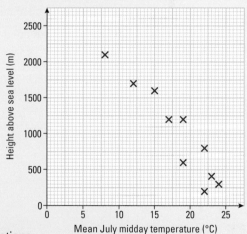

The table shows this information for two more weather stations.

Height of weather station above sea level	1000	500
Mean July midday temperature (°C)	20	22

a Plot this information on the scatter graph.

b What type of correlation does this scatter graph show?

c Draw a line of best fit on a copy of the scatter graph.

A weather station is 1800 metres above sea level.

d Estimate the mean July midday temperature for this weather station.

At another weather station the mean July midday temperature is 18°C.

e Estimate the height above sea level of this weather station.

June 2008

C

7 The scatter graph shows some information about the ages and values of fourteen cars. The cars are the same make and type.

  a Describe the relationship between the age of a car and its value in pounds.

  b Draw a line of best fit on a copy of the scatter graph.

A car is 3 years old.

  c Use your line of best fit to find an estimate of its value.

A car has a value of £3500.

  d Use your line of best fit to find an estimate of its age.

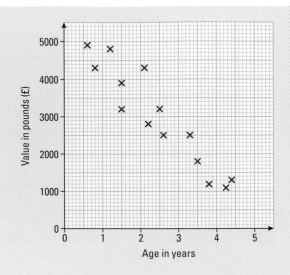

March 2008

8 Jake recorded the weight, in kg, and the height, in cm, of each of ten children. The scatter graph shows information about his results.

  a Describe the relationship between the weight and the height of these children.

  b Draw a line of best fit on a copy of the scatter graph.

  c Use your line of best fit to estimate the height of a child whose weight is 47 kg.

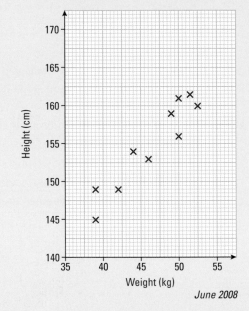

June 2008

9 A garage sells motorcycles. The scatter graph shows information about the price and age of the motorcycles.

  a What type of correlation does the scatter graph show?

  b Draw a line of best fit on a copy of the scatter graph.

Mae buys a motorcycle from this garage for £1500.

  c Use your line of best fit to estimate the age of the motorcycle.

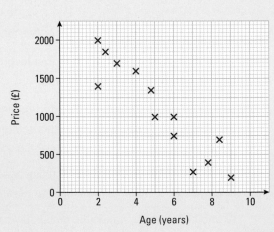

June 2007

10 The table gives the age and price of 10 second-hand Minis.

a Estimate the approximate price of a new Mini.
b Which Mini is different to the others? Suggest reasons why
   it is priced like it is.

Age (years)	Price
5	£5400
2	£7000
4	£5600
6	£3400
7	£5200
3	£6800
8	£1700
1	£10 500
8	£2000
6	£3200

26 PROBABILITY

Weather forecasters use probability to predict the weather. The US Storm Prediction Center monitors all the storms in 'Tornado Alley' to predict which ones will become tornadoes. In 1925 the Tri-State Tornado, the longest and deadliest twister in US history, killed almost 700 people. Today, better forecasting of tornadoes has reduced the average annual death toll to just 50.

Objectives

In this chapter you will:
- learn to represent on a probability scale how likely it is that an event will happen
- learn to write probabilities as numbers
- find probabilities from a sample space diagram
- discover how to estimate a probability from the results of an experiment
- find probabilities from a two-way table.

Before you start

You need to know:
- your tables and the rules of basic arithmetic
- how to use a number line
- how to add and subtract fractions and decimals.

26.1 The probability scale

Objectives

- You can represent how likely it is that an event will happen on a probability scale.
- You can use words to describe probabilities.

Why do this?

How likely is it that it will rain tomorrow? A knowledge of probability helps us to decide whether to wear a coat or not.

Get Ready

1. Will Christmas Day always follow Christmas Eve? Is it possible to throw a 7 on an ordinary dice?

Key Points

- An **event** that is **certain** to happen has a probability of 1.
- An event that is **impossible** has a probability of 0.
- The **probability** that an event will happen is always less than or equal to 1, or greater than or equal to 0. This can be written as $0 \leqslant$ probability $\leqslant 1$.

Example 1

Represent how likely each of the following events is on a **probability scale**.

a   If today is Monday, tomorrow will be Tuesday.

b   A human will grow to be 10 metres tall.

c   It will be sunny every day for a week in January in London.

d   If you spin an ordinary coin it will land on a head.

> Draw a probability scale from 0 to 1.

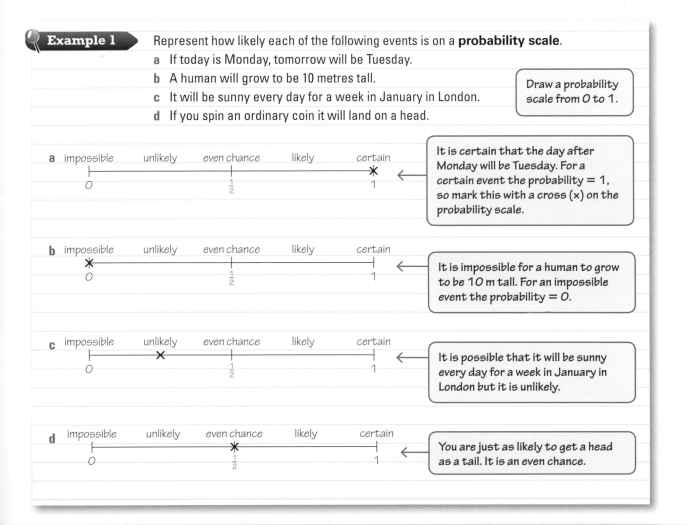

a   It is certain that the day after Monday will be Tuesday. For a certain event the probability = 1, so mark this with a cross (x) on the probability scale.

b   It is impossible for a human to grow to be 10 m tall. For an impossible event the probability = 0.

c   It is possible that it will be sunny every day for a week in January in London but it is unlikely.

d   You are just as likely to get a head as a tail. It is an even chance.

F

⚙ **Exercise 26A**

Questions in this chapter are targeted at the grades indicated.

1 How likely is each of the following events? In each case, represent your answer on a probability scale.

a The sun will rise tomorrow.

b A pet cat will live for ever.

c The next baby born will be a girl.

d You will use a mobile phone today.

e It will snow on Christmas Day in Manchester.

2 Give two examples of events that you think:

a are impossible

b are certain

c have about an even chance

d are possible but unlikely

e are likely.

3 If you take 7 letters at random from a bag of 100 letters, how likely is it that you will be able to use the letters to make a 7-letter word? Represent your answer on a probability scale.

4 How likely do you think it is that the next national election will be won by:

a Conservatives

b Labour

c Liberal Democrats?

Represent your answers on the same probability scale.

26.2 Writing probabilities as numbers

◎ **Objective**

● You can use a number to represent a probability.

◈ **Why do this?**

If you write probabilities as numbers it is easier to compare them to make a decision, for example which raffle prize you are more likely to win at a school fair.

◈ **Get Ready**

1. Copy and complete this table.

Fraction	$\frac{3}{10}$		$\frac{3}{8}$	
Decimal		0.6		
Percentage				65%

🔖 **Key Points**

● **Outcomes** are **mutually exclusive** when they cannot happen at the same time. For example, rolling a 3 and rolling a 4 on a dice are mutually exclusive outcomes – you cannot roll a 3 and a 4 at the same time.

● For **equally likely** outcomes the probability that an event will happen is

$$\text{Probability} = \frac{\text{number of successful outcomes}}{\text{total number of possible outcomes}}$$

● A probability can be written as a fraction, a decimal or a percentage.

Example 2

A fair dice is rolled.

Work out the probability of getting an even number.

$P(even) = \dfrac{number\ of\ successful\ outcomes}{total\ number\ of\ possible\ outcomes}$

$= \dfrac{3}{6}$ ← There are three successful outcomes: 2, 4 and 6.

$= \dfrac{1}{2}$ ← There are six possible outcomes: 1, 2, 3, 4, 5 or 6.

Simplify the fraction. Divide top and bottom by 3.

So $\dfrac{3 \div 3}{6 \div 3} = \dfrac{1}{2}$

ResultsPlus
Examiner's Tip

A fair dice is one that is not biased. Each number has an equal chance of being thrown. So the outcomes are equally likely.

Exercise 26B

1 Here are a number of shapes.

One of these shapes is chosen at random. Work out the probability that the shape will be:

  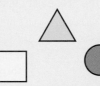

a a square b a triangle c a square or a triangle.

2 A game consists of spinning a pointer on a dial.
Work out the probability that the pointer will stop on

a 0 b 1 c 2 d 3

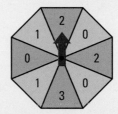

3 A bag contains 10 counters. 3 of the counters are red and 7 of the counters are blue.
A counter is taken at random from the bag. Write down the probability that it will be:

a red b blue c yellow.

4 A letter is chosen at random from the word STATISTICS. Write down the probability that it will be:

a A b S c I d S or T e G.

5 A fair dice is rolled. Work out the probability of getting:

a a 5 b a 1 or a 2 c an odd number

d a number less than 5 e a number greater than 6 f a prime number.

6 There are 120 raffle tickets in a hat. Of these, 67 are yellow and the rest are green.
A raffle ticket is taken at random from the hat. Work out the probability that it will be:

a yellow b green.

7 A box of sweets contains 3 mints, 7 toffees and 5 lemon drops.
A sweet is taken at random from the box. Write down the probability that it will be:

a a mint b a toffee c a lemon drop

d a mint or a toffee e not a toffee f not a lemon drop or a toffee.

F

E

26.3 The probability that something will *not* happen

◎ Objective

○ You can work out the probability that something will not happen if you know the probability that it will happen.

◈ Why do this?

You could work out the probability of not being hit by lightning because there is a known probability for the chance of being hit by lightning.

◈ Get Ready

1. $1 - 0.7 = ?$

2. $1 - ? = 0.6$

3. $1 - \frac{1}{2} = ?$

4. $1 - ? = \frac{3}{4}$

🔍 Key Point

◉ The sum of the probabilities of all the mutually exclusive outcomes of an event is 1.

◉ If the probability of an event happening is P then the probability of it not happening is therefore $1 - P$.

🔍 Example 3

The probability that it will rain tomorrow is 0.2.
Work out the probability that it will not rain tomorrow.

$P(\text{not rain}) = 1 - P(\text{rain})$
$= 1 - 0.2$
$= 0.8$

← If the probability that it will rain tomorrow is 0.2, then the probability it will not rain is $1 - 0.2$.

⚙ Exercise 26C

E

1 The probability that it will snow on New Year's Day is 0.3.
Work out the probability that it will not snow on New Year's Day.

2 The probability that my bike will get a puncture on the way to school today is 0.15.
Work out the probability that my bike will not get a puncture on the way to school today.

3 The probability that Stephanie will have chips for dinner tonight is $\frac{2}{3}$.
Work out the probability that Stephanie will not have chips tonight.

4 The probability that my lottery ticket will win a prize is $\frac{3}{53}$.
Work out the probability that my lottery ticket will not win a prize.

5 A weather forecaster says that there is a 70% chance of getting rain tomorrow.
What is the probability that it will not rain tomorrow?

6 The probability that Tracy will not eat all her carrots at lunch is 0.64.
Work out the probability that she will eat all her carrots at lunch.

7 An insurance broker says that when she receives a claim the probability that it will not be for an accident in the home is 0.325. Work out the probability that the next claim received by the insurance broker will be for an accident in the home.

8 The probability that Kimberley's train will be late is 0.32. Kimberley says that the probability that her train will not be late is 0.78. She is wrong. Explain why.

Example 4

The diagram shows a 4-sided spinner that is biased.

The spinner is spun.

This table gives the probability that it will land on B, C or D.

	A	B	C	D
Probability		0.13	0.45	0.25

Work out the probability that the spinner will land on A.

$$P(A) + 0.13 + 0.45 + 0.25 = 1$$ ← It is certain that the spinner will land on A or B or C or D. So the probabilities add up to 1.

$$P(A) + 0.83 = 1$$ ← Add the decimals: $0.13 + 0.45 + 0.25 = 0.83$

$$P(A) = 1 - 0.83$$
$$= 0.17$$ ← Take 0.83 from both sides of the equation.

Exercise 26D

1 Heather's bus can either be early, on time or late.

This table gives the probability that her bus will be early and the probability that her bus will be late.

	early	on time	late
Probability	0.1		0.6

Work out the probability that Heather's bus will be on time.

2 A biased dice is rolled.

This table gives the probability that it will land on each of the numbers 1, 2, 3, 4 and 5.

	1	2	3	4	5	6
Probability	0.2	0.1	0.3	0.2	0.1	

Work out the probability that the dice will land on 6.

3 Imran has some coloured cards. Each card is either red or yellow or blue or green. One of these cards is taken at random. This table gives the probability that the card will be red or blue or green.

Colour	red	yellow	blue	green
Probability	0.36		0.19	0.28

Work out the probability that the card will:

a be yellow

b not be blue

c be a red card or a green card.

D

D

4 A bag contains a number of balls. Each ball is either red or blue or yellow or green.
A ball is taken at random from the bag.

The probability that the ball will be red is 0.25.
The probability that the ball will be blue is 0.3.
The probability that the ball will be yellow is 0.2.
Work out the probability that the ball will be green.

5 When Kasha spins this 3-sided spinner, the probability that it will land on 1 is $\frac{2}{9}$.
The probability that it will land on 3 is $\frac{4}{9}$.
Work out the probability that it will land on 2.

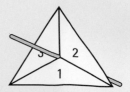

6 When Weymouth Wanderers play football they can win, lose or draw the game.
The probability that they will win the game is $\frac{3}{5}$. The probability that they will lose the game is $\frac{3}{10}$.
Work out the probability that they will draw the game.

26.4 **Sample space diagrams**

◉ Objective

○ You can record all the possible outcomes of an experiment in a sample space diagram.

◈ Why do this?

When two things happen at the same time a sample space diagram is a good way to show all the possible outcomes.

◈ Get Ready

1. Write down all the possible outcomes when you:
a spin a coin **b** roll a dice **c** play a game of football

🔑 Key Point

◉ A **sample space diagram** can be used to find a **theoretical probability**.

🔎 Example 5

An ordinary dice is rolled and a 4-sided spinner is spun.
a Draw a sample space diagram to show all the possible outcomes.
b Work out the probability of getting a total score of 7.

a

		1	2	3	4	5	6
	4	(1, 4)	(2, 4)	(3, 4)	(4, 4)	(5, 4)	(6, 4)
	3	(1, 3)	(2, 3)	(3, 3)	(4, 3)	(5, 3)	(6, 3)
Spinner	2	(1, 2)	(2, 2)	(3, 2)	(4, 2)	(5, 2)	(6, 2)
	1	(1, 1)	(2, 1)	(3, 1)	(4, 1)	(5, 1)	(6, 1)
		1	2	3	4	5	6

Dice

A sample space diagram shows all the possible outcomes, e.g. (6, 4) is the outcome of rolling a 6 on the dice and spinning a 4 on the spinner.

Identify all the possible ways of getting a total score of 7: (3, 4), (4, 3), (5, 2) and (6, 1).

b $P(7) = \dfrac{\text{number of successful outcomes}}{\text{total number of possible outcomes}}$

$= \dfrac{4}{24}$ ←

> There are four outcomes which give a total score of 7.

$= \dfrac{1}{6}$ ←

> There are a total of 24 possible outcomes.

> This is the theoretical probability. The probability you expect to get for a fair dice and a fair spinner.

Exercise 26E

1 A coin is spun, and an ordinary dice is rolled. Show all the possible outcomes.

A02
A03

2 An ordinary dice is rolled and a 4-sided spinner is spun.
Use the sample space diagram in Example 5 to work out the probability of getting:
a a total score of 4
b the same number on the dice and the spinner
c a total score less than 6.

3 The ace, king, queen and jack of clubs and the ace, king, queen and jack of diamonds are put into two piles. The sample space diagram shows all the possible outcomes when a card is taken from each pile.

	J	AJ	KJ	QJ	JJ
	Q	AQ	KQ	QQ	JQ
Clubs	K	AK	KK	QK	JK
	A	AA	KA	QA	JA
		A	K	Q	J

Diamonds

Work out the probability that:
a both cards will be aces   b the cards will be a pair
c only one of the cards will be a jack   d at least one card will be a king
e one card will be a diamond   f neither card will be a queen
g both cards will be diamonds.

4 Two 3-sided spinners are spun.
Draw a sample space diagram to show all the possible outcomes.
One possible outcome is (2, 3).

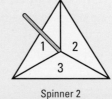

Spinner 1   Spinner 2

D A02

5   Mandy has some sheets and pillowcases in a drawer. The colours of the sheets are either white or yellow or blue or green. The colours of pillowcases are either white or green or orange. Mandy takes at random a sheet and a pillowcase from the drawer.

   a Draw a sample space diagram to show all the possible combinations of colours for the sheets and pillowcases.

   b Work out the probability that Mandy takes a sheet and a pillowcase of:

   i the same colour   **ii** different colours.

A03

6   Simon is going to spin a 3-sided spinner and a 4-sided spinner. The spinners are fair.
What is the most likely total score.
Give a reason for your answer.

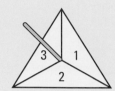
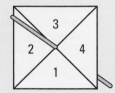

26.5 Relative frequency

◉ Objectives

○ You can find an estimate of a probability from the results of an experiment.

○ You can compare experimental and theoretical probabilities.

◈ Why do this?

Insurance companies use relative frequency to estimate risk. The greater the number of claims the greater the risk.

◈ Get Ready

1. Simplify these fractions.

   a $\frac{2}{4}$ **b** $\frac{6}{9}$ **c** $\frac{8}{12}$ **d** $\frac{50}{100}$

◈ Key Points

◉ You can use **relative frequency** to find an estimate for a probability.

◉ Estimated probability $= \dfrac{\text{number of successful trials}}{\text{total number of trials}}$

◉ The estimated or **experimental probability** may be different from the theoretical probability.

◉ Generally, the more **trials** you undertake the nearer your estimate will be to the actual probability.

Example 6   Samina spins a fair coin 50 times.
She gets 21 heads.
Write down the estimated and theoretical probability of getting a head.

Estimated probability $= \dfrac{\text{number of successful trials}}{\text{total number of trials}} = \dfrac{21}{50}$

Theoretical probability $= \dfrac{1}{2}$

E

Exercise 26F

1   Roll a dice 60 times and record your results in a frequency table like this.

Number	Tally	Frequency
1		
2		
3		
4		
5		
6		
	Total	60

   a   Use the results in your table to work out the estimated probability of getting:
      i a 6
      ii an odd number
      iii a number bigger than 4.
   b   Write down the theoretical probability of getting:
      i a 6   ii an odd number   iii a number bigger than 4.
   c   Do you think your dice is fair? Give a reason for your answer.

2   a   Write down the theoretical probability of getting a head when you spin an ordinary coin.
   b   Now spin a coin 50 times and record your results in a frequency table.
   c   Use your results to write down the estimated probability of getting a head. Comment on your answer.

3   Throw a drawing pin 50 times and record whether it lands on its head or on its side.
   a   Use your results to write down the estimated probability of getting a head.
   b   How could you improve on your answer to part **a**?

Head   Side

4   Make a 4-sided dice of your own out of card.
   Test the dice to see whether it is fair or not.

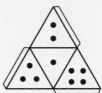

Net of a 4-sided dice

5   A letter is chosen at random from the words in a book.
   a   Work out an estimate of the probability of getting the letter k.
   b   Is your estimate affected by the language in which the book is written? Explain your answer.

A02
A03

26.6 Two-way tables

Objective

● You can find probabilities from a two-way table.

Why do this?

Two-way tables help travel agents decide where people are most likely to go on holiday at certain times of the year. For example, beach holidays are more popular in the summer and ski holidays are more popular in winter.

Get Ready

1. Can you find the missing numbers?
   a $3 + 4 + ? = 10$ b $9 + ? + 11 = 25$ c $? + 16 + 9 = 35$

Key Points

● For a two-way table, sum of row totals = sum of column totals.

Example 7

Carmen asks 20 people where they went for their summer holidays.
This two-way table gives some of the information from her results.

	France	Italy	Spain	Total
Boys	4	5		11
Girls	1			9
Total		7		20

a Copy and complete the table.

b Carmen picks one of the 20 students at random.
   Write down the probability that this student:

   i will be a girl   ii went to Italy   iii will be a girl who went to France.

a

> This number is 6 because:
> $1 + 2 + 6 = 9$

> This number is 2 because
> $4 + 5 + 2 = 11$

	France	Italy	Spain	Total
Boys	4	5	2	11
Girls	1	2	6	9
Total	5	7	8	20

> This number is 9 because
> $11 + 9 = 20$

> This number is 5 because:
> $4 + 1 = 5$

> This number is 2 because:
> $5 + 2 = 7$

> This number is 8 because
> $2 + 6 = 8$ or $5 + 7 + 8 = 20$

b   i From the table above, it can be seen that 9 students are girls.
       So, $P(\text{girl}) = \frac{9}{20}$

   ii The table shows that 7 students went to Italy.
       So, $P(\text{Italy}) = \frac{7}{20}$

   iii From the table, it can be seen that 1 girl
        went to France. So, $P(\text{girl, France}) = \frac{1}{20}$

> Use $P = \dfrac{\text{number of successful outcomes}}{\text{total number of possible outcomes}}$
>
> Here 9 students are girls, so the number of successful outcomes = 9.
> There are 20 students altogether, so the total number of possible outcomes = 20.

Exercise 26G

D

1 Kumar counted the number of butterflies and the number of moths in his garden in May and June.
The following two-way table provides some of the information from his results.

	Butterflies	Moths	Total
May	9	4	
June			
Total		7	25

Copy and complete the table.

2 The table below gives some information about how some students travel to school.

	Walk	Bus	Cycle	Total
Boys	4		3	12
Girls	7			
Total		9		25

a Copy and complete the table.

b One of the students is picked at random. Work out the probability that this student is:
   i a girl   ii a girl who walks to school
   iii a boy who cycles to school   iv a student who comes by bus.

3 45 students each went to one activity on Saturday night.
The following two-way table shows some information about where they went.

	Cinema	Club	Bowling	Total
Boys	5			23
Girls			4	
Total	14	15		45

a Copy and complete the table.

b One of the students is picked at random. Write down the probability that this student:
   i will be a boy   ii went to the cinema   iii will be a girl who went bowling.

4 Some students each had one drink and one snack in the school canteen.
The table below gives some information about what the students had to eat and drink.

	Orange	Lemonade	Milk	Total
Sandwiches	5			13
Biscuits	4		5	
Crisps			1	16
Total	17	18	7	

a Copy and complete the table.

b One of the students is picked at random.
Use your table to write down the probability that this student had:
   i lemonade   ii crisps
   iii orange and biscuits   iv lemonade and biscuits.

c John says that the probability of picking someone who had milk and biscuits is the same as picking someone who had orange and sandwiches. Is he right? Give a reason for your answer.

26.7 Predicting outcomes

⊕ **Get Ready**

1. Flip a coin 10 times and keep a tally of how many times the outcome is heads.

🔍 **Key Points**

◉ Predicted number of outcomes = probability × number of trials

🔍 **Example 8**

A fair 4-sided spinner is spun 100 times.
Find an estimate for the number of times it will land on a 3.

The theoretical probability that the spinner will land on a 3 is $\frac{1}{4}$.

So, when the spinner is spun 100 times we expect it to land on a 3.

$\frac{1}{4} \times 100 = 25$ times.

← Use P(3) = …
Here there is one outcome which is successful (i.e. '3'), and the total number of possible outcomes = 4 (i.e. 1, 2, 3 or 4).

← Predicted number of outcomes = probability × number of trials
$\frac{1}{4} \times 100 = 100 \div 4 = 25$

⚙ **Exercise 26H**

E

A03

1 An ordinary coin is spun 100 times. How many times do you expect it to land on a head?

A03

2 Piers spins an ordinary 5-sided spinner (numbered 1 to 5) 150 times.
How many times can he expect it to land on 4?

A03

3 Harry is going to roll an ordinary dice 90 times. Work out an estimate for the number of times it will land on:

    a 6   **b** an even number   **c** 1 or 2.

4 The table gives information about the probability that Tom will win, draw or lose a game of Go.

	Win	Draw	Lose
Probability	0.4	0.25	0.35

Tom is going to play 40 games of Go.
Find an estimate for the number of games he will:

    a win

    b lose.

5 A bag contains 3 red balls and 5 blue balls.

A ball is taken at random from the bag and its colour is recorded. The ball is now put back into the bag and another ball is taken at random from the bag. This is repeated 60 times.

Find an estimate for the total number of:

a red balls taken from the bag

b blue balls taken from the bag.

A03 E

6 The probability that an insurance company will get a claim for an accident in the home is 0.72.

If the insurance company gets 340 claims during the next month, find an estimate for the number of these claims that will not be for an accident in the home.

A03

* 7 The diagram shows part of Holly's design for a game. In her game a player pays x pence to spin a star. When the star stops spinning the player wins the amount shown by the arrow.

Holly wants to gain an average of 5p each time the game is played.

Show how this can be done by adding six more numbers to the star and finding a suitable value for x.

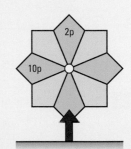

A03 C

Chapter review

- An **event** that is **certain** to happen has a probability of 1.
- An event that is **impossible** has a probability of 0.
- The **probability** that an event will happen is always less than or equal to 1, or greater than or equal to 0. This can be written as $0 \leqslant$ probability $\leqslant 1$.
- **Outcomes** are **mutually exclusive** when they cannot happen at the same time.
- For **equally likely** outcomes the probability that an event will happen is

  Probability $= \dfrac{\text{number of successful outcomes}}{\text{total number of possible outcomes}}$

- A probability can be written as a fraction, a decimal or a percentage.
- If the probability of an event happening is p then the probability of it not happening is $1 - p$.
- You can list all possible outcomes of an experiment in a **sample space diagram**.
- You can use **relative frequency** to find an estimate for a probability.
- Estimated probability $= \dfrac{\text{number of successful trials}}{\text{total number of trials}}$
- Generally, the more trials you undertake the nearer your estimate will be to the actual probability.
- For a two-way table, sum of row totals = sum of column totals.
- Predicted number of outcomes = probability $\times$ number of **trials**.

Review exercise

1 Tom throws an ordinary coin once.

   a On a copy of the probability scale, mark with a cross (×) the probability that the coin will show tails.

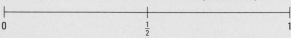

   0   ½   1

Tom rolls an ordinary dice once.

   b On a copy of the probability scale, mark with a cross (×) the probability that he will score a number less than 6.

   0   ½   1

Tom takes a Maths test.

   c On a copy of the probability scale, mark with a cross (×) the probability that he will score more than 0.

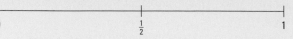

   0   ½   1

Nov 2008

2 Lucy uses some letter cards to spell the word "NOVEMBER".

Lucy takes one of these cards at random.

Write down the probability that Lucy takes a card with a letter E.

Nov 2009

3 Here are some statements.

On a copy, draw an arrow from each statement to the word which best describes its likelihood.

One has been done for you.

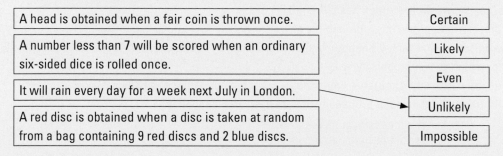

June 2008

4 Iqbal eats in a cafe.

He can choose **one** main course and **one** piece of fruit.

Main Course	Fruit
Fish	Apple
Lamb	Banana
Salad	Pear

One possible combination is (Fish, Pear).

Write down all the possible combinations that Iqbal can choose.

March 2008

5 The diagram shows some 3-sided, 4-sided and 5-sided shapes.
The shapes are black or white.

a Copy and complete the two-way table.

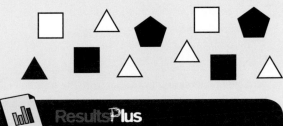

	Black	White	Total
3-sided shape		4	5
4-sided shape	2		
5-sided shape		0	
Total			11

Exam Question Report

92% of students did very well on part **a** of this question.

Ed takes a shape at random.

b Write down the probability the shape is white **and** 3-sided.

March 2008

6 Ishah spins a fair 5-sided spinner.
She then throws a fair coin.

a List all the possible outcomes she could get.

Ishah spins the spinner once and throws the coin once.

b Work out the probability that she will get a 1 and a head.

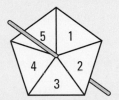

March 2009 adapted

7 80 children went on a school trip.
They went to London or to York.
23 boys and 19 girls went to London.
14 boys went to York.

a Use this information to complete a copy of the two-way table.

	London	York	Total
Boys			
Girls			
Total			

One of these 80 children is chosen at random.

b What is the probability that this child went to London?

March 2009

8 There are three beads in a bag.
One bead is blue, one bead is yellow and one bead is green.
Zoe takes a bead at random from the bag.

a On a copy of the probability scale, mark with the letter B the probability that she takes a blue bead.

0        $\frac{1}{2}$ 1

Zoe now throws a coin.
One possible outcome for the bead and the coin is (green, heads).

b List all the possible outcomes for the bead and the coin.
One has already been done for you.

May 2008

9 There are 3 red pens, 4 blue pens and 5 black pens in a box.
Sameena takes a pen, at random, from the box.
Write down the probability that she takes a black pen.

June 2008

E

10 There are 8 pencils in a pencil case.
1 pencil is red.
4 pencils are blue.
The rest are black.
A pencil is taken at random from the pencil case.
Write down the probability that the pencil is black.   *June 2008*

11 Emily has a bag of 20 fruit flavour sweets.
7 of the sweets are strawberry flavour,
11 are lime flavour,
2 are lemon flavour.

Emily takes at random a sweet from the bag.
Write down the probability that Emily

a takes a strawberry flavour sweet

b does **not** take a lime flavour sweet

c takes an orange flavour sweet.   *June 2006*

D

12 A bag contains only red, green and blue counters.
The table shows the probability that a counter chosen at random from the bag will be red or will be green.

Colour	Red	Green	Blue
Probability	0.5	0.3	

Mary takes a counter at random from the bag.
a Work out the probability that Mary takes a blue counter.
The bag contains 50 counters.
b Work out how many green counters there are in the bag.   *March 2009*

13 Here is a 5-sided spinner.
The sides of the spinner are labelled 1, 2, 3, 4 and 5.
The spinner is biased.
The probability that the spinner will land on each of the numbers
1, 2, 3 and 4 is given in the table.

Number	1	2	3	4	5
Probability	0.15	0.05	0.2	0.25	x

Work out the value of x.   *Nov 2008*

14 Here is a 4-sided spinner.
The sides of the spinner are labelled Red, Blue, Green and Yellow.
The spinner is biased.
The table shows the probability that the spinner will land on each
of the colours Red, Yellow and Green.

Colour	Red	Blue	Green	Yellow
Probability	0.2		0.3	0.1

Work out the probability the spinner will land on Blue.   *May 2008*

15 Marco has a 4-sided spinner.

The sides of the spinner are numbered 1, 2, 3 and 4.

The spinner is biased.

The table shows the probability that the spinner will land on each of the numbers 1, 2 and 3.

Number	1	2	3	4
Probability	0.20	0.35	0.20	

a Work out the probability that the spinner will land on the number 4.

Marco spins the spinner 100 times.

b Work out an estimate for the number of times the spinner will land on the number 2.

16 Here are the ages, in years, of 15 teachers.

35 52 42 27 36 23 31 41 50 34 44 28 45 45 53

One of these teachers is picked at random.

Work out the probability that the teacher is more than 40 years old.

May 2008 adapted

17 Two spinners are each numbered 1 to 4.

When they are both spun, the score is found by adding the two numbers.

E.g. a 1 and a 4 scores 5.

Three friends are playing with these spinners and devise a set of rules.

If Alice gets a score of 6, 7 or 8 she wins.

If Robbie scores 4 or 5 he wins.

If Megan scores 1, 2 or 3 she wins.

Who should win the most games?

 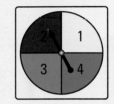

18 You need two coins to play this game. Toss the coins and record the results.

If you get two heads the red horse moves one space.

If you get two tails the blue horse moves one space.

If you get a head and a tail then the yellow horse moves one space.

The first horse to move eight spaces wins the race.

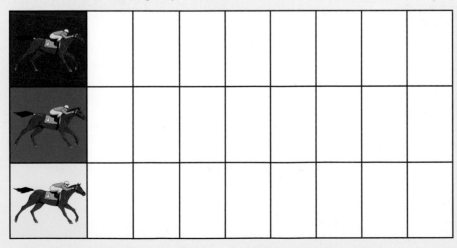 

Play the game three times and record your results.

Compare your results with a partner.

Can you explain why one horse has more of a chance than the others?

A03

* 19 You need two dice to play this game.

Roll the dice and add up the numbers that are showing on the dice.

If you roll a 4 and a 5 then Horse 9 would move one space.

The first horse to move eight spaces is the winner.

Horse 1								
Horse 2								
Horse 3								
Horse 4								
Horse 5								
Horse 6								
Horse 7								
Horse 8								
Horse 9								
Horse 10								
Horse 11								
Horse 12								

One horse has no chance of winning. Can you state which one it is and why?

Play the game twice and describe the results of the race.

Compare your results with a partner.

Use probability to explain what is happening.

A02

* 20 Devise and carry out an experiment to estimate the number of times a drawing pin would land pin up if you threw it 1000 times.

27 PYTHAGORAS' THEOREM

The diagram shows the dimensions of the Parthenon Temple in Athens. By using right-angled triangles as the basis for the construction, Pythagoras' Theorem could be used to check that the right angles were accurately constructed. This is an example of maths being used in architecture that dates back to 447 BCE.

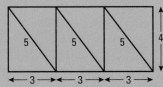

◎ Objective

In this chapter you will:
- ◉ understand and use Pythagoras' Theorem.

◈ Before you start

You need to know:
- ◉ how to square numbers using a calculator
- ◉ how to find the square root of numbers using a calculator
- ◉ how to work out 4^2, 3.5^2, $4^2 + 6^2$
- ◉ how to work out $\sqrt{49}$, $\sqrt{12.25}$.

27.1 Finding the length of the hypotenuse of a right-angled triangle

⊙ Objectives

- ○ You understand Pythagoras' Theorem.
- ○ You can use Pythagoras' Theorem to find the hypotenuse.

⊘ Why do this?

Three thousand years ago the Egyptians and Babylonians used knotted rope to make a 90° angle using a 3, 4, 5 triangle. They used the right angle in the triangle to make their buildings have square corners. These days, builders sometimes use pieces of wood with length 3 feet, 4 feet and 5 feet to do the same thing.

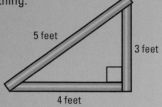

⬆ Get Ready

1. Find the value of
   a 4^2 b 2.3^2 c $7^2 + 24^2$

2. Find the value of
   a $\sqrt{25}$ b $\sqrt{169}$ c $\sqrt{0.25}$

3. Make a copy of these right-angled triangles and mark the hypotenuse (longest side).

a 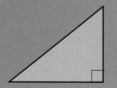 b  c  d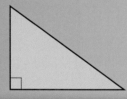

🔍 Key Points

● For a right-angled triangle
  $c^2 = a^2 + b^2$ or $a^2 + b^2 = c^2$

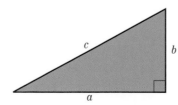

● To find the length of the **hypotenuse** (c):
  square each of the other sides (a) and (b), add the squares and then square root the sum.

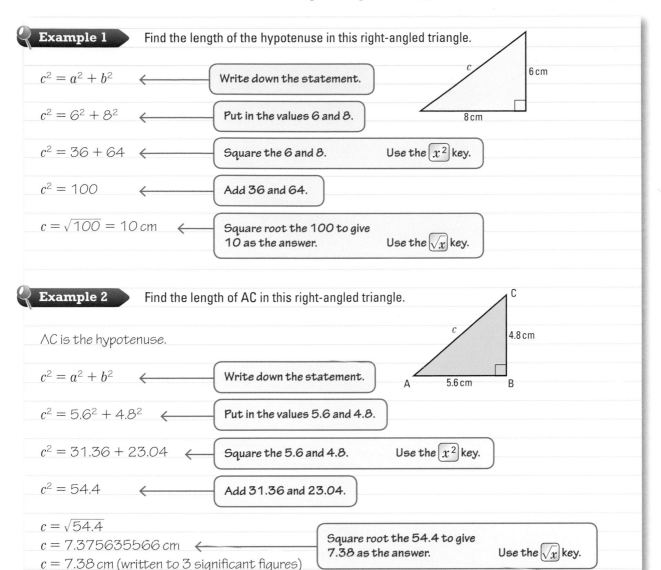

Example 1

Find the length of the hypotenuse in this right-angled triangle.

$c^2 = a^2 + b^2$ ← Write down the statement.

$c^2 = 6^2 + 8^2$ ← Put in the values 6 and 8.

$c^2 = 36 + 64$ ← Square the 6 and 8.   Use the x^2 key.

$c^2 = 100$ ← Add 36 and 64.

$c = \sqrt{100} = 10\,\text{cm}$ ← Square root the 100 to give 10 as the answer.   Use the $\sqrt{x}$ key.

Example 2

Find the length of AC in this right-angled triangle.

AC is the hypotenuse.

$c^2 = a^2 + b^2$ ← Write down the statement.

$c^2 = 5.6^2 + 4.8^2$ ← Put in the values 5.6 and 4.8.

$c^2 = 31.36 + 23.04$ ← Square the 5.6 and 4.8.   Use the x^2 key.

$c^2 = 54.4$ ← Add 31.36 and 23.04.

$c = \sqrt{54.4}$
$c = 7.375635566\,\text{cm}$ ← Square root the 54.4 to give 7.38 as the answer.   Use the $\sqrt{x}$ key.
$c = 7.38\,\text{cm}$ (written to 3 significant figures)

Exercise 27A

Questions in this chapter are targeted at the grades indicated.

C

1. Calculate the length of the hypotenuse in these right-angled triangles. Give your answers correct to 3 significant figures.

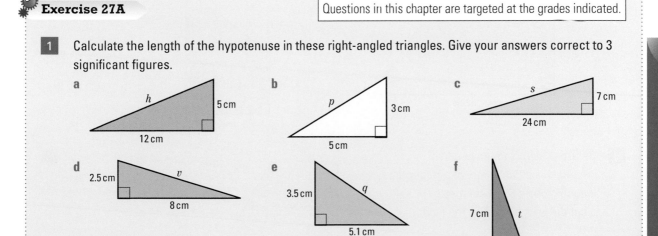

C

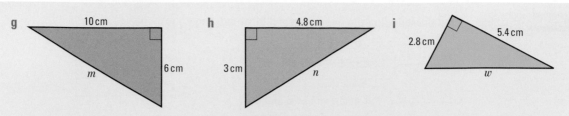

g 10 cm

m

6 cm

h 4.8 cm

3 cm

n

i 5.4 cm

2.8 cm

w

2 Calculate the length of the missing side in these right-angled triangles. Give your answers correct to 3 significant figures.

a

B

7 cm

A

14 cm

C

b

Q

2 cm

P

6 cm

R

c

X

9 cm

Y

25 cm

Z

d

E

4.5 cm

F

10 cm

G

e

H

5.5 cm

J

8.1 cm

K

f

L

9 cm

M

4.4 cm

N

g R 13 cm S

4 cm

T

h U 6.8 cm V

5 cm

W

i

G

3.8 cm

7.4 cm

F

H

A02
A03

3 Find the perimeter of these trapeziums.
Give your answer correct to 3 significant figures.

a

8 cm

6 cm

12 cm

b

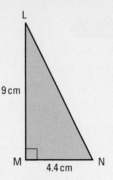

9 cm

4.5 cm

6 cm

A02
A03

4 A farmer wants to fence a field.
The field is in the shape of a trapezium.
Fencing costs £5.50 per metre.
Find the cost of fencing the field.

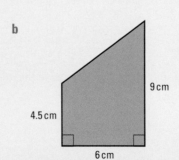

45 m

65 m

82 m

27.2 Finding the length of one of the shorter sides of a right-angled triangle

◎ Objective

◉ You can use Pythagoras' Theorem to find one of the shorter sides of a right-angled triangle.

⬦ Get Ready

1. Find the value of
   a $5^2 - 4^2$ **b** $5^2 - 3^2$ **c** $5.6^2 - 2.3^2$ **d** $13^2 - 5^2$ **e** $25^2 - 7^2$

2. Find the value of
   a $\sqrt{9}$ **b** $\sqrt{16}$ **c** $\sqrt{169}$ **d** $\sqrt{6.25}$ **e** $\sqrt{576}$

3. Make a copy of these right-angled triangles and mark the two shorter sides.

   a   **b**   **c**   **d**

⬙ Key Points

◉ For a right-angled triangle
$$c^2 = a^2 + b^2 \quad \text{or} \quad a^2 + b^2 = c^2$$
so $\quad a^2 = c^2 - b^2 \quad \text{or} \quad b^2 = c^2 - a^2$

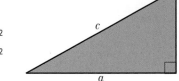

◉ To find the length of one of the shorter sides (a or b):
square each of the other sides (c) and (a or b), subtract the squares and then square root the sum.

🔍 Example 3

Find the length of BC in this right-angled triangle.

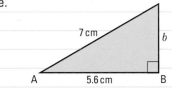

AC is the hypotenuse.
BC is one of the shorter sides.

$c^2 = a^2 + b^2$ ⟵   Write down the statement.

$7^2 = 5.6^2 + b^2$ ⟵   Put in the values 5.6 and 7.

$b^2 = 7^2 - 5.6^2$ ⟵   Square the 5.6 and 7.   Use the $\boxed{x^2}$ key.

$b^2 = 49 - 31.36$ ⟵   Subtract 31.36 from 49.

$b^2 = 17.64$
$b = \sqrt{17.64}$ ⟵   Square root the 17.64 to give 4.2 as the answer.   Use the $\boxed{\sqrt{x}}$ key.
$b = 4.2\text{ cm}$

Exercise 27B

C

1 Calculate the length of the shorter side marked with a letter in these right-angled triangles. Give your answers correct to 3 significant figures.

a

13 cm
5 cm
x

b

5 cm
y
4 cm

c

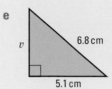

25 cm
s
24 cm

d

3.5 cm
12 cm
r

e

v
6.8 cm
5.1 cm

f

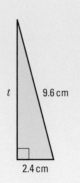

t
9.6 cm
2.4 cm

g

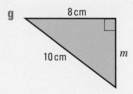

8 cm
10 cm
m

h

n
3 cm
4.8 cm

i

2.8 cm
w
5.6 cm

2 Calculate the length of the missing side in these right-angled triangles. Give your answers correct to 3 significant figures.

a

B
13 cm
A
12 cm
C

b

Q
5 cm
3 cm
P
R

c

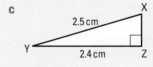

X
2.5 cm
Y
2.4 cm
Z

d

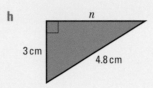

E
12.5 cm
F
8 cm
G

e

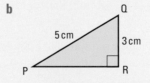

H
6 cm
3.6 cm
J
K

f

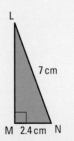

L
7 cm
M
2.4 cm
N

g

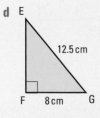

R
S
6 cm
12 cm
T

h

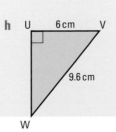

U
6 cm
V
9.6 cm
W

i

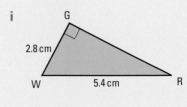

G
2.8 cm
W
5.4 cm
R

3 A 7.5 metre-long ladder leans against a vertical wall.
The foot of the ladder is 1.5 metres from the base of the wall.
How far up the wall does the ladder reach?

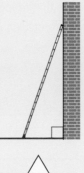

C

4 A farmer has a field in the shape of an equilateral triangle.
Each side of the field is of length 500 metres.
He sells the field at 50p per square metre.
How much money does he sell the field for?
Give your answer correct to the nearest pound.

A03

27.3 Checking to see if a triangle is right-angled or not

Objective

- If you know the lengths of all the sides of a triangle, you can use Pythagoras' Theorem to show whether the triangle is right-angled or not.

Why do this?

In buildings, engineers need to check whether an angle is a right angle or not to ensure that the walls, doors or windows are straight.

Get Ready

Which of these are correct?
1. $5.4^2 + 3.6^2 = 42.12$ 2. $3.5^2 + 3.1^2 = 19.86$ 3. $4.8^2 + 3.2^2 = 33.28$

Key Point

- If the length of the longest side of a triangle squared is equal to the sum of the squares of the other two sides then the triangle has a right angle.

Example 4   Prove that the triangle ABC is right-angled.

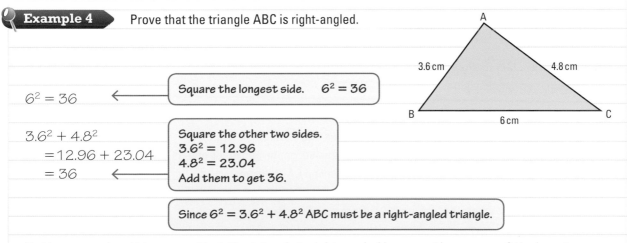

$6^2 = 36$ ← | **Square the longest side.** $6^2 = 36$ |

$3.6^2 + 4.8^2$
$\quad = 12.96 + 23.04$
$\quad = 36$ ← | **Square the other two sides.**
$3.6^2 = 12.96$
$4.8^2 = 23.04$
Add them to get 36. |

| **Since $6^2 = 3.6^2 + 4.8^2$ ABC must be a right-angled triangle.** |

Both are equal so this proves that the triangle is right-angled because the square of the hypotenuse is equal to the sum of the squares of the other two sides.

Example 5

Josie says that triangle PQR is right-angled because $8^2 = (6 + 2)^2$.
Josie is wrong. Explain why.

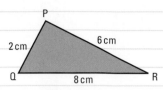

> Square the longest side.
> $8^2 = 64$

$8^2 = 64$

$6^2 + 2^2$
$\quad = 36 + 4$
$\quad = 40$

> Square the other two sides.
> $6^2 = 36$
> $2^2 = 4$

> Since $8^2 \neq 6^2 + 2^2$ ABC is not a right-angled triangle.

64 and 40 are not equal so triangle is not right-angled.

Exercise 27C

1 Check if these triangles have right angles.

a   Triangle ABC where AB = 5 cm, BC = 12 cm and CA = 13 cm

b   Triangle PQR where PQ = 5 cm, QR = 10 cm and RP = 12 cm

c   Triangle XYZ where XY = 4 cm, YZ = 6 cm and ZX = 7 cm

d   Triangle FGH where FG = 3.5 cm, GH = 4.5 cm and HF = 5.5 cm

e   Triangle RST where RS = 6 cm, ST = 8 cm and TR = 10 cm

f   Triangle JKL where JK = 5 cm, KL = 5 cm and LJ = 7 cm

2 Jenny says that triangle PQR is right-angled because $12^2 = (6 + 6)^2$.
Jenny is wrong. Explain why.

3 Jason says that triangle PQR is not right-angled because $10^2 \neq (6 + 8)^2$.
Jason is wrong. Explain why.

4 An acute-angled triangle has three acute angles.
An obtuse-angled triangle has two acute angles and one obtuse angle.
Investigate whether these triangles are right-angled, acute-angled or obtuse-angled.

a   Triangle ABC where AB = 5 cm, BC = 10 cm and CA = 10 cm

b   Triangle PQR where PQ = 5 cm, QR = 8 cm and RP = 12 cm

c   Triangle XYZ where XY = 4 cm, YZ = 6 cm and ZX = 7 cm

d   Triangle FGH where FG = 3.5 cm, GH = 4.5 cm and HF = 6.5 cm

e   Triangle RST where RS = 7.5 cm, ST = 10 cm and TR = 12.5 cm

f   Triangle JKL where JK = 5 cm, KL = 5 cm and LJ = 5 cm

27.4 Finding the length of a line segment

⊙ Objective

⊙ You can use Pythagoras' Theorem to find the length of the line segment between two coordinates.

⟐ Why do this?

Sat navs use Pythagoras' Theorem to calculate the shortest distance between two places.

⟐ Get Ready

Find the lengths of the missing sides in these right-angled triangles, ABC.

1. AC where AB = 3.2 cm and BC = 4.3 cm
2. BC where AC = 7.5 cm and AB = 2.9 cm
3. AB where AC = 9.8 cm and BC = 1.8 cm

🔑 Key Point

⊙ To find the distance between two points on a coordinate grid:
   - ⊙ subtract the x-coordinates and square
   - ⊙ subtract the y-coordinates and square
   - ⊙ add the results
   - ⊙ square root the answer.

🔍 Example 6

A has coordinates (1, 1). B has coordinates (4, 5).
Find the length of the line segment AB.

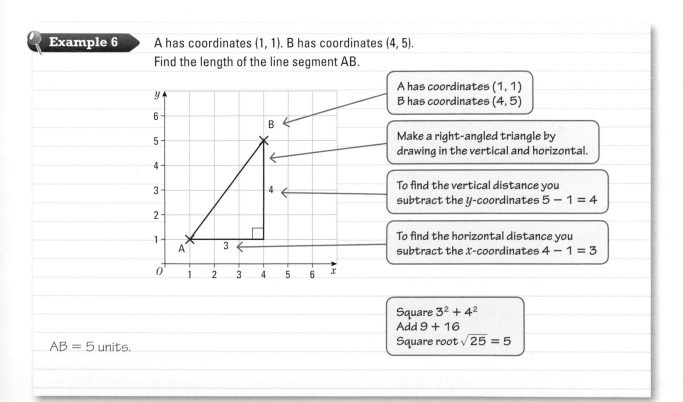

A has coordinates (1, 1)
B has coordinates (4, 5)

Make a right-angled triangle by drawing in the vertical and horizontal.

To find the vertical distance you subtract the y-coordinates $5 - 1 = 4$

To find the horizontal distance you subtract the x-coordinates $4 - 1 = 3$

Square $3^2 + 4^2$
Add $9 + 16$
Square root $\sqrt{25} = 5$

AB = 5 units.

Example 7   Find the length of the line segment ST.

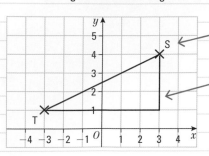

S has coordinates (3, 4)
T has coordinates (−3, 1)

Make a right-angled triangle by drawing in the vertical and horizontal.

To find the horizontal distance you subtract the x-coordinates 3 − −3 = 3 + 3 = 6

To find the vertical distance you subtract the y-coordinates 4 − 1 = 3

Square $6^2 + 3^2$
Add 36 + 9
Square root $\sqrt{45} = 6.708$

ST = 6.71 units (to 2 d.p.).

Exercise 27D

C

1   Work out the length of each of the line segments shown on the grid. Give your answers correct to 3 significant figures.

   a OA   **b** BC   **c** DE   **d** FG   **e** HJ

   f KL   **g** MN   **h** PQ   **i** ST   **j** UV

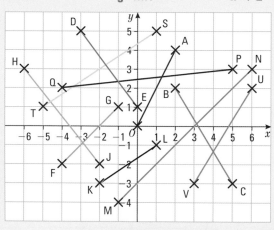

2   Work out the lengths of each of these line segments. Give your answers correct to 3 significant figures.

   a   **b**   **c**   **d**

   a B (4, 3), A (−1, −2)

   b Q (7, 4), P (1, −2)

   c R (−2, 5), S (5, −3)

   d U (2, 8), V (5, −3)

3 Work out the lengths of each of these line segments. Give your answers correct to 3 significant figures.
  a AB when A is $(-1, -1)$ and B is $(9, 9)$
  b PQ when P is $(2, -4)$ and Q is $(-6, 9)$
  c ST when S is $(5, -8)$ and T is $(-2, 1)$
  d CD when C is $(1, 7)$ and D is $(-7, 2)$
  e UV when U is $(-2, 3)$ and V is $(6, -8)$
  f GH when G is $(-2, -6)$ and H is $(7, 3)$

Chapter review

- For a right-angled triangle:
  $c^2 = a^2 + b^2$ or $a^2 + b^2 = c^2$

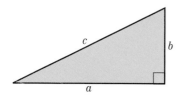

- To find the length of the **hypotenuse** (c):
  square each of the other sides (a) and (b), add the squares and then square root the sum.
- To find the length of one of the shorter sides $(a$ or $b)$:
  square each of the other sides (c) and $(a$ or $b)$, subtract the squares and then square root the sum.
- If the length of the longest side of a triangle squared is equal to the sum of the squares of the other two sides then the triangle has a right angle.
- To find the distance between two points on a coordinate grid:
  - subtract the x-coordinates and square
  - subtract the y-coordinates and square
  - add the results
  - square root the answer.

Review exercise

1 ABC is a right-angled triangle.

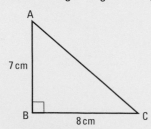

AB = 7 cm,
BC = 8 cm.
Work out the length of AC.
Give your answer correct to 2 decimal places.

ResultsPlus
Exam Question Report

83% of students answered this question poorly. The most common wrong answer was 15, obtained by either adding the two sides $(7 + 8)$ or subtracting $(8^2 - 7^2)$.

June 2008

2 **a** Work out the area of the triangle.

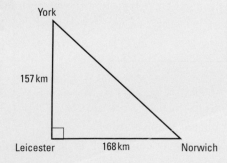

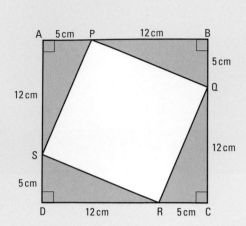

4 copies of the triangle and the quadrilateral PQRS are used to make the square ABCD.

b Work out the area of the quadrilateral PQRS. *Nov 2007*

3 In the triangle XYZ

XY = 5.6 cm

YZ = 10.5 cm

angle XYZ = 90°

Work out the length of XZ.

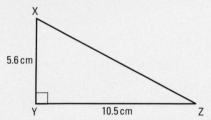

Nov 2007

4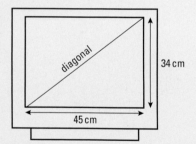

A rectangular television screen has a width of 45 cm and a height of 34 cm.

Work out the length of the diagonal of the screen.

Give your answer correct to the nearest centimetre. *June 2007*

5 The diagram shows three cities.

Norwich is 168 km due east of Leicester.

York is 157 km due north of Leicester.

Calculate the distance between Norwich and York.

Give your answer correct to the nearest kilometre.

Nov 2006

6

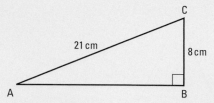

In triangle ABC
angle ABC = 90°
BC = 8 cm
AC = 21 cm
Work out the length of AB.
Give your answer correct to 3 significant figures.

ResultsPlus
Exam Question Report

82% of students answered this question poorly.
The majority of students did not recognise that
they were required to use Pythagoras' Theorem.

March 2007

7 ABC is a right-angled triangle.
AC = 6 cm
BC = 9 cm
Work out the length of AB.
Give your answer correct to 3 significant figures.

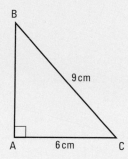

* **8** Prove that triangle ABC is a right-angled triangle.

A03

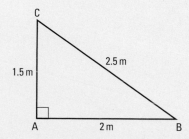

9 A has coordinates (1, 7). B has coordinates (6, 8).
Find the length of the line segment AB.
Give your answer correct to 2 decimal places.

10 P has coordinates (9, 3). Q has coordinates (2, 0).
Find the length of the line segment PQ.
Give your answer correct to 1 decimal place.

June 2006

11 Paul flies his helicopter from Ashwell to Birton.
He flies due west from Ashwell for 4.8 km. He then flies due south for 7.4 km to Birton.
Calculate the shortest distance between Ashwell and Birton.

A03

12 ABC is a right-angled triangle.
Calculate the area of the triangle ABC.

A02

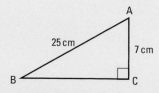

13 PQR is an isosceles triangle with PQ = PR.

PQ = PR = 10 cm

QR = 6 cm

Calculate the area of triangle PQR.

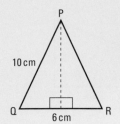

14 A farmer has a field in the shape of an isosceles triangle.

The sides of the field have lengths 500 metres, 500 metres and 600 metres.

He sells the field at 45p per square metre.

For how much money does he sell the field?

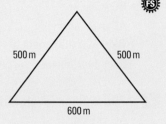

15 A is the point with coordinates (2, 5).

B is the point with coordinates (8, 13).

Calculate the length AB.

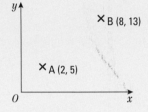

Diagram **NOT** accurately drawn

2006

You could use a formula to work out the distance you could drive before running out of fuel. It might be good idea if you were driving somewhere remote like Death Valley in the USA. Temperatures have been known to reach 57°C (134°F) in July and there is no shade. It is still known for people to die if they break down or run out of fuel in one of the most remote areas.

⊙ Objectives

In this chapter you will:
- use formulae
- write word formulae
- substitute numbers into expressions
- use and write algebraic formulae
- find the value of a term which is not the subject of the formula
- change the subject of a formula.

◈ Before you start

You need to know:
- the order of operations
- the difference between expressions, variables and terms.

28.1 Using word formulae

◈ Get Ready

Work out

1. $15 \times 25p$ **2.** $36 \times £5.90$ **3.** 3.14×12

✎ Key Points

● A **word formula** uses words to represent a relationship between two quantities.
● When using formulae you need to remember the order of operations, BIDMAS (see Chapter 9).
● Formulae are used when calculating:
  ● area of triangles (see Section 14.3)
  ● angles in polygons (see Section 7.2)
  ● area of circles (see Chapter 17).
● Make sure the units of measure are consistent when calculating using formulae.

Example 1 ▶ This word formula can be used to work out the perimeter of an equilateral triangle.

Perimeter = 3 × length of side

Work out the perimeter of an equilateral triangle with sides of 9 cm.

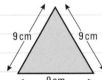

9 cm   9 cm

9 cm

Perimeter = 3 × 9 = 27 cm ← Put 9 into the word formula for length of side.

Example 2 ▶ Suki is paid £6.50 per hour.
  a Write a word formula for her weekly pay.
  b Work out her pay for a week when she works 24 hours.

a Weekly pay = hourly rate × hours worked per week ← A word formula must be in the form something = something else.

b Pay = £6.50 × 24 ← Put 24 and £6.50 into the formula. Do not forget the units.
       = £156

Exercise 28A

Questions in this chapter are targeted at the grades indicated.

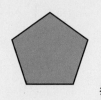

1 This word formula can be used to work out the perimeter of a regular pentagon.

Perimeter = 5 × length of side

Work out the perimeter of a regular pentagon with sides of 7 cm.

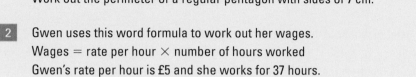

2 Gwen uses this word formula to work out her wages.

Wages = rate per hour × number of hours worked

Gwen's rate per hour is £5 and she works for 37 hours.

Work out her wages.

3 This word formula can be used to work out the area of a triangle.

Area = $\frac{1}{2}$ base × vertical height

A triangle has base 8 cm and vertical height 10 cm.

Work out the area of the triangle.

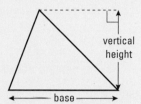

4 This word formula can be used to work out the perimeter of a rectangle.

Perimeter = 2 × length + 2 × width

Work out the perimeter of a rectangle with a length of 8 cm and width of 3 cm.

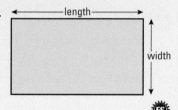

5 Owen uses this word formula to work out his phone bill.

Total bill = cost per minute × number of minutes + monthly charge

The cost per minute is 6p. Owen made 70 minutes of calls.

The monthly charge is £15. Work out his total bill.

6 Kirsty buys 12 stamps at 42p each.

a Write a word formula for the total cost of the stamps.

b Work out the cost of the stamps.

7 a Write a word formula to work out the number of packets of crisps left in a machine after a number have been sold.

b Use your formula to work out the number of packets left in a machine that holds 56 packets, when 27 have been sold.

8 Georgina uses this word formula to work out her take-home pay.

Take-home pay = rate per hour × number of hours worked − deductions

Georgina's rate per hour is £7. She worked for 40 hours and her deductions were £96.

Work out her take-home pay.

9 For each part write a word formula, then use it to calculate the answers.

a Alex shared a bag of sweets equally between herself and her three brothers. There were 84 sweets in the bag. How many did each person have?

b At a buffet lunch there were 36 slices of pizza. Every person at the lunch had 3 slices. All the slices were eaten. How many people were at the lunch?

F

10 This word formula can be used to work out the average speed for a journey.

Average speed $= \dfrac{\text{total distance travelled}}{\text{time taken}}$

Raj travels 215 miles in 5 hours. Work out his average speed in miles per hour (mph).

11 This word formula can be used to work out the angle sum, in degrees, of a polygon.
Angle sum = (number of sides − 2) × 180
Work out the angle sum of a polygon with 7 sides.

12 This word formula can be used to work out the size, in degrees, of each exterior angle of a regular polygon.

Exterior angle $= \dfrac{360}{\text{number of sides}}$

Work out the size of each exterior angle of a regular polygon with 8 sides.

13 This word formula can be used to work out the area inside a circle.
Area = π × radius × radius
Work out the area of a circle with a radius of 4 cm.
Give your answer to the nearest whole number.

28.2 Substituting numbers into expressions

◉ Objectives

- You can work out the value of an algebraic expression by substituting in the values of each letter or letters.
- You can substitute negative numbers into expressions.

⟨?⟩ Why do this?

You could work out pocket or birthday money if you get a certain amount of money per year depending on your age.

⟨↗⟩ Get Ready

1. Use the word formula perimeter = 2 × length + 2 × width to work out the perimeter of a rectangle with:
   a length 4 cm, width 2 cm   **b** length 5 cm, width 3 cm   **c** length 7 cm, width 5 cm.

⟨◐⟩ Key Point

- You can **substitute** values into an algebraic expression in the same way as you substitute values into a word formula.

Example 3 $a = 4$ and $b = 5$

Work out the value of

    a $3a - b$ **b** $3ab$ **c** $a^2 + 2b$

a $3a - b = 3 \times 4 - 5$ ← The numbers replace the letters so $3a = 3 \times 4$ and $b = 5$.

    $= 12 - 5$ ← Remember: multiplication before subtraction.

    $= 7$

b $3ab = 3 \times 4 \times 5 = 60$ ← $3ab = 3 \times a \times b$

c $a^2 + 2b = 4 \times 4 + 2 \times 5$ ← $a^2 = a \times a = 4 \times 4$

    $= 16 + 10$

    $= 26$

Exercise 28B

1 $p = 3$, $q = 2$, $r = 5$ and $s = 0$

Work out the value of the following expressions.

a $p + r$	**b** rs	**c** $5p - 2r$
d pqr	**e** $pr - pq$	**f** $6pq + 3qr$
g $4(p + 7)$	**h** $p(q + 4)$	**i** $p(r - q)$
j $r^2 + 1$	**k** $(q + 1)^2$	**l** $2p^3$
m $(p + r)^2$	**n** $(p - q)^3$	

2 $p = \frac{3}{4}$, $q = \frac{1}{4}$, $r = 2$ and $s = 1$

Work out the value of the following expressions.

a $4q$	**b** $7qr$	**c** $4p - 6q$
d qrs	**e** $qr - pq$	**f** $4qr - 5pq$
g $6(r - 2)$	**h** $r(r - 2)$	**i** p^2
j $3r^3 - 2$	**k** $(r - 4)^2$	**l** $6q^3$
m $(p + q)^3$		

3 $p = 0.5$, $q = 2$, $r = 3$ and $s = 1.25$

Work out the value of the following expressions.

a pq	**b** $6p + 4q$	**c** $7p + 8s$
d $pr + 4q$	**e** $qr + rs$	**f** $6pr - 7qs$
g $5(p + q)$	**h** $q(p + r)$	**i** $5p^2$
j $5q^2 + 7$	**k** r^3	**l** $p^2 + r^2$
m $p^3 - q^3$		

D

Example 4 $p = 2$, $q = -3$ and $r = -5$

Work out the value of

| a $p + r$ | b $q - r$ | c pq | d qr |
| e $p(q + r)$ | f $r^2 + 6r$ | g $(p + q)^2$ | h $4r^3$ |

a $p + r = 2 + (-5)$ ← Adding a negative number is the same as subtracting a positive number.

$\quad = 2 - 5$

$\quad = -3$

b $q - r = -3 - (-5)$ ← Subtracting a negative number is the same as adding a positive number.

$\quad = -3 + 5$

$\quad = 2$

c $pq = 2 \times -3$

$\quad = -6$ ← When you multiply two numbers which have different signs the answer is negative. Remember: 2 means $+2$.

d $qr = -3 \times -5$

$\quad = 15$ ← When you multiply two numbers which have the same sign the answer is positive.

e $p(q + r) = 2 \times (-3 + -5)$ ← A bracket means times. Work out the value of the bracket first.

$\quad = 2(-8)$

$\quad = -16$

f $r^2 + 6r = (-5)^2 + 6 \times -5$ ← Square and multiply before adding.

$\quad = (-5 \times -5) - 30$

$\quad = 25 - 30$

$\quad = -5$

g $(p + q)^2 = (2 + (-3))^2$

$\quad = (2 - 3)^2$ ← Work out the value of the bracket first.

$\quad = (-1)^2$ ← Square.

$\quad = -1 \times -1$

$\quad = 1$

h $4r^3 = 4 \times (-5)^3$ ← $(-5)^3$ means cube of -5. Cube before you multiply.

$\quad = 4 \times (-5 \times -5 \times -5)$

$\quad = 4 \times -125$

$\quad = -500$

Exercise 28C

In this exercise $a = -5$, $b = 6$, $c = -2$, $d = \frac{1}{2}$ and $e = 1$.
Work out the value of the following expressions.

1 $a + b$	**2** $a - b$	**3** $b - a$	**4** $a - c$
5 $b - c$	**6** $a + b + c$	**7** $3a + 7$	**8** $4a + 3b$
9 $2b + 5c$	**10** $2a - 5c$	**11** $3b - 2a$	**12** ab
13 acd	**14** $3bc$	**15** $bd - 1$	**16** $ab - bc$
17 $2ab + 3ac$	**18** $3ac - 2bc$	**19** $abcd$	**20** $3d(a + 1)$
21 $b(c - a)$	**22** $c(a + b)$		
23 $5(c - 1)$	**24** $a(b + c)$	**25** a^2	**26** $3c^2d$
27 $4a^2 - 3$	**28** $5c^2 + 3c$	**29** $2a^2 - 3a$	**30** $(a + 1)^2$
31 $(c + 3)^2$	**32** $(a + b)^2$	**33** $(a + c)^2$	**34** $(c - a)^2$
35 $2b^3$	**36** $3a^3$	**37** $6c^3$	**38** $2(b + c)^2$
39 $a^2 - b^2$	**40** $(a - c)^3$	**41** $(e - d)^2$	**42** $(c - 5d)^2$

E

D

C

28.3 Using algebraic formulae

◉ Objective

○ You can substitute number values for the letters in a formula to work out a quantity.

◈ Why do this?

If you are given an algebraic formula to work out your pay, you need to know how to use it!

◈ Get Ready

$a = -5$, $b = 6$, $c = \frac{1}{2}$
Work out

1. $a + (b \times c)$ **2.** abc **3.** $\dfrac{(a + b)}{c}$

◆ Key Point

◉ An **algebraic formula** uses letters to show a relationship between quantities, e.g. $A = lw$.
The letter that appears on its own on one side of the $=$ sign and does not appear on the other side is called the **subject** of the formula.
In the formula above, A is the subject of the formula.

Example 5

The formula for the perimeter of this isosceles triangle is $P = 2a + b$.

Work out the value of P when $a = 8$ and $b = 5$.

$P = 2 \times 8 + 5 = 16 + 5 = 21$ ← Put $a = 8$ and $b = 5$ into the formula to find P.

A03

Example 6

Harry's pay is worked out using the formula $P = hr + b$

where P = pay

h = hours worked

r = rate of pay

b = bonus.

Work out Harry's pay when $h = 35$, $r =$ £10 and $b =$ £45.

$P = hr + b$

$P = 35 \times 10 + 45$

$P = 350 + 45$ ← Substitute $h = 35$, $r = 10$ and $b = 45$.

$P =$ £395

Exercise 28D

F

1 $y = 2x + 3$ is the equation of a straight line.

Work out the value of y when

a $x = 4$ **b** $x = 6$ **c** $x = 10$ **d** $x = 7.5$

E

2 The formula for the area of a parallelogram is $A = bh$.

Work out the value of A when

a $b = 7$ and $h = 3$ **b** $b = 9$ and $h = 7$

c $b = 6$ and $h = 3.7$ **d** $b = 8.4$ and $h = 4.5$

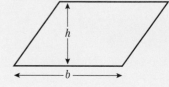

3 The formula for the circumference of a circle is $C = \pi d$.

Work out, to the nearest whole number, the value of C when

a $d = 2$ **b** $d = 5$ **c** $d = 3.7$ **d** $d = 9.3$

4 Euler's formula for the number of edges of a solid is $E = F + V - 2$, where E is the number of edges, F is the number of faces and V is the number of vertices.

Work out the value of E when

a $F = 6$ and $V = 8$ **b** $F = 16$ and $V = 19$

5 The formula $F = 1.8C + 32$ can be used to convert a temperature from degrees Celsius to degrees Fahrenheit. Work out the value of F when

a $C = 10$ **b** $C = 100$ **c** $C = -30$ **d** $C = 0$

D

6 The formula for the volume of a cuboid is $V = lwh$.
Work out the value of V when
  a $l = 5, w = 4$ and $h = 2$
  b $l = 8, w = 5$ and $h = 3.5$
  c $l = 10, w = 6$ and $h = 4$
  d $l = 9.3, w = 4.2$ and $h = 5.1$

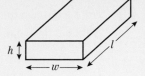

7 The formula $v = u + at$ can be used to work out velocity.
Velocity means speed in a particular direction.
Work out the value of v when
  a $u = 8, a = 4$ and $t = 3$
  b $u = 0, a = 10$ and $t = 2$
  c $u = 7, a = 2.6$ and $t = 5$
  d $u = 12, a = 10$ and $t = 4.7$

8 The formula $T = 15(W + 1)$ can be used to work out the time needed to cook a turkey.
Work out the value of T when
  a $W = 5$ b $W = 8$ c $W = 12$ d $W = 18$

9 The formula for the volume of a cuboid is $V = l^2h$. Work out the value of V when
  a $l = 2$ and $h = 5$ b $l = 7$ and $h = 9.8$ c $l = 2.5$ and $h = 4.3$

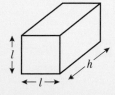

28.4 Writing an algebraic formula to represent a problem

◎ Objective

● You can form a simple formula
including squares, cubes and roots.

◈ Why do this?

You could write a formula to work out the total cost of your holiday
including air fares, hotel costs and entertainment costs, then you
could alter any one of these to find how it affects the total cost.

◈ Get Ready

Use the formula $C = \pi d$ to work out C when
1. $d = 3$ 2. $d = 8$ 3. $d = 10$

◉ Key Point

● You can use information given in words to write an algebraic formula to solve a problem.

🔍 Example 7

Florence's pay is worked out using the formula
pay = number of hours work × rate per hour + commission.

a Write this as an algebraic formula.
b Work out Florence's pay when she works 30 hours at £7.50 per hour and gets a
   commission of £20.

a $P = nr + c$ ←————————— ┌───┐
                            │ Use letter for pay = P, number of hours worked = n, │
                            │ rate per hour = r and commission = c. │
                            └───┘

b $P = nr + c$ ←———————
   $P = 30 \times 7.50 + 20$ ┌───┐
                                  │ Substitute $n = 30, r =$ £7.50 and $c =$ £20. │
   $P = 225 + 20$ └───┘
   $P = £245$

E

⚙ **Exercise 28E**

1 Write a formula for the perimeter P of this regular hexagon, with side l.
Work out the value of P when
  a $l = 3$ b $l = 7$
  c $l = 29$ d $l = 8.6$

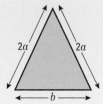

2 Write a formula for the perimeter of this isosceles triangle.
Work out the perimeter when
  a $a = 6$ and $b = 4$ b $a = 12$ and $b = 7$
  c $a = 5.3$ and $b = 3.4$ d $a = 4.7$ and $b = 8.5$

3 a Write an algebraic formula for the price of a number of pens that cost 70p each.
   b Use your formula to work out the cost of:
        i 4 pens   ii 6 pens   iii 12 pens.

4 Write a formula for the volume of this cube.
Work out the volume when
  a $s = 2$ cm   b $s = 4.5$ cm

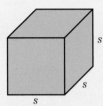

5 Write a formula for the surface area of the cube in question 4.
Work out the surface area when
  a $s = 2$ cm   b $s = 4.5$ cm

D

6 Write a formula for the length of the side of a square given the area.
Work out the length of the side when the area is
  a 4 cm^2 b 1.44 cm^2

28.5 Finding the value of a term which is not the subject of a formula

◎ **Objective**

○ You can work out a value in a formula when it is not the subject of the formula.

⬦ **Get Ready**

Solve these equations.
1. $2x + 4 = 10$ 2. $24 = 2a - 8$ 3. $6x = 30$

Key Point

◉ To find out the value of a term which is not the subject of a formula, put the given values into the formula and then solve the resulting equation.

Example 8

$P = 2a + b$

Work out the value of a when $P = 25$ and $b = 7$.

$25 = 2a + 7$ ← Solve this equation.

$2a = 18$ ← Subtract 7 from both sides.

$a = 9$ ← Divide both sides by 2.

ResultsPlus
Examiner's Tip

Substitute the value of the term back into the formula to check your answer.

Exercise 28F

1 $P = 6l$
Work out the value of l when
a $P = 24$ b $P = 54$ c $P = 138$ d $P = 35.4$

2 $A = bh$
a Work out the value of h when
   i $A = 45$ and $b = 5$ ii $A = 54$ and $b = 3$
b Work out the value of b when
   i $A = 35$ and $h = 7$ ii $A = 120$ and $h = 8$

3 $E = F + V - 2$
a Work out the value of F when
   i $E = 9$ and $V = 5$ ii $E = 21$ and $V = 9$
b Work out the value of V when
   i $E = 15$ and $F = 10$ ii $E = 30$ and $F = 12$

4 $y = 2x + 3$
Work out the value of x when
a $y = 15$ b $y = 27$ c $y = -10$ d $y = -3$

5 $P = 2a + b$
a Work out the value of b when
   i $P = 15$ and $a = 6$ ii $P = 23$ and $a = 4.5$
b Work out the value of a when
   i $P = 11$ and $b = 5$ ii $P = 19$ and $b = 8$

6 $y = 4x - 5$
Work out the value of x when
a $y = 3$ b $y = -31$ c $y = 75$ d $y = -6$

F

E

D

D

7 $V = hwl$

Work out the value of h when

    a $V = 24$, $l = 3$ and $w = 2$ **b** $V = 60$, $l = 6$ and $w = 2$ **c** $V = 80$, $l = 4$ and $w = 5$

8 $v = u + at$

    a Work out the value of u when

        i $v = 19$, $a = 7$ and $t = 2$ **ii** $v = 25$, $a = 6$ and $t = 3$

    b Work out the value of a when $v = 17$, $u = 5$ and $t = 2$.

    c Work out the value of t when $v = 31$, $u = 3$ and $a = 7$.

9 $y = \dfrac{x}{5}$

Work out the value of x when

    a $y = 4$ **b** $y = 17$ **c** $y = 7.4$ **d** $y = 0$

10 $t = \dfrac{d}{s}$

Work out the value of d when

    a $t = 3$ and $s = 5$ **b** $t = 9$ and $s = -8$ **c** $t = 7.5$ and $s = 6$ **d** $t = 5.6$ and $s = -10.4$

28.6 Changing the subject of a formula

Objective

○ You can rearrange a formula to make a different variable the subject of the formula.

Why do this?

If you wanted to find out the distance that you had left to travel but the subject of the formula was time, you could rearrange it to find out the distance.

Get Ready

Solve these equations.

1. $4 = 5 + 3x$ **2.** $18 = 2(x + 4)$ **3.** $24 = \frac{1}{3} \times 8x$

Key Point

◉ You can change the subject of a formula by carrying out the same operations on both sides of the equals sign.

Example 9

Make t the subject of the formula $v = u + at$.

$v = u + at$

$v - u = at$ ← Subtract u from both sides.

$t = \dfrac{v - u}{a}$ ← Divide both sides by a.

Example 10 ▶ Make l the subject of the formula $P = 2(l + b)$.

$P = 2(l + b)$

$P = 2l + 2b$ ← Multiply out the brackets.

$2l = P - 2b$ ← Subtract $2b$ from both sides.

$l = \dfrac{P}{2} - b$ ← Divide both sides by 2.

Alternatively, you can divide both sides by 2

$\dfrac{P}{2} = l + b$

and then subtract b from both sides

$l = \dfrac{P}{2} - b$

Example 11 ▶ Make h the subject of the formula $V = \frac{1}{3}Ah$.

$V = \frac{1}{3}Ah$

$3V = Ah$ ← Multiply both sides by 3.

$h = \dfrac{3V}{A}$ ← Divide both sides by A.

$h = \dfrac{V}{\frac{1}{3}A}$ is also correct but it is best not to have a fraction within another fraction.

Exercise 28G

Rearrange each formula to make the letter in square brackets the subject.

1	$P = 5d$	$[d]$	**2**	$P = IV$	$[I]$	**3**	$A = lw$	$[w]$
4	$C = \pi d$	$[d]$	**5**	$V = lwh$	$[h]$	**6**	$A = \pi rl$	$[r]$
7	$y = 4x - 3$	$[x]$	**8**	$t = 3n + 5$	$[n]$	**9**	$P = 2x + y$	$[y]$
10	$y = mx + c$	$[m]$	**11**	$v = u - gt$	$[u]$	**12**	$v = u - gt$	$[t]$
13	$A = \frac{1}{2}bh$	$[b]$	**14**	$I = \dfrac{PRT}{100}$	$[T]$	**15**	$T = \dfrac{D}{V}$	$[V]$
16	$\dfrac{PV}{T} = k$	$[V]$	**17**	$\dfrac{PV}{T} = k$	$[T]$	**18**	$I = m(v - u)$	$[v]$
19	$A = \frac{1}{2}(a + b)h$	$[b]$	**20**	$y = \frac{1}{3}x - 2$	$[x]$	**21**	$y = 2(x - 1)$	$[x]$
22	$x = 3(y + 2)$	$[y]$	**23**	$H = 17 - \dfrac{A}{2}$	$[A]$	**24**	$3x - 2y = 6$	$[x]$
25	$3x - 2y = 6$	$[y]$	**26**	$P = 6(q - 7) - 5(q - 6)$	$[q]$	**27**	$4y^2 - 2x = 6(x - 8y)$	$[x]$

D

C

Chapter review

- A **word formula** uses words to represent a relationship between two quantities.
- When using formulae you need to remember the order of operations, BIDMAS.
- Formulae are used when calculating:
  - area of triangles (see Section 14.3)
  - angles in polygons (see Section 7.2)
  - area of circles (see Chapter 17).
- Make sure the units of measure are consistent when calculating using formulae.
- You can **substitute** values into an algebraic expression in the same way as you substitute values into a word formula.
- An **algebraic formula** uses letters to show a relationship between quantities, e.g. $A = lw$.
  The letter that appears on its own on one side of the $=$ sign and does not appear on the other side is called the **subject** of the formula. In the formula above, A is the subject.
- You can use information given in words to write an algebraic formula to solve a problem.
- To find out the value of a term which is not the subject of a formula, put the given values into the formula and then solve the resulting equation.
- You can change the subject of a formula by carrying out the same operations on both sides of the equals sign.

Review exercise

1 Nathan is three years younger than Ben.
  a Write down an expression for Nathan's age.
  Daniel is twice as old as Ben.
  b Write down an expression for Daniel's age.
  Nov 2008

2 Amy, Bryony and Christina each collect bracelet charms.
  Bryony has twice as many charms as Amy.
  a Write down an expression for the number of charms that Bryony has.
  Christina has 7 charms less than Amy.
  b Write down an expression for the number of charms that Christina has.

3 Adam, Brandon and Charlie each buy some stamps.
  Brandon buys three times as many stamps as Adam.
  a Write down an expression for the number of stamps Brandon buys.
  Charlie buys 5 more stamps than Adam.
  b Write down an expression for the number of stamps Charlie buys.

4 You can use this rule to work out the cost, in pounds, of hiring a carpet cleaner.

| Multiply the number of days' hire by 6 |
| Add 4 to your answer |

  Jill hires the carpet cleaner for 3 days.
  a Work out the cost.
  Carlos hires the carpet cleaner. The cost is £52.
  b Work out for how many days Carlos hires the carpet cleaner.
  June 2009

5 Navjeet uses this rule to work out his pay.

> Pay = Number of hours worked × rate of pay per hour

This week Navjeet worked for 10 hours.
His rate of pay per hour was £4.50.
a Use this rule to work out his pay.
Last week Navjeet's pay was £66.
He worked for 12 hours.
b Work out Navjeet's rate of pay per hour last week.

June 2006 A03

6 This word formula can be used to work out the total cost, in pounds, of running a car.

$$\text{Total cost} = \text{fixed costs} + \frac{\text{number of miles travelled}}{6}$$

a Flora's fixed costs were £500 and she travelled 9000 miles. Work out her total cost.
b Harry's total cost was £2700 and he travelled 12 000 miles. Work out his fixed costs.
c Ali's total cost was £1600 and his fixed costs were £400. Work out the number of miles he travelled.

7 To calculate the cost of printing leaflets for a school fair, the printer uses the formula:
$C = 40 + 0.05n$
where C is the cost in pounds and n is the number of leaflets printed.
a How much would it cost to print 200 leaflets?
b Can you suggest what 40 and 0.05 represent?

8 If $a = 1$, $b = 2$ and $c = -3$, find the value of

a $\dfrac{c - ab}{c + ab}$ b $3(a + b)^2 - 2(b - c)^2$

9 $a = 4$, $b = \frac{1}{4}$, $c = -3$

Work out

a $\dfrac{5a}{b} + 7$ b $6a + \dfrac{2c}{3}$ c $3a - 6b + c$ d $a(a - 8b)$

10 The formula $v = u - gt$ can be used to work out velocity.
a Work out the value of v when $u = 45$, $g = 10$ and $t = 3$.
b Work out the value of u when $v = 8$, $g = 10$ and $t = 2$.

11

a Write a formula for the area of a square of side l.
b Work out the area when $l = 9$.

D

12

a Write a formula for the perimeter of a rectangle.
b Work out the perimeter when
  i $l = 9$ and $w = 4$
  ii $l = 6.7$ and $w = 3.4$

13 The cost, C in £, of buying t trees and b bushes together with delivery is given by the formula
$$C = 10t + 6b + 15.$$
Greg has £315 to spend and needs 35 bushes.
How many trees can he afford?

14 $s = ut + \frac{1}{2}at^2$ is a formula for working out the distance, s, moved by an object.
Work out s when
a $u = 4.2$, $a = 10$ and $t = 3$
b $u = 5$, $a = -10$ and $t = 5.7$
c $u = -3$, $a = -32$ and $t = 6$

15 $A = \frac{1}{2}bh$ is the formula for working out the area of a triangle. Work out the area of a triangle when
a $b = 30$ cm and $h = 20$ cm
b $b = 15$ cm and $h = 26$ cm
c $b = 7.3$ cm and $h = 2.9$ cm
d $b = 2.3$ cm and $h = 1.3$ cm

16 $v = u + at$ is a formula for finding the speed of an object.
Find v when
a $u = 6$, $a = 10$ and $t = 5$
b $u = 8$, $a = -10$ and $t = 6$
c $u = 20$, $a = -32$ and $t = 4\frac{1}{2}$

17 $p = 2$
Work out the value of $5p^2$.

18 $v = u + 10t$
Work out the value of v when
a $u = 10$ and $t = 7$
b $u = -2.5$ and $t = 3.2$

19 **a** Work out the value of $2a + ay$ when $a = 5$ and $y = -3$.
b Work out the value of $5t^2 - 7$ when $t = 4$.

20 Tom the plumber charges £35 for each hour he works at a job, plus £50.
The amount Tom charges, in pounds, can be worked out using this rule.

> Multiply the number of hours
> he works by 35
> Add 50 to your answer

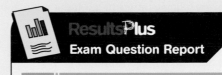
ResultsPlus
Exam Question Report

86% of students did very well on part **a** of this question.

Tom charged a customer £260 for a job.
a How many hours did Tom work?
Tom works h hours at a job.
He charges P pounds.
b Write down a formula for P in terms of h.

June 2007

21 The cost of hiring a car can be worked out using this rule.

> Cost = £90 + 50p per mile

A03

Bill hires a car and drives 80 miles.
a Work out the cost.
The cost of hiring a car is C pounds.
b Write the formula for C.

Nov 2007

22 The formula used to convert temperatures in Fahrenheit, F, into Celsius, C, is given by:
$$C = \frac{5(F - 32)}{9}$$
a Find C when $F = 77$.
b Use the formula to find the freezing point of water in Fahrenheit.
A newspaper headline read 'Phew, what a scorcher! Temperature soars into the 100s.'
c What temperature unit are they saying? What is its equivalent in the other unit?

23 There are many factors that determine how long it would take to climb a mountain.
a Name as many as you can.
In 1892 a Scottish mountaineer named William Naismith devised a rule which stated that you must allow 1 hour for every 5 km you walk forward, plus $\frac{1}{2}$ hour for every 300 metres of ascent.
Estimate, using Naismith's rule, how long it would take to
b walk 10 km with 900 m of ascent   c walk 20 km with 300 m of ascent
d walk 12 km with 1000 m of ascent   e walk 18 km with 450 m of ascent.

24 The formula $d = \frac{a + b}{3}$ can be used to work out the distance apart two bushes should be planted.
a Work out the value of d when $a = 50$ and $b = 43$.
b Work out the value of b when $d = 29$ and $a = 59$.

25 $F = 1.8C + 32$ is a formula which links temperatures in degrees Fahrenheit (F) with temperatures in degrees Celsius (C).
a Use the formula to convert:
    i 20°C   ii 45°C   iii 70°C into °F.
b Use the formula to convert:
    i 212°F   ii 122°F   iii 77°F into °C.

C

26 The width of a rectangle is x centimetres.
The length of the rectangle is $(x + 4)$ centimetres.
   a Find an expression, in terms of x, for the perimeter of the rectangle.
   Give your expression in its simplest form.
The perimeter of the rectangle is 54 centimetres.
   b Work out the length of the rectangle.

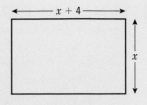

27 The diagram shows a trapezium.
All the lengths are in centimetres.
The perimeter of the trapezium is P cm.
Find a formula, in terms of a and b, for P.
Give your answer in its simplest form.

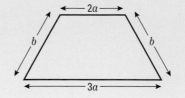

Diagram **NOT**
accurately drawn

28 $V = \frac{1}{3}\pi r^2 h$ is the formula for finding the volume of a cone.
Work out the volume of a cone with:
   a $r = 5$ cm and $h = 10$ cm   **b** $r = 7$ cm and $h = 15$ cm
   c $r = 4.6$ cm and $h = 9.2$ cm

29 **a** $S = 2p + 3q$
   Work out the value of S when $p = -4$ and $q = 5$.
   b $T = 2m + 20$
   Work out the value of m when $T = 30$.

30 Make b the subject of the formula $P = 2a + 2b$.

In questions 31–35 rearrange the formula to make the letter in square brackets the subject.

31 $P = 2x + y$ $[x]$

32 $s = \dfrac{a + b + c}{2}$ $[a]$

33 $T = \dfrac{D}{V}$ $[D]$

34 $A = \frac{1}{2}(a + b)h$ $[h]$

35 $y = \dfrac{5 - x}{2}$ $[x]$

Answers

Chapter 1 Answers

1.1 Get ready

1 508, 510
2 13
3 56

Exercise 1A

1 Students' numbers; the digits that must be as in the examples below are shown in **bold**.

	Ten thousand	Thousands	Hundreds	Tens	Units
a		**4**	1	2	3
b				**3**	9
c	8	9	**1**	5	5
d			5	3	**9**
e		2	4	**0**	8
f	6	7	**4**	5	5
g		**7**	7	3	**7**
h	3	**6**	5	9	**6**

2 For example:
Teacher on left: 51, 52, 53, 54, 55
Teacher in middle: 134, 154, 384, 494, 514
Teacher on right: 1200, 2235, 4289, 8276, 3244
3 **a** 60, sixty   **b** 600, six hundred
   c 60 000, sixty thousand
   d 6, six   **e** 6000, six thousand

1.2 Get ready

1 **a** Forty   **b** Four   **c** Four hundred

Exercise 1B

1 **a** 325   **b** 1718   **c** 6204   **d** 19 420
2 **a** two hundred and thirty-seven
   b three hundred and twenty-one
   c one thousand seven hundred and ninety-two
   d six thousand five hundred and two
   e one thousand and fifty-three
3 **a** 73, 179, 183, 190, 235   **b** 970, 2015, 2105, 2439, 2510
   c 2998, 3000, 3003, 3033, 30 300
   d 56 321, 56 745, 56 762, 59 342
4 **a** 69, 1010, 2306   **b** 76 152, 70 363, 151 400
5 **a** ten million four hundred and sixty-seven thousand five hundred and forty-two

b seven hundred and ninety-three thousand nine hundred and sixty-three
c one million three hundred and forty thousand four hundred and fifteen
d sixty-four million three hundred and fifty-one thousand
e ten million six hundred and twenty-seven thousand two hundred and fifty

6 **a**

Peugeot 505	seven thousand nine hundred and ninety-five pounds
Focus	eleven thousand four hundred and ninety-five pounds
Ka	four thousand eight hundred and thirty-five pounds
Mini	six thousand five hundred and forty-nine pounds
Sharan	thirteen thousand two hundred and five pounds

b

Sharan	£13 205
Focus	£11 495
Peugeot 505	£7995
Mini	£6549
Ka	£4835

1.3 Get ready

1 £3645, £4190, £5250, £5490
2 462 690, 10 348 276, 10 524 145, 40 280 780, 60 424 213
3 70 363, 76 152, 150 400

Exercise 1C

1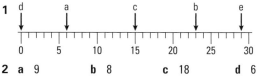
2 **a** 9   **b** 8   **c** 18   **d** 6
   e 25   **f** 7
3 **a** 4, increase   **b** 6, increase   **c** 7, decrease
   d 5, decrease   **e** 22, increase   **f** 17, decrease

1.4 Get ready

1 **a** 62   **b** 7   **c** 17

Exercise 1D

1 43 **2** 80   **3** 331   **4** 270 marks
5 45 **6** 96 fish   **7** 84 passengers   **8** 216 songs
9 40 **10** 416 people

Exercise 1E

1 305   **2** 3166   **3** 1003   **4** 12 CDs
5 **a** 19   **b** 17   **c** 36
6 **a** 88   **b** 59   **c** 133

1.5 Get ready

1 a 169   **b** 0

Exercise 1F

1 a i 270   **ii** 2700   **iii** 27 000
  b i 80   **ii** 800   **iii** 8000
  c i 3010   **ii** 30 100   **iii** 301 000
  d i 600   **ii** 6000   **iii** 60 000
  e i 50 200   **ii** 502 000   **iii** 5 020 000
2 a 700   **b** 5200   **c** 3660
  d 21 500   **e** 4600   **f** 642 000
3 a 455   **b** 159   **c** 1884
  d 4184   **e** 1185   **f** 1596
4 a 408   **b** 975   **c** 1749
  d 5024   **e** 24 581   **f** 14 144
  g 7738   **h** 1102
5 923   **6** 4494
7 782 miles   **8** 1449 supporters
9 1548 matches   **10** 384 tins

Exercise 1G

1 a 366   **b** 43   **c** 900   **d** 87
2 a 24   **b** 23   **c** 14   **d** 16
  e 160   **f** 113   **g** 18   **h** 89
  i 92   **j** 91   **k** 204   **l** 340
3 a 16   **b** 44   **c** 16   **d** 41
  e 21   **f** 34   **g** 40   **h** 300
4 a 21   **b** 15   **c** 40   **d** 20
  e 14
5 £480
6 a 10 finalists   **b** 30 finalists   **c** 36 finalists
7 a 7 trips   **b** 13 trips   **c** 36 trips
8 a 12 cases   **b** 41 cases   **c** 80 cases

1.6 Get ready

1 £38
2 £5.00

Exercise 1H

1 a 60   **b** 60   **c** 190
  d 190   **e** 990   **f** 2410
2 a 300   **b** 700   **c** 2400
  d 3100   **e** 8800   **f** 29 500
3 a 2000   **b** 36 000   **c** 29 000
  d 322 000   **e** 717 000   **f** 2 247 000
4

	Length (ft)	Cruising speed (mph)	Takeoff weight (lb)
Airbus A310	150	560	36 100
Boeing 737	90	580	130 000
Saab 2000	90	400	50 300
Dornier 228	50	270	12 600
Lockheed L1011	180	620	496 000

1.7 Get ready

1 266, 403, 557, 577, 615

Exercise 1I

1 a i 5, −10   **ii** −10, −3, 0, 4, 5
  b i 0, −13   **ii** −13, −9, −7, −2, 0
  c i 13, −15   **ii** −15, −6, −3, 6, 13
  d i −2, −21   **ii** −21, −20, −13, −5, −2
2 a 0, −1   **b** −2, −5   **c** 3, 7   **d** −7, −12
  e −15, −24   **f** −1, 2
3 a 7   **b** −3   **c** −1   **d** −5
  e −6   **f** 3   **g** 0   **h** −7
  i −5   **j** −2
4 a −40   **b** −70   **c** 30   **d** −30
  e −160   **f** 270   **g** −30   **h** 0
  i 120   **j** −400
5 a Minsk   **b** Minsk, 49°C   **c** Tripoli, 26°C
6 22°C
7 6°C

1.8 Get ready

1 Civetta
2 Val d'Isere

Exercise 1J

1 a 5°C   **b** 3°C   **c** 6°C   **d** 10°C
  e 10°C   **f** 8°C   **g** 5°C   **h** 13°C
2 a −2°C   **b** −3°C   **c** −7°C   **d** 3°C
  e 3°C   **f** −6°C   **g** −9°C

1.9 Get ready

1 a −2°C   **b** −10°C

Exercise 1K

1 a −7   **b** 4   **c** 10   **d** 9
  e −1   **f** 2   **g** −2   **h** −10
2 11 metres
3 −12°C

Exercise 1L

1 a −3   **b** −3   **c** 4
  d 12   **e** −4   **f** −12
2 a −90   **b** 4   **c** 5
  d −14   **e** −2   **f** 12
3 a −20   **b** 2   **c** −20
  d 6   **e** 9   **f** −42
4 a 24   **b** −15   **c** −4
  d 27   **e** −40   **f** 3
5 a 10   **b** −56   **c** 36
  d 21   **e** −3   **f** −42

1.10 Get ready

1 a 42   **b** 63   **c** 64   **d** 18

Exercise 1M

1 42, 18, 1110, 73 536, 500 000
2 105, 537, 811, 36 225
3 a 14, 16   **b** 30, 32   **c** 198, 200

4 a 3 **b** 29 **c** 199
5 5674, 6574, 7564, 5764, 6754, 7654, 4576, 4756, 5476, 5746, 7456, 7546
6 a 8732 **b** 2387
7 4, 8, 12, 12, 14, 14, 16, 16, 16, 22, 22, 42, 31, 23, 17, 17, 15, 15, 9, 9, 3, 3, 1

Exercise 1N

1 a 1, 3, 5, 15 **b** 1, 2, 4, 5, 10, 20
  c 1, 2, 3, 4, 6, 8, 12, 24 **d** 1, 2, 3, 6, 9, 18
  e 1, 13
  f 1, 2, 3, 5, 6, 9, 10, 15, 18, 30, 45, 90
2 a 2, 4, 6, 8, 10 **b** 5, 10, 15, 20, 25
  c 10, 20, 30, 40, 50 **d** 7, 14, 21, 28, 35
  e 13, 26, 39, 52, 65
3 Students' multiples of 10 greater than 50, e.g. 60, 70, 80
4 a 1, 4, 6, 8, 12 **b** 5, 20 **c** 1, 4, 8, 16 **d** 6, 9, 12
5 31, 37
6 97
7 5432
8 a 12, 18, 24, 27, 36, 54, 108, 216 – although a bunch of roses is unlikely to have more than 24 roses.
  b 14, 15, 21, 30, 42, 70, 105, 210 – although a bunch of roses is unlikely to have more than 21 roses.

Exercise 1O

1 a 1, 2 **b** 1, 5 **c** 1, 2, 3, 4, 6, 12
  d 1, 2, 5, 10 **e** 1, 2, 4, 8 **f** 1, 5
  g 1, 2, 4 **h** 1, 2, 3, 6 **i** 1
2 a 2, 3, 5 **b** 5 **c** 2, 3, 7
  d 3, 13 **e** 3, 5, 7
3 a $3 \times 3 \times 5$
  b $2 \times 2 \times 3 \times 3$
  c $2 \times 2 \times 7$
  d $2 \times 2 \times 2 \times 2 \times 5$
  e $2 \times 2 \times 2 \times 3 \times 3$

1.11 Get ready

1 a $3 \times 3 \times 11$
  b $2 \times 2 \times 3 \times 3 \times 3 \times 3$
  c $5 \times 5 \times 7$

Exercise 1P

1 a 4 **b** 3 **c** 6 **d** 2 **e** 7
2 a 12 **b** 12 **c** 60 **d** 144 **e** 850
3 a 36, 6 **b** 360, 60 **c** 168, 12
  d 910, 13 **e** 288, 24 **f** 120, 20
4 120 seconds or 2 minutes
5 20 seconds

1.12 Get ready

1 a 121 **b** 169 **c** 27

Exercise 1Q

1 a 9 **b** 36 **c** 100 **d** 4 **e** 400
2 a 8 **b** 125 **c** 343 **d** 8000 **e** 1728
3 a 25 **b** 216 **c** 81 **d** 27

Review exercise

1 For example, 4**2** 98**3**
2 a Fifty **b** Five thousand
  c Fifty thousand **d** Fifty **e** Five
3 a Three thousand seven hundred and twenty-three
  b One hundred and seven
  c Two thousand and seven
  d Fifteen thousand and seventy-one
4 a 21 231 **b** 507 **c** 70 203
5 a 2005 **b** 2001 **c** 2003 **d** 2005
6 £92 534
7 583 girls
8 a 6, 12 **b** 4, 16 **c** 3, 4, 6, 12 **d** 8, 27
9 a 5, 9, 27, 35, 37 **b** 9, 12, 27, 36 **c** 4, 8, 12, 16
  d 5, 37 **e** 4, 9, 16, 36 **f** 8, 27
10 a Any number using all the digits once only is a multiple of 3
  b 7416, 4716, 7164 or 1764

11

Number of packs	Number of cans
1	24
2	48
3	72
4	96
5	120

12 factor
13 a 8 **b** 7
14 15, 30, 45, 60, 75, 90
15 a 30 **b** 300 **c** 2000
  d 78 940 **e** 8 000 000 **f** 80
16 644
17 £103.74
18 £738
19 a 9°C **b** −5°C
20 a −4 **b** 2 **c** −15 **d** 10 **e** −2
21 a −21 **b** 20 **c** −8 **d** 5 **e** −7
22 For example, 9, 8 and 3
23 7 cm
24 77
25 £3.50
26 a −4°C **b** 7°C **c** Leeds
27 2 multiplied by another prime number will give an even number.
28 1000
29 a 48 **b** 17 **c** 13
30 63 (both 9)
31 a 152 **b** 7 coaches
32 For example, the carton could have dimensions 12 cm by 20 cm by 20 cm
33 5 times (every 12 seconds)
34 a $2 \times 2 \times 3 \times 3 \times 3$
  b 12
35 a i $2 \times 2 \times 3 \times 5$ **ii** $2 \times 2 \times 2 \times 2 \times 2 \times 3$
  b 12
  c 480
36 3 packs of doughnuts, 4 packs of cakes
37 a 5 **b** 3 **c** 3
38 120 000 miles

Chapter 2 Answers

2.1 Get ready

1 **a** 90 **b** 180 **c** 45 **d** 30

Exercise 2A

1 **a** 180° **b** 90° **c** 360°
2 **a** 90° **b** 135° **c** 90°
3 **a** 180° **b** 90° **c** 360°
4 NE
5 300°

2.2 Get ready

1 **a** 90° **b** 135° **c** 180°

Exercise 2B

1 **b**
2 **a** obtuse **b** acute **c** right angle
 d reflex **e** acute **f** obtuse
 g right angle **h** reflex **i** obtuse

2.3 Get ready

1 **c**
2 Students' acute angles
3 Obtuse angle

Exercise 2C

1 a = angle ABC b = angle ACB c = angle DFE
 d = angle EDF e = angle HGI f = angle HJI
 g = angle GIJ h = angle LKO i = angle MNO
 j = angle KLM k = angle QPS l = angle PQR
 m = angle PSR n − angle TVU o − angle TUV
2 **a** angle BCD, angle EAB
 b angle CDE **c** angle DEA
 d angle ABC **e** DC **f** AE
3 Students' drawings

2.4 Get ready

1 **a** 180° **b** 90° **c** 270°

Exercise 2D

1 60° 2 30°
3 155° 4 120°
5 45° 6 120°

2.5 Get ready

1 **a** About 60° **b** About 130° **c** About 40°

Exercise 2E

1 **a** 51° **b** 134°
2 **a** 112° **b** 61° **c** 115°
3 **a** 98° **b** 88° **c** 124°
4 Mild steel

2.6 Get ready

Students' drawings

Exercise 2F

1 **a**   **b**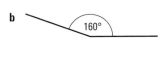
 160°

c   **d**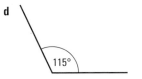
 55° 115°

e   **f**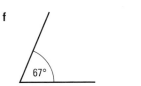
 43° 67°

g 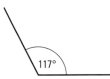 **h**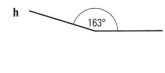
 117° 163°

i   **j**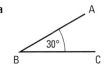
 17° 84°

2 **a** 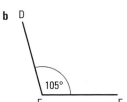 **b**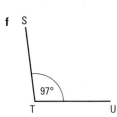

c 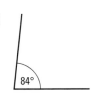 **d**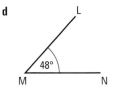

e   **f**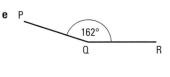

2.7 Get ready

1 Students' accurate drawings
2 60°, 60°, 5 cm
3 The sides and angles are equal.

Exercise 2G

1 a 70° b 20°
c 43° d 70°
e 46° f 71°
2 a 70° b 60° c 90°
3 a isosceles
b equilateral
c right-angled
4 55°
5 angle DEF = 25°, angle FDE = 130°

2.8 Get ready

1 a 140
b 113
c 48
2 a 313
b 252
c 113

Exercise 2H

1 $a = 60°$ $b = 85°$ $c = 90°$
$d = 100°$ $e = 49°$ $f = 34°$
2 $g = 32°$ $h = 109°$ $i = 30°$
$j = 47°$ $k = 133°$ $l = 133°$
$m = 119°$ $n = 61°$ $p = 61°$
3 $a = 62°$, angles on a straight line = 180°
$b = 169°$, angles around a point = 360°
$c = 38°$, vertically opposite angles
$d = 142°$, angles on a straight line = 180°
$e = 142°$, vertically opposite angles
$f = 127°$, angles on a straight line = 180°
$g = 53°$, vertically opposite angles
$h = 127°$, vertically opposite angles
$l = 64°$, angles around a point = 360°, $2l = 128°$
$m = 60°$, angles on a straight line = 180°, $3m = 180°$
$n = 45°$, angles on a straight line = 180°, $4n = 180°$
$o = 45°$, vertically opposite angles
$p = 135°$, vertically opposite angles
$q = 90°$, angles on a straight line = 180°
$r = 60°$, angles on a straight line = 180°, $3r = 180°$

Review exercise

1 7 cm
2 Acute angle
3

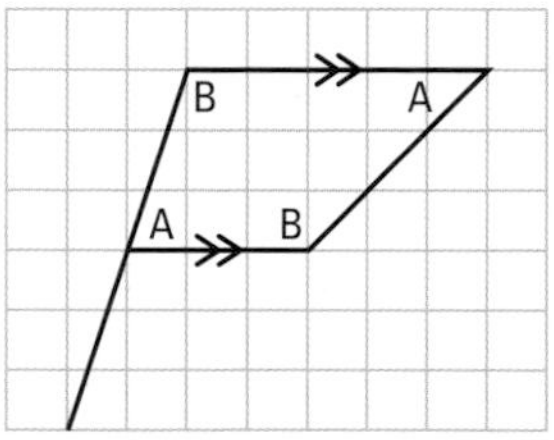

4

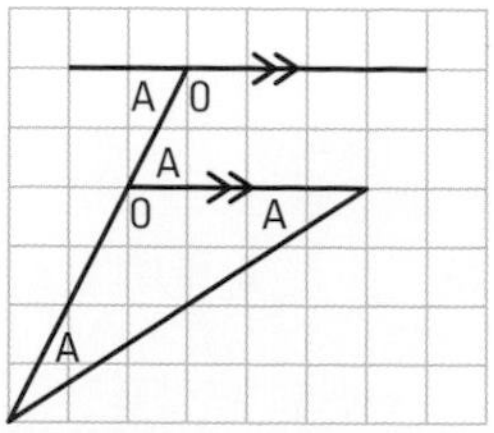

5 Students' drawings
6 a i 30° ii Vertically opposite angles
b Angles around a point should add up to 360°. These add up to 385°.
7 140°
8 a 4 cm b 30°
9 a 5.5 cm b 42°
10 a About 55°
b About 145°
11 Students' drawings
12 Angles on a straight line should add up to 180°. These add up to 170°.
13 a 69° b 52° c 70°

Chapter 3 Answers

3.1 Get ready

1 a Survey of classmates b e.g. internet
c e.g. internet

Exercise 3A

1 Specify the problem.
Decide what information to collect.
Collect the information.
Present and display the information.
Interpret the findings.
Act on the conclusions drawn from the findings.
2 a Type of house
b Garden size, Price
c Number of bedrooms
3 a Number of hours boys spend watching television
Number of hours girls spend watching television
b Continuous data

3.2 Get ready

1 a 3 b 6 c 10

Exercise 3B

1 a

Colour	Tally	Frequency
Silver	~~||||~~ ~~||||~~ |	11
White	|	1
Blue	~~||||~~	5
Black	||||	4
Red	~~||||~~ ||||	9
Total		30

b Silver

2 a

Age (years)	Tally	Frequency
13	||||	4
14	ⅢⅡ ||	7
15	ⅢⅡ ||||	9
16	ⅢⅡ ⅢⅡ	10
Total		30

b 11 members

3 a

Length (cm)	Tally	Frequency
5–6	ⅢⅡ |	6
7–8	ⅢⅡ |	6
9–10	ⅢⅡ |||	8
11–12	||||	4
Total		24

b 6 worms

3.3 Get ready

1 Tally chart

Exercise 3C

1 It does not allow for more than 4 times. The boxes overlap.
2 It does not give categories for the answer. Some people might not want to give their precise age.
3 It does not allow for 'No'.
  The question is biased, as it starts 'Do you agree …'.

3.4 Get ready

1 Depends on class size.
  For a class of 30 students it takes 450 seconds or 7.5 minutes.

Exercise 3D

1 The sample is too small.
  It may not include people who shop on other days.
  It may not include people who play sport on Saturday mornings.
2 The sample only includes students from one year.
3 The sample is small.
  The people may not read the magazine.

3.5 Get ready

1 A 12
  B 6

Exercise 3E

1

	Music	Drama	Total
Art	7	6	13
PE	5	12	17
Total	12	18	30

2

	Large	Small	Total
Vanilla	8	10	18
Chocolate	6	4	10
Total	14	14	28

3 a

	Junior	Senior	Family	Total
Full week	14	36	24	74
Weekends	28	56	20	104
Total	42	92	44	178

b 28 weekend junior members

4 a

	Pizza	Salad	Pasta	Total
Gateau	12	10	3	25
Ice cream	10	10	20	40
Fruit	4	2	1	7
Total	26	22	24	72

b 20 people
c 72 people

Exercise 3F

1 a 50.7 mm   **b** 1.2°C   **c** October
2 a 63 moons   **b** −200°C   **c** Mercury
  d Venus   **e** 4 planets
3 a 90 minutes
  b Newcastle
  c London to Birmingham
  d London to Inverness

Review exercise

1 a

Country	Tally	Frequency
Australia	|	1
France	||||	4
Italy	ⅢⅡ	5
Spain	|||	3
USA	ⅢⅡ ||	7

b 20

2 a

Length (cm)	Tally	Frequency
13					3
14								6
15						4
16					3
17				2

b 3

3 There is no set period of time and no option for 0 texts. The sample may not be suitable as it may not be representative of the whole population.

4 a 10 centime
b 1 dollar

5 a Volvo
b Mazda

6 i The colours of walls in each classroom
ii Number of school meals sold;
the cost of a school outing
iii Heights of students

7 The question doesn't specify a length of time and the options for the response boxes overlap. The sample is biased as her class may not be representative of the whole population.

8 The question doesn't specify a length of time. It is not clear what the options for the response boxes represent. Students' questions, for example:
How many times do you visit the cinema each month?
☐ Never ☐ 1 or 2 ☐ 3 or 4 ☐ 5 or more

9 How many times do you shop at the supermarket each month?
☐ Never ☐ 1 or 2 ☐ 3 or 4 ☐ 5 or more

10 How many emails do you send per week?
☐ None ☐ 1 to 3 ☐ 4 to 6 ☐ 7 or more

11 People leaving the cinema are more likely to be frequent cinema goers than the population in general.
She should also question men.

12

Preferred animal	Tally	Frequency
Lion		
Tiger		
Elephant		
Monkey		
Giraffe		

13

Country	Tally	Frequency
France	||||	5
Spain	||||| ||	7
Italy	||||	4
England	||||	4

14 a £2510
b £1500

15

	Small	Medium	Large	Total
Pine	7	12	4	23
Oak	10	16	8	34
Yew	3	8	2	13
Total	20	36	14	70

16 a Other classes may be voting differently.
b 80 votes

17 a 1400
b The 6 acres field may not be representative of the whole farm.

Chapter 4 Answers

4.1 Get ready

1 $5a$

Exercise 4A

1 $4a$
2 $6a$
3 $3p$
4 $5x$
5 $6j$

Exercise 4B

1 a $p + 3$ **b** $x + 4$ **c** $q - 5$ **d** $g - 5$
e $h + 4$ **f** $k - 6$ **g** $j - 6$ **h** $a + 3$
i $y - 4$ **j** $m - 3$ **k** $p + 6$ **l** $h + 7$
2 $c + 12$
3 $a - 3$
4 $d + 12$
5 $g - 7$
6 $x + y$

Exercise 4C

1 $4f + 12t$ **2** $4f + 10t$ **3** $4a + 9b$
4 $50g + 20s$ **5** $5p + 3m + 4n$ **6** $6x + 12y$
7 $12x + 6y$ **8** $90r + 10d$

4.2 Get ready

1 $s + b$ **2** $h - 12$ **3** $r + g$

Exercise 4D

1 a a, b **b** x, y **c** a, t
d x, y **e** t, d **f** a, s
g b **h** g **i** t
j a, b
2 a $3a, 4b$ **b** $x, 4y$ **c** $5a, 4t$
d x, y **e** $2t, 5d$ **f** $2a, 5s, 8$
g $4b, 6h$ **h** $9g, 6r, 4$ **i** $5t, 7s, 3$
j $2a, 5b$
3 a a **b** y **c** t
d x **e** d **f** a, s
g b **h** g **i** t
j a, b
4 Students' expressions, e.g. $3a + 4y$, $x + 9g + 6$
5 Students' terms, e.g. $8a, 7b, 5x$

4.3 Get ready

1 a 8 apples   **b** 5 bananas $-$ 1 pear
  c 6 apples $+$ 5 bananas

Exercise 4E

1	$5t$	**2**	$2c$	**3**	$4x$
4	$3a$	**5**	$6y$	**6**	$8a$

Exercise 4F

1	$5a$	**2**	$2p$	**3**	$8s$
4	$3x$	**5**	$6b$	**6**	$5k$
7	$9a$	**8**	$6x$	**9**	$9b$
10	p	**11**	$10n$	**12**	$4p$
13	$4x$				

Exercise 4G

1	$7a + 9b$	**2**	$9m + 7n$	**3**	$7p + 8q$
4	$2e + 2f$	**5**	$2g + 7h$	**6**	$2p + 5r$
7	$3j + 3k$	**8**	$6m + 8n$	**9**	$8a + 5b$
10	$3a + 3b$	**11**	$3m + 3n$	**12**	$7p + 2q$
13	$6e + f$	**14**	$2g + 3h$	**15**	$6p + r$
16	$9j + 3k$	**17**	$8n$	**18**	$3a + 2b$
19	$3p + 3j$	**20**	$4t$	**21**	$6x$
22	0	**23**	$4g$	**24**	m

Exercise 4H

1	$7a + 9$	**2**	$9m + 7$	**3**	$5p + 3q + 7$
4	$2e + 2$	**5**	$7h + 2$	**6**	$g + 4$
7	$3j + 3$	**8**	$6m + 1$	**9**	$8a + 5b + 1$
10	$3a + c + 3$	**11**	$6m + 5n + 9$	**12**	$3p + 6q + 2$
13	$4e + f + 2$	**14**	$3p + 4r + 4$	**15**	$p + r$

Exercise 4I

1	$2x^2$	**2**	$5y^2$	**3**	$2a^2$
4	$7a^2 + 9b^3$	**5**	$9m^2 + 7n$	**6**	$7p^3 + 2pq$
7	$4ef$	**8**	$2g^2 + 7h^3$	**9**	$2pq + 7r^3 - 2r^2$
10	$6jk$	**11**	$6m^3 + 8n$	**12**	$7a^2 + 4b^2$
13	$2a^3 + 3b^2$	**14**	$2m + 3n^2$	**15**	$7pq + 5p^3 - 3q^2$
16	$6e^3 + f^2$	**17**	$gh + 2h^3$	**18**	$7pqr$

Exercise 4J

1	$-3a$	**2**	$-2m + 5n$	**3**	$-3p - 4q$
4	$-5e - 1$	**5**	$2g - 3h$	**6**	$2p^2 - 5r^3$
7	$-3k$	**8**	$-m^3 - 8n$	**9**	$-2a - 5b$
10	$-ab$	**11**	$-2m - 2$	**12**	$-3p - 6$
13	$e - f$	**14**	$-2g^2 - 2$	**15**	$-4p + 2r + 2$
16	$-2j + 3k$	**17**	$-8n - 5$	**18**	$3a - 8b$
19	$-3p - 7j$	**20**	-8	**21**	$-4x + 1$
22	0	**23**	$-3g^3$	**24**	$-8mn + 2m^2$

4.4 Get ready

1 a 144   **b** 80   **c** 800

Exercise 4K

1	ab	**2**	xy	**3**	b^2
4	d^3	**5**	rst	**6**	abc
7	g^3	**8**	$2ef$	**9**	$3jk$
10	h^2	**11**	$5s^2$	**12**	$6t^3$
13	rt	**14**	xyt	**15**	$3mn$
16	$7abc$				

Exercise 4L

1	$6ab$	**2**	$20xy$	**3**	$6b^2$
4	$12d^3$	**5**	$35rs$	**6**	$12bc$
7	$15g^2$	**8**	$14ef$	**9**	$24jk$
10	$20h^2$	**11**	$25s^2$	**12**	$12t^3$
13	$12rt$	**14**	$35xy$	**15**	$18mn$
16	$30abc$	**17**	$4g^2$	**18**	$49h^2$
19	$8x^3$	**20**	$25n^2$	**21**	$30fgh$
22	$24jk$	**23**	$36hi$	**24**	$4a^2b$

4.5 Get ready

1 a 3   **b** $\frac{3}{2}$ **c** $\frac{3}{4}$

Exercise 4M

1	$4p$	**2**	p	**3**	4
4	$4n$	**5**	$2t$	**6**	15
7	$6k$	**8**	$2a$	**9**	$2x$
10	3	**11**	$5x$	**12**	8
13	$2p$	**14**	2	**15**	$4c$
16	2				

4.6 Get ready

1 $2a + 6$ **2** $3p + 6$

Exercise 4N

1	$2a + 8$	**2**	$3b + 6$	**3**	$4c + 24$
4	$5a - 20$	**5**	$3b - 15$	**6**	$5x + 15$
7	$2y - 4$	**8**	$6n + 12$	**9**	$15 + 3g$
10	$10 - 2x$	**11**	$6 - 3y$	**12**	$20 - 5h$
13	$10a + 50$	**14**	$3g + 21$	**15**	$4s - 20$
16	$21 - 3w$				

Exercise 4O

1	$6a + 8$	**2**	$15b + 12$	**3**	$20c + 24$
4	$6a - 15$	**5**	$15b - 21$	**6**	$10x + 25$
7	$6y - 8$	**8**	$12n + 42$	**9**	$15 + 9g$
10	$10 - 4x$	**11**	$6 - 15y$	**12**	$20 - 15h$
13	$40a + 30$	**14**	$15g + 21$	**15**	$12s - 20$
16	$21 - 12w$				

Exercise 4P

1	$a^2 + 4a$	**2**	$b^2 + 2b$	**3**	$ac + 6a$
4	$2a^2 - 4a$	**5**	$b^2 - 5b$	**6**	$x^2 + 3x$
7	$2y^2 - 2y$	**8**	$3n^2 + 2n$	**9**	$5g + g^2$
10	$5x - 2x^2$	**11**	$2y - 3y^2$	**12**	$4h - 5h^2$
13	$2a^2 + 2ab$	**14**	$2g^2 + 14g$	**15**	$4s^2 + 4st$
16	$21w - 3w^2$	**17**	$10p^2 + 15p$	**18**	$15x^2 - 5xy$
19	$2gh + 6h^2$	**20**	$20p - 10p^2$		

4.7 Get ready

1 a $6p + 38$ **b** $26 - 14s$ **c** $55 + 5g$

Answers

Exercise 4Q

1	$2(a + 3)$	**2**	$2(n + 4)$	**3**	$2(a - 6)$
4	$3(k + 2)$	**5**	$3(f - 3)$	**6**	$5(p - 2)$
7	$5(r + 4)$	**8**	$3(x - 4)$	**9**	$7(w + 2)$
10	$3(m - 5)$	**11**	$4(q + 2)$	**12**	$2(s + 1)$
13	$5(a - 5)$	**14**	$6(x + 5)$	**15**	$8(p - 5)$
16	$5(y - 1)$				

Exercise 4R

1	$a(a + 2)$	**2**	$a(a + 8)$	**3**	$y(y^2 + 2)$
4	$j(j^2 - 3)$	**5**	$s(s - 9)$	**6**	$x(x^2 - 5)$
7	$p(p + 6)$	**8**	$a(a - 1)$	**9**	$p(p^2 + 1)$
10	$m^2(m - 1)$	**11**	$c(c^2 + 8)$	**12**	$a(2 + a)$
13	$x^2(x - 2)$	**14**	$x(x + 7)$	**15**	$p(p^2 - 1)$
16	$y(y - 5)$				

Exercise 4S

1	$2a(3a + 1)$	**2**	$3a(a + 3)$	**3**	$2y(2y^2 + 1)$
4	$3j(2j^2 - 1)$	**5**	$3s(s - 3)$	**6**	$5x(2x^2 - 1)$
7	$3p(p + 2)$	**8**	$2a(2a - 1)$	**9**	$5p(p^2 + 2)$
10	$3m^2(2m - 1)$	**11**	$4c(c^2 + 2)$	**12**	$6a(2 + a)$
13	$2x^2(3x - 1)$	**14**	$5x(x + 6)$	**15**	$4p(2p^2 - 1)$
16	$5y(5y - 1)$				

4.8 Get ready

1 Students' expressions

Exercise 4T

1	Formula	**2**	Expression
3	Equation	**4**	Equation
5	Formula	**6**	Equation
7	Expression	**8**	Formula
9	Formula	**10**	Formula

4.9 Get ready

1 a 11   **b** 1   **c** 14

Exercise 4U

1	4	**2**	15	**3**	12
4	3	**5**	6	**6**	15
7	20	**8**	10	**9**	9
10	18	**11**	13	**12**	8
13	19	**14**	7	**15**	10
16	6	**17**	27	**18**	10
19	0	**20**	3		

Exercise 4V

1	10	**2**	−6	**3**	12
4	7	**5**	−6	**6**	15
7	20	**8**	−5	**9**	28
10	11	**11**	7	**12**	−7
13	31	**14**	8	**15**	20
16	−11	**17**	16	**18**	6
19	24	**20**	5		

Exercise 4W

1	14	**2**	24	**3**	20
4	15	**5**	36	**6**	40
7	−8	**8**	−5	**9**	24
10	78	**11**	5	**12**	22
13	35	**14**	2	**15**	76
16	4	**17**	102	**18**	20
19	−36	**20**	−20		

Review exercise

1 If Luke has x pounds, the total is $4x + 9$ pounds.

2 a

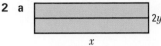

b

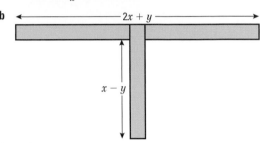

3 $8a + 4x$

4 a $10x - 20y$ **b** 210

5 a $3c$ **b** $3e + 2f$ **c** $5a$
   d $4xy$ **e** $2a + 7b + 8$

6 a $3bc$ **b** $2x + 5y$ **c** m^3 **d** $6np$

7 $x(x + 4)$

8 $12a - 28$

Chapter 5 Answers

5.1 Get ready

1 a two hundred and seventy three
   b four thousand and seventy six
   c three thousand seven hundred and fifty three

Exercise 5A

1

	Hundreds	Tens	Units	•	Tenths	Hundredths	Thousandths
a		4	1	•	6		
b			4	•	1	6	
c	7	3	4	•	6		
d			1	•	4	6	3
e			0	•	6	4	3
f			1	•	0	0	5
g			5	•	0	1	
h			0	•	0	8	6

2 a 5 units   **b** 5 tenths   **c** 4 tenths
   d 4 hundredths   **e** 1 ten   **f** 9 thousandths
   g 3 hundreds   **h** 7 ten-thousandths
   i 0 hundredths   **j** 3 tenths   **k** 1 thousandth
   l 2 hundredths

5.2 Get ready

1 a One tenth   **b** One hundredth
   c One   **d** Ten   **e** One thousandth

Exercise 5B

1 £1.09, £1.13, £1.18, £1.20, £1.29, £1.31
2 a 0.9, 0.76, 0.71, 0.68, 0.62
   b 3.75, 3.4, 3.12, 2.13, 2.09
   c 0.42, 0.407, 0.3, 0.09, 0.065
   d 6.52, 6.08, 3.7, 3.58, 3.0
   e 0.13, 0.105, 0.06, 0.024, 0.009
   f 2.2, 2.09, 1.3, 1.16, 1.1087, 1.08

3

Rascini	52.037 seconds
Killim	53.027 seconds
Ascarina	53.072 seconds
Bertollini	53.207 seconds
Silverman	53.702 seconds
Alloway	54.320 seconds

5.3 Get ready

1 a 61
   b 45

Exercise 5C

1 6.1		**2** 3.25		**3** 68.9
4 26.02		**5** 1.0		**6** 18.725
7 19.8		**8** 11.001		**9** 1.914
10 118.17		**11** 31.97		**12** 28.71
13 19.122		**14** 18.326		**15** 11.064
16 31.006		**17** 15.0976		**18** 178.585

Exercise 5D

1 a £6.20   **b** £4.14   **c** £14.10   **d** £97.30
   e £0.11   **f** £0.35   **g** £6.19   **h** £11.14
   i £0.90   **j** £4.07   **k** £14.03   **l** £13.25
2 a 1.225   **b** 11.649   **c** 2.254   **d** 168.58

5.4 Get ready

1 a 1856   **b** 21 425   **c** 2856

Exercise 5E

1 a £13.50   **b** £5.48   **c** £5.20   **d** £1.20
2 a 4.5   **b** 45   **c** 450
   d 2.03   **e** 20.3   **f** 203
3 a 30.4   **b** 3.04   **c** 0.304
   d 11.25   **e** 1.125   **f** 0.1125
   g 1.125   **h** 0.01125   **i** 0.001 125

4 a 64.2   **b** 642   **c** 6.42
   d 562.3   **e** 56.23   **f** 0.5623
5 a 172.2 kg   **b** 0.0945 seconds   **c** 0.0144 m
   d 0.04 miles   **e** 0.9 litres   **f** 0.0012 hours
6 a £116.25   **b** £167.40   **c** £255.75
7 a £117.75   **b** £196.25   **c** £337.55
8 a 68.25 litres   **b** 113.75 litres   **c** 295.75 litres

5.5 Get ready

1 a 9   **b** 36   **c** 121   **d** 64

Exercise 5F

1 a 9.61   **b** 17.64   **c** 28.09   **d** 4.1209   **e** 0.16
2 a 3.375   **b** 15.625   **c** 32.768   **d** 0.008   **e** 0.125

5.6 Get ready

1 a 41   **b** 65   **c** 6

Exercise 5G

1 a 3.45   **b** 0.345   **c** 0.0345
   d 207.1   **e** 0.2701   **f** 0.027 01
   g 6.5   **h** 0.65   **i** 0.065
2 a 16.12   **b** 0.633   **c** 14.84
   d 34.221   **e** 5.027   **f** 0.0046
3 a 7.5   **b** 5.75   **c** 1.125   **d** 1.75
   e 1.2   **f** 0.85   **g** 6.2   **h** 1.8
4 £15.40
5 0.625 kg or 625 g
6 a 15.5   **b** 13.2   **c** 21   **d** 3.5
   e 1.07   **f** 6.2   **g** 41.4   **h** 32

5.7 Get ready

1 4
2 5
3 a 56   **b** 3

Exercise 5H

1 a 8   **b** 13   **c** 14   **d** 6
   e 11   **f** 20   **g** 1   **h** 20
   i 1   **j** 100   **k** 20   **l** 2
2 a 3.6   **b** 5.3   **c** 0.1   **d** 9.3
   e 10.7   **f** 8.0   **g** 2.1   **h** 0.5
   i 2.5   **j** 125.7   **k** 0.1   **l** 9.9
3 a 14 mm   **b** 80 m   **c** 1 kg   **d** £204
   e 4 lb   **f** 0 tonne   **g** 11 g   **h** 8 min

Exercise 5I

1 a i 4.226   **ii** 4.23   **b i** 9.787   **ii** 9.79
   c i 0.416   **ii** 0.42   **d i** 0.058   **ii** 0.06
2 a i 10.517   **ii** 10.52   **b i** 7.503   **ii** 7.50
   c i 21.730   **ii** 21.73   **d i** 9.089   **ii** 9.09
3 a i 15.598   **ii** 15.60   **b i** 0.408   **ii** 0.41
   c i 7.247   **ii** 7.25   **d i** 6.051   **ii** 6.05
4 a i 29.158 cm   **ii** 29.16 cm   **b i** 0.055 kg   **ii** 0.05 kg
   c i 13.379 km   **ii** 13.38 km   **d i** £5.998   **ii** £6.00
5 a 5.617   **b** 0.0   **c** 0.9240   **d** 0.9
   e 9.7   **f** 1.01

Answers

5.8 Get ready

1 **a** 20   **b** 50   **c** 200

Exercise 5J

1 **a** 40   **b** 700   **c** 300   **d** 0.3
   e 20 000   **f** 0.007   **g** 1000   **h** 5
   i 10   **j** 20
2 100 medals
3 30 000 spectators

5.9 Get ready

1 Manchester United 80 000
   Chelsea 40 000
   Real Madrid 80 000
   Newcastle United 50 000
   Liverpool 40 000
   Barcelona 70 000

Exercise 5K

1 **a** **i** 0.062   **ii** 0.0618   **b** **i** 0.16   **ii** 0.165
   c **i** 96   **ii** 96.3   **d** **i** 41   **ii** 41.5
2 **a** **i** 730   **ii** 735   **b** **i** 0.079   **ii** 0.0795
   c **i** 5.7   **ii** 5.69   **d** **i** 590   **ii** 586
3 **a** **i** 0.015   **ii** 0.0148   **b** **i** 2200   **ii** 2220
   c **i** 76   **ii** 76.2   **d** **i** 0.38   **ii** 0.380
4 **a** **i** 8.4   **ii** 8.38   **b** **i** 36   **ii** 36.0
   c **i** 190   **ii** 187   **d** **i** 0.067   **ii** 0.0666
5 **a** **i** 220 000   **ii** 219 000   **b** **i** 4 000 000   **ii** 3 990 000
   c **i** 310 000   **ii** 307 000   **d** **i** 26 000   **ii** 25 600
6 £1.61
7 83.3 seconds

5.10 Get ready

1 600   2 1200   3 0.8

Exercise 5L

1 **a** $70 \times \frac{60}{30} = 140$

   b $\frac{200 \times 300}{200} = 300$

   c $\frac{90 \times 30 \times 100}{300} = 90$

   d $\frac{200}{10 \times 100} = 0.2$

   e $\frac{500}{10 \times 50} = 1$

   f $\frac{100 \times 90}{20 \times 30} = 15$

2 $50 \times 100 = 5000$ seats
3 $70 \times 50 = 3500$ tins
4 $30 \times £5 = £150$
5 **a** $20 \times \frac{0.2}{4} = 1$

   b $6 \times \frac{30}{1} = 180$

   c $\frac{900}{20} \times 0.5 = 22.5$

5.11 Get ready

1 **a** 60   **b** 600   **c** 6000
2 **a** 30   **b** 3   **c** 0.3

Exercise 5M

1 **a** 1792   **b** 1792   **c** 17.92
2 **a** 146.4   **b** 1.464   **c** 0.1464
3 **a** 726   **b** 0.726   **c** 7.26
4 **a** 0.64   **b** 0.64   **c** 64

Review exercise

1 **a** 4.09, 4.85, 5.16, 5.23, 5.9
   b 0.021, 0.07, 0.34, 0.37, 0.4
   c 5, 5.007, 5.01, 7.07, 7.23
   d 0.06, 0.23, 1.001, 1.08, 1.14
2 Thiamin   0.0014 g
   Riboflavin   0.0015 g
   Vitamin B6   0.002 g
   Iron   0.014 g
   Sodium   0.02 g
   Fibre   1.5 g
3 **a** 8.5   **b** 18.57
   c 0.1172   **d** 11.1432
   e 58.4   **f** 15.58
   g 46.4   **h** 0.0118
4 Yes, it weighs 19.9 kg.
5 211.6 kg
6 **a** 11.7   **b** 1.44   **c** 0.12
   d 40.96   **e** 5.1   **f** 20.35
7 360 000
8 1.35 m
9 **a** 1000   **b** 1   **c** 1000   **d** 0.1
10 **a** £2.45
   b £2.85
   c £2.45
11 0.85 kg
12 146.2 km
13 She buys 13 bottles and has 16p left.
14 **a** 57   **b** 0.103   **c** 600
15 **a** 23.5   **b** 1.8   **c** 0.3   **d** 150.0
16 **a** 7.26   **b** 73.04   **c** 0.042   **d** 0.721
17 **a** 8300   **b** 20 100   **c** 0.5   **d** 20.9
18 $40 \times 8 = £320$
19 $6 \div 20 = £0.30$ or 30p
20 **a** £153.90
   b £59.85
21 $30 \times 6 = £180$
22 **a** $(800 \times 5000)/3000 = 1333$
   b $(4 \times 5)/10 = 2$
   c $(2 \times 8)/(4 \times 2) = 2$
23 **a** 1632   **b** 16.32   **c** 3.4
24 **a** 15.456   **b** 0.15456   **c** 3220
25 $2000 \div 50 \div 0.4 = 100$ miles
26 £2.81
27 **a** Energy consumption is less for A (0.95 kWh/cycle).
   b 160 kWh, £19.20
   c Yes, because a washing machine should last at least 3 years.

Chapter 6 Answers

6.1 Get ready

1 **a** right-angled
   b obtuse
   c acute

Exercise 6A

1 **A** and **1** and **3**, **B** and **4** and **6**, **C** and **2** and **6**, **D** and **2** and **3**,
   E and **2** and **5**, **F** and **5** and **1**
2 **a** **B** and **D**   **b** **A** and **B**
   c **C** and **E**   **d** **C, D** and **E**
3 Students' sketch of an obtuse-angled triangle that is also
   isosceles.
4 This is not an isosceles triangle as all of the lengths and
   angles are different. An isosceles triangle has 2 equal
   angles and 2 sides of the same length.
5 Gerry is correct. Because all of the angles are 60°, they are
   therefore all less than 90°.

6.2 Get ready

1   **2**   **3**

Exercise 6B

1 **A** and **5**   **B** and **2**   **C** and **4**   **D** and **1**
   E and **2**   **F** and **3**   **G** and **4**   **H** and **6**
2 **a** **A** = Parallelogram   **B** = Trapezium   **C** = Square
        D = Rhombus   **E** = Kite   **F** = Rectangle
        G = Square
   b **A, C, D, F** and **G**   **c** **E**
3 **a** Square
   b Octagon
   c Trapezium
   d Right-angled triangle
4 **a** True   **b** False   **c** True   **d** True
5 **a** FG   **b** Yes   **c** Angle EHG   **d** Yes
   e Isosceles triangle
6 Diagram **b**
7 Rhombus and kite

6.3 Get ready

1 **a** and **d**

Exercise 6C

1 **A** and **I**, **B** and **F**, **D** and **G**.
2 **a** **i** **B** and **D**   **ii** **A** and **C**
   b **i** **A** and **D**   **ii** All
   c **i** **B** and **C**   **ii** All
3

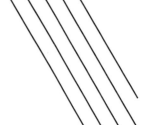

6.4 Get Ready

1 **a**
   b
   c
   d
   e

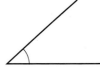

2 **a**   **b**

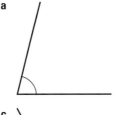

c   **d**

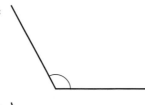

e

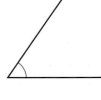

Answers

Exercise 6D

1 Students' accurate drawings

Exercise 6E

1 Students' accurate drawings
2 a Students' accurate drawing of quadrilateral
  b 6.2 cm   **c** 68°
3 a

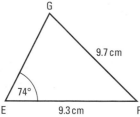

b

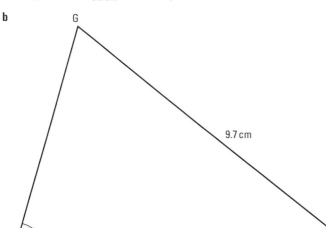

  c 6.3 cm
4 The two sides do not meet as the largest side
  (10.6 cm) is bigger than the sum of the other
  two sides (4.3 cm + 5.1 cm = 9.4 cm)
5 Students' drawings of parallel lines

Exercise 6F

1 diameter   **2** tangent   **3** chord
4 radius   **5** segment   **6** sector

6.6 Get ready

1 a   **b**   **c**

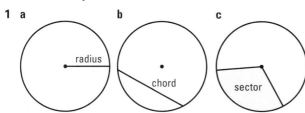

Exercise 6G

Students' accurate drawings

6.7 Get ready

1 a Yes   **b** No   **c** No

Exercise 6H

1 a   **b**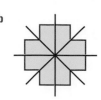

  c 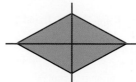   **d**

  e 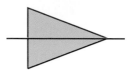   **f**

2 a Yes, 1 line
  b No
  c Yes, 1 line
  d Yes, 6 lines
  e Yes, 2 lines
  f Yes, 4 lines

3 a 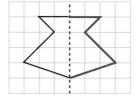   **b**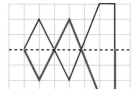

  c

4 a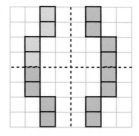

b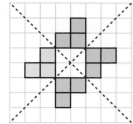

6.8 Get ready

1 a 360   **b** 90   **c** 180

Exercise 6I

1 a Yes, order 2   **b** No
  c Yes, order 4   **d** Yes, order 2
2 a 5   **b** 2
  c 8   **d** 4

3

4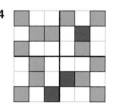

Exercise 6J

1

Shape	Name of shape	Number of lines of symmetry	Order of rotational symmetry
	Rectangle	2	2
	Equilateral triangle	3	3
	Rhombus	2	2
	Regular hexagon	6	6
	Parallelogram	0	2

Review exercise

  1 Isosceles triangle
  2 a A and C
    b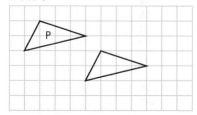

3

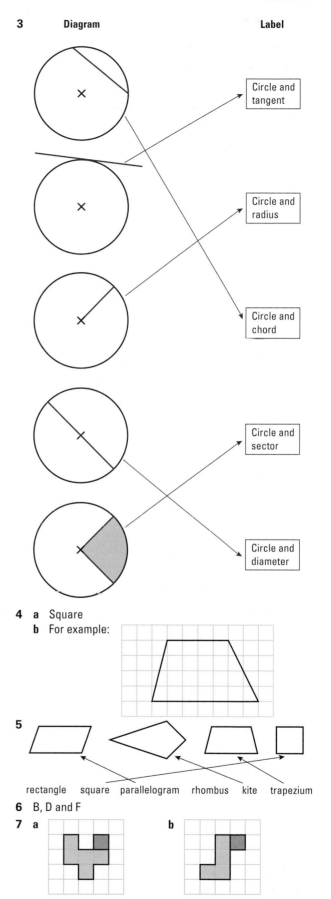

  4 a Square
    b For example:

  5

rectangle   square   parallelogram   rhombus   kite   trapezium

  6 B, D and F
  7 a   **b**

8 **i** D
   ii B
   iii A

9 **a**

Mirror line

  b

10

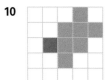

11 **a**

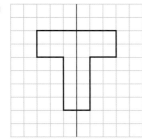

  b 2

12 **a**   **b**

13

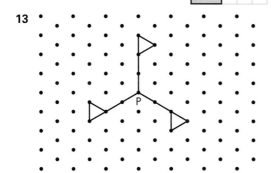

14 **a**

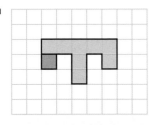

b

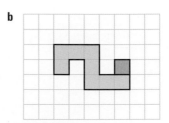

15 **a** A and D
   b B and C

16

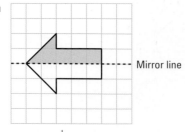

17 Students' accurate drawings
18 Students' accurate drawings
19 Students' accurate drawings

Chapter 7 Answers

7.1 Get ready

1 **a** $a = 113°$ **b** $b = 63°, c = 117°, d = 63°$ **c** $e = 58°$

Exercise 7A

1 110° **2** 50° **3** 90°
4 144° **5** 98°

7.2 Get ready

1 Equilateral triangle
2 A square is a quadrilateral with 4 <u>equal</u> sides and 4 <u>equal</u> angles.

Exercise 7B

1 **a** pentagon **b** hexagon **c** octagon
2 **a** 720° **b** 1080° **c** 2340° **d** 3240°
3 **a** 7 **b** 11 **c** 22

7.3 Get ready

1 $x = 53°$
2 $x = 152$
3 $x = 80$

Exercise 7C

1 $a = 102°$ $b = 55°$ $c = 93°$ $d = 85°$
   $e = 95°$ $f = 138°$ $g = 42°$
2 **a** 72° **b** 120° **c** 18°
3 **a** 108° **b** 60° **c** 162°
4 30 sides
5 36°, 10 sides

7.4 Get ready

1 **a** yes **b** yes **c** no

Exercise 7D

1 a

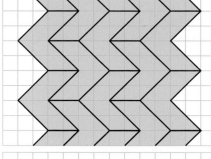

or

b

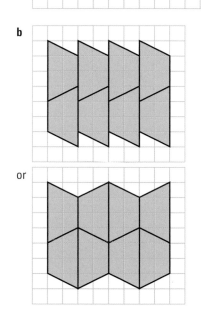

or

c

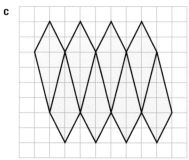

d

or

e

or

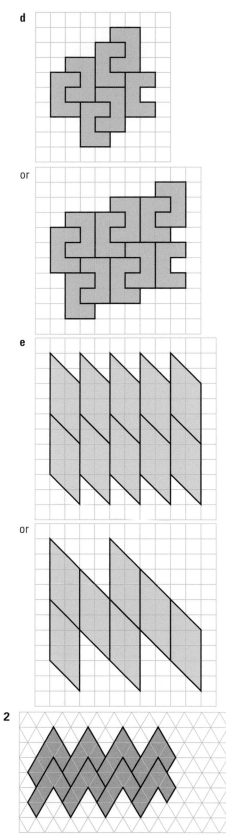

2

Exercise 7E

1 a A,B; C,E; D,F; G,H; I,J; K,L   **b** A,G; A,H; B,G; B,H; C,K;
                                              C,L; E,K; E,L

2 Students' drawings

7.6 Get ready

1 127° (angles on a straight line)
2 59° (angles around a point)
3 38° (vertically opposite)

Exercise 7F

1 **a** *a* and *d*, *b* and *e*   **b** *b* and *d*, *c* and *f*
2 **a** *a* = 25° corresponding
   b *b* = 110° alternate
   c *a* = 111° corresponding
        b = 111° vertically opposite/alternate
   d *a* = 148° corresponding
        b = 32° corresponding
        c = 148° angles on straight line/alternate/vertically opposite
        d = 32° alternate/angles on straight line/vertically opposite
   e *a* = 61° alternate
        b = 119° angles on a straight line
        c = 61° corresponding/vertically opposite
        d = 119° vertically opposite
   f *a* = 113° angles on a straight line
        b = 113° alternate
        c = 67° angles on a straight line
3 **a** *a* = 125°   *b* = 55°   *c* = 125°   *d* = 55°
   b *e* = 108°   *f* = 72°   *g* = 72°   *h* = 108°

7.7 Get ready

1 *x* = 40°
2 *a* = 27° (corresponding) *b* = 27° (vertically opposite/alternate)
3 ADC = 81°   BCD and BAD = 99°

7.8 Get ready

1
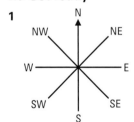

2 *a* = 24°
3 Students' drawings
4 *a* = 68°   *b* = 112°

Exercise 7G

1 **a** East   **b** North-east   **c** North-west
2 **a** 072°   **b** 225°   **c** 133°
   d 307°   **e** 220°   **f** 087°
3 **a** 080°   **b** 111°   **c** 291°

Exercise 7H

1 **a** 124°   **b** 320°   **c** 230°
2 Students' drawings
3 **a** 293°   **b** 075°   **c** 203°
4 **a** 054°   **b** 190°   **c** 263°

5
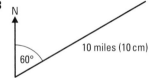

6 235°
7 348°
8 123°
9 295°
10 177°

7.9 Get ready

1 **a** 5600 cm   **b** 560 cm
2 **a** 12 : 24   **b** 27 : 9
   c 12 : 24   **d** 24 : 12

Exercise 7I

1 11 km
2 **a** 1.4 km
   b 17.4 cm

3
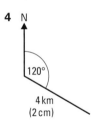

4
(includes: N, 120°, 4 km (2 cm))

5 **a**
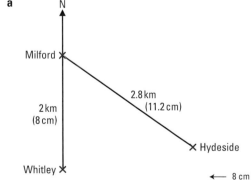

   b **i** 2.3 km
        ii 80°
6 19.9 km, 344°
7 108 yards, 017°
8 024°, 124 km
9 Students' accurate drawings

Review exercise

1 **a** For example:

P

   b For example:

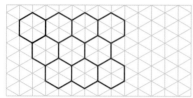

L

2 Students' drawings
3 110°
4 60°
5 **a** **i** 25°
      ii 130°
   b 65°
6 122°
7

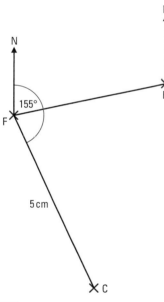

8 **a** 078°
   b

9 **a** 30°
   b 48°
10 2.25 cm
11 1.46 m
12 **i** 127°
    ii Alternate angles
13 **i** 58°
    ii Alternate angles
14 **i** 120°
    ii Corresponding angles
15 *x* is 130° because angles on a straight line add up to 180°.
    y is 50° because it is a corresponding angle to angle PBC.

16 **a** Hexagon
    b **i** 120°
        ii Angles on a straight line add up to 180°.
17 **a** **i** 63°
        ii Angles in a triangle add up to 180°. Angles ABC
            and ACB are the two equal angles of the isosceles
            triangle.
    b 117°
18 **a** 330°
    b

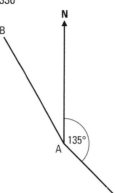

19 **a** 60°
    b 120°
20 **a** *a* = 35° (angles on a straight line add up to 180°,
            corresponding angles)
        b = 35° (vertically opposite angles)
    b *a* = 123° (alternate angles)
        b = 57° (angles on a straight line add up to 180°)
21 **a** 130°
    b 250 km
    c

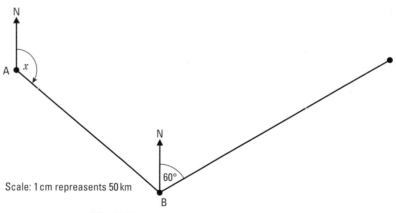

Scale: 1 cm repreasents 50 km

22 144°
23 You can draw parallel lines around any triangle:

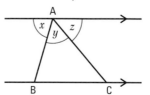

Angle ABC = *x* (alternate angles)
Angle ACB = *z* (alternate angles)
So the angle sum of the triangle is *x* + *y* + *z*.
But *x* + *y* + *z* = 180° (angles on a straight line).
So the angle sum of the triangle is also 180°.

Answers

Chapter 8 Answers

8.1 Get ready

1 a 4   **b** 20   **c** 7   **d** 6

Exercise 8A

1

Shape	Fraction shaded	Fraction not shaded
circle	$\frac{1}{2}$	$\frac{1}{2}$
square grid	$\frac{1}{4}$	$\frac{3}{4}$
pentagon	$\frac{2}{5}$	$\frac{3}{5}$
rectangle grid	$\frac{3}{10}$	$\frac{7}{10}$
octagon	$\frac{5}{8}$	$\frac{3}{8}$
triangle	$\frac{4}{9}$	$\frac{5}{9}$

2 a Any single box shaded, e.g.   **b** Any 3 boxes shaded, e.g.

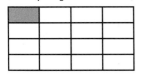 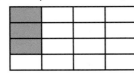

c Any 8 boxes shaded, e.g.   **d**

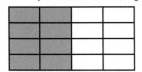 

3 a Any single sector shaded, e.g.

b Any 3 sectors shaded, e.g.

c Any 3 sectors shaded, e.g.

4 a $\frac{15}{28}$ **b** $\frac{13}{28}$

5 a 75 surfers   **b** $\frac{47}{75}$ **c** $\frac{28}{75}$

8.2 Get ready

1 a 8   **b** 9   **c** 6

Exercise 8B

1 Students' fractions, e.g.

   a e.g. $\frac{6}{8}, \frac{3}{4}$ **b** e.g. $\frac{2}{6}, \frac{1}{3}$ **c** e.g. $\frac{4}{8}, \frac{1}{2}$

   d e.g. $\frac{4}{8}, \frac{2}{4}, \frac{1}{2}$

2 a $\frac{3}{4} = \frac{6}{8} = \frac{9}{12} = \frac{12}{16} = \frac{15}{20} = \frac{18}{24}$

   b $\frac{2}{7} = \frac{4}{14} = \frac{6}{21} = \frac{8}{28} = \frac{10}{35} = \frac{12}{42}$

   c $\frac{4}{5} = \frac{8}{10} = \frac{12}{15} = \frac{16}{20} = \frac{20}{25} = \frac{24}{30}$

   d $\frac{1}{3} = \frac{3}{9} = \frac{6}{18} = \frac{9}{27} = \frac{12}{36} = \frac{15}{45}$

3 a $\frac{3}{18}$ **b** $\frac{6}{14}$ **c** $\frac{18}{48}$ **d** $\frac{12}{21}$

   e $\frac{30}{36}$ **f** $\frac{6}{9}$ **g** $\frac{24}{54}$ **h** $\frac{40}{56}$

   i $\frac{90}{100}$ **j** $\frac{84}{144}$ **k** $\frac{49}{56}$ **l** $\frac{18}{81}$

4 a $\frac{3}{6}, \frac{2}{6}$

   b **i** $\frac{12}{30}, \frac{15}{30}$ **ii** $\frac{7}{70}, \frac{10}{70}$ **iii** $\frac{3}{12}, \frac{10}{12}$

     iv $\frac{5}{10}, \frac{6}{10}$ **v** $\frac{16}{24}, \frac{3}{24}$ **vi** $\frac{15}{20}, \frac{12}{20}$

8.3 Get ready

1 a 20   **b** 70   **c** 12

Exercise 8C

1 a $\frac{8}{20}, \frac{5}{20} . \frac{1}{4}$ **b** $\frac{10}{20}, \frac{16}{20} . \frac{2}{4}$

   c $\frac{8}{12}, \frac{9}{12} . \frac{2}{3}$ **d** $\frac{6}{10}, \frac{7}{10} . \frac{3}{5}$

2 a $\frac{3}{6}$ **b** $\frac{1}{7}$ **c** $\frac{5}{6}$

   d $\frac{3}{5}$ **e** $\frac{2}{3}$ **f** $\frac{3}{4}$

3 a $\frac{1}{2}, \frac{2}{3}, \frac{3}{4}$ **b** $\frac{7}{15}, \frac{4}{5}, \frac{5}{6}$

   c $\frac{1}{2}, \frac{3}{4}, \frac{4}{5}$ **d** $\frac{5}{14}, \frac{3}{7}, \frac{1}{2}, \frac{4}{7}$

4 $\frac{7}{8}, \frac{3}{4}, \frac{1}{2}, \frac{2}{5}, \frac{2}{10}$

8.4 Get ready

1 a $5\frac{3}{4}$ **b** 18   **c** 35

Exercise 8D

1 a $2\frac{1}{2}$ **b** $1\frac{3}{4}$ **c** $1\frac{2}{7}$ **d** $1\frac{3}{8}$

   e $1\frac{1}{8}$ **f** $3\frac{1}{5}$ **g** $2\frac{3}{10}$ **h** $4\frac{4}{5}$

   i $2\frac{2}{7}$ **j** $2\frac{2}{5}$ **k** $6\frac{2}{3}$ **l** $1\frac{7}{9}$

   m $9\frac{3}{4}$ **n** $5\frac{2}{5}$ **o** $2\frac{8}{9}$ **p** $1\frac{7}{10}$

2 a $\frac{3}{2}$ **b** $\frac{11}{2}$ **c** $\frac{11}{4}$ **d** $\frac{5}{3}$

   e $\frac{13}{4}$ **f** $\frac{22}{5}$ **g** $\frac{37}{10}$ **h** $\frac{26}{5}$

   i $\frac{31}{4}$ **j** $\frac{9}{4}$ **k** $\frac{19}{10}$ **l** $\frac{28}{3}$

   m $\frac{17}{6}$ **n** $\frac{43}{8}$ **o** $\frac{29}{8}$ **p** $\frac{109}{100}$

8.5 Get ready

1 a 78   **b** 133   **c** 105

Exercise 8E

1 a $\frac{3}{8}$ b $\frac{3}{32}$ c $\frac{8}{25}$ d $\frac{9}{32}$
  e $\frac{5}{36}$ f $\frac{21}{40}$ g $\frac{9}{50}$ h $\frac{4}{9}$
  i $\frac{3}{16}$ j $\frac{8}{15}$ k $\frac{4}{21}$ l $\frac{4}{15}$
  m $\frac{2}{35}$ n $\frac{10}{21}$ o $\frac{9}{8} = 1\frac{1}{8}$ p $\frac{1}{5}$

2 a $\frac{2}{5}$ b $\frac{3}{5}$ c $\frac{1}{2}$ d $\frac{6}{25}$
  e $\frac{5}{8}$ f $\frac{1}{8}$ g $\frac{4}{15}$ h $\frac{4}{7}$
  i $\frac{2}{7}$ j $\frac{5}{14}$ k $\frac{1}{5}$ l $\frac{1}{6}$
  m $\frac{1}{7}$ n $\frac{2}{5}$ o $\frac{7}{2} = 3\frac{1}{2}$ p $\frac{13}{20}$

3 a $\frac{7}{2} = 3\frac{1}{2}$ b $\frac{10}{3} = 3\frac{1}{3}$ c $\frac{24}{5} = 4\frac{4}{5}$ d 6
  e 14   f 6   g 4   h 10

4 a £4   b £5   c £7   d £9
  e 7 cm   f 21 cm   g 44 kg   h 18 kg

5 a $\frac{8}{9}$ b $\frac{14}{15}$ c $\frac{3}{8}$ d $\frac{13}{8} = 1\frac{5}{8}$
  e $\frac{17}{6} = 2\frac{5}{6}$ f $\frac{10}{9} = 1\frac{1}{9}$ g $\frac{7}{4} = 1\frac{3}{4}$

6 $82\frac{1}{2}$ minutes = 1 hour $22\frac{1}{2}$ minutes

7 a $\frac{21}{4} = 5\frac{1}{4}$ b $\frac{133}{24} = 5\frac{13}{24}$ c $\frac{21}{5} = 4\frac{1}{5}$ d $\frac{21}{5} = 4\frac{1}{5}$
  e $\frac{5}{8}$ f $\frac{28}{15} = 1\frac{13}{15}$ g 16   h 3

8.6 Get ready

1 a 24   b 30   c 70

Exercise 8F

1 a $\frac{4}{3} = 1\frac{1}{3}$ b $\frac{3}{4}$ c $\frac{3}{2} = 1\frac{1}{2}$ d $\frac{5}{7}$
  e $\frac{10}{3} = 3\frac{1}{3}$ f $\frac{15}{8} = 1\frac{7}{8}$ g $\frac{10}{9} = 1\frac{1}{9}$ h $\frac{7}{8}$
  i $\frac{4}{9}$ j $\frac{8}{15}$ k $\frac{9}{16}$ l 2

2 a 16   b 16   c 10   d $\frac{64}{7} = 9\frac{1}{7}$
  e 5   f $\frac{12}{7} = 1\frac{5}{7}$ g 15   h 24

3 a 5   b $\frac{26}{20} = 1\frac{3}{10}$ c $\frac{5}{3} = 1\frac{2}{3}$ d $\frac{39}{76}$
  e $\frac{1}{2}$ f 2   g $\frac{17}{27}$ h $\frac{21}{40}$

4 a $\frac{3}{32}$ b $\frac{5}{12}$ c $\frac{1}{10}$ d $\frac{4}{25}$
  e $\frac{1}{3}$ f $\frac{13}{24}$ g $\frac{17}{60}$ h $\frac{1}{6}$
  i $\frac{2}{9}$ j $\frac{1}{9}$ k $\frac{7}{6} = 1\frac{1}{6}$ l $\frac{7}{4} = 1\frac{3}{4}$

8.7 Get ready

1 a 56   b 15   c 6

Exercise 8G

1 a $\frac{7}{8}$ b $\frac{7}{9}$ c 1   d $\frac{8}{9}$
  e $\frac{1}{2}$ f $\frac{6}{7}$ g 1   h $\frac{8}{5} = 1\frac{3}{5}$
  i $3\frac{2}{9}$ j $2\frac{2}{3}$ k $1\frac{3}{4}$ l $1\frac{7}{8}$

2 a $\frac{3}{4}$ b $\frac{5}{8}$ c $\frac{11}{8} = 1\frac{3}{8}$ d $\frac{5}{6}$
  e $\frac{7}{6} = 1\frac{1}{6}$ f $\frac{7}{10}$ g $\frac{4}{3} = 1\frac{1}{3}$ h $\frac{11}{10} = 1\frac{1}{10}$

3 a $\frac{29}{24} = 1\frac{5}{24}$ b $\frac{17}{20}$ c $\frac{31}{36}$ d $\frac{71}{40} = 1\frac{31}{40}$
  e $\frac{17}{30}$ f $\frac{13}{12} = 1\frac{1}{12}$ g $\frac{23}{24}$ h $\frac{19}{18} = 1\frac{1}{18}$
  i $\frac{7}{8}$ j $\frac{19}{24}$ k $\frac{17}{16} = 1\frac{1}{16}$

4 a $\frac{5}{6}$ b $\frac{17}{30}$ c $\frac{33}{40}$ d $\frac{31}{36}$
  e $\frac{53}{42} = 1\frac{11}{42}$ f $\frac{83}{70} = 1\frac{13}{70}$ g $\frac{41}{30} = 1\frac{11}{30}$ h $\frac{31}{35}$
  i $\frac{23}{40}$ j $\frac{11}{30}$ k $\frac{20}{21}$

5 a $\frac{29}{8} = 3\frac{5}{8}$ b $\frac{53}{8} = 6\frac{5}{8}$ c $\frac{65}{16} = 4\frac{1}{16}$ d $\frac{35}{8} = 4\frac{3}{8}$
  e $\frac{67}{16} = 4\frac{3}{16}$ f $\frac{89}{30} = 2\frac{29}{30}$ g $\frac{145}{42} = 3\frac{19}{42}$ h $\frac{167}{42} = 3\frac{41}{42}$
  i $\frac{88}{15} = 5\frac{13}{15}$ j $\frac{26}{9} = 2\frac{8}{9}$

6 $7\frac{1}{12}$ miles

7 a $\frac{23}{4} = 5\frac{3}{4}$ b $\frac{19}{6} = 3\frac{1}{6}$ c $\frac{33}{8} = 4\frac{1}{8}$ d $\frac{109}{12} = 9\frac{1}{12}$
  e $\frac{83}{16} = 5\frac{3}{16}$ f $\frac{44}{12} = 3\frac{2}{3}$ g $\frac{43}{6} = 7\frac{1}{6}$ h $\frac{109}{15} = 7\frac{4}{15}$

Exercise 8H

1 a $\frac{2}{11}$ b $\frac{2}{9}$ c $\frac{3}{4}$ d $\frac{1}{6}$
  e $\frac{1}{2}$ f $\frac{1}{4}$ g $\frac{1}{2}$ h $\frac{3}{7}$

2 $\frac{3}{5}$

3 a $\frac{1}{4}$ b $\frac{1}{8}$ c $\frac{1}{8}$ d $\frac{5}{8}$
  e $\frac{1}{2}$ f $\frac{1}{4}$ g $\frac{1}{2}$ h $\frac{1}{5}$
  i $\frac{1}{8}$ j $\frac{3}{8}$ k $\frac{1}{6}$

4 a $\frac{1}{6}$ b $\frac{7}{24}$ c $\frac{1}{30}$ d $\frac{13}{30}$
  e $\frac{2}{15}$ f $\frac{3}{30}$ g $\frac{11}{30}$ h $\frac{3}{20}$
  i $\frac{103}{20} = 5\frac{3}{20}$ j $\frac{43}{6} = 7\frac{1}{6}$

5 $\frac{9}{16}$

6 a $\frac{19}{8} = 2\frac{3}{8}$ b $\frac{5}{4} = 1\frac{1}{4}$ c $\frac{11}{5} = 2\frac{1}{5}$ d $\frac{27}{10} = 2\frac{7}{10}$
  e $\frac{9}{10}$ f $\frac{3}{4}$ g $\frac{14}{5} = 2\frac{4}{5}$ h $\frac{77}{24} = 3\frac{5}{24}$
  i $\frac{22}{9} = 2\frac{4}{9}$ j $\frac{137}{40} = 3\frac{17}{40}$ k $\frac{111}{35} = 3\frac{6}{35}$

8.8 Get ready

1 a 0.375   b 0.5   c 0.04

Exercise 8I

1 a 0.6   b 0.5   c 0.7   d 0.35
  e 0.16   f 0.06   g 0.875   h 0.45
  i 0.76   j 0.3125   k 0.125   l 0.54
  m 0.09   n 0.065   o 0.6   p 0.95

2 a $\frac{3}{10}$ b $\frac{37}{100}$ c $\frac{93}{100}$ d $\frac{137}{1000}$
  e $\frac{293}{1000}$ f $\frac{7}{10}$ g $\frac{59}{100}$ h $\frac{3}{1000}$
  i $\frac{3}{100\,000}$ j $\frac{13}{10\,000}$ k $\frac{77}{100}$ l $\frac{77}{1000}$
  m $\frac{39}{100}$ n $\frac{41}{10\,000}$ o $\frac{19}{1000}$ p $\frac{31}{1000}$

3 a 0.8   b 0.75   c 1.125   d 0.19
  e 3.6   f 0.52   g 0.625   h 3.425
  i 0.14   j 4.1875   k 3.15   l 4.3125
  m 0.007   n 1.28   o 15.9375   p 2.35

4 a $\frac{12}{25}$ b $\frac{1}{4}$ c $1\frac{7}{10}$ d $3\frac{203}{500}$
  e $4\frac{3}{1000}$ f $2\frac{1}{40}$ g $\frac{49}{1000}$ h $4\frac{7}{8}$
  i $3\frac{3}{4}$ j $10\frac{101}{1000}$ k $\frac{5}{8}$ l $2\frac{64}{125}$
  m $\frac{13}{16}$ n $14\frac{7}{50}$ o $9\frac{3}{16}$ p $60\frac{13}{200}$

Review exercise

1 a $\frac{1}{4}$ b $\frac{2}{3}$ c $\frac{3}{4}$ d $\frac{2}{5}$

2 Students' equivalent fractions, e.g.

Answers

a $\frac{8}{10}, \frac{12}{15}, \frac{16}{20}$ **b** $\frac{4}{14}, \frac{6}{21}, \frac{8}{28}$ **c** $\frac{3}{4}, \frac{6}{8}, \frac{9}{12}$

d $\frac{4}{5}, \frac{8}{10}, \frac{12}{15}$ **e** $\frac{3}{10}, \frac{6}{20}, \frac{9}{30}$ **f** $\frac{2}{3}, \frac{4}{6}, \frac{6}{9}$

3 a 0.25 **b** 0.375 **c** 0.7

d 0.6 **e** 0.12 **f** 0.74

4 a $\frac{17}{50}$ **b** $\frac{1}{8}$ **c** $\frac{3}{10}$

d $\frac{1}{40}$ **e** $\frac{3}{20}$ **f** $3\frac{1}{10}$

5 $\frac{3}{20}$

6 a $\frac{3}{8}$ **b** $\frac{1}{3}$ **c** $\frac{1}{12}$ **d** $\frac{1}{24}$ **e** $\frac{1}{6}$

7 a $\frac{2}{3}$ **b** $\frac{4}{5}$ **c** $\frac{3}{4}$ **d** $\frac{4}{9}$

8 $\frac{1}{3}, \frac{3}{10}, \frac{29}{100}, \frac{2}{7}, \frac{4}{15}$

9 $\frac{4}{5}$

10 $\frac{450}{1000}, 0.6, \frac{7}{10}, \frac{3}{4}$

11 a $\frac{2}{15}$ **b** $\frac{1}{9}$ **c** $\frac{3}{4}$

12 $\frac{7}{25}$

13 a $\frac{2}{3}$ **b** $\frac{5}{9}$ **c** $\frac{39}{40}$ **d** $1\frac{1}{36}$ **e** $\frac{4}{9}$

14 $3\frac{2}{15}$ m

15 a $\frac{18}{24} = \frac{3}{4}$ of the bracelet is gold

b $\frac{1}{3}$

c 30 g

d 50 g

e 10 carat

16 a $8\frac{1}{2}$ **b** $1\frac{5}{16}$ **c** $\frac{98}{125}$

17 $3\frac{1}{3}$

18 a $1\frac{5}{16}$ **b** $1\frac{3}{4}$ **c** $2\frac{7}{8}$

d $2\frac{11}{15}$

19 $\frac{4}{15}$

20 Yes, it is $6\frac{2}{16}$ cm long

21 15 glasses

Chapter 9 Answers

9.1 Get ready

1 a 4 **b** 125 **c** 512 **d** 1 000 000

Exercise 9A

1 a 64 **b** 256 **c** 1

d 10 000 **e** 625 **f** 7776

2 a 2^4 **b** 4^5 **c** 1^6

d 8^3 **e** $3^2 \times 8^4$ **f** $2^3 \times 4^4$

3 a 16 **b** 243 **c** 216

d 25 **e** 512 **f** 11 664

g 65 536 **h** 10 125 **i** 31 104

j 256

4

Power of 10	Index	Value	Value in words
10^3	3	1000	One thousand
10^2	2	100	One hundred
10^6	6	1 000 000	One million
10^1	1	10	Ten
10^5	5	100 000	One hundred thousand

5 a 6400 **b** 400 **c** 6000

d 4 **e** 125 **f** 16

6 a 3 **b** 4 **c** 6

d 4 **e** 2 **f** 3

g 4 **h** 2

9.2 Get ready

1 a 4 **b** 64 **c** 2401 **d** 10 000

Exercise 9B

1 a 6^{11} **b** 8^8 **c** 2^6

2 a 4 **b** 6^3 **c** 7^4

3 a 4^5 **b** 5^2 **c** 3

4 a 5^{13} **b** 2^{11}

5 a 10^5 **b** $9^0 = 1$

6 a 6^{11} **b** 5^6

7 a 3^8 **b** 4^6

8 a 6^9 **b** 5^5 **c** 4^{12}

9 a 5^6 **b** 7^8

9.3 Get ready

1 a 7^7 **b** 8^2 **c** 5^6

Exercise 9C

1 a x^{10} **b** y^{11} **c** x^{14}

2 a a^8 **b** b^6 **c** d^{11}

3 a p^3 **b** q^{10} **c** t^4

4 a j^6 **b** k **c** n^2

5 a x^9 **b** y^9 **c** z^{10}

6 a $6x^5$ **b** $15y^{29}$ **c** $24z^{10}$

7 a $3p^5$ **b** $5q^2$ **c** $2r^3$

8 a d^{12} **b** e^{10} **c** f^9 **d** g^{63}

9 a g^{24} **b** h^4 **c** 1 **d** 1

10 a $2187d^{14}$ **b** $64e^3$ **c** 1

11 a 1 **b** b^4 **c** c^{10}

12 a $8d^{10}$ **b** $2e^4$ **c** $16f^4$

9.4 Get ready

1 a 9 **b** 11 **c** 17

Exercise 9D

1 a 9 **b** 1 **c** 25

d 13 **e** 21 **f** 23

g 6 **h** 4 **i** 8

j 5 **k** 6 **l** 2

m 1 **n** 5 **o** 4

2 a $4 + 5 = 9$ **b** $4 \times 5 = 20$

c $(2 + 3) \times 4 = 20$ **d** $(3 - 2) \times 5 = 5$

e $(5 - 2) \times 3 = 9$ **f** $4 \div 2 + 8 = 10$

g $5 \times 4 + 5 + 2 = 27$ **h** $5 \times 4 + 5 - 2 = 23$

3 a 49 **b** 25 **c** 243 **d** 123

e 72 **f** 17 **g** 22 **h** 7

i 7 **j** 0 **k** 7 **l** 0

9.5 Get ready

1 a 48 **b** 3 **c** 120

Exercise 9E

1 $5x + 14$	**2** $20x + 17$
3 $23x + 14$	**4** $-8x + 12$ or $12 - 8x$
5 $9 - 21x$	**6** $15 - 23x$
7 $12x - 22y$	**8** $22y - 2x$
9 $-4x - 21y$	**10** $x + 4y$
11 $7y + 2$	**12** 27
13 $x - 3$	**14** $7y - 16x + 6$
15 $21x - 19y$	**16** $21y - 27x$
17 $10x - 7y$	**18** $20x - 32y$
19 $8xy + 3x$	**20** $8xy + 2x + y$
21 $6x - 4y - 3xy$	**22** $10xy - 20x^2 - 2y^2$

9.6 Get ready

1 a 8 **b** 4 **c** ab

Exercise 9F

1 a $2(x + 3)$ **b** $2(3y + 1)$ **c** $5(3b - 1)$
  d $2(2r - 1)$ **e** $x(3 + 5y)$ **f** $4(3x + 2y)$
  g $4(3x - 4)$ **h** $3(3 - x)$ **i** $3(3 + 5g)$
2 a $x(3x + 4)$ **b** $y(5y - 3)$ **c** $a(2a + 1)$
  d $b(5b - 2)$ **e** $c(7 - 3c)$ **f** $d(d + 3)$
  g $m(6m - 1)$ **h** $x(4y + 3)$ **i** $n^2(n - 8)$
3 a $4x(2x + 1)$ **b** $3p(2p + 1)$ **c** $3x(2x - 1)$
  d $3b(b - 3)$ **e** $3a(4 + a)$ **f** $5c(3 - 2c)$
  g $7x^3(3x + 2)$ **h** $4y^2(4y - 3)$ **i** $2d^2(3d^2 - 2)$
4 a $ax(x + 1)$ **b** $pr(r - 1)$ **c** $ab(b - 1)$
  d $q(r^2 + q)$ **e** $ax(a + x)$ **f** $by(b - y)$
  g $3a^2(2a - 3)$ **h** $4x^3(2 - x)$ **i** $6x^3(3 + 2x^2)$
5 a $6ab(2a + 3b)$ **b** $2xy(2x - y)$
  c $4ab(a + 2b + 3)$ **d** $2xy(2x + 3y - 1)$
  e $3ax(4x + 2a - 1)$ **f** $abc(a + b + c)$
6 a $5(x + 4)$ **b** $2(6y - 5)$
  c $x(3x + 5)$ **d** $y(4 - 3y)$
  e $2a(4 + 3a)$ **f** $4b(3b - 2)$
  g $cy(y + 1)$ **h** $3dx(x - 2)$
  i $3cd(3c + 5d)$

Review exercise

1 a 6^3 **b** 11^2 **c** 2^6
2 a 625 **b** 128 **c** 1000 **d** $100\,000$
3 a 144 **b** 784 **c** 400 **d** $30\,000$
4 a $2^2 \times 3^3$ **b** $5^2 \times 7^2$ **c** $4^2 \times 8^4$ **d** $2^3 \times 6^3$
5 a 512 **b** $10\,000$ **c** 125 **d** 144 **e** 800
6 a 5 **b** 5 **c** 3
7 a $5(x + 3y)$ **b** $3(5p - 3q)$ **c** $c(d + e)$
8 6 or -6
9 2
10 2^{14}
11 a $5(p - 4)$ **b** $x = 3.5$
12 a 2^7 **b** 5^5 **c** 3^5 **d** 7^3
  e 9^4 **f** 8^2 **g** 7^3 **h** 6^5
13 a x^9 **b** x^3 **c** x^{15} **d** x
  e x^5 **f** x^{12} **g** 1 **h** x^6
  i x^{11} **j** x^9 **k** x^3 **l** x^{11}
14 a $4x^8$ **b** $15x^8$ **c** $21x^5$ **d** $4x^4$
  e $8x^5$ **f** $9x$ **g** x^{10} **h** x^9
15 a a^7 **b** $15x^3y^4$

16 a $9ab - 6a^2 - 4b^2$ **b** $11pq + 18p^2$
  c $13c^2 + 12cd$ **d** $a^2 + 2ab + b^2$
  e $5ab + ac - bc$ **f** $-4ab - 4ac - 9bc$
17 a $x(x - 7)$ **b** $t(t + a)$ **c** $x(bx - 1)$
  d $p(3p + y)$ **e** $a(q^2 - t)$
18 a $x(x - 5)$ **b** $10\,500$

Chapter 10 Answers

10.1 Get ready

1 0.25 **2** 0.1 **3** 0.01

Exercise 10A

1 a 0.75 **b** 0.125 **c** 0.375
  d 0.0625 **e** 0.5625
2 a 0.64 **b** 0.175 **c** 0.65
  d 0.46 **e** 0.5875

Exercise 10B

1 a $0.\dot{6}$ **b** $0.\dot{2}$ **c** $0.\dot{4}$
  d $0.\dot{7}$ **e** $0.\dot{8}$
2 a $0.1\dot{6}$ **b** $0.4\dot{5}$ **c** $0.8\dot{1}$
  d $0.41\dot{6}$ **e** $0.7\dot{7}\dot{2}$
3 a $0.\dot{1}4285\dot{7}, 0.\dot{2}8571\dot{4}, 0.\dot{4}2857\dot{1}, 0.\dot{5}7142\dot{8}, 0.\dot{7}1428\dot{5}, 0.\dot{8}5714\dot{2}$,
  b the same six digits repeat

10.2 Get ready

1 0.25 **2** 2.5 **3** 0.04

Exercise 10C

1 a 0.1 **b** 0.25 **c** 0.125
  d 0.2 **e** $0.\dot{1}$
2 a 3 **b** 4 **c** $1\frac{1}{2}$
  d $1\frac{1}{5}$ **e** $3\frac{1}{3}$
3 a 0.4 **b** 0.02 **c** 0.0625
  d 0.0125 **e** 5 **f** 2
  g 20 **h** 8 **i** 25
  j 100
4 1000
5 a 0.025 **b** 1
6 a 0.01 **b** 1

Mixed Exercise 10D

1 a 0.625 **b** 0.4375 **c** $0.\dot{5}$
  d $0.6\dot{3}$ **e** $0.08\dot{3}$
2 $\frac{1}{9} = 0.\dot{1}$
3 No, $\frac{1}{3} = 0.\dot{3}$
4 $\frac{1}{25}$ or 0.04
5 1.25

10.3 Get ready

1 £6.68   **2** £75   **3** £4.74

Exercise 10E

1 £8.90   **2** £7.10   **3** £2.30
4 £26.60   **5** £56.70   **6** £44.04

7 14 **8** 9 **9** 8
10 34
11 17 days 12 hours
12 6p
13 Current plan:
    September   £25
    October   £25
    November   £25
    New plan:
    September $15 + (40 \times 0.20) + (30 \times 0.12) = £26.60$
    October $15 + (45 \times 0.20) = £24$
    November $15 + (35 \times 0.20) + (60 \times 0.12) = £29.20$
    Raja should not switch to the new monthly plan because in September and November he would have spent more per month and in October he would have spent only slightly less.

10.4 Get ready

1 36   **2** 8   **3** 5

Exercise 10F

1 a 6.25   **b** 10.24   **c** 2209
   d 3.24   **e** 320.41
2 a 216   **b** 10 648   **c** 9.261
   d 39.304   **e** 3.375
3 a 32   **b** 14 641   **c** 243
   d 20 736   **e** 15 625
4 a 299   **b** 20.76   **c** 39   **d** 12.71
5 a 11.221   **b** 12.149   **c** 8.335   **d** 1.2

Exercise 10G

1 a 23   **b** 17   **c** 34   **d** 25
2 a 1.6   **b** 4.7   **c** 3.7   **d** 9.4
3 a 12.2   **b** 8.9   **c** 11.1   **d** 15.5
4 a 6   **b** 9   **c** 7   **d** 12
5 a 3.4   **b** 5.8   **c** 4.9   **d** 4.4
6 a 12   **b** 7.7   **c** 0.2   **d** 3
7 James : boat A
    Matthew : boats A and B

10.5 Get ready

1 24   **2** 13.5   **3** 2

Exercise 10H

1 a 62.41   **b** 7.2361   **c** 79.507   **d** 2.744
2 a 313   **b** 1372   **c** 1199   **d** 1120
3 a 5.6   **b** 16   **c** 4   **d** 1.5
4 a 4   **b** 5.2   **c** 4.5   **d** 5.2
5 a 1.02415(...)   **b** 5.27441(...)
   c 2.85209(...)   **d** 1.55266(...)

Review Exercise

1 £3.80
2 80
3 14
4 a £1.90   **b** 27
5 a 8.41   **b** 1728   **c** 13.69   **d** 10.648

6 a 7.2   **b** 28   **c** 8   **d** 2.5
7 a 13.7   **b** 2.2   **c** 16.8   **d** 7.4
8 $\frac{5}{6}$ and $\frac{1}{7}$
9 a £2.35   **b** £4.25   **c** £2.05
10 31
11 26
12 £268.65
13 24.4
14 a 40.1956   **b** 6.5   **c** 28.812
15 a 4.05975(...)   **b** 0.10053(...)
   c 1.24531(...)
16 a 0.57901(...)   **b** 1.24148(...)
17 Recommended daily calorie intakes:
    Sophie $655 + (9.6 \times 68) + (1.8 \times 165) - (4.7 \times 32) = 1454.4$
    Chelsea $655 + (9.6 \times 55) + (1.8 \times 175) - (4.7 \times 47) = 1277.1$
    Kenny $66 + (13.7 \times 98) + (5 \times 191) - (6.8 \times 27) = 2180$
    Hassan $66 + (13.7 \times 117) + (5 \times 182) - (6.8 \times 38) = 2320.5$
    Hassan has the greatest daily calorie requirement.
    Chelsea has the smallest daily calorie requirement.
18 a $1\frac{3}{5}$ **b** 0.4   **c** 4
19 a 0.85   **b** 0.075
20 11.94117(...)
21 Large costs $380 \div 200 = 1.9$p per gram
    Regular costs $350 \div 175 = 2$p per gram
    So Rob is correct. The large tub is better value for money.
22 2.26541(...)
23 a Single burger and regular fries
   b £4.81

Chapter 11 Answers

11.1 Get ready

1 26.5 cm by 19.5 cm

Exercise 11A

1 a 2.5 cm, 25 mm   **b** 4 cm, 40 mm   **c** 5 cm, 50 mm
   d 6.5 cm, 65 mm   **e** 8.2 cm, 82 mm   **f** 4.4 cm, 44 cm
   g 7.6 cm, 76 mm   **h** 6.8 cm, 68 mm   **i** 5.7 cm, 57 mm
   j 3.3 cm, 33 mm   **k** 4.8 cm, 48 mm   **l** 6.2 cm, 62 mm

2 a

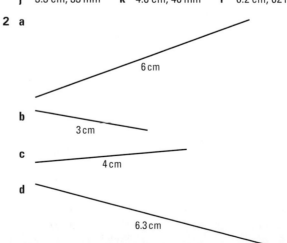

b 3 cm
c 4 cm
6 cm
d 6.3 cm

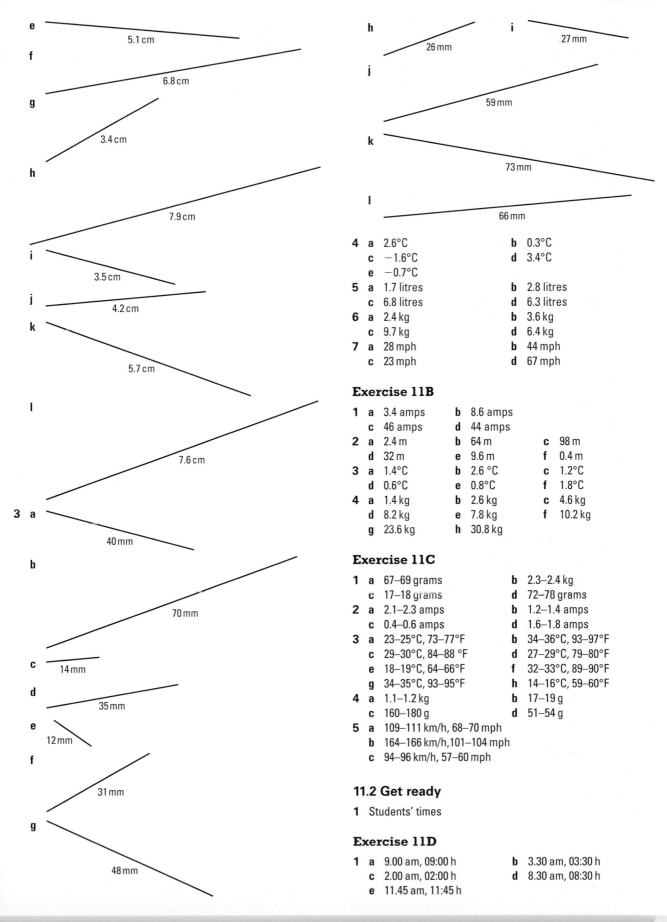

e 5.1 cm

f 6.8 cm

g 3.4 cm

h 7.9 cm

i 3.5 cm

j 4.2 cm

k 5.7 cm

l 7.6 cm

h 26 mm   i 27 mm

j 59 mm

k 73 mm

l 66 mm

3 a 40 mm

b 70 mm

c 14 mm

d 35 mm

e 12 mm

f 31 mm

g 48 mm

4 **a** 2.6°C **b** 0.3°C
 c −1.6°C **d** 3.4°C
 e −0.7°C
5 **a** 1.7 litres **b** 2.8 litres
 c 6.8 litres **d** 6.3 litres
6 **a** 2.4 kg **b** 3.6 kg
 c 9.7 kg **d** 6.4 kg
7 **a** 28 mph **b** 44 mph
 c 23 mph **d** 67 mph

Exercise 11B

1 **a** 3.4 amps **b** 8.6 amps
 c 46 amps **d** 44 amps
2 **a** 2.4 m **b** 64 m **c** 98 m
 d 32 m **e** 9.6 m **f** 0.4 m
3 **a** 1.4°C **b** 2.6 °C **c** 1.2°C
 d 0.6°C **e** 0.8°C **f** 1.8°C
4 **a** 1.4 kg **b** 2.6 kg **c** 4.6 kg
 d 8.2 kg **e** 7.8 kg **f** 10.2 kg
 g 23.6 kg **h** 30.8 kg

Exercise 11C

1 **a** 67–69 grams **b** 2.3–2.4 kg
 c 17–18 grams **d** 72–78 grams
2 **a** 2.1–2.3 amps **b** 1.2–1.4 amps
 c 0.4–0.6 amps **d** 1.6–1.8 amps
3 **a** 23–25°C, 73–77°F **b** 34–36°C, 93–97°F
 c 29–30°C, 84–88 °F **d** 27–29°C, 79–80°F
 e 18–19°C, 64–66°F **f** 32–33°C, 89–90°F
 g 34–35°C, 93–95°F **h** 14–16°C, 59–60°F
4 **a** 1.1–1.2 kg **b** 17–19 g
 c 160–180 g **d** 51–54 g
5 **a** 109–111 km/h, 68–70 mph
 b 164–166 km/h,101–104 mph
 c 94–96 km/h, 57–60 mph

11.2 Get ready

1 Students' times

Exercise 11D

1 **a** 9.00 am, 09:00 h **b** 3.30 am, 03:30 h
 c 2.00 am, 02:00 h **d** 8.30 am, 08:30 h
 e 11.45 am, 11:45 h

Answers

2 a 8.15 pm, 20:15 h   **b** 1.15 pm, 13:15 h
   c 5.00 pm, 17:00 h   **d** 7.00 pm, 19:00 h
   e 5.15 pm, 17:15 h

3 a 08:00 h   **b** 11:15 h   **c** 15:40 h
   d 08:20 h   **e** 20:55 h   **f** 15:25 h
   g 02:30 h   **h** 17:25 h   **i** 22:15 h
   j 07:20 h   **k** 09:45 h   **l** 13:15 h
   m 23:25 h   **n** 02:50 h   **o** 13:50 h
   p 12:20 h

4 a 11.10 am   **b** 8.20 am   **c** 7.40 am
   d 11.35 pm   **e** 2.17 pm   **f** 9.35 am
   g 6.16 pm   **h** 5.25 pm   **i** 1.20 pm
   j 1.10 pm   **k** 8.30 am   **l** 1.35 pm
   m 3.42 am   **n** 10.16 pm   **o** 9.17 am
   p 1.37 pm

5 a 11.15 am   **b** 12:10 h   **c** 2.30 pm
   d 10.45 am   **e** 11:30 h   **f** 5.40 am
   g 1.50 pm   **h** 10:30 h   **i** 16:25 h
   j 08:40 h   **k** 9.45 pm   **l** 00:45 h

6 a 156 weeks   **b** 210 minutes   **c** 300 seconds
   d 60 months   **e** 96 hours   **f** 9 hours
   g 6 years   **h** 150 seconds   **i** $1\frac{1}{2}$ minutes
   j 208 weeks   **k** 480 minutes   **l** 60 hours
   m 182 weeks   **n** 730 days

7 a 156   **b** 20   **c** 12   **d** 168

Exercise 11E

1 a 25 mins   **b** 50 mins   **c** 2 h 10 min
   d 10 h 30 min   **e** 9 hours   **f** 4 h 10 min
   g 3h 38 min   **h** 18 h 32 min   **i** 14 h 45 min
   j 3 h 50 min   **k** 22 h 50 min   **l** 3 h 15 min

2 16 h 20 min
3 8 h 10 min
4 6 h 50 min

5

Flight number	Departure time	Arrival time	Flight time
BA52	2220	0445	**6 h 25 min**
XA160	0542	0914	**3 h 32 min**
FC492	1415	**1855**	4 h 40 min
TC223	1002	**1425**	4 h 23 min
AL517	**0222**	0759	5 h 37 min
AB614	1917	0521	**10 h 4 min**
FX910	0243	**0634**	3 h 51 min
BI451	**0945**	1217	2 h 32 min
AE105	**2056**	0225	5 h 29 min
DA452	1539	**2227**	6 h 48 min

Ella should take the flight that departs at 09:45h and arrives at 12:17h.

Exercise 11F

1 a 31 min   **b** 2 min
   c 19:40 h   **d** 07:35 h
   e 30 min   **f** 25 min
   g 4   **h** 27
   i 10 min   **j** 45 min
   k 07:52 h   **l** 11:13 h

2 a i 37 min   **ii** 18 min
    iii 26 min   **iv** 20 min
   b 41 min
   c

Manchester Victoria	09:05	11:35	Wigan Wallgate	08:30	11:55
Salford	09:08	11:38	Ince	08:35	12:00
Salford Crescent	09:10	11:40	Hindley	08:38	12:03
Swinton	09:19	11:49	Westhoughton	08:42	12:07
Moorside	09:22	11:52	Bolton	08:50	12:15
Walkden	09:26	11:56	Moses gate	08:53	12:18
Atherton	09:32	12:02	Farnworth	08:56	12:21
Hag Fold	09:35	12:05	Kearsley	08:59	12:24
Daisy Hill	09:37	12:07	Clifton	09:06	12:31
Hindley	09:42	12:12	Salford Crescent	09:10	12:35
Ince	09:45	12:15	Salford	09:12	12:37
Wigan Wallgate	09:50	12:20	Manchester Victoria	09:15	12:40

11.3 Get ready

1 Students' examples of objects

Exercise 11G

1 ml   **2** mm/cm
3 kg   **4** tonnes
5 litres   **6** metres
7 mm/cm   **8** kg
9 grams   **10** mm
11 litres   **12** mg/grams
13 km   **14** grams
15 cm/m

Exercise 11H

1 a 400 cm   **b** 5 cm   **c** 800 cm
   d 1300 cm   **e** 20 cm   **f** 3500 cm
   g 7.4 cm   **h** 12.2 cm

2 a 30 mm   **b** 60 mm   **c** 220 mm
   d 400 mm   **e** 2000 mm   **f** 54 mm
   g 137 mm   **h** 51.5 mm

3 a 6000 m   **b** 5 m   **c** 20 000 m
   d 30 m   **e** 45 000 m   **f** 800 m
   g 1400 m   **h** 2450 m

4 a 2000 g   **b** 30 000 g   **c** 400 000 g
   d 250 000 g   **e** 2 000 000 g   **f** 55 000 g
   g 120 g   **h** 4200 g

5 a 4 l   **b** 7 l   **c** 20 l
   d 45 l   **e** 2.5 l   **f** 3.7 l
   g 6.52 l   **h** 3.13 l

6 a 3000 ml   **b** 20 000 ml   **c** 200 000 ml
   d 450 000 ml   **e** 35 000 ml   **f** 7500 ml
   g 400 ml   **h** 1430 ml

7 a 3 km   **b** 8 km   **c** 30 km
   d 68 km   **e** 4.2 km   **f** 5.6 km
   g 5.41 km   **h** 2.14 km

8 a 5 t   **b** 6 t   **c** 40 t
   d 57 t   **e** 3.6 t   **f** 4.5 t
   g 7.63 t   **h** 4.25 t

9   **a** 4 kg   **b** 2000 kg   **c** 20 kg
     d 15 000 kg   **e** 200 kg   **f** 3700 kg
     g 6.4 kg   **h** 1.23 kg

10 500
11 90 litres
12 25
13 750 kg
14 20
15 50 days

Exercise 11I

1 6 mm, 3 cm, 60 mm, 30 cm, 4 m, 4 km
2 400 ml, 700 ml, 1 l, 3000 ml , 6 l
3 450 g, 0.5 kg, 600 g, 0.62 kg
4 0.6 cm, 370 mm, 40 cm, 55 cm, 600 mm, 1.4 m
5 0.2 cm, 0.4 cm, 9 mm, 55 mm, 6 cm, 77 mm, 46 cm
6 75 ml, 0.08 l , 260 ml, 0.3 l, 450 ml , 600 ml

11.4 Get ready

1 Students' measurements in feet and inches, and in centimetres

Exercise 11J

1 **a** 3 feet   **b** 4 gallons   **c** 28 inches
    d 64 ounces   **e** 63 inches   **f** 7 yards
    g 52 inches   **h** 48 pints   **i** 49 pounds
    j 18 pints   **k** 21 stones 6 pounds   **l** 12 feet
    m 9 inches   **n** 2 gallons 6 pints   **o** 224 ounces
2 5 foot 3 inches
3 9 stone 3 pounds
4 6 foot 2 inches

Exercise 11K

1 **a** 24 km   **b** 22 pounds   **c** 7 pints
    d 15 cm   **e** 270 cm   **f** 30 miles
    g 5 kg   **h** 100 inches (or 97.5 inches)   **i** 120 km
2 7.2 m   **3** 35 cm
4 4 gallons   **5** 3 bottles
6 640 km   **7** 8.75 pints
8 93.75 miles   **9** 11
10 4 kg (A)

11.5 Get ready

1 Explanation of how speed is measured in car

Exercise 11L

1 30 mph   **2** 8 mph
3 4 mph   **4** 2 mph
5 50 mph   **6** 32 km/h
7 60 mph   **8** 50 mph
9 80 km/h   **10** 400 km/h

Exercise 11M

1 80 miles   **2** 4 hours
3 105 miles   **4** $1\frac{1}{2}$ hours
5 $2\frac{1}{2}$ hours   **6** 4 hours
7 12.5 km   **8** $2\frac{1}{2}$ hours
9 1400 miles   **10** 160 miles

Exercise 11N

1 87.5 miles   **2** 6.5 miles
3 80 km/h   **4** 3 hours 12 min
5 82 km   **6** 60 mph
7 7 hours 48 min   **8** 297 miles
9 50 km/h
10 **a** 34 m   **b** 306 km/h

11.6 Get ready

1 **a** 7   **b** 52   **c** 14
2 **a** 300 cm   **b** 7 cm
3 **a** 220 mm   **b** 47 mm

Exercise 11O

1 12.5 cm
2 44.5 g
3 3.5 litres
4 **a** 9.65 cm   **b** 9.75 cm
5 **a** 1.585 m   **b** 1.595 m
6 The pencil could be as long as 105 mm, which is longer than the shortest possible length of the pencil case, 101.5 mm
7 The cupboard could as narrow as 81.5 cm, and the gap as wide as 81.75 cm

Review exercise

1 **a** metres, grams, litres
    b 400 cm
    c 1.5 kg
2 **a** 17.8 cm   **b** −2°C   **c** 2.9 kg
3 **a** **i** metres   **ii** kilograms
    b 20 mm
4 **a** **i** kilometres   **ii** litres
    b **i** 50 mm   **ii** 4 kg
5 **a** 32
    b   110 120 130 140 150 160
    c 4.4
    d   3 3.1 3.2 3.3 3.4 3.5
6 **a** 1.8 m   **b** 7 m
7 **a** 1.8 m   **b** 7 m
8 1.5 m
9 No, 1.5 km is 1500 m
10 **a** Huntingdon   **b** 3 minutes   **c** 10: 05
11 **a** 4.6 kg
    b **i** 2.2 lb   **ii** 11 lb
12 **a** 09: 50
    b **i** 15 minutes   **ii** 09: 45
    c 1 hour 20 minutes
13 **a** 07: 37   **b** 10 minutes   **c** 18 minutes
14

	Metric	Imperial
Distance from London to Cardiff	km	miles
Weight of a bag of potatoes	kg	pounds
Volume of fuel in a car's fuel tank	litres	gallons

Answers

15 a 57 minutes

b

	Time
Zoe leaves the hotel	08:53
Zoe Leaves Sa Pobla on a train	09:23
Train arrives at Inca (Zoe gets off)	09:41
Zoe leaves Inca on another train	11:41
Zoe arrives at Palma	12:20

16 432 mph
17 120 litres
18 85 kg
19 80 km/h
20 a 126.5 g   **b** 127.5 g

Chapter 12 Answers

12.1 Get ready

1 Discrete quantitative
2 Qualitative data
3 Continuous quantitative data

Exercise 12A

1 a Thursday **b** 100 emails **c** 80 emails **d** 70 emails
2 a Juice **b** 60 drinks **c** 35 drinks **d** 210 drinks
3 a

Margherita	⊕ ⊕ ◔
BBQ Chicken	⊕ ⊕ ⊕ ⊕
Hawaiian	⊕ ⊕ ◖
Meat feast	⊕ ⊕ ◖

b Margherita: 18 pizzas; BBQ chicken 32 pizzas; Hawaiian 22 pizzas; Meat feast 20 pizzas

4

English	👤 👤 👤
Mathematics	👤 👤 👤 👤
Spanish	👤 👤
Science	👤 👤 👤 👤 👤
Technology	👤 👤

Key
👤 represents two students

5

Walk	⊞ ▯
Bus	⊞ ⊞ ⊞ ⊞ ⊞ ⊞ ⊞ ⊞ ⊞ ⊞
Train	▯
Cycle	⊞ ⊞ ⊞ ▯
Car	▯

Key
⊞ represents four students

12.2 Get ready

1 270°   **2** $\frac{1}{20}$ **3** 70°

Exercise 12B

1

City	Reading	Swindon	Bristol	Cardiff
Angle	$\frac{6}{40} \times 360$ $= 54°$	$\frac{5}{40} \times 360$ $= 45°$	$\frac{9}{40} \times 360$ $= 81°$	$\frac{20}{40} \times 360$ $= 180°$

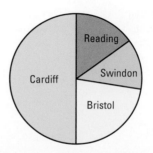

2

Fruit	Strawberry	Apple	Banana	Grape	Orange
Angle	$\frac{25}{60} \times 360 =$ $150°$	$\frac{15}{60} \times 360$ $= 90°$	$\frac{6}{60} \times 360$ $= 36°$	$\frac{9}{60} \times 360$ $= 54°$	$\frac{5}{60} \times 360$ $= 30°$

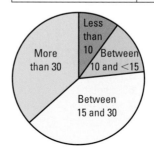

3

Time in minutes	Angle
Less than 10	$\frac{3}{30} \times 360 = 36°$
Between 10 and <15	$\frac{4}{30} \times 360 = 48°$
Between 15 and 30	$\frac{12}{30} \times 360 = 144°$
More than 30	$\frac{11}{30} \times 360 = 132°$

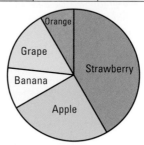

4

Category	Thriller	Classic	Romance	Non-fiction
Angle	$\frac{120}{540} \times 360$ $= 80°$	$\frac{60}{540} \times 360$ $= 40°$	$\frac{270}{540} \times 360$ $= 180°$	$\frac{90}{540} \times 360$ $= 60°$

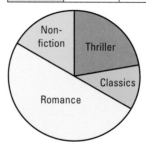

5

Type of car	Ford	Nissan	Toyota	Renault
Angle	$\frac{24}{90} \times 360$ $= 96°$	$\frac{30}{90} \times 360$ $= 120°$	$\frac{27}{90} \times 360$ $= 108°$	$\frac{9}{90} \times 360$ $= 36°$

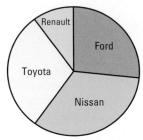

6

Day	Monday	Tuesday	Wednesday	Thursday	Friday
Angle	$\frac{7}{36} \times 360$ $= 70°$	$\frac{8}{36} \times 360$ $= 80°$	$\frac{10}{36} \times 360$ $= 100°$	$\frac{6}{36} \times 360°$ $= 60°$	$\frac{5}{36} \times 360$ $= 50°$

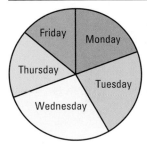

Exercise 12C

1 a Going out with friends **b** Reading
  c $\frac{1}{3}$ **d** 40 adults
2 a Spain **b** 4° **c** 20 girls
  d 108° **e** 27 girls

3

Activity	Angle	Number of hours
Sleep	120°	8
School	105°	7
Play	30°	2
Watch TV	60°	4
Eat	15°	1
Homework	30°	2

4 a $\frac{1}{4}$ **b** 300 people
  c $3\frac{1}{3}$ people **d** 160 people
5 a 2° **b** 180 people
  c Vanilla and chocolate. They have the largest angles.
6 a $\frac{1}{5}$ **b** 60 students

12.3 Get ready

1 25
2 84
3 720
4 140

Exercise 12D

1 a 7 students **b** 26 students
2 a 5 cars **b** Garage F **c** 33 cars
3 a Sam **b** 3 hours **c** Caitlin and Laura
  d Owen

4

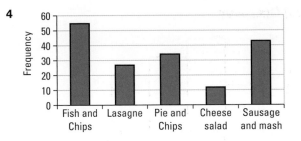

12.4 Get ready

1 a Red **b** Blue

Exercise 12E

1 a 30°C **b** 33°C
  c Resort G **d** Resorts C and F
  e Resort A
2 a Factory A **b** 45 males
  c 45 people **d** 130 people
  e Factory A. It has more workers.
3 a Shortbread **b** Protein
  c 25%

4

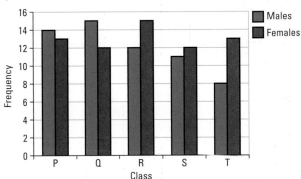

12.5 Get ready

1 a C **b** E **c** A and D

Exercise 12F

1 a 4 times **b** Number 6 **c** 41 times
2 a 10 cars **b** 2 people **c** 100 cars **d** 65 people
3 a Because it represents continuous data.
  b 5 runners **c** 6 runners **d** 20 runners

4

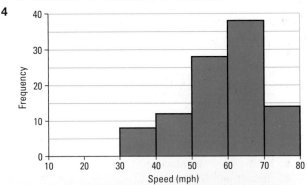

12.6 Get ready

1

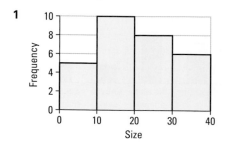

Exercise 12G

1 a, b

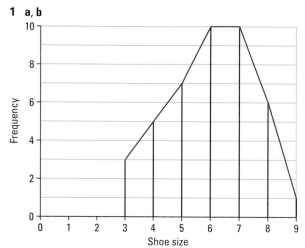

2 a, b

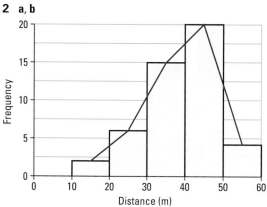

3 a Once **b** 4 times **c** $\frac{1}{4}$
  d The train, because the blue line is above the red line for the higher time values.

Review exercise

1 a i 8 **ii** 10
  b

Asad	⊞ ⊞
Betty	⊞ ⊞ ⊞ ▫
Chris	⊞ ⊞ ⊟
Diana	⊞ ⊞ ⊞
Erikas	⊞ ⊞ ▫

2 a 7 TVs **b** Monday
  c Tuesday and Wednesday

3 a 3 students **b** Cat **c** 22 students
4 a 20 plates **b** 15 plates
  c

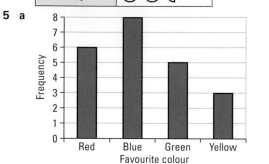

5 a

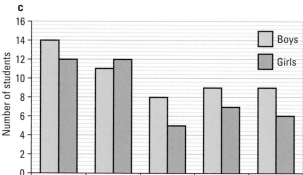

  b Blue
6 a Wednesday and Friday
  b 320 minutes or 5 hours 20 minutes
7 a 14 boys **b** 5 girls
  c

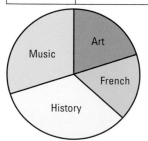

  d Tuesday
8 a $\frac{1}{4}$ **b** 60 students
9

Subject	Number of students	Angle
Art	12	72°
French	10	60°
History	20	120°
Music	18	108°

10

Drink	Frequency	Angle
Hot chocolate	20	80°
Soup	15	60°
Coffee	25	100°
Tea	30	120°

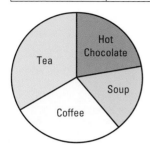

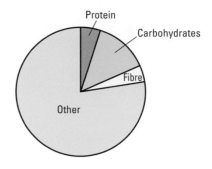

11 **a** 9 am
  b José's temperature steadily reduced.
12 **a** £126 billion
  b £35 billion
  c £700 billion
13

14 **a** $\frac{5}{18}$
  b **i** 16 students
    ii 72 students
  c The pie chart doesn't give frequencies, only proportions.
15

16

Protein	Carbohydrate	Fibre	Other
6 g	16 g	5 g	93 g
$\frac{6}{120} \times 360$ $= 18°$	$\frac{16}{120} \times 360$ $= 48°$	$\frac{5}{120} \times 360$ $= 15°$	$\frac{93}{120} \times 360$ $= 279°$

17 Students' composite bar charts

Chapter 13 Answers

13.1 Get ready

1 2, 4, 6, 8, 10, 12, 14, 16, 18, 20
2 1, 3, 5, 7, 9, 11, 13, 15, 17, 19
3 Check

Exercise 13A

1 **a** 12, 15; add 3   **b** 15, 19; add 4   **c** 25, 30; add 5
  d 22, 27 ; add 5   **e** 13, 16; add 3   **f** 13, 15; add 2
  g 23, 28; add 5   **h** 16, 19; add 3   **i** 18, 22; add 4
  j 40, 50; add 10
2 **a, b** **i** 21, 25; add 4   **ii** 17, 20; add 3   **iii** 23, 27; add 4
    iv 24, 28; add 4   **v** 20, 23; add 3   **vi** 29, 35; add 6
    vii 22, 26; add 4   **viii** 31, 37; add 6   **ix** 43, 51; add 8
    x 25, 29; add 4
3 question 1
  a 30   **b** 39   **c** 50   **d** 47
  e 28   **f** 23   **g** 48   **h** 31
  i 38   **j** 100
  question 2
  i 37   **ii** 29   **iii** 39   **iv** 40
  v 32   **vi** 59   **vii** 38   **viii** 55
  ix 75   **x** 41
4 **a**

Week	1	2	3	4	5
Money in piggy bank (£)	2	4	6	8	10

  b 10 weeks

Exercise 13B

1 **a** 12, 10; subtract 2   **b** 9, 7; subtract 2
  c 35, 30; subtract 5   **d** 22, 17; subtract 5
  e 10, 7; subtract 3   **f** 11, 9; subtract 2
  g 17, 10; subtract 7   **h** 13, 10; subtract 3
  i 13, 9; subtract 4   **j** 50, 40; subtract 10
2 **a, b** **i** 25, 21; subtract 4   **ii** 15, 12; subtract 3
    iii 43, 39; subtract 4   **iv** 22, 19; subtract 3
    v 18, 15; subtract 3   **vi** 37, 31; subtract 6
    vii 14, 12; subtract 2   **viii** 31, 26; subtract 5
    ix 36, 29; subtract 7   **x** −4, −6; subtract 2
3 question 1
  a 2   **b** −1   **c** 10   **d** −3
  e −5   **f** 1   **g** −18   **h** −2
  i −7   **j** −10

Answers

question 2

i 5   ii 0   iii 23   iv 7

v 3   vi 7   vi 4   viii 6

ix 1   x −10

4

Day	M	Tu	W	Th	F
Money left at end of day (£)	17	14	11	8	5

£5

Exercise 13C

1 a 16, 32; multiply by 2   **b** 256; multiply by 4

c 25, 625; multiply by 5   **d** 1000, 10 000; multiply by 10

e 48, 96; multiply by 2   **f** 54, 162; multiply by 3

g 128, 512; multiply by 4   **h** 20 000, 200 000; multiply by 10

i 250, 1250; multiply by 5   **j** 375; multiply by 5

2 a, b   **i** 32, 64; multiply by 2

ii 243, 729; multiply by 3

iii 1024, 4096; multiply by 4

iv 3125, 15 625; multiply by 5

v 80, 160; multiply by 2

vi 324, 972; multiply by 3

vii 810, 2430; multiply by 3

viii 50 000, 500 000; multiply by 10

ix 160, 320; multiply by 2

x 7776, 46 656; multiply by 6

3 question 1

a 512   **b** 262 144   **c** 1 953 125

d 1 000 000 000   **e** 1536   **f** 39 366

g 524 288   **h** 2 000 000 000

i 3 906 250   **j** 5 859 375

question 2

i 1024   **ii** 59 049   **iii** 1 048 576

iv 9 765 625   **v** 2560   **vi** 78 732

vii 196 830   **viii** 5 000 000 000

ix 5120   **x** 60 466 176

4 a

Month	1	2	3	4	5
Number of rabbits	2	4	8	16	32

b 1024 rabbits

Exercise 13D

1 a 4, 2; divide by 2   **b** 16; divide by 4

c 25, 5; divide by 5   **d** 100, 10; divide by 10

e 12, 6; divide by 2   **f** 6; divide by 3

g 64, 32; divide by 2   **h** 300, 30; divide by 10

i 50, 10; divide by 5   **j** 2, 0.2; divide by 10

2 a, b   **i** 1, $\frac{1}{2}$; divide by 2   **ii** 3, 1; divide by 3

iii 8, 4; divide by 2   **iv** 1, $\frac{1}{5}$ (0.2); divide by 5

v 5, 2.5; divide by 2   **vi** 12, 4; divide by 3

vii 30, 10; divide by 3   **viii** 5, 0.5; divide by 10

ix 10, 5; divide by 2   **x** 1, $\frac{1}{6}$; divide by 6

3 question 1

a 0.5   **b** 0.0625   **c** 0.04   **d** 0.01

e 1.5   **f** $\frac{2}{9}$ **g** 8   **h** 0.03

i 0.08   **j** 0.0002

question 2

i $\frac{1}{8}$ **ii** $\frac{1}{9}$ **iii** 1   **iv** 0.008

v 0.625   **vi** $\frac{4}{9}$ **vii** $\frac{10}{9}$ **viii** 0.005

ix 1.25   **x** $\frac{1}{216}$

---

4 a

Years	0	10	20	30	40
Number of atoms	2560	1280	640	320	160

b 2.5 atoms

Exercise 13E

1 a i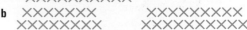

ii Add 1 cross to each row

iii

b

ii Add 2 crosses to each row

iii

c i

ii Add 6 matches

iii

d i

ii Add 2 matches

iii

e

ii Add 1 cross to each row

iii

2 a 5, 9, 13

b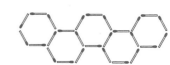

c Add 4 matches

d 41 matches

3 a 6, 11, 16

b

c Add 5 matches

d 51 matches

13.2 Get ready

1 a 8

b 6

c 14

d 20

2 a 4

b 0

c −4

Exercise 13F

1 a

Term number	Term
1	3
2	6
3	9
4	12

b Multiply the term number by 3   **c** Add 3

2 a

Term number	Term
1	7
2	14
3	21
4	28

b Multiply the term number by 7   **c** Add 7

3 a

Term number	Term
1	4
2	8
3	12
4	16

b Multiply the term number by 4   **c** Add 4

4 a

Term number	Term
1	2
2	4
3	6
4	8

b Multiply the term number by 2   **c** Add 2

5 a

Term number	Term
1	8
2	16
3	24
4	32

b Multiply the term number by 8   **c** Add 8

6 a

Term number	Term
1	10
2	20
3	30
4	40

b Multiply the term number by 10   **c** Add 10

7 a

Term number	Term
1	12
2	24
3	36
4	48

b Multiply the term number by 12   **c** Add 12

8 a

Term number	Term
1	50
2	100
3	150
4	200

b Multiply the term number by 50
c Add 50

Exercise 13G

1 a

Term number	1	2	3	4	5
Term	5	8	11	14	17

b Multiply by 3 and add 2
c Add 3

2 a

Term number	1	2	3	4	5
Term	3	5	7	9	11

b Multiply by 2 and add 1
c Add 2

3 a

Term number	1	2	3	4	5
Term	2	5	8	11	14

b Multiply by 3 and subtract 1
c Add 3

4 a

Term number	1	2	3	4	5
Term	1	5	9	13	17

b Multiply by 4 and subtract 3
c Add 4

5 a

Term number	1	2	3	4	5
Term	3	8	13	18	23

b Multiply by 5 and subtract 2
c Add 5

6 a

Term number	1	2	3	4	5
Term	7	10	13	16	19

b Multiply by 3 and add 4
c Add 3

7 a

Term number	1	2	3	4	5
Term	8	11	14	17	20

b Multiply by 3 and add 5
c Add 3

8 a

Term number	1	2	3	4	5
Term	4	9	14	19	24

b Multiply by 5 and subtract 1
c Add 5

9 a

Term number	1	2	3	4	5
Term	7	11	15	19	23

b Multiply by 4 and add 3
c Add 4

10 a

Term number	1	2	3	4	5
Term	1	6	11	16	21

b Multiply by 5 and subtract 4
c Add 5

Answers

Exercise 13H

1

Term number	Term
1	4
2	7
3	10
4	13
5	16
↓	↓
10	31
↓	↓
11	34

2

Term number	Term
1	1
2	3
3	5
4	7
5	9
↓	↓
10	21
↓	↓
13	25

3

Term number	Term
1	8
2	13
3	18
4	23
5	28
↓	↓
10	53
↓	↓
15	78

4

Term number	Term
1	1
2	5
3	9
4	13
5	17
↓	↓
10	37
↓	↓
12	45

5

Term number	Term
1	11
2	21
3	31
4	41
5	51
↓	↓
10	101
↓	↓
15	151

6

Term number	Term
1	2
2	7
3	12
4	17
5	22
↓	↓
10	47
↓	↓
14	67

7 a 39   **b** 12   **8 a** 49   **b** 14
9 a 35   **b** 15

13.3 Get ready

1 a 5   **b** -5 **c** 3   **d** 4   **e** -3
2 a 50   **b** -5 **c** 31   **d** 43   **e** 23

Exercise 13I

1 (1) $3n + 1$ (2) $2n - 1$ (3) $5n + 3$
2 a $2n - 1$; 39   **b** $2n + 1$; 41   **c** $3n - 1$; 59
   d $3n + 2$; 62   **e** $4n - 3$; 77   **f** $4n - 2$; 78
   g $5n - 3$; 97   **h** $5n - 1$; 99   **i** $5n + 3$; 103
   j $2n + 3$; 43   **k** $45 - 5n$; -55 **l** $40 - 2n$; 0
   m $38 - 3n$; -22 **n** $22 - 2n$; -18 **o** $21 - 2n$; -19
   p $200 - 10n$; 0
3 a

b

Pattern number	1	2	3	4	5	6
Number of sticks	6	10	14	18	22	26

   c $4n + 2$ **d** 82 sticks

13.4 Get ready

1 a $3n + 1, 61$ **b** $5n - 2, 98$ **c** $2n + 11, 51$

Exercise 13J

1 The sequence is of all the odd numbers, so 21 is a member
   of the number pattern because it is an odd number.
   34 is not in the pattern as it is not an odd number.

2 The sequence is of all the odd numbers above 1, so 63 is a member of the number pattern because it is an odd number higher than 1.
86 is not in the pattern as it is not an odd number.

3 The nth term is $3n - 1$. If 50 is in the pattern, $50 = 3n - 1$, giving $n = 17$, so 50 is in the pattern.
None of the terms are a multiple of 3. 66 is a multiple of 3, so 66 is not in the pattern.

4 The nth term is $3n + 2$. If 50 is in the pattern, $50 = 3n + 2$, giving $n = 16$, so 50 is in the pattern.
The nth term is $3n + 2$. If 62 is in the pattern, $62 = 3n + 2$, giving $n = 20$, so 62 is in the pattern.

5 The nth term is $4n - 3$. If 101 is in the pattern, $101 = 4n - 3$, giving $n = 26$, so 101 is in the pattern.
The nth term is $4n - 3$. If 150 is in the pattern, $150 = 4n - 3$, giving $n = 38.25$, so 150 is not in the pattern.

6 All the terms are even, so 101 is not in the pattern.
The nth term is $4n - 2$. If 98 is in the pattern, $98 = 4n - 2$, giving $n = 25$, so 98 is in the pattern.

7 The sequence contains all the numbers ending in 2 or 7, so 97 is in the pattern.
All the numbers end in 2 or 7, so 120 is not in the pattern.

8 All the numbers end in 4 or 9, so 168 is not in the pattern.
The sequence contains all the numbers ending in 4 or 9, so 169 is in the pattern.

9 The sequence contains all the numbers below 45 ending in 0 or 5, so 85 is not in the pattern, as it is higher than 40.
All the numbers end in 0 or 5, so 4 is not in the pattern.

10 All the terms are even, so 71 is not in the pattern.
The sequence contains all the even numbers below 40, so 82 is not in the pattern as it is higher than 38.

11 All the terms are odd, so 46 is not in the pattern.
The nth term is $4n - 1$. If 79 is in the pattern, $79 = 4n - 1$, giving $n = 20$, so 79 is in the pattern.

12 The nth term is $6n - 1$. If 119 is in the pattern, $119 = 6n - 1$, giving $n = 20$, so 119 is in the pattern.
All the terms are odd, so 72 is not in the pattern.

Review exercise

1 a

b 41 squares

2 a 116   **b** 112
   c No. 9 is an odd number and all the terms in the sequence are even.

3

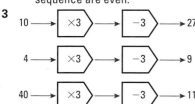

4 a

Input	Output
4	−2
2	−6
−3	−16

   b $2(n - 5)$

5 No. When $n = 9$, $n^2 + 4 = 85$, which is not a prime number.

6 a 8, 13   **b** 55

7 Dylan could be right if you are doubling the previous term. Evie could be right if the difference between the terms was increasing by one.

8 The nth term is $3n + 2$. If 140 is in the sequence, $140 = 3n + 2$, giving $n = 46$, so 140 is in the sequence.

9 $x = 4.5$

10 a 31   **b** $4n - 1$

Chapter 14 Answers

14.1 Get ready

1 3 cm, 2 cm   **2** 4 cm, 5 cm, 5 cm

Exercise 14A

1	**A** 16 cm		**B** 20 cm		**C** 20 cm	
2	**a** 26 cm		**b** 15 cm		**c** 13.7 cm	
3	**a** 12 cm		**b** 10.5 cm		**c** 11 cm	
4	9 m					

Exercise 14B

1 40 cm

2 a 24 cm   **b** 102 mm   **c** 32 cm

3 34 cm

4 7.2 cm

14.2 Get ready

1 a 4   **b** 5   **c** 3

Exercise 14C

1 12 cm²

2 A 9 cm²   B 15 cm²   C 22 cm²   D 22 cm²

3 T_1 8 cm² T_2 9 cm² T_3 10 cm² T_4 12 cm²

4 F 12 cm²   G 18 cm²   H 22 cm²

5 P 25 cm²   Q 13 cm²   R 28 cm²

Mixed Exercise 14D

1 a 18 cm   **b** 14 cm²

2 21 cm²   **3** 24 mm   **4** 13 m

14.3 Get ready

1 6 cm and 2 cm   **2** 4.5 cm and 5 cm   **3** 2 cm and 3 cm

Exercise 14E

1 a 15 cm²   **b** 14 mm²   **c** 40 m²

2 a 9 cm²   **b** 20.25 cm²   **c** 13.69 cm²

3 19.76 m²

Exercise 14F

1 a 6 cm²   **b** 48 m²   **c** 15 cm²
   d 21.6 cm²   **e** 16.74 mm²

2 30 cm²

3 4500 cm²

Answers

Exercise 14G

1 **a** 10 cm² **b** 12 cm² **c** 12 cm²
2 24 000 cm²

Exercise 14H

1 15 cm²   55 cm²   48 m²   30 m²
2 **a** 14 cm² **b** 40 cm² **c** 15 cm² **d** 64 m²
3 4 cm

14.4 Get ready

1

Exercise 14I

1 **a** 100 cm² **b** 36 cm² **c** 64 cm²
2 150
3 £103.80
4 **a** 2 **b** 220 g
5 **a** 88 cm² **b** 22.5 m² **c** 108 m²
6 **a** 150 cm² **b** 36 cm² **c** 55.5 mm²

Review exercise

1 **a** 14 cm **b** 6 cm²
2 **a** 60 cm **b** 200 cm²
3 60 m
4 7.8 m by 2.9 m
5 28 cm²
6 **a** 11.25 cm² **b** 18 cm
7 264 paving stones
8 3 tins needed for 63.5 m²
9 45 cm²
10 45 cm²
11 56 cm²
12 £31 425

Multiplication

1 £249.50
2 Nan must give £35.80
3 £48.46

Number

1 An even number of 5p coins will give a even total. The total will remain even when you add 2p coins, so an odd total such as 23p cannot be made.
2 3 fish and chips, 2 sausage and chips
3 One 5 kg pack and four 3 kg packs
4 4%
5 20% off the price is the best value for the customer
6 Prices are increased by 12%, as 0.8 × 1.4 = 1.12

2D Shapes

1 $x = 30°$ 2 65 3 3.53 km
4 3.66 m 5 24 cm

Interpreting and displaying data

1
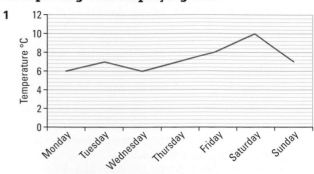

2
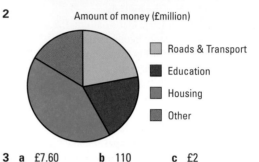

Amount of money (£million)

3 **a** £7.60 **b** 110 **c** £2
   d When a scatter graph is drawn there is a positive correlation between the number of pages and the price of the book. A line of best fit can be drawn and estimates made.
   The estimate for part **c** is the least reliable estimate, as it is outside the range of data used to draw the graph.

4
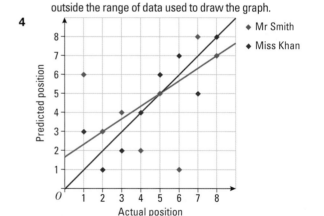

Miss Khan was closest to predicting the actual positions.

Area

1 Any two shapes made of 6 squares where some of the squares are joined at more than one edge.
2 £210
3 £248
4 Some possibilities are 74 cm, 40 cm, 30 cm and 26 cm
5 £8

Average and range

1 Hatton: mean = 36.5; range = 8
   Georgie: mean = 37.5; range = 24
   Hatton's team were more consistent with a lower range but due to two high-scoring members the mean for Georgie's team was higher.

2 Any 3 different prices with mean of £8.97
e.g. £8.96   £8.97   £8.98

3 **a** e.g. 1, 2, 3, 4, 5; 1, 1, 4, 4, 5
b e.g. 1, 2, 3, 4, 5; 2, 2, 2, 4, 5
c e.g. 1, 2, 3, 4, 5; 1, 1, 1, 5, 7

4 50 minutes

5 Farmer Pearce: mean = 1477.5 g; range = 1000 g
Farmer Hicken has the larger mean but also the larger range so his lambs are less consistent. Some may be very large and others very small.

6

	Florida	Varadero	Red Sea
Average Max Temp °C	27	26.8	26
Average Min Temp °C	15	19.8	16
% humidity	7.2	14	0

Florida has the highest temperature during the day but it also has the lowest temperature at night.

Varadero's maximum temperature is only slightly lower but the temperatures don't drop as much at night. It has the highest humidity of the three resorts.

The best choice for most people is the Red Sea. The day and night temperatures are less extreme, the maximum temperature is only 1° lower than Florida and humidity is zero.

Probability

1 The modes of transport are not equally likely. The probability of travelling to school by bus is $\frac{7}{30}$.

2 **a** He is only correct if each question has only two answers and he chooses at random.
b Most tests have four or five answers so he is unlikely to be correct.

3 The possible outcomes are HTT, HHT, HTH, HHH, THT, TTH, THH, TTT. So the probability of HHH or TTT is $\frac{2}{8}$ or $\frac{1}{4}$.

4 $\frac{1}{9}$

5 0.3

6 $\frac{23}{100}$ or 23% or 0.23

Holiday	July	Aug	Sept	Total
Hotel	18	23	16	57
Self Catering	14	19	10	43
Total	32	42	26	100

2012 Olympics

1 7.17 am

2 750 euros

3

	Athlete 1	Athlete 2	Athlete 3
Swimming	843	677	1060
Shooting	1216	1252	916
Running	1272	1104	984
Show Jumping	1116	1200	1172
Fencing	1048	832	832
Total	5495	5065	4964

Athlete 1 has the highest total score.

Learning to drive

1

Speed (mph)	Thinking distance (m)	Breaking distance (m)	Total distance (m)
20	6	6	12
30	9	14	23
40	12	24	36
50	15	38	53
60	18	55	73
70	21	75	96

2 Minimum cost = £757.49

3 £1920

Healthy living

1 Samantha ate 34.2 g of fat and 3 g of salt. Daisy ate 44.4 g of fat and 3.5 g of salt. Darren ate 40.8 g of fat and 2.9 g of salt. Samantha ate the least fat. Daisy ate the most fat and salt. Darren ate the least salt.

2 There are roughly 123 000 obese children in Year 6 and 57 000 obese children in Reception.

There are roughly 91 000 overweight children in Year 6 and 78 000 overweight children in Reception.

There are more obese children than overweight children in Year 6. There are more overweight children than obese children in Reception. There are more obese and overweight children in Year 6 than in Reception.

3 She should spend 12 minutes (0.2 hours) on flexibility, 24 minutes (0.4 hours) on aerobic exercise and 84 minutes (1.4 hours) on anaerobic exercise.

Money management

1 £142.88

2 £5.28 per week = £15.84 for three weeks

3 £6545 in income tax and £3526.60 in NIC   Total £10,071.60

Kitchen Design

1 No set answer but all conditions must be met for full marks

2 Using 305 mm square £132   Using 330 mm square £130

3

	White	Coloured	Cost
10 × 10	54	36	£7.98 + £2 + £9.98 = £19.96
15 × 15	24	16	£8.99 + £8 = £16.99

University

1 Elaine 221 square units = £245.26
Ryan 117 square units = £129.85
Danielle 117 square units = £129.85
Rashid 331 square units = £367.34
Saria 225 square units = £249.70

2 375 g beef mince, 2 onions, 10 mushrooms, 2 cans tomatoes, 125ml tomato puree, 750 g spaghetti

3 Elaine needs a further 18.7 marks to reach 40%.

She could get this entirely from Unit 6 with a mark of $\frac{33}{60}$.
So even getting zero in Unit 5 would give her a chance of passing.

Chapter 15 Answers

15.1 Get ready

1 a One square to the right and one down
b Two squares to the right
c One square up

Exercise 15A

1 a i Ghost Ride   **ii** Big Dipper   **iii** Helter Skelter
   iv Exit   **v** Dodgems
b i (3, 0)   **ii** (4, 3)   **iii** (0, 6)
   iv (1, 2)   **v** (4, 5)
2 a i B   **ii** F   **iii** A
   iv E   **v** O
b i (0, 0)   **ii** (6, 2)   **iii** (3, 0)
   iv (2, 7)   **v** (0, 4)
3 a i U   **ii** T   **iii** S
   iv P   **v** V
b i (3, 1)   **ii** (6, 2)   **iii** (5, 4)
   iv (5, 1)   **v** (3, 7)

4
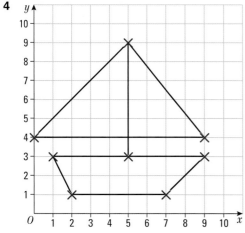

5
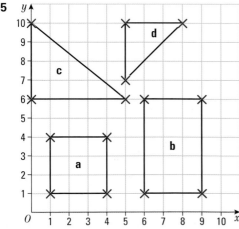

6 a, b
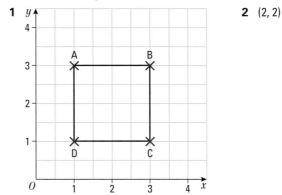
c (1, 5)

7 a, b
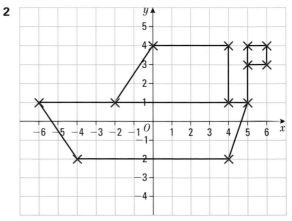
c (3, 5)

15.2 Get ready

1

2 (2, 2)

Exercise 15B

1 A (5, 2)   B (2, 4)   C (0, 1)   D (−2, 4)
   E (−4, 2)   F (−3, 0)   G (−2, −3)   H (4, −1)
   I (5, 0)   J (−4, −2)   K (0, −2)   L (2, −1)

2

3 Students' pictures and coordinates

15.3 Get ready

1 **a** 25   **b** 35   **c** 5   **d** 6
   e 1   **f** 2   **g** 1   **h** 1.5

Exercise 15C

1 **a** (2, 3)   **b** (4, 3)   **c** (7, 2.5)
   d (3.5, 3.5)   **e** (2.5, 2)
2 **a** (4, 4)   **b** (4, 6)   **c** (5.5, 5)
   d (4.5, 4.5)
3 **a** (5, 5)   **b** (4, 6.5)   **c** (3.5, 4.5)
   d (4, 4.5)   **e** (4, 5.5)   **f** (4.5, 4.5)

Exercise 15D

1 **a** (1, 2)   **b** (3.5, −0.5)   **c** (−1.5, 3)
   d (−2.5, −0.5)   **e** (−4, 0.5)   **f** (−0.5, −2)
   g (2.5, −0.5)   **h** (0.5, 2.5)   **i** (−2, 3)
   j (4.5, −0.5)
2 **a** (1.5, 0.5)   **b** (4, 1)
   c (1.5, 1)   **d** (3.5, 2.5)
3 **a** (4, 4)   **b** (−2, 2.5)   **c** (1.5, −3.5)
   d (−3, 4.5)   **e** (2, −2.5)   **f** (2.5, −1.5)

15.4 Get ready

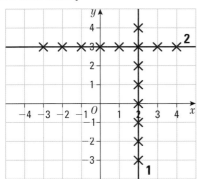

3 (2, 3)

Exercise 15E

1 **a** $y = 3$ **b** $y = 1$ **c** $y = -1$ **d** $y = -3$
2 **a** $x = -4$ **b** $x = -2$ **c** $x = 1$ **d** $x = 4$
3

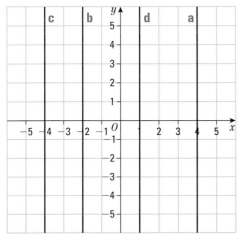

4

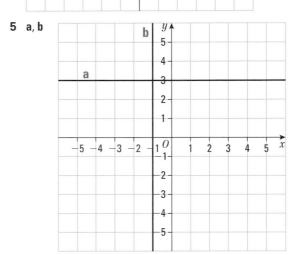

5 **a, b**

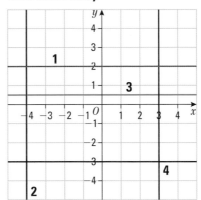

   c (−1, 3)

15.5 Get ready

Exercise 15F

1 a i

x	−3	−2	−1	0	1	2	3
$y = x - 1$	−4	−3	−2	−1	0	1	2

ii

x	−3	−2	−1	0	1	2	3
$y = 2x - 4$	−10	−8	−6	−4	−2	0	2

iii

x	−3	−2	−1	0	1	2	3
$y = 3x + 1$	−8	−5	−2	1	4	7	10

iv

x	−3	−2	−1	0	1	2	3
$y = x + 4$	1	2	3	4	5	6	7

v

x	−2	−1	0	1	2
$y = 4x + 1$	−7	−3	1	5	9

b

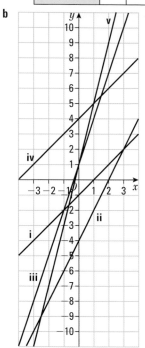

4

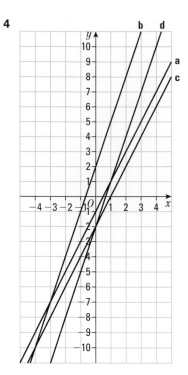

5

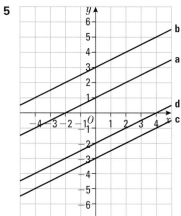

2

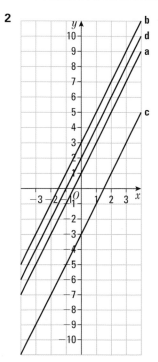

3

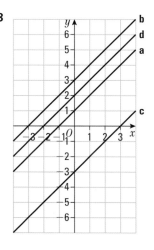

Exercise 15G

1 a i

x	−3	−2	−1	0	1	2	3
$y = -x - 1$	2	1	0	−1	−2	−3	−4

ii

x	−3	−2	−1	0	1	2	3
$y = -2x - 4$	2	0	−2	−4	−6	−8	−10

iii

x	−3	−2	−1	0	1	2	3
$y = -3x + 1$	10	7	4	1	−2	−5	−8

iv

x	−3	−2	−1	0	1	2	3
$y = -x + 4$	7	6	5	4	3	2	1

v

x	-2	-1	0	1	2
$y = -4x + 1$	9	5	1	-3	-7

b

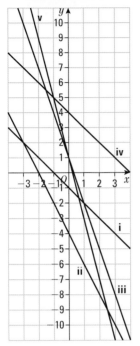

2

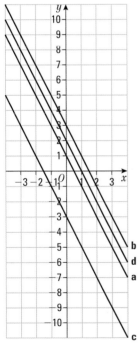

3

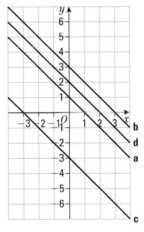

4

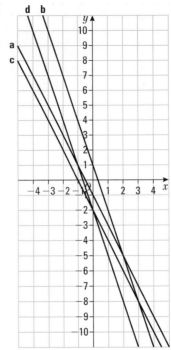

5

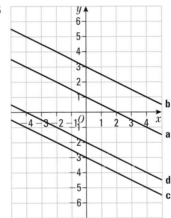

15.6 Get ready

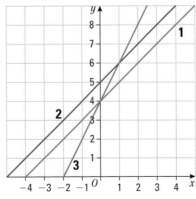

They both cross the y-axis at $y = 4$.

Exercise 15H

1
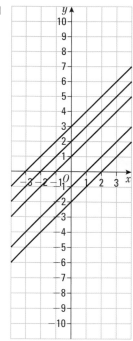

2
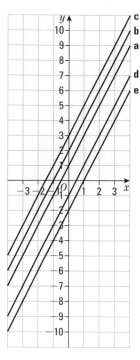

5
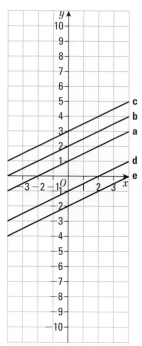

For each question, the lines are parallel.

3
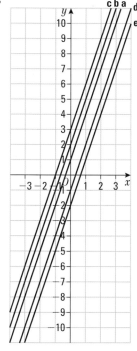

4
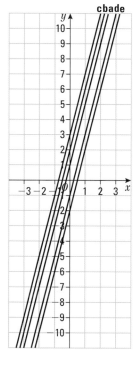

Exercise 15I

1
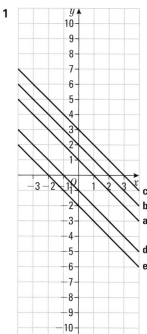

2
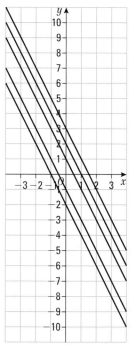

3

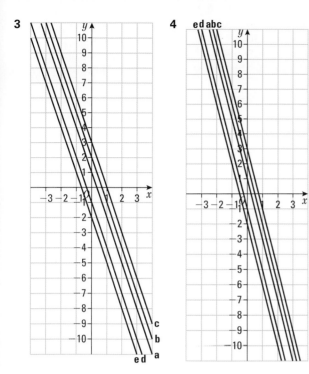

4 ed abc

5

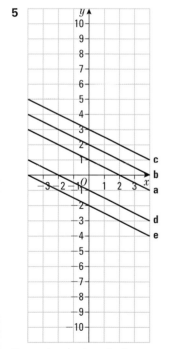

For each question, the lines are parallel.

1

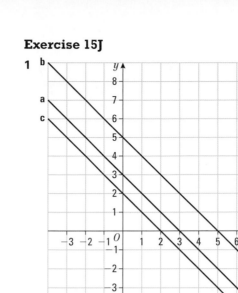

2

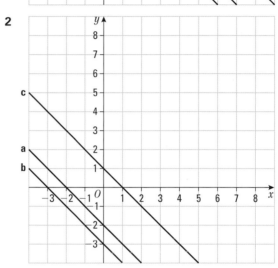

3

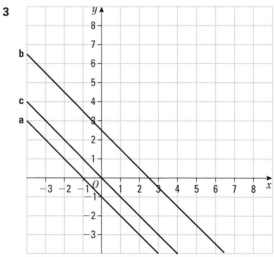

Answers

15.7 Get ready

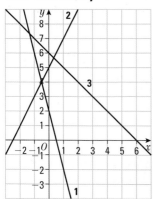

Exercise 15K

1 a $y = x + 3$ **b** $y = 2x + 1$ **c** $x + y = 4$
  d $y = -\frac{4}{3}x + 1$

2 a $y = 3$ **b** $y = 2x + 3$ **c** $y = -3x + 1$
  d $x = -3$ **e** $x + y = 2$

3 a $y = -1$ **b** $y = 3x + 2$ **c** $y = -\frac{1}{2}x + 1$
  d $x = 3$ **e** $y = -2x - 1$

4 a $y = 1$ **b** $y = 2x + 1$ **c** $y = -4x + 1$
  d $x = 1$ **e** $x + y = 1$

Review exercise

1 a $(3, 3)$ **b** $(4, 0)$
  c

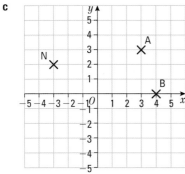

  d $(-3, 0)$

2 i B   **ii** C   **iii** D   **iv** A

3 a

x	-1	0	1	2	3
$y + 2x = 6$	8	6	4	2	0

  b

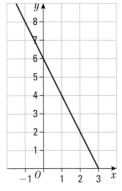

  c $y = 1.2$ **d** $x = -0.3$

4

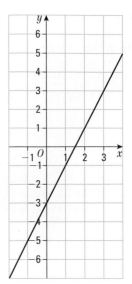

5

6

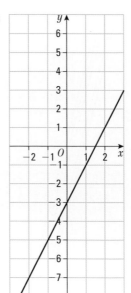

  c $x = 2.67$

7 a i $(-6, 4)$ **ii** $(3, 4)$ **iii** $(3, -2)$
  b i

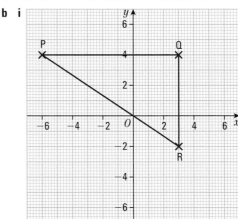

  ii right-angled triangle
  c i $(3, 1)$ **ii** $(-1.5, 4)$ **iii** $(-1.5, 1)$
  d $(-6, -2)$

8 a i $(2, 3)$ **ii** $(2, -1)$
  b $(-4, -1)$
  c $(2, 1)$

9 A $y = 2x + 1$
  B $y = -3$
  C $y = 3x - 3$

Chapter 16 Answers

16.1 Get ready

1 a 284 296 302 325 336
  b 0.55 0.59 0.6 0.61 0.625
2 a 38 times   **b** Green   **c** Two more times

Exercise 16A

1 a 2   **b** 10 and 12   **c** No mode   **d** Dog
2 16
3 Silver
4 £7.20 and £8.20

Exercise 16B

1 a 5   **b** 7   **c** 6.5   **d** 4   **e** 5.5
2 1.2°C
3 39.5 thousands of tonnes
4 a £26 100   **b** £22 000–£27 500

Exercise 16C

1 a 6   **b** 5   **c** 7.5
2 a 4 meals   **b** about 100
3 a 190 cm   **b** 186 cm   **c** 185 cm
4 a No mode   **b** 131 emails   **c** 131.5 emails

16.2 Get ready

1 a 48   **b** 54   **c** 58.9 (3 s.f.)

Exercise 16D

1 The value of the mean will be affected by the extreme value of the most expensive jumper.
2 a The mode
  b No. The mode is the lowest of the prices of cars.
3 a 2 emails   **b** 6 emails   **c** 6 emails
  d The value of the mode is low compared to the number of junk emails received on many days. The value of the mean is affected by the extreme value of 17. The median is the best average to represent these data, though in this case it is the same as the mean.
4 Mode = 10 and 13 minutes
  Median = 13 minutes
  Mean $= \dfrac{10 + 10 + 13 + 13 + 16 + 17 + 40}{7} = 17$ minutes
  The data is numeric and there is one extreme value, so the median is the best average to use.

16.3 Get ready

1 a 13, 2
  b 29, 19
  c 161, 130

Exercise 16E

1 a 19   **b** 11   **c** 6
2 No. Petra has used the first and last scores, which are not the lowest and highest ones.
  Range = 25 − 11 = 14

3 a Economics: range = 44; Psychology: range = 15
  b Psychology has the most consistent marks, because the range is smaller. This may be because the students are not being stretched.
4 a Machine 1: range = 3 ml; Machine 2: range = 6 ml
  b Machine 1: mean = 30 ml; Machine 2: mean = 30 ml
  c Both machines filled the bottles with the same mean amount of water. Machine 1 had a smaller range. The amounts of water were less spread out which shows that the machine was more consistent.
5 Max Rangers:
  Mean =
  $$\dfrac{170 + 172 + 180 + 190 + 184 + 179 + 176 + 183 + 186 + 190 + 170}{11}$$
  = 180 cm
  Range = 190 − 170 = 20
  Red United:
  Mean =
  $$\dfrac{179 + 190 + 187 + 170 + 180 + 182 + 163 + 188 + 181 + 190 + 179}{11}$$
  = 180.818181… = 180.8 cm (to 1 d.p.)
  Range = 190 − 163 = 27
  The mean heights are very close but Red United had a wider spread due to an extreme value (163).

16.4 Get ready

1 26 35 42 58   **2** 151 152 153 154
3 0.1 0.2 0.4 0.5   **4** 0.1 0.11 0.12 0.9

Exercise 16F

1

0	1	2	3	4	6		
1	2	2	2	4	6		
2	2	3	4	5	7	9	
3	0	2	3	4	5	6	8
4	2	3	6	7			

Key: 1|2 stands for 12

2

0	6	8	9	9				
1	0	2	4	5	7	7	7	8
2	1	2	2	5	6	7		
3	1	9						

Key: 1|2 stands for 12

3 a 21 weeks   **b** 26 cars
  c 26 cars   **d** 31 cars

4

0	2	2	3	3	5	5	7	8	
1	0	2	4	5	6	6	6	8	8
2	0	2	3	3	3	4			
3	0								

  b 30 − 2 = 28 mins
  c mode = 16 and 23
  median $= \dfrac{15 + 16}{2} = 15.5$ mins
  mean $= \dfrac{335}{24} = 14$ mins (2 s.f.)

16.5 Get ready

1 a 5   **b** 30

Exercise 16G

1 a 1 car   **b** 2 cars   **c** 2 cars
   d 130 spaces. The average household (mean and median) has two cars.
2 a 65 g   **b** 65 g   **c** 65.2 g
3 a 7 peas   **b** 7 peas   **c** 6.5 peas

16.6 Get ready

1 a Discrete   **c** Discrete   **c** Continuous
2 a False   **c** True   **c** False

Exercise 16H

1 a 6 to 8   **b** 6 to 8
2 a 11 to 15   **b** 11 to 15
3 a $1.1 \leqslant d < 1.3$ **b** $1.3 \leqslant d < 1.5$

16.7 Get ready

1 a 7   **b** 60   **c** 75.5   **d** 0.8   **e** 125 000

Exercise 16I

1 7.4 packets
2 1.5 breakdowns a week
3 11.125 minutes

Review exercise

1 a

Score	Tally	Frequency
1					3
2						4
3									7
4							5
5						4
6			1

   b 3
2 a £379   **b** £400   **c** £150   **d** £65
3 a 6   **b** 6.2
4 30 mm
5 a 3.5   **b** 3.6   **c** 4
6 a 65 kg   **b** 34 kg
7 1, 4, 5, 5, 5; 2, 3, 5, 5, 5; 1, 3, 5, 5, 6; 2, 2, 5, 5, 6; 1, 2, 5, 5, 7; 1, 1, 5, 5, 8
8 a 29   **b** 46   **c** 46   **d** 45
9 a £22   **b** £34   **c** £46.70
   d The median is the best average here. The mode is the lowest value and the mean is distorted by the expensive dress (£180).
10 a Eating apple: mean 138.9 g, range 18 g
      Cooking apple: mean 148 g, range 26 g
   b On average the cooking apples are heavier, but they vary in weight more than the eating apples.

11 a

```
4 | 6 8
5 | 1 2 8
6 | 0 3 4 6 8
7 | 4 7 8 9
8 | 7
```

      4|5 stands for 45 kg
   b 5 people
   c 41 kg
12 a 15 tracks   **b** 13.5 tracks
13

```
2 | 3 7 8
3 | 1 4 5 6
4 | 1 2 4 5 5
5 | 0 2 3
```

   4|5 stands for 45 years old
14 As some (injured!) spiders will have less than eight legs, the mean number of legs will be less than eight. So the statement can be true.
15 a The mode is 7 rooms because this has the highest frequency.
   b 6.3 rooms, to 1 d.p.
16 The mean for males is 48 and the range is 48.
     The mean for females is 61.1 and the range is 70.
     On average the females get higher scores, but their scores vary more.
17 a Expression
   b 84
   c This is a suitable number as there are already two spare bottles per class.
18 Meena must score $70 \times 6 = 420$ marks in total on the 6 tests to get an average score of 70.
     So far she has $64 \times 5 = 320$ marks.
     Therefore she needs $420 - 320 = 100$ marks on the sixth test.
19 If you want to show that the average is high use the mean (£30 000). If you want to show that the average is low use the mode or the median (both £10 000).
20 a 10 to 12 and 13 to 15 are both modal classes.
   b 13 to 15
   c £13.50
21 a $31 \leqslant h < 33$
   b $29 \leqslant h < 31$
   c 30.36 cm
22 19 minutes
23 18.4 minutes
24 a $10\,000 < x \leqslant 14\,000$
   b $14\,000 < x \leqslant 16\,000$

Chapter 17 Answers

17.1 Get ready

1 a diameter   **b** radius   **c** circumference
2 a 2   **b** $\frac{1}{2}$

Exercise 17A

1 a 39.6 m   b 59.7 cm   c 26.4 cm
   d 78.5 mm   e 184.7 cm
2 a 21.7 cm   b 31.7 mm   c 16.7 cm
   d 30.5 cm   e 15.7 m
3 11.3 m
4 100.5 cm
5 a 4.49 m   b 0.94 m   c 4.78

Exercise 17B

1 a 14.5 m   b 6.4 cm   c 18.5 cm
   d 11.8 mm   e 31.8 cm
2 a 4.8 cm   b 11.4 mm   c 10.2 cm
   d 14.8 cm   e 7.9 m
3 59.8 cm

17.2 Get ready

1 78.54   2 28.27
3 483.61   4 799.44

Exercise 17C

1 a 81.7 m^2 b 28.3 cm^2 c 238 cm^2
   d 726 mm^2 e 278 cm^2
2 a 707 cm^2 b 468 mm^2 c 43.0 cm^2
   d 119 cm^2 e 50.3 m^2
3 155 m^2
4 58.1 cm^2
5 a 380 cm^2 b 81 cm^2 c 299 cm^2
6 411 cm^2

17.3 Get ready

1 76.97   2 48.69
3 123.37   4 16.49

Exercise 17D

1 a i 46.3 cm   ii 127 cm^2
   b i 25.7 m   ii 39.3 m^2
   c i 30.8 cm   ii 56.5 cm^2
   d i 14.3 cm   ii 12.6 cm^2
   e i 22.1 cm   ii 30.2 cm^2
   f i 57.8 m   ii 103 m^2
2 a 3.21 m^2 b £89.96
3 a 28.5 cm^2 b 35.7 cm

Review exercise

1 30.9 cm^2
2 a 2 cm   b 175.9 cm^2
3 50.27 cm
4 157 cm^2
5 314 cm^2
6 2.04 m
7 20.57 cm
8 a 7854 cm^2 b 125.7 cm
9 £134.40
10 £36.80

Chapter 18 Answers

18.1 Get ready

1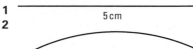

2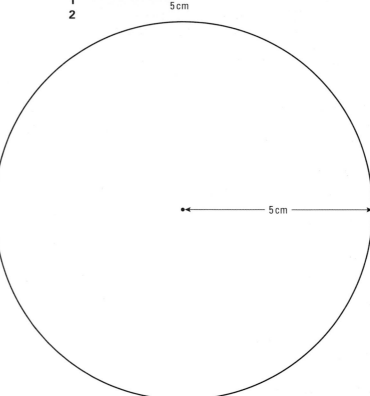

Exercise 18A

1 a 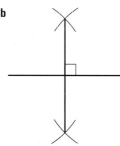   b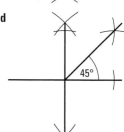

c 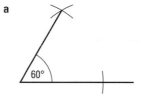   d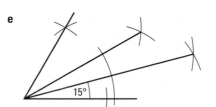

e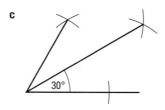

Answers

2 a

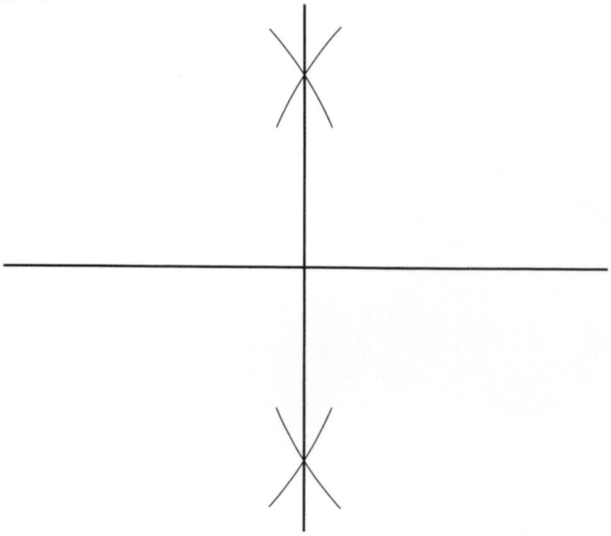

b

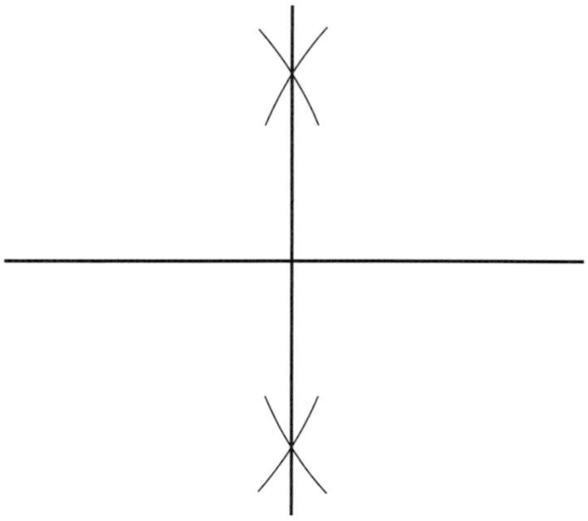

c

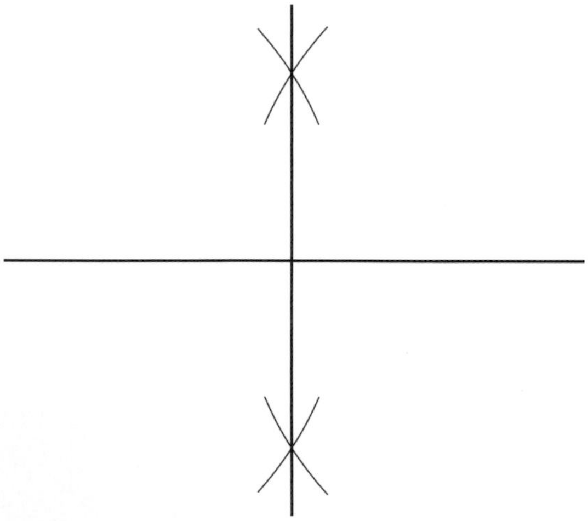

d

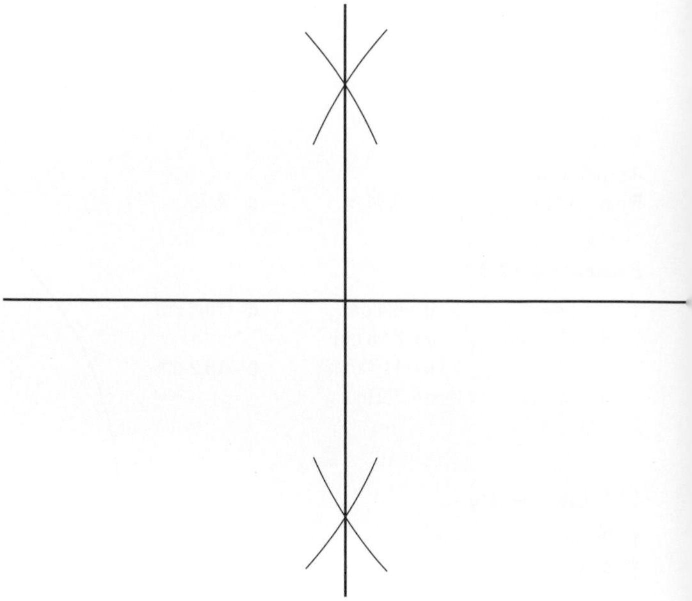

3 a

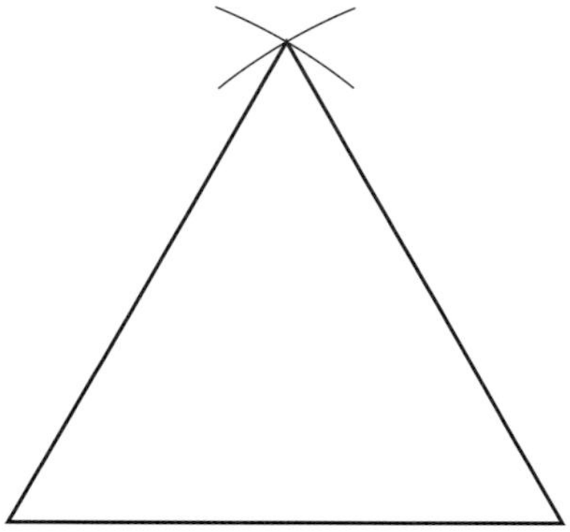

b

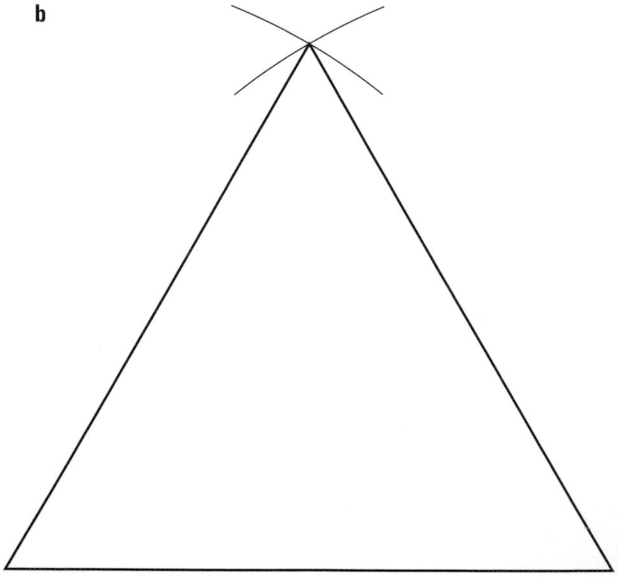

c

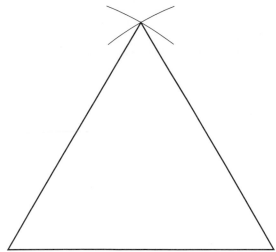

b

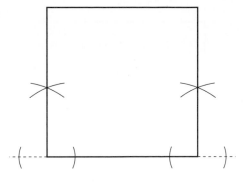

d

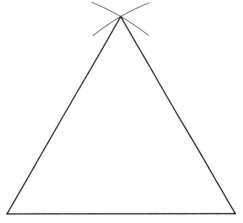

4 a

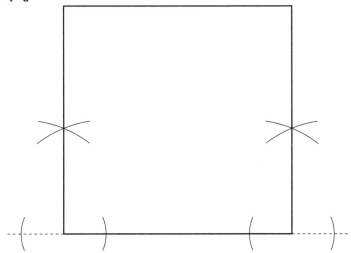

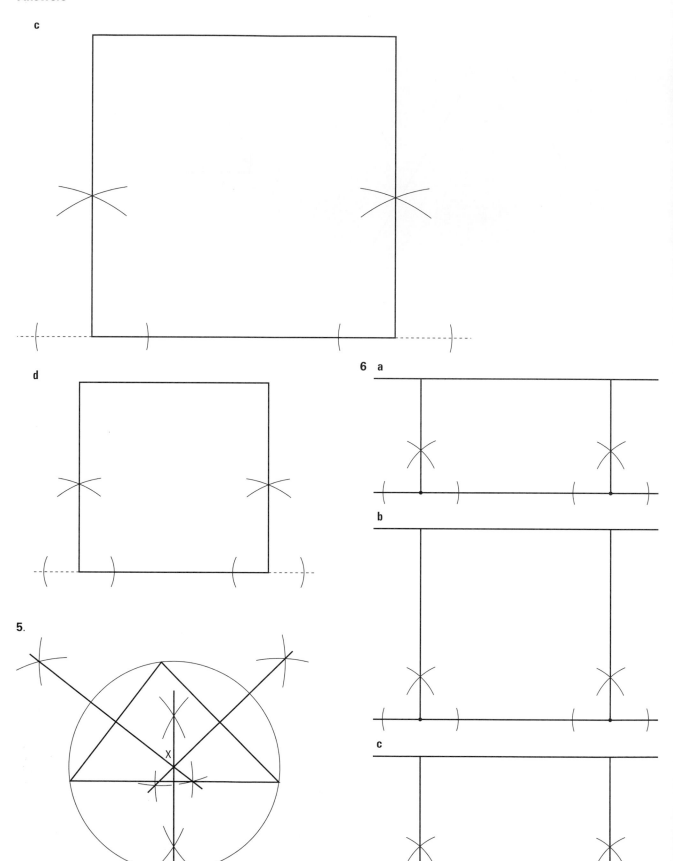

c

d

5.

6 a

b

c

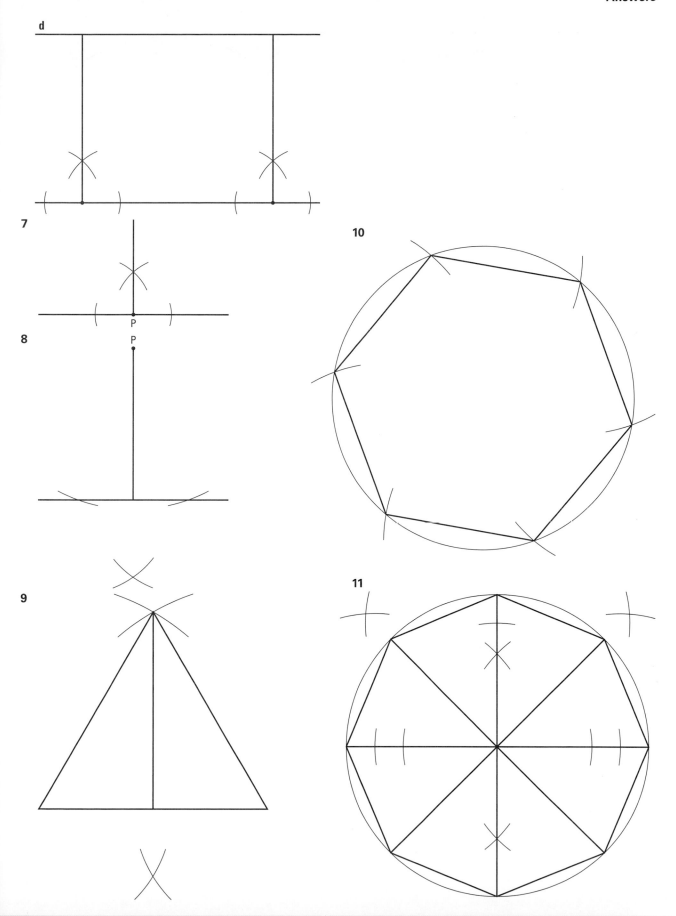

d

7

8

9

10

11

12

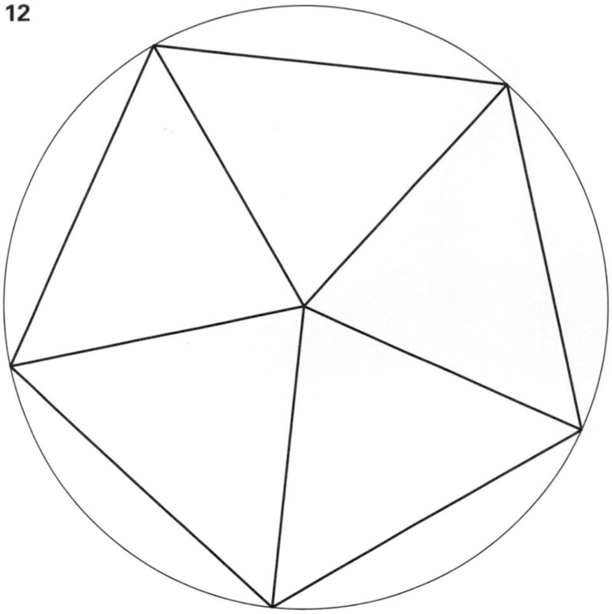

2

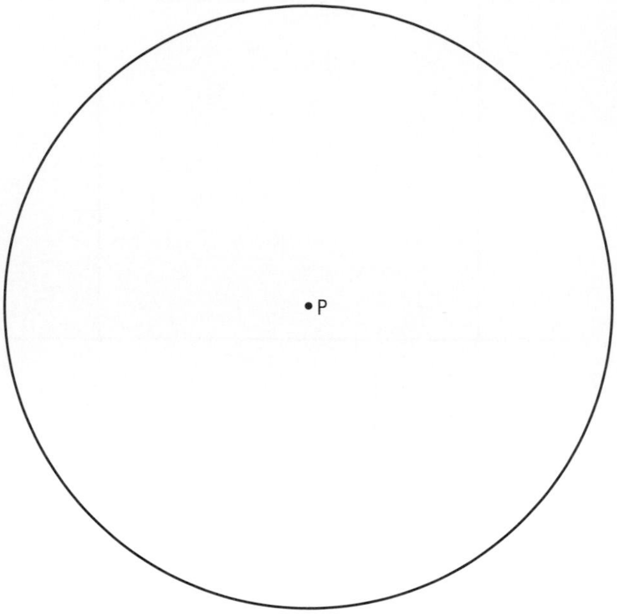

18.2 Get ready

1

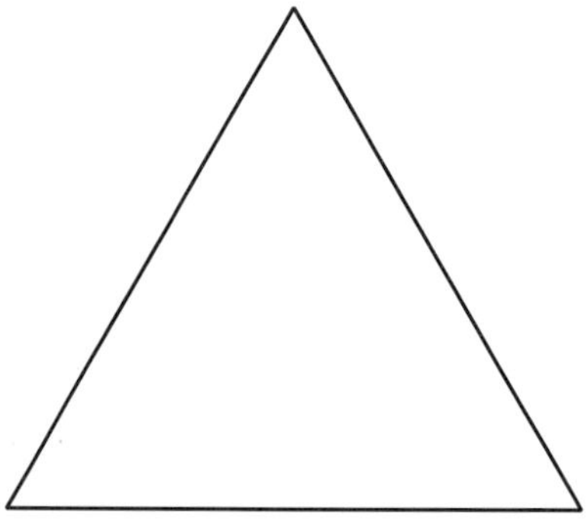

3

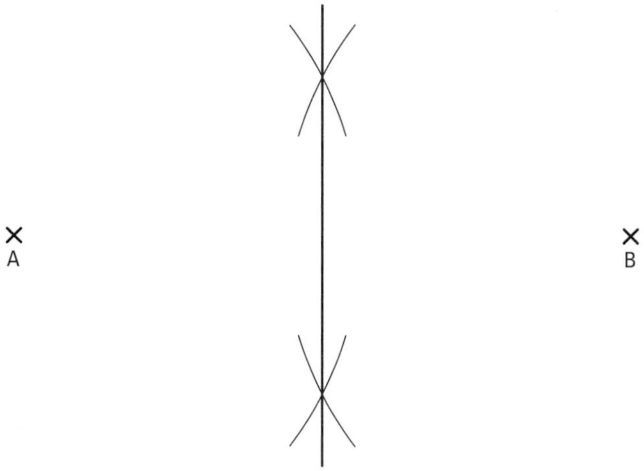

Exercise 18B

1

4

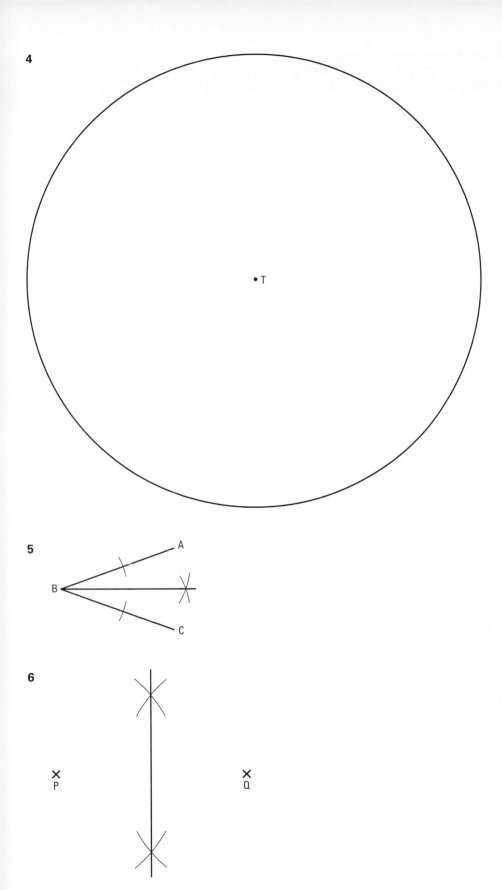

5

6

7

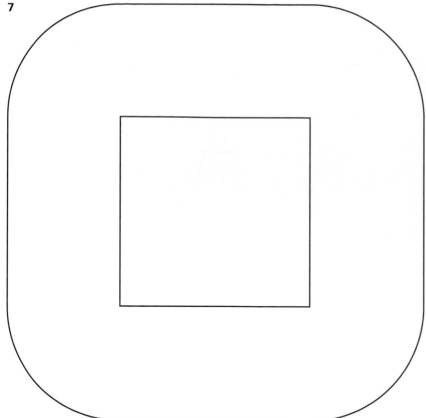

8

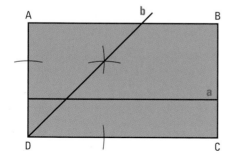

9

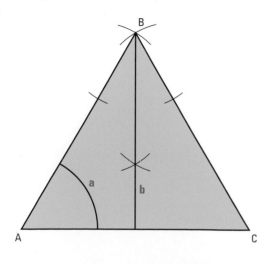

10

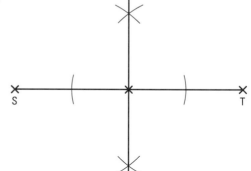

11

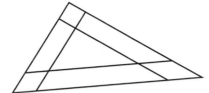

12

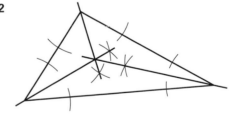

18.3 Get ready

1

Exercise 18C

1

2

3

4

5

6

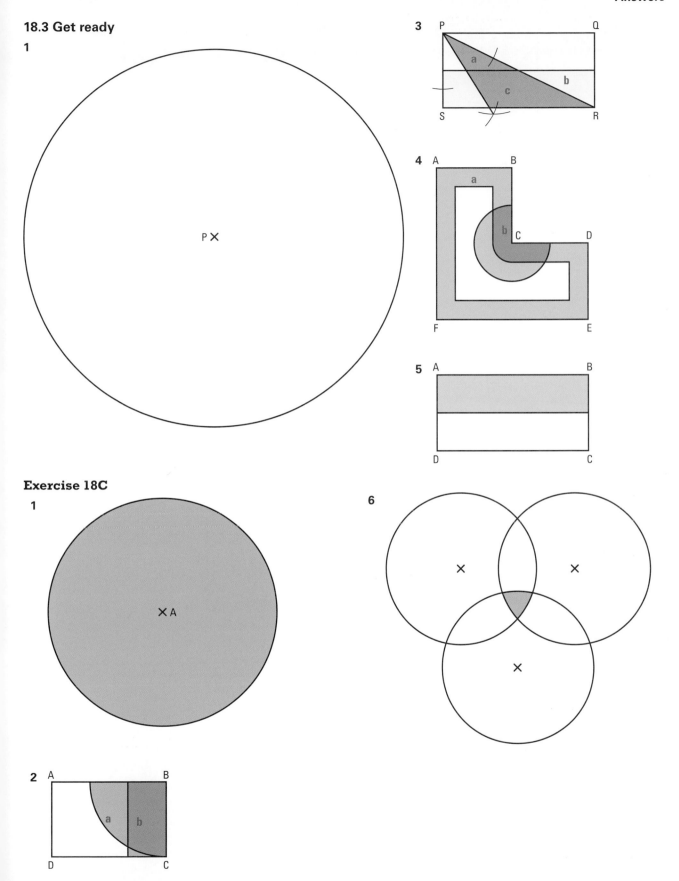

7

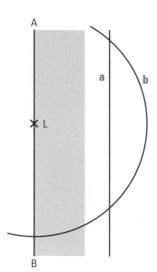

10

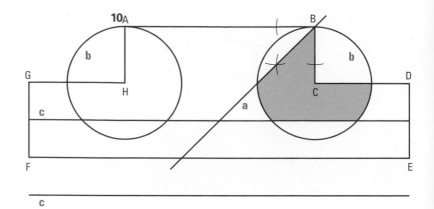

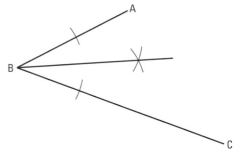

8

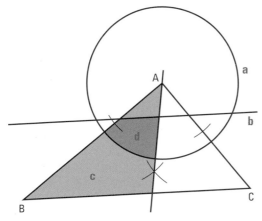

Review exercise

1

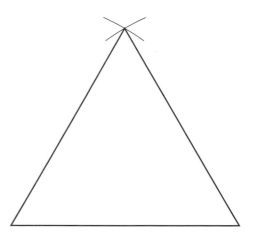

2

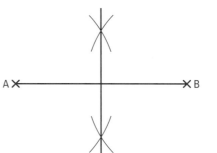

3

9

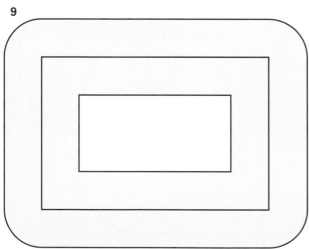

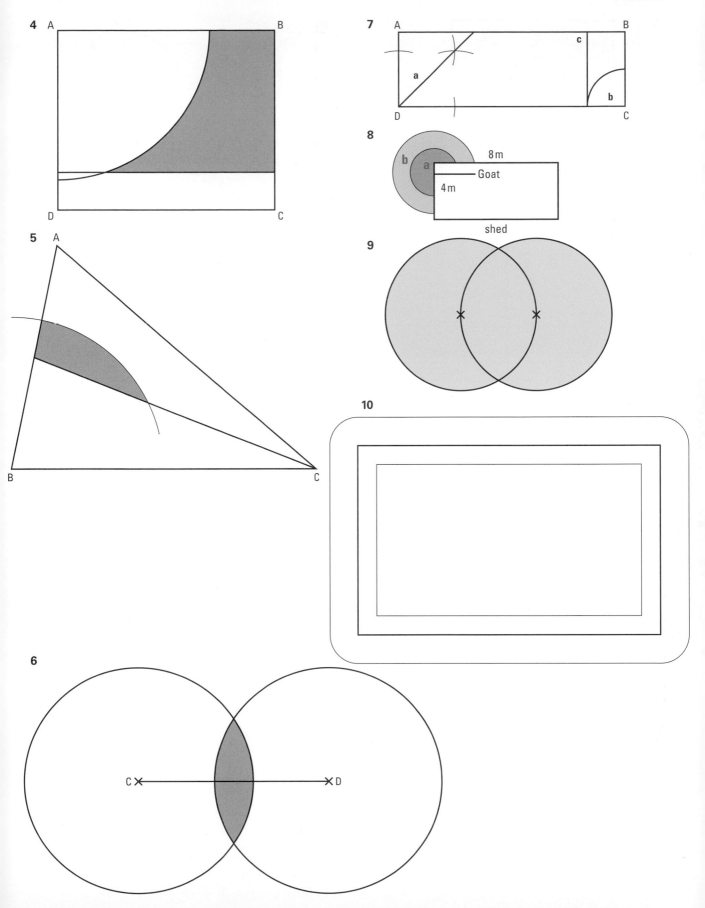

11

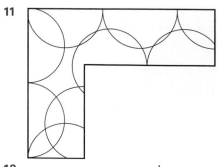

12

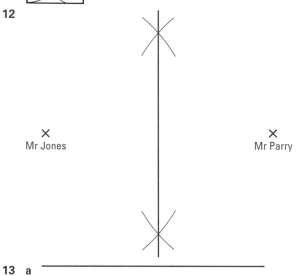

Mr Jones   Mr Parry

13 a ———————————————————

b

Chapter 19 Answers

19.1 Get ready

1 $\frac{4}{5}$ **2** $\frac{2}{5}$ **3** 0.27

Exercise 19A

1 a 70%   **b** 36%   **c** 50%   **d** 75%
   e 40%   **f** 60%
2 a 30%   **b** 64%   **c** 50%   **d** 25%
   e 60%   **f** 40%
3 a any 7 squares shaded
   b any 2 squares shaded
4 40%
5 75%
6 27%
7 65%

Exercise 19B

1 a 0.5   **b** 0.45   **c** 0.62   **d** 0.95
   e 0.29   **f** 0.3   **g** 0.03   **h** 0.07
2 1.25
3 0.125
4 0.032

Exercise 19C

1 a $\frac{3}{5}$ **b** $\frac{3}{4}$ **c** $\frac{7}{20}$ **d** $\frac{9}{10}$
   e $\frac{1}{20}$ **f** $\frac{4}{5}$ **g** $\frac{21}{25}$ **h** $\frac{8}{25}$
2 $\frac{16}{25}$
3 $\frac{6}{25}$
4 $\frac{11}{20}$
5 a $\frac{1}{8}$ **b** $\frac{1}{40}$ **c** $\frac{3}{8}$ **d** $\frac{7}{40}$

Exercise 19D

1 a 0.23   **b** 0.25   **c** $\frac{1}{4}$
2 a 0.74   **b** 0.7   **c** 74%
3 a 0.45, 48%, $\frac{1}{2}$ **b** 0.53, 55%, $\frac{6}{10}$
   c 68%, 0.7, $\frac{3}{4}$ **d** 0.2, 27%, $\frac{3}{10}$
4 $\frac{7}{40}$
5 a 30%, $\frac{1}{3}$, 0.4, 45%, $\frac{1}{2}$ **b** $\frac{1}{20}$, 10%, 0.12, 15%, $\frac{1}{5}$
   c 0.6, 0.63, $\frac{13}{20}$, $\frac{2}{3}$, 68%   **d** $\frac{3}{5}$, 62%, 0.65, $\frac{27}{40}$, 70%

Mixed Exercise 19E

1 20%
2 85%
3 a $\frac{3}{10}$ **b** 30%   **c** 5 more squares shaded
4 0.7
5 $\frac{37}{100}$
6 $\frac{7}{25}$
7 0.25, 0.3, $\frac{1}{3}$, 35%, $\frac{3}{8}$
8 Ryan, as Sam only got 40%

19.2 Get ready

1 8   **2** 16   **3** 4.5

Exercise 19F

1 a £12   **b** 40 kg   **c** 8 m   **d** 38 p
   e £50   **f** 9 cm   **g** $30   **h** £23
2 a £6   **b** 7 km   **c** 14 km   **d** £36
   e £16   **f** 3 kg   **g** 210 m*l*   **h** £14
3 No, he has saved £3540
4 150 g   **5** £5400   **6** 34p
7 £72   **8** £42   **9** £16 280

Exercise 19G

1 a £4.80   **b** 38.7 kg   **c** £199.80   **d** 236.8 km
   e 158.4 m*l*   **f** $20.70   **g** £13.60   **h** 1380 m
2 115
3 £48
4 54
5 £1440
6 £7.50
7 a £14.70   **b** £22.75
8 £15
9 Cheaper to pay cash. The credit plan costs £705.50.
10 a 38%   **b** 95 g
11 £5.60 profit

19.3 Get ready

1 15   **2** 27   **3** 14.8

Exercise 19H

1 600 g		**2** £27 295		**3** £96.57	
4 £611.60		**5** £56.40		**6** £89.30	
7 £3726		**8** £1560		**9** £7200	

10 £1.47

11 **a** £59.50

    b £153

    c £11.90

12 £108

13 Sample calculation for 2010:

Item	Cost before discount	% discount	Cost after discount
Labour	£120	12.5%	£105
Parts	£84	5%	£79.80
		Total before VAT	£184.80

Riverside Garage	Total before VAT	£184.80
	VAT at $17\frac{1}{2}$ %	£32.34
	Total with VAT	£217.14

19.4 Get ready

1 $\frac{7}{10}$ **2** $\frac{2}{3}$ **3** 30

Exercise 19I

1 **a** 70%   **b** 46%   **c** 65%

    d 12%   **e** 60%

2 72%

3 90%

4 80%

5 40%

6 36%

Exercise 19J

1 **a** 70%   **b** 7%   **c** 36%

    d 65%   **e** 36%

2 40%

3 18.75%

4 18%

5 **a** 15.5%   **b** 4.5%

6 55%

7 8 buckets

Review exercise

1 70%

2 **a** 0.1   **b** 0.04   **c** $\frac{13}{50}$

3 **a** 0.25   **b** $\frac{1}{4}$

4 **a** $\frac{3}{5}$ **b** 45%

5 **a** £30   **b** 5 m

6 30%

7 75%

8 **a** 0.92   **b** $\frac{3}{100}$ **c** 20 g

9 **a** 62.5%   **b** $\frac{1}{4}$

10 **a** 60 g   **b** 20 g

11 22.5 g

12 **a** £249   **b** £357   **c** £6600

13 60 480

14 Emma, Majda got 67.5%

15 Sheds For U $\frac{75}{100} \times 320 = £240$

    Garden World $17\frac{1}{2}$% of £210 = £36.75

                      210 + 36.75 = £246.75

    Ed's Sheds $\frac{2}{3} \times 345 = £230$

    Jack should buy his shed at Ed's Sheds.

16 £51.75

17 £17

18 £17.04

19 **a** 55%   **b** 60%   **c** 56.25%

20 After 1 year = $1 \times 1.1 = 1.1$

    After 2 years = $1 \times 1.05 = 1.155 = 115.5\%$

    Ziggy is not correct. Rachael had a 15.5% pay rise over the two years.

21 **a** 45%   **b** 53.6%

Chapter 20 Answers

20.1 Get ready

1 Octagon, pentagon, regular hexagon

Exercise 20A

1 Cuboid   **2** Cone

3 Pentagonal prism   **4** Tetrahedron

5

	Shape	Object
1	sphere	football
2	sphere	tennis balls
3	cuboid	cereal cartons
4	cuboid	biscuit tin
5	cuboid	cardboard box
6	cube	choc box
7	cube	dice
8	cylinder	biscuit packet
9	cylinder	flour bin
10	cylinder	rubbish bin
11	pyramid	food cover
12	triangular prism	grater
	cone	ice cream cone
	Etc.	

Exercise 20B

1

	Shape	Faces	Edges	Vertices
A	Cube	6	12	8
B	Pentagonal prism	7	15	10
C	Triangular prism	5	9	6
D	Square-based pyramid	5	8	5
E	Cuboid	6	12	8
F	Tetrahedron	4	6	4
G	Octagonal prism	10	24	16

2 Triangle

Answers

3

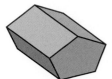

4 a hexagon   **b** hexagonal-based pyramid

20.2 Get ready

1   **2**   **3**

Exercise 20C

1 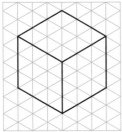   **2**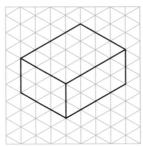

3

4

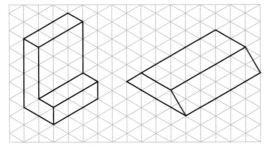

Exercise 20D

1 a 1C, 3A, 4B
   b Net of cuboid

2 a

b

3 a Cuboid but others are possible — most paper is rectangular
   Many different possibilities.

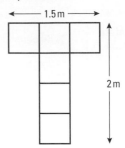

b Cylinder — oil is a liquid and a cylinder will hold more, also less corners for it to get stuck in.

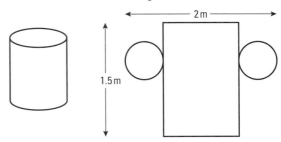

20.3 Get ready

1 A and T, B and X, C and Z

Exercise 20E

1

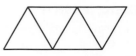

2 a Front elevation   Plan   Side elevation

b Front elevation   Plan   Side elevation

c Front elevation   Plan   Side elevation

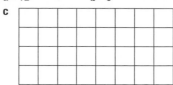

3 a   **b**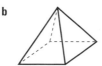

20.4 Get ready

1   12 cm^2
2   4 cm^2
3   27 cm^2
4 **a** 10 cm^2 **b** 21 cm^2 **c** 10 cm^2 **d** 12.57 cm^2

Exercise 20F

1 **a** 24 cm^3 **b** 6 cm^3 **c** 4 cm^3
2 **a** 3 cm^3 **b** 96 cm^3
3 **a** 224 m^3 **b** 612 mm^3
4   343 cm^3
5   5 cm
6   One possibility 40 cm $\times$ 60 cm $\times$ 70 cm

Exercise 20G

1   15 m^3
2   144 cm^3
3   24 cm^3
4 **a** 750 cm^3 **b** 60 cm^3 **c** 12   **d** 30 cm^3

Exercise 20H

1 **a** 565 cm^3 **b** 188 cm^3 **c** 198 cm^3
2   54 cm^3
3   432 cm^3
4 **a** 32   **b** 0.696 m^3

20.5 Get ready

1 **a** 22.94 cm^2 **b** 8.88 m^2 **c** 14.5 cm^2

Exercise 20I

1   122 cm^2
2   148 cm^2
3   3600 cm^2
4   204 cm^2
5   2 cm
6   3 cm

Exercise 20J

1 **a** 15.71 cm   **b** 141.37 cm^2 **c** 19.63 cm^2
   d 180.64 cm^2
2 **a** 94.25 cm^2 **b** 226.19 cm^2
3   251.33 cm^2

20.6 Get ready

1   24 cm^2 2   8.4 m^2 3   28.27 m^2

Exercise 20K

1 **a** 12   **b** 8
  c

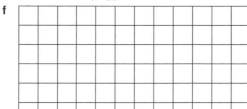

  d 24   **e** 32
  f

  g 36   **h** 72
  i   **i** $\frac{24}{12} = 2$ **ii** $\frac{36}{12} = 3$ These give the scale factor.
  j   **i** $\frac{32}{8} = 4$ **ii** $\frac{72}{8} = 9$ These are the scale factor squared.
  k   **i** 96   **ii** 512
2 **a** 16 cm and 20 cm   **b** 72 cm   **c** 320 cm^2
3   180 cm^2.
4 **a** 60 m   **b** 2500 m^2.
5   54 cm^2.
6   80 cm and 384 cm^2.

Exercise 20L

1 **a** 2   **b** 480 cm^3
2   540 cm^3
3   268 cm^3
4   2187 cm^3
5   30 cm $\times$ 24 cm $\times$ 12 cm

20.7 Get ready

1 **a** 2.5 m   **b** 35 cm   **c** 630 cm
  d 0.35 km   **e** 1500 m   **f** 36 mm

Exercise 20M

1 **a** 30 000 cm^2 **b** 45 000 cm^2 **c** 3 cm^2
  d 0.34 cm^2
2 **a** 6 000 000 m^2 **b** 400 000 m^2 **c** 2 m^2
  d 0.345 m^2
3   1.45 m^2
4   2.25 m^2
5 **a** 1.4 m^2 **b** 14 000 cm^2
6   200
7   11.7 m^2
8   1350 cm^2

Exercise 20N

1 **a** 4 000 000 cm^3 **b** 4.5 cm^3 **c** 0.4 cm^3
  d 3000 cm^3
2 **a** 0.4 litres   **b** 5.6 litres
  c 1 000 l **d** 0.0035 litres
3   1 000 000
4 **a** 144 cm^3 **b** 144 000 mm^3

Answers

5 42 000 cm³
6 100
7 675 000 litres
8 2100
9 2.1 cm
10 962

Review exercise

1 12 cm³
2 i Cylinder ii Cone
3 20 cm³
4 i 6 ii 12 iii 8
5 i 6 ii 12 iii 8
6 420 cm³
7

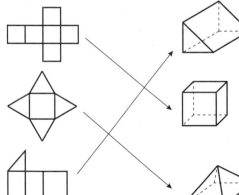

8

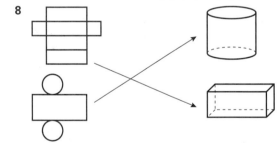

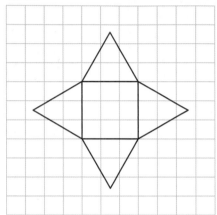

9 a

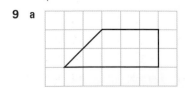

b

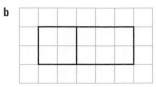

10 a

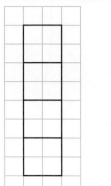

b

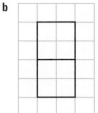

11

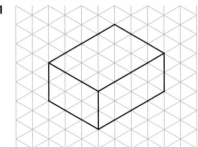

12 5 cm
13 A
14

15

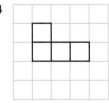

16 150
17 150
18 a 5 cm
    b i 19 **ii** 8
19 a

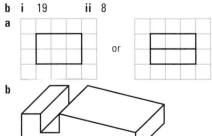

or

b

c 27 cm³
20 503 cm³
21 4.27 m²

22 135 cm^3
23 9.3 cm
24 **a** 994   **b** 361 cm^3
25 Area of floor = $5 \times 3 = 15$ m^2
Cost of carpet = £18.50 $\times$ 15 = £277.50
Cost of underlay = £3.50 $\times$ 15 = £52.50
Room perimeter = $2(5 + 3) = 16$ m
Height of room = 2.5 m
Therefore 9 rolls of wallpaper are needed.
Cost of wallpaper = £25 $\times$ 9 = £225
Total cost = 277.50 + 52.50 + 225 = £555
26 **a** The cup can hold 159 ml
    b 3 bottles, £3.75
27 2430 cm^3
28 **a** 150 cm^2 **b** 125000 mm^3
29 162 cm^2
30 8000000 cm^3

Chapter 21 Answers

21.1 Get ready

1 **a** 78   **b** 133   **c** 105

Exercise 21A

1 $x - 7 = 9$
2 $x + 3 = 11$
3 $8x = 32$
4 $9 + x = 20$
5 $7x - 3 = 32$
6 $3x = 21$
7 $3(x + 1) = 24$
8 $4x + 5 = 13$
9 $2(5 + x) = 16$
10 $7 + 5x = 27$
11 **a** $x + 7 = 11$ **b** 4
12 **a** $5x = 45$ **b** 9
13 **a** $4(x + 2) = 24$ **b** 4
14 **a** $2x - 5 = 13$ **b** 9
15 **a** $6x + 7 = 31$ **b** 4

21.2 Get ready

1 $3x - 4 = 14$ 2 $6x + 2 = 26$ 3 $7x - 4 = 17$

Exercise 21B

1 $a = 1$
2 $y = 2$
3 $h = 7$
4 $p = 9$
5 $q = 10$
6 $d = 8$
7 $x = 0$
8 $t = 4$
9 $r = 3$
10 $k = 1$
11 $n = 1$
12 $x = 5$
13 $m = 5$
14 $y = 16$
15 $w = 0$
16 $q = 12$
17 $p = 2$
18 $t = 0$
19 $a = 12$
20 $x = 0$
21 $p = 38$
20 $a = 4$
23 $b = 1$
24 $v = 0$

Exercise 21C

1 $a = 5$
2 $p = 11$
3 $q = 8$
4 $x = 8$
5 $y = 13$
6 $s = 3$
7 $x = 22$
8 $y = 21$
9 $s = 27$
10 $a = 1$
11 $p = 11$
12 $c = 4$
13 $a = 1$
14 $p = 5$
15 $q = 0$
16 $a = 2$
17 $b = 2$
18 $c = 15$
19 $p = 5$
20 $y = 21$
21 $t = 5$
22 $p = 0$
23 $p = 24$
24 $p = 0$

Exercise 21D

1 $a = 2$
2 $p = 2$
3 $p = 3$
4 $s = 3$
5 $k = 5$
6 $u = 4$
7 $g = 7$
8 $l = 7$
9 $j = 2$
10 $f = 4$
11 $r = 9$
12 $v = 9$

Exercise 21E

1 $a = 10$
2 $b = 20$
3 $s = 12$
4 $c = 30$
5 $t = 24$
6 $s = 72$
7 $h = 72$
8 $f = 28$
9 $d = 45$
10 $a = 75$
11 $b = 40$
12 $r = 52$
13 $a = 60$
14 $b = 32$
15 $k = 48$

Mixed Exercise 21F

1 $a = 1$
2 $b = 3$
3 $c = 5$
4 $p = 9$
5 $q = 4$
6 $d = 8$
7 $p = 3$
8 $r = 2$
9 $s = 2$
10 $r = 3$
11 $e = 1$
12 $p = 0$
13 $a = 12$
14 $b = 60$
15 $s = 30$

21.3 Get ready

1 $q = 7$ 2 $x = 16$ 3 $r = 6$ 4 $p = 28$

Exercise 21G

1 $a = 2$
2 $a = 3$
3 $a = 2$
4 $a = 3$
5 $p = 0$
6 $p = 2$
7 $q = 12$
8 $r = 2$
9 $t = 5$
10 $f = 3$
11 $r = 13$
12 $a = 1$
13 $a = 0$
14 $d = 3$
15 $c = 4$
16 $a = 3$
17 $z = 5$
18 $r = 18$
19 $s = 12$
20 $b = 18$
21 $c = 24$
22 $f = 27$
23 $h = 4$
24 $x = 15$

Exercise 21H

1 $a = 1\frac{1}{2}$
2 $a = 3\frac{1}{2}$
3 $a = 2\frac{2}{3}$
4 $a = 4\frac{1}{3}$
5 $p = 1\frac{3}{5}$
6 $p = 4\frac{2}{5}$
7 $e = 0$
8 $t = 1\frac{1}{2}$
9 $j = 1\frac{1}{2}$
10 $c = 1\frac{4}{7}$
11 $k = \frac{1}{4}$
12 $d = 3\frac{1}{3}$
13 $u = \frac{2}{9}$
14 $q = 2\frac{1}{4}$
15 $y = 1\frac{2}{7}$

Exercise 21I

1 $a = -1$
2 $a = -2$
3 $a = -4$
4 $a = -1$
5 $a = -2$
6 $p = -2$
7 $s = -5$
8 $p = -2$
9 $k = -1$
10 $h = -1$
11 $y = -5$
12 $e = -9$
13 $t = 0$
14 $w = -1$
15 $c = -2$
16 $a = 0$

Mixed Exercise 21J

1 $s = 3$
2 $d = 3$
3 $m = 5$
4 $h = 4$
5 $k = 9$
6 $y = 2$
7 $p = 1\frac{2}{5}$
8 $f = 3\frac{1}{4}$
9 $s = 3\frac{2}{3}$
10 $g = -2\frac{2}{7}$
11 $f = 4\frac{1}{4}$
12 $k = 3\frac{3}{5}$
13 $s = -5\frac{2}{3}$
14 $j = 3\frac{2}{3}$
15 $b = -\frac{5}{9}$
16 $r = 3\frac{1}{2}$
17 $t = -3\frac{2}{5}$
18 $y = -\frac{6}{7}$

Answers

19 $e = -\frac{1}{3}$ **20** $f = -1\frac{1}{4}$ **21** $g = -\frac{2}{5}$
22 $h = -1$ **23** $c = -1\frac{2}{3}$ **24** $s = -\frac{5}{8}$
25 $z = 4$ **26** $x = 25$ **27** $p = 4$
28 $c = -18$ **29** $a = 48$ **30** $e = -24$

21.4 Get ready
1 $x = 9$ **2** $b = 12$ **3** $x = -\frac{1}{2}$

Exercise 21K
1 $a = 19$ **2** $b = 0$ **3** $c = 24$
4 $d = 10$ **5** $e = 6$ **6** $f = 16$
7 $g = 6$ **8** $h = 21$ **9** $m = 7$
10 $p = \frac{1}{3}$ **11** $q = 0$ **12** $v = \frac{4}{5}$
13 $x = -6$ **14** $y = 1\frac{2}{3}$ **15** $c = -1$
16 $b = 4\frac{1}{2}$ **17** $d = 7$ **18** $n = 15$
19 $t = -4$ **20** $c = \frac{2}{3}$

21.5 Get ready
1 $g = 22$ **2** $a = 24$ **3** $b = 6$

Exercise 21L
1 $a = -4$ **2** $c = 5$ **3** $p = 6$
4 $b = 1$ **5** $q = 3$ **6** $x = 3$
7 $d = 5$ **8** $y = 3$ **9** $n = 7$
10 $k = 0$ **11** $u = 2\frac{1}{2}$ **12** $r = 2\frac{2}{5}$
13 $v = 4\frac{2}{3}$ **14** $t = \frac{4}{5}$ **15** $m = 2\frac{1}{2}$
16 $g = \frac{5}{6}$ **17** $b = \frac{1}{2}$ **18** $h = 1\frac{1}{3}$
19 $e = 4\frac{1}{2}$ **20** $f = \frac{2}{3}$

21.6 Get ready
1 $a = 1$ **2** $b = 4$ **3** $c = 3$

Exercise 21M
1 $x = 2$ **2** $x = 4$ **3** $x = 13$
4 $x = 2$ **5** $x = 1$ **6** $x = 3$
7 $x = 1$ **8** $x = 1$ **9** $x = 8$
10 $x = 2$ **11** $x = 1$ **12** $x = 4$
13 $x = -3$ **14** $x = 0$ **15** $x = 2\frac{1}{2}$
16 $x = \frac{3}{5}$ **17** $x = -1$ **18** $x = 1\frac{2}{3}$
19 $x = -2$ **20** $x = -\frac{2}{3}$

21.7 Get ready
1 $20°$ **2** $160°$ **3** $10\,\text{cm}$

Exercise 21N
1 8   **3** 12
2 $a = 40°$, largest angle is $80°$ **4** $130°, 80°, 150°$
5 8 cm, 9 cm, 7 cm   **6** 11
7 $y = 7$ **8** 52 years
9 15 cm   **10** $x = 6, y = 4$

21.8 Get ready
1 20   **2** 12 years   **3** 9 cm

Exercise 21O
1 $x = 1.83$ **2** $x = 8.73$ **3** $x = 4.4$
4 $x = 7.9$ **5** $x = 3.11$ **6** $x = 2.81$
7 $x = 3.3$

21.9 Get ready
1 $-19, -15, 1.34, 5, 24$

Exercise 21P
1 a $4 < 6$ **b** $5 > 2$ **c** $12 > 8$
   d $6 = 6$ **e** $15 > 8$ **f** $3 < 24$
   g $10 > 3$ **h** $0 < 0.1$ **i** $6 > 0.7$
   j $4.5 = 4.5$ **k** $0.2 < 0.5$ **l** $4.8 > 4.79$
2 a True   **b** False, $2 < 6$
   c False, $6 = 6$ **d** False, $6 < 8$
   e False, $6 > 5$ **f** False, $8 < 14$
   g False, $7 > 6.99$ **h** False, $6 < 6.01$
   i False $7 > 0$ **j** False, $4 = 4$
   k False, $6 > 4$ **l** True
3 a 5   **b** 4, 5, 6, 7
   c 0, 1, 2, 3   **d** 4, 5
   e 2, 3, 4   **f** 3, 4, 5
   g 4, 5, 6   **h** $-2, -1, 0, 1, 2, 3$
   i 0, 1, 2, 3, 4   **j** $-1, 0, 1, 2, 3, 4, 5, 6$
   k $-3, -2, -1, 0, 1, 2$ **l** $-4, -3, -2, -1, 0, 1, 2$
   m 1, 2, 3, 4   **n** 0, 1, 2, 3, 4
   o $-5, -4, -3, -2, -1$ **p** $-3, -2, -1, 0, 1, 2, 3$

21.10 Get ready
1 $-3, -2, -1, 0, 1, 2, 3, 4$
2 $-5, -4, -3, -2, -1, 0, 1$
3 $-6, -5, -4, -3, -2, -1, 0, 1, 2, 3, 4, 5, 6, 7$

Exercise 21Q
1 a

b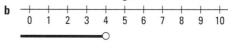

c

d

e

f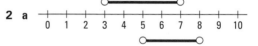

2 a

b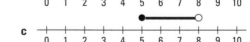

c

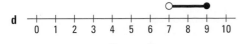

d

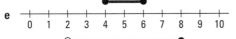

e

f

g

h

i

j

3 a

b

c

d

e

f

g

h

i

j

4 a $2 < x \leq 6$ **b** $7 \leq x \leq 8$ **c** $1 < x < 6$
   d $3 \leq x < 7$ **e** $-3 < x \leq 1$ **f** $-1 \leq x \leq 3$
   g $-4 < x < 1$ **h** $-3 \leq x < 4$

21.11 Get ready

1

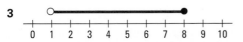

2

3

Exercise 21R

1 $x < 4$ **2** $x \geq 6$ **3** $x \leq 6$
4 $x > 6$ **5** $x < 9$ **6** $x > 4$
7 $x \geq 0$ **8** $x \leq 5$ **9** $x \geq 3$
10 $x < 2$ **11** $x > 3$ **12** $x \geq 1$
13 $x > 2\frac{3}{4}$

14 $x \leq \frac{1}{2}$

15 $x \geq -2$

16 $x > 1\frac{1}{4}$

17 $x < -1\frac{1}{2}$

18 $x \leq \frac{3}{4}$

19 $2, 3, 4$ **20** $-3, -2, -1, 0, 1$
21 $-2, -1, 0, 1$ **22** $0, 1, 2, 3$
23 $-3, -2, -1, 0$ **24** $1, 2$
25 $-1, 0, 1, 2, 3$ **26** $-2, -1, 0, 1, 2$
27 $-3, -2, -1$ **28** $x < 2\frac{1}{2}$
29 $x \geq \frac{3}{4}$ **30** $x > -3$
31 $x \geq -2\frac{2}{3}$ **32** $x > -8$
33 $x > 3\frac{1}{2}$ **34** $x \geq 2\frac{3}{4}$
35 $x \leq -\frac{2}{3}$ **36** $x > \frac{7}{8}$
37 $x > 1$ **38** $x \geq 2$
39 $x \leq -1$ **40** $x > -1\frac{2}{3}$
41 $x < 2\frac{1}{2}$ **42** $x \leq 3$
43 $x \geq 7$ **44** $x > 0$
45 $x < 3$ **46** $x \geq 6$
47 $x < 2$ **48** $x > -\frac{4}{5}$
49 $x \leq 1$ **50** $x \leq 3$
51 $x < 2\frac{1}{5}$ **52** $x \geq 2$
53 $x \geq \frac{1}{2}$ **54** $x < -3$
55 $x > -4\frac{2}{3}$, smallest integer satisfying inequality is -4
56 $x \leq -\frac{3}{5}$, largest integer satisfying inequality is -1

Review Exercise

1 a $a = 5$ **b** $c = 10$ **c** $p = 7$ **d** $d = 12$
   e $x = 3$ **f** $b = 4$ **g** $a = -4$ **h** $b = -6$
   i $c = -1$ **j** $e = 5\frac{1}{2}$ **k** $h = -2$ **l** $m = -\frac{3}{4}$
   m $p = -2\frac{5}{6}$ **n** $q = -\frac{2}{3}$
2 a $r = 3$ **b** $x = 5$ **c** $c = 5$ **d** $b = \frac{1}{2}$
   e $d = 1\frac{2}{3}$ **f** $y = \frac{4}{5}$ **g** $t = 2\frac{1}{3}$ **h** $w = 2\frac{1}{2}$
   i $u = -3$ **j** $w = -6\frac{1}{2}$ **k** $y = -2\frac{1}{2}$

3 **a** $x = 2$ **b** $y = \frac{1}{2}$
4 **a** $x = 4$ **b** $y = 3\frac{1}{2}$ **c** $2t + 23$
5 $x = -1\frac{1}{2}$
6 **a** $x = 7$ **b** $y = 1\frac{1}{2}$
7 **a** $a = 3$ **b** $b = 7$ **c** $c = 5$ **d** $d = 7$
  e $e = 6$ **f** $f = 3\frac{3}{7}$ **g** $m = 1$ **h** $t = 3\frac{1}{4}$
  i $a = 3$ **j** $b = 4$ **k** $d = \frac{1}{2}$ **l** $g = 1\frac{1}{3}$
  m $p = -4\frac{2}{3}$
8 48 cm
9 **a** $4x + 10$ **b** $x = 6$
10 **a** $x = 2.35$ **b** $x = 4.55$ **c** $x = 4.27$
  d $x = 3.76$ **e** $x = 2.4$ **f** $x = 1.77$
11 $x = 1.7$
12 **a** 55.927 125
  b $x = 3.65$ gives an answer that is too small. $x^3 + 2x$ gets bigger as x gets bigger, so the solution must be greater than 3.65. Yousef is correct.
13 **a** $5x + 60 = 360$ **b** $x = 60$
14 Uzma £18, Hajra £38, Mabintou £76
15 A 2.5 cm, B 7.5 cm, C 30 cm
16 A £8, B £12, C £4
17 **a** If Emil has x CDs, $4x + 12 = 32$.
  b 18 CDs
18 **a** $x = 3.6$ **b** 7.6 cm
19 16 m
20 **a**
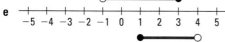
  b
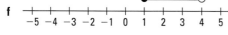
  c
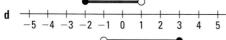
  d
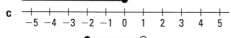
  e
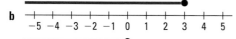
  f
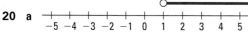
21 **a** $x \leqslant -1$ **b** $x > 3$ **c** $0 \leqslant x < 2$
  d $-3 < x < 3$ **e** $-2 < x \leqslant 0$ **f** $-1 \leqslant x \leqslant 3$
22 **a** $-3, -2, -1, 0$ **b** $1, 2, 3$
  c $-2, -1, 0, 1, 2, 3, 4$ **d** $-3, -2$
23 **a** $x > 10$ **b** $x \leqslant 5$ **c** $x < 4\frac{1}{2}$
  d $x \leqslant -2$ **e** $x \geqslant -1\frac{1}{2}$ **f** $x < 0$
  g $x > -\frac{1}{2}$ **h** $x < 6\frac{1}{2}$ **i** $x \geqslant -2$
24 **a** $x < 2\frac{1}{2}$

  b $x \geqslant -\frac{1}{2}$

  c $x > 1\frac{2}{3}$

d $x \leqslant -1$
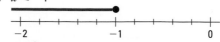
e $x < 2$
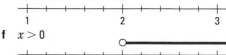
f $x > 0$

25 **a** $x < 3\frac{1}{2}$ **b** $y^2 + 7y + 12$
26 **a** $x = \frac{3}{5}$ **b** $-2, -1, 0, 1, 2, 3$
27 **a** $x > 3$ **b** $y \geqslant 6$
28 **a** A is x B is $x + 4$ C is $2(x + 4)$
  $L = x + x + 4 + 2(x + 4)$
  $L = 2x + 4 + 2x + 8$
  $L = 4x + 12$
  b $4x + 12 < 50$ **c** $0 < x < 9.5$
29 $9\frac{1}{3}$ kg
30 $0 < x < 4$

Chapter 22 Answers

22.1 Get ready

1

Number of bars	1	2	3	4	5	6	7	8	9	10
Cost in pence	20	40	60	80	100	120	140	160	180	200

Exercise 22A

1 **a**
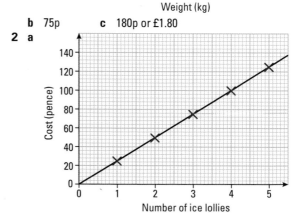
  b 75p **c** 180p or £1.80
2 **a**

  b **i** 200p or £2 **ii** 150p or £1.50

3 a

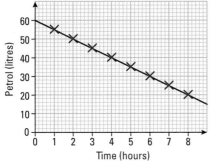

b 60 litres
c 32.5 litres

4 a

Distance travelled in km	0	5	10	15	20	25
Petrol used in litres	0	2	4	6	8	10

b

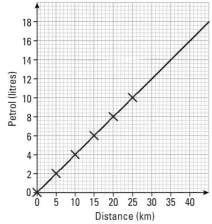

c 1.6 litres
d 37.5 km

5 a

Weeks	0	1	2	3	4	5	6	7	8
Expected depth of water in m	144	140	136	132	128	124	120	116	112

b

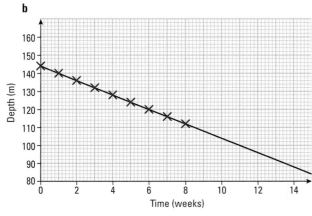

c 104 m
d 12 weeks

6 a

Usage time in minutes	Cost in pounds
0	0
5	2
10	4
15	6
20	8
25	10
30	12
35	14
40	16
45	18
50	20

b

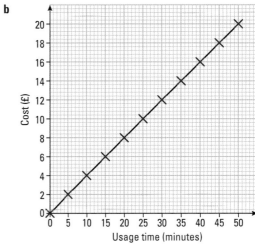

c £12.80
d 21 minutes

7 a

Usage time in minutes	Cost in pounds
0	15
5	15.5
10	16
15	16.5
20	17
25	17.5
30	18
35	18.5
40	19
45	19.5
50	20

Answers

b

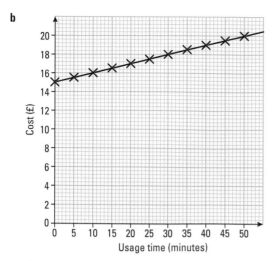

c £18.20
d 20 minutes

8 a

Units used	Cost in pounds
0	0
10	5
20	10
30	15
40	20
50	25
60	30
70	35
80	40
90	45
100	50

b

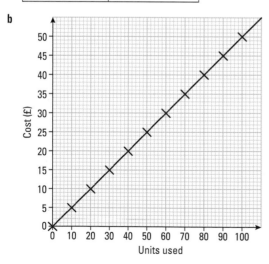

c £16
d 90 units

9 a

Units used	Cost in pounds
0	20
10	22.5
20	25
30	27.5
40	30
50	32.5
60	35
70	37.5
80	40
90	42.5
100	45

b

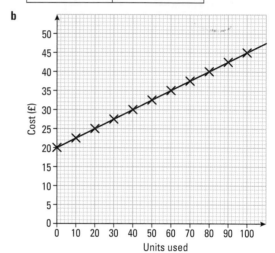

c £28
d 72 units

10 a Tariff 1   B
     Tariff 2   A
     Tariff 3   C

b Tariff 2 would be cheapest because the cost of £24 is the lowest of the three tariffs.

11

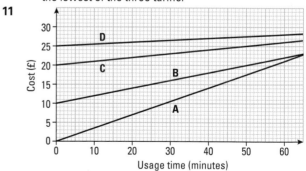

Jodie should choose Tariff A.

Exercise 22B

1 A 1, B 5, C 2, D 4, E 3
2 A 3, B 1, C 4, D 2, E 5
3 A 2, B 4, C 3, D 5, E 1
4 A 4, B 5, C 2, D 1, E 3

5 a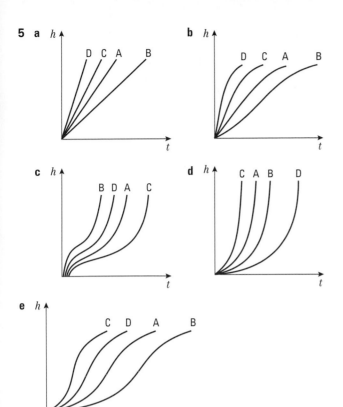

b h

c h

d h

e h

6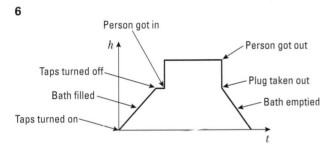

Person got in
Taps turned off
Person got out
Bath filled
Plug taken out
Taps turned on
Bath emptied

7 The taps were turned on and the bath filled at a constant rate for 2 minutes, to a depth of 20 cm.
The taps were turned on further (e.g. more hot water) for 1 minute, after which the bath was 35 cm deep.
The taps were turned off and the bath left for 1 minute.
The person got into the bath and the depth increased to 50 cm.
After 3 minutes, more hot water was added for 1 minute, increasing the depth to 54 cm, and the person stayed in a further 0.4 minute.
At 8.4 minutes, the person got out of the bath and the depth decreased to 40 cm.
The plug was taken out and the water drained out of the bath in 2.6 minutes.

8 The balloon climbed rapidly for 2 minutes to a height of 175 m.
It then climbed more slowly for 2 minutes to a height of 225 m.
It stayed at this height for 2 minutes, before descending slowly for 4 minutes to 175 m.
The burner was used for 2 minutes to make the balloon ascend rapidly to 290 m, where it stayed for 2 minutes.
The balloon then descended slowly for 4 minutes to 220 m and then more rapidly for 4 minutes to ground level.

22.2 Get ready

1 a £4   **b** £8   **c** £16   **d** £80
2 a €2   **b** €10   **c** €30   **d** €20

Exercise 22C

1 a i HK$120   **ii** HK$60   **iii** HK$96
   iv HK$1200   **v** HK$2400
   b i £5   **ii** £2.50   **iii** £7.50
   iv £50   **v** £100

2

°C	5	20	27	28	10	38	35	80	93	40
°F	41	68	80	82	50	100	95	176	200	104

3 a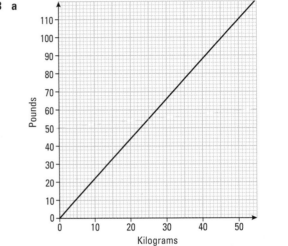

b

Kilograms	0	4.5	9	45	30	15	22.5	6	35	50
Pounds	0	10	20	99	66	33	50	14	77	110

4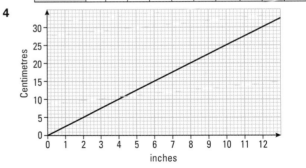

Inches	0	1	2	4	6	8	9	8	10	12
Centimetres	0	2.5	5	10	15	20	22.5	20	25	30

5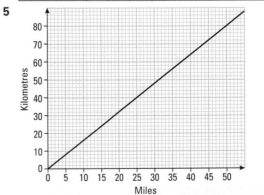

Answers

Miles	0	5	10	40	22.5	30	45	12.5	24	50
Kilometres	0	8	16	64	36	48	72	20	38.4	80

6

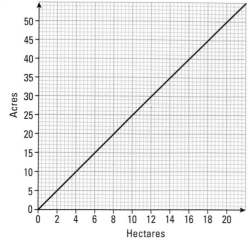

Hectares	0	8	12	12	15	17	9.6	18	3	20
Acres	0	20	30	30	37.5	42.5	24	45	7.5	50

22.3 Get ready

1 30 mph **2** 100 miles **3** 2 hours

Exercise 22D

1 a 10 minutes **b** 300 m **c** 15 minutes
 d 15 minutes **e** 30 m per minute = 1.8 km per hour
 f 20 m per minute = 1.2 km per hour
2 a Students' stories, e.g.
   David travelled for 1 hour, stopped for a cup of tea for
   half an hour and then drove faster for another hour,
   when he reached his aunt's house. He stayed at his
   aunt's for 1 hour and then drove straight home in half an
   hour.
 b 1st section: 20 km per hour
   2nd section: 0 km per hour
   3rd section: 40 km per hour
   4th section: 0 km per hour
   5th section: 120 km per hour
3 a Tom set off at 08:00, travelled 45 km in 1 hour, stopped for
   20 minutes and travelled another 45 km in 40 minutes. He
   reached Sarah's house at 10:00. He then travelled back
   90 km in 1 hour.
 b Sarah set off at 08:20 and drove 30 km in 40 minutes. At
   09:00 she stopped for 10 minutes and then drove 60 km in
   40 minutes, reaching Tom's house at 09:50.
 c 09:20, 45 km from London
4 A→B: The depth of the water goes up 10 cm to 30 cm in 5
   minutes.
   B→C: The depth stays at 30 cm for 5 minutes.
   C→D: Imran gets into the bath and the depth increases to
   60 cm.
   D→E: The depth stays constant at 60 cm for 15 minutes.
   E→F: Imran takes the plug out and the depth decreases
   from 60 cm to zero in 5 minutes.

5

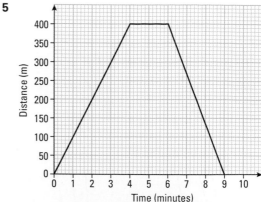

6

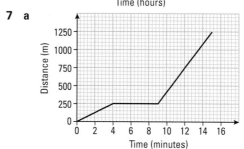

7 a

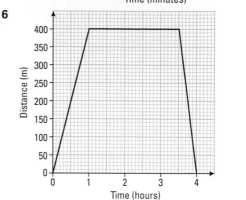

 b 167 m per minute = 10 km per hour

8

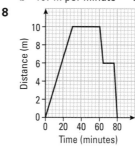

9 a i 12 m **ii** 45 m **b i** 2.8 seconds **ii** 2.2 seconds
10 a i 1985 m **ii** 1800 m **iii** 1525 m
 b i 11:02:05 **ii** 11:02:40 **iii** 11:03:12
11 a **b** 15.5 m per second
   c 16.5 m

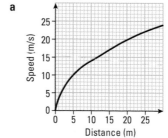

22.4 Get ready

1 a 9
  b 45
  c 15
  d 45
  e −6
  f −12

Exercise 22E

1 a i

x	−3	−2	−1	0	1	2	3
$y = x^2 + 2$	11	6	3	2	3	6	11

ii

x	−3	−2	−1	0	1	2	3
$y = -x^2 - 2$	−11	−6	−3	−2	−3	−6	−11

iii

x	−3	−2	−1	0	1	2	3
$y = -x^2 + 1$	−8	−3	0	1	0	−3	−8

iv

x	−3	−2	−1	0	1	2	3
$y = -x^2 + 4$	−5	0	3	4	3	0	−5

v

x	−3	−2	−1	0	1	2	3
$y = x^2 + 3$	12	7	4	3	4	7	12

b

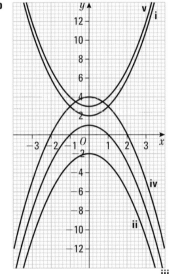

2 a i

x	−3	−2	−1	0	1	2	3
$y = 2x^2 + 1$	19	9	3	1	3	9	19

ii

x	−3	−2	−1	0	1	2	3
$y = -2x^2 + 1$	−17	−7	−1	1	−1	−7	−17

iii

x	−2	−1	0	1	2
$y = -3x^2 + 1$	−11	−2	1	−2	−11

iv

x	−3	−2	−1	0	1	2	3
$y = 2x^2 - 1$	17	7	1	−1	1	7	17

v

x	−3	−2	−1	0	1	2	3
$y = -2x^2 - 1$	−19	−9	−3	−1	−3	−9	−19

b

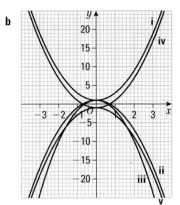

3

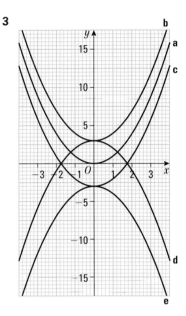

4

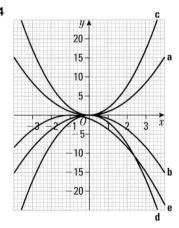

5

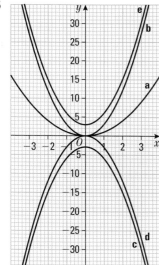

Exercise 22F

1 a i

x	-3	-2	-1	0	1	2	3
x^2	$+9$	$+4$	$+1$	0	$+1$	$+4$	$+9$
$2x$	-6	-4	-2	0	$+2$	$+4$	$+6$
$+1$	$+1$	$+1$	$+1$	$+1$	$+1$	$+1$	$+1$
$y = x^2 + 2x + 1$	4	1	0	1	4	9	16

ii

x	-3	-2	-1	0	1	2	3
x^2	$+9$	$+4$	$+1$	0	$+1$	$+4$	$+9$
$3x$	-9	-6	-3	0	$+3$	$+6$	$+9$
$+2$	$+2$	$+2$	$+2$	$+2$	$+2$	$+2$	$+2$
$y = x^2 + 3x + 2$	2	0	0	2	6	12	20

iii

x	-3	-2	-1	0	1	2	3
x^2	$+9$	$+4$	$+1$	0	$+1$	$+4$	$+9$
$2x$	-6	-4	-2	0	$+2$	$+4$	$+6$
-5	-5	-5	-5	-5	-5	-5	-5
$y = x^2 + 2x - 5$	-2	-5	-6	-5	-2	3	10

iv

x	-3	-2	-1	0	1	2	3
x^2	$+9$	$+4$	$+1$	0	$+1$	$+4$	$+9$
$-2x$	$+6$	$+4$	$+2$	0	-2	-4	-6
$+3$	$+3$	$+3$	$+3$	$+3$	$+3$	$+3$	$+3$
$y = x^2 - 2x + 3$	18	11	6	3	2	3	6

v

x	-3	-2	-1	0	1	2	3
x^2	$+9$	$+4$	$+1$	0	$+1$	$+4$	$+9$
$-2x$	$+6$	$+4$	$+2$	0	-2	-4	-6
-3	-3	-3	-3	-3	-3	-3	-3
$y = x^2 - 2x - 3$	12	5	0	-3	-4	-3	0

b

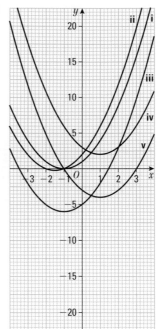

2

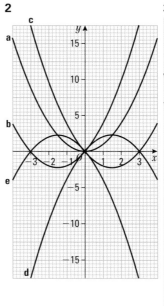

3

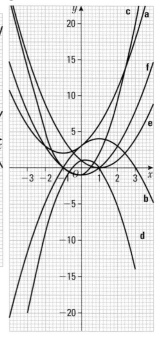

4

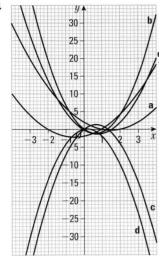

22.5 Get ready

1

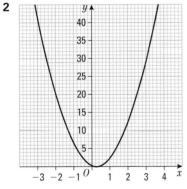

2

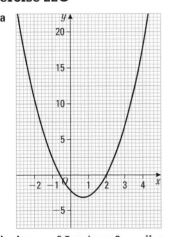

Exercise 22G

1 a

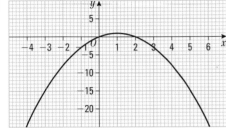

    b **i** $x = -0.5$ and $x = 2$ **ii** $x = -1.8$ and $x = 3.3$

2 a

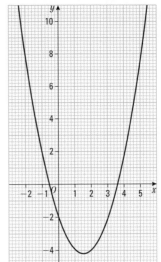

b
i $x = -0.6$ and $x = 3.6$
ii $x = -1.5$ and $x = 4.5$

3 a

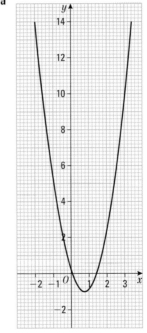

b
i $x = 0$ and $x = 1.5$
ii $x = -0.9$ and $x = 2.4$

4 a

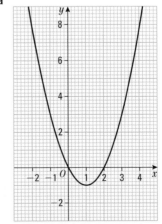

b
i $x = 0$ and $x = 2$
ii $x = -1$ and $x = 3$

5 a

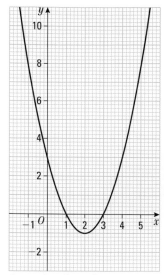

b **i** $x = 1$ and $x = 3$
  ii $x = 0.3$ and $x = 3.7$

Review exercise

1 A: The temperature remains steady
  B: The temperature is steadily increasing
  C: The temperature steadily increases, then rapidly decreases
  D: The temperature is steady to start with then steadily decreases
  E: The temperature rises to begin with, then remains the steady for a period and finally rises again.
  F: The temperature rises steadily to at the start, then remains steady for a period, then falls steadily

2 a

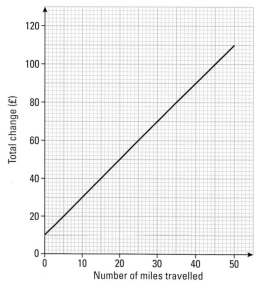

b Up to 20 miles

3

C	0	20	40	60	80	100
F	30	70	110	150	190	230

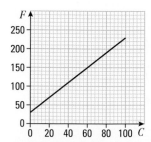

4 a $C = 6d + 14$
  b £74
5 a Tariff A: $C = 20$
    Tariff B: $C = 0.04m + 10$
  b Tariff A has a flat charge of £20, so calls don't cost any extra. Tariff B has a cost of £10 plus 4p per minute.
  c Josh should use Tariff B unless he uses more than 250 minutes a month, in which case Tariff A would be cheaper.

6

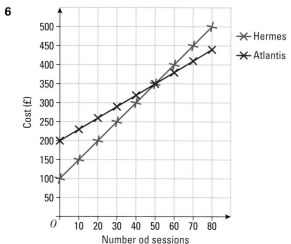

Atlantis is cheaper if she goes to more than 50 sessions

7 a

x	−1	0	1	2	3	4	5
y	3	−2	−5	−6	−5	−2	3

b

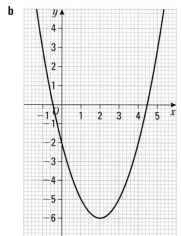

c $x = -0.4$ and $x = 4.4$
8 a $x = 0$ and $x = 2$
  b $x = -0.7$ and $x = 2.7$
  c $x = 1$

9 $x = 3$

10 a

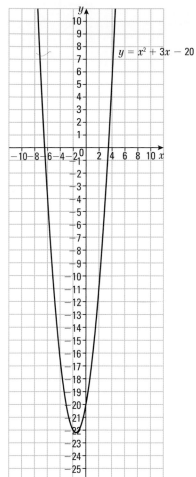

$y = x^2 + 3x - 20$

b $x = 3.2$

Chapter 23 Answers

23.1 Get ready

1 Symmetrical letters:
A B C D E H I K M N O S T U V W X Y Z

Exercise 23A

1 Rotation		**2** Translation	
3 Enlargement		**4** Reflection	
5 Enlargement		**6** Reflection	
7 Rotation		**8** Translation	
9 Translation		**10** Rotation	
11 Enlargement		**12** Reflection	

23.2 Get ready

1 4 right and 5 up
2 right, 2 up, 2 right, 3 up
3 right, 5 up, 1 right

Exercise 23B

1 a 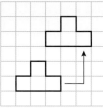 **b**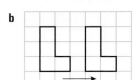

c 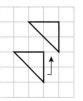 **d**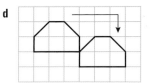

e   **f**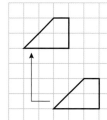

g 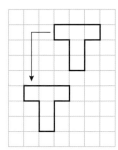 **h**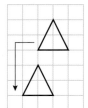

2

3 a $\begin{pmatrix} 5 \\ -4 \end{pmatrix}$ **b** $\begin{pmatrix} -1 \\ -4 \end{pmatrix}$ **c** $\begin{pmatrix} -4 \\ 3 \end{pmatrix}$ **d** $\begin{pmatrix} 2 \\ 5 \end{pmatrix}$

e $\begin{pmatrix} 6 \\ -3 \end{pmatrix}$ **f** $\begin{pmatrix} 5 \\ 2 \end{pmatrix}$ **g** $\begin{pmatrix} -2 \\ -3 \end{pmatrix}$ **h** $\begin{pmatrix} 3 \\ 2 \end{pmatrix}$

4 a $\begin{pmatrix} 6 \\ -2 \end{pmatrix}$ **b** $\begin{pmatrix} 1 \\ 5 \end{pmatrix}$ **c** $\begin{pmatrix} -7 \\ -4 \end{pmatrix}$ **d** $\begin{pmatrix} 7 \\ 4 \end{pmatrix}$

e $\begin{pmatrix} -6 \\ 2 \end{pmatrix}$ **f** $\begin{pmatrix} 5 \\ -7 \end{pmatrix}$

23.3 Get ready

1 a 3 **b** 2 **c** 2 **d** 4

Answers

Exercise 23C

1 a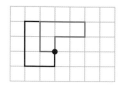

b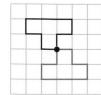

c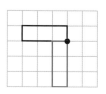

d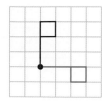

d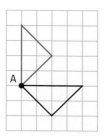

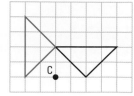

2 a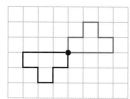

b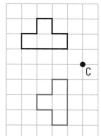

c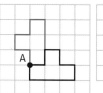

d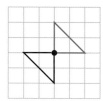

4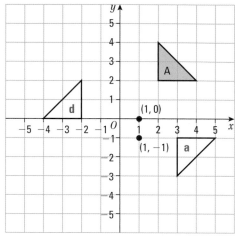

3 a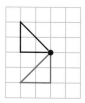

b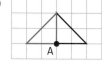

c

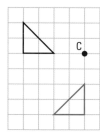

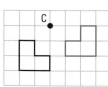

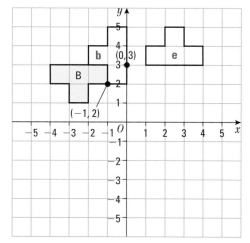

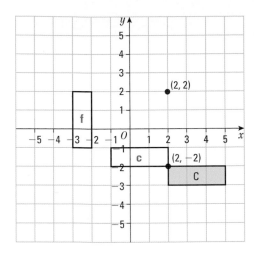

Exercise 23D

1 a Rotation of 90° clockwise about (−1, 0)
  b Rotation of 90° clockwise about (0, 0)
  c Rotation of 180° about (0, 0)
  d Rotation of 90° anticlockwise about (0, 0)
2 a Rotation of 90° anticlockwise about (−2, 2)
  b Rotation of 180° about (2, 1)
  c Rotation of 90° clockwise about (0, −1)
  d Rotation of 180° about (0, 2)

23.4 Get ready

1 a 1   **b** 0   **c** 1   **d** 6

Exercise 23E

1 a   **b**

c   **d**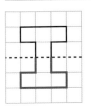

e   **f**

g 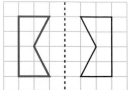   **h**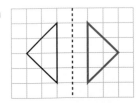

2 a 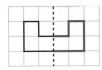   **b**

c 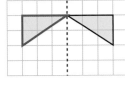   **d**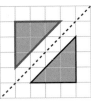

e 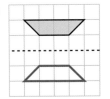   **f**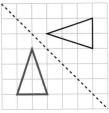

g 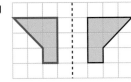   **h**

Exercise 23F

1 a Reflection in the x-axis
  b Reflection in the y-axis
  c Reflection in the x-axis
  d Reflection in the line $x = 2$
  e Reflection in the line $y = -2$
  f Reflection in the line $x = -1$
2 a Reflection in the y-axis
  b Reflection in the line $y = 2$
  c Reflection in the line $x = 3$
  d Reflection in the line $y = x$
  e Reflection in the line $y = 3\frac{1}{2}$
  f Reflection in the line $y = -x$
  g Reflection in the line $y = -1\frac{1}{2}$
3 a Reflection in the line $y = 1$
  b Reflection in the line $y = x$
  c Reflection in the line $y = -x$
  d Reflection in the line $x = 1\frac{1}{2}$
  e Reflection in the line $y = -x$
  f Reflection in the line $y = x$

23.5 Get ready

1

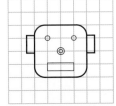

Answers

Exercise 23G

1

2
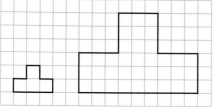

3
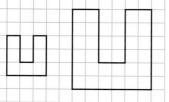

4
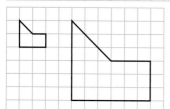

5
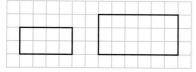

6

7
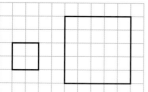

8
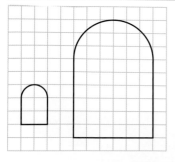

9
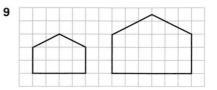

Exercise 23H

1 a
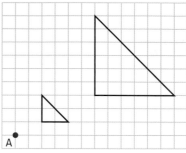

b
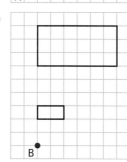

c

d **e**
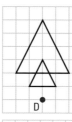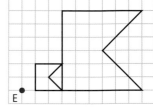

f
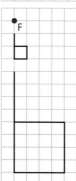

2 a
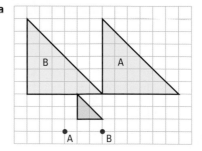

b

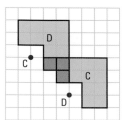

c

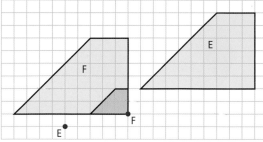

3 a

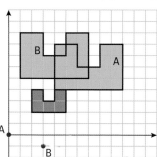

b

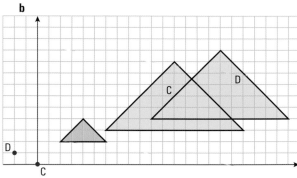

c

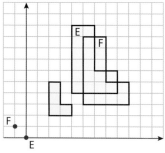

Exercise 23I

1 a Enlargement sf 3 centre (1, 3)
  b Enlargement sf 2 centre (9, 0)
  c Enlargement sf 3 centre (2, 2)
2 a Enlargement sf 2 centre (8, 8)
  b Enlargement sf 3 centre (0, 7)
  c Enlargement sf 2 centre (9, 4)

23.6 Get ready

1

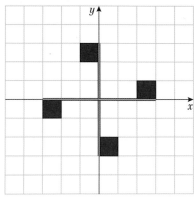

Exercise 23J

1 ab

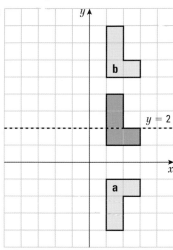

  c Translation by $\binom{0}{4}$

2 ab

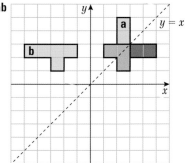

  c Reflection in the y-axis

3 ab

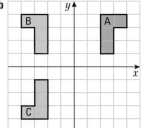

  c Rotation of 180° about point (0, 0)

Answers

4 ab

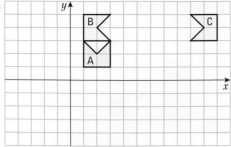

c Rotation of 90° anticlockwise about (6, 6)

5 a

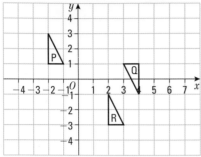

b Translation $\begin{pmatrix} 4 \\ -4 \end{pmatrix}$

Review exercise

1 a

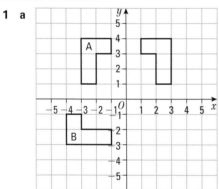

b Rotation 90° anticlockwise about (0, 0)

2

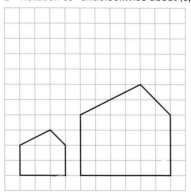

3

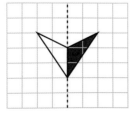

4

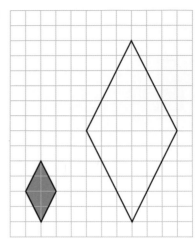

5

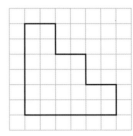

6 a, b

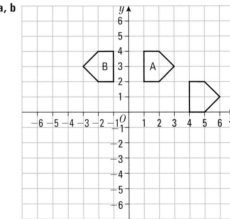

7 a, b

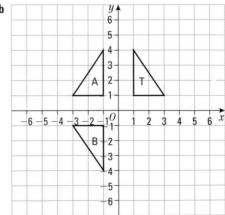

c Enlargement scale factor 3, centre O

8

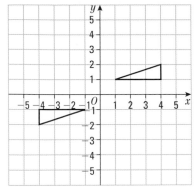

9 Reflection in the y-axis

10 a

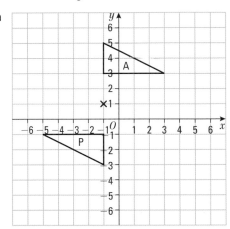

b

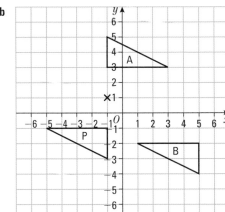

c

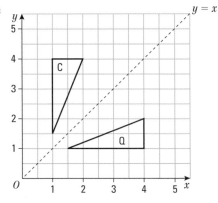

11 Rotation 90° clockwise about $(-2, 3)$

12 a Reflection in the line $y = x$

b

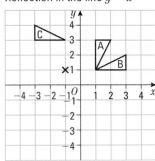

13 a

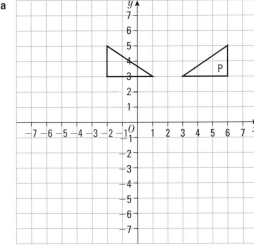

b Translation by $\begin{pmatrix} 5 \\ -4 \end{pmatrix}$

14 Rotation 180° about (1, 0)

Chapter 24 Answers

24.1 Get ready

1 $\frac{5}{8}$ **2** $\frac{3}{4}$ **3** $\frac{4}{5}$

Exercise 24A

1 a 3:7   **b** 8:7   **c** 4:11
2 a 16:13   **b** 13:16
3 a 2:3   **b** 2:5   **c** 5:3   **d** 2:3:5

Exercise 24B

1 $\frac{1}{3}$
2 a $\frac{2}{5}$ **b** $\frac{3}{5}$
3 a $\frac{5}{8}$ **b** $\frac{3}{8}$
4 a $\frac{1}{10}$ **b** $\frac{7}{10}$
5 1:1
6 1:2

Exercise 24C

1 a 1:2   **b** 1:3   **c** 7:5   **d** 2:3
   e 4:3   **f** 1:3   **g** 8:15   **h** 1:4
2 2:3

3 4:3
4 5:9:6
5 3:4

Exercise 24D

1 **a** 1:3 **b** 2:5 **c** 3:10 **d** 4:1
2 **a** 3:10 **b** 20:3 **c** 8:3 **d** 1:4
3 9:20
4 3:16
5 **a** 3:1 **b** 1:3

Exercise 24E

1 **a** 1:2.5 **b** 1:2.4 **c** 1:0.3 **d** 1:1.25
2 1:18
3 1:14.5
4 **a** 1:5 **b** 1:20 **c** 1:0.2 **d** 1:6

Mixed Exercise 24F

1 1:2
2 **a** $\frac{4}{9}$ **b** $\frac{5}{9}$
3 **a** 8:3 **b** 1:4 **c** 3:8:2
4 **a** 3:1 **b** 2:5
5 3:20
6 **a** 2:3 **b** 1:1.5
7 1:2.25
8 7 adults

24.2 Get ready

1 4 **2** $\frac{2}{8}, \frac{3}{12}$ **3** 0.0365 m

Exercise 24G

1 20
2 **a** 4 **b** 6 **c** 20
3 **a** 6 kg **b** 60 kg
4 **a** **i** 40 g **ii** 200 g **iii** 300 g
   b **i** 90 g **ii** 150 g **iii** 375 g
5 **a** 200 m*l* **b** 750 m*l*
6 15
7 1.25 litres

Exercise 24H

1 400 cm
2 720 cm
3 240 m
4 25 cm
5 **a** 360 cm **b** 26 cm
6 6 km
7 1.75 km
8 12 cm

24.3 Get ready

1 $\frac{4}{7}$ **2** 2:3 **3** 1:15

Exercise 24I

1 £16, £64

2 £9, £15
3 £18, £27
4 Alex 30 sweets, Ben 10 sweets
5 copper 680 g, nickel 120 g
6 15
7 £5, £15, £20
8 flour 450 g, sugar 150g, butter 300 g
9 £6
10 40 *l*

Mixed Exercise 24J

1 14.4 m
2 100 g
3 160 cm
4 25
5 13.5 km
6 £6
7 manganese 25 kg, nickel 5 kg
8 600 cm : 240 cm : 360 cm

24.4 Get ready

1 £1.35
2 17p
3 £1.95

Exercise 24K

1 £1.20
2 £2.24
3 £57.50
4 96 m
5 £3.75
6 £74.90
7 17.5 g
8 500 sheets

Exercise 24L

1 200 g flour, 100 g margarine, 150 g cheese
2 150 g pastry, 150 g bacon, 112.5 g cheese, 3 eggs, 225 ml milk
3 **a** 350 g **b** 75 g **c** 75 g

Exercise 24M

1 216 euros
2 26 970 roubles
3 **a** $426 **b** £250
4 £75
5 £40

Mixed Exercise 24N

1 300 g
2 £43.75
3 £42.50
4 **a** 472 euros **b** £200
5 250 g margarine, 250 g caster sugar, 5 eggs, 562.5 g flour, 75 m*l* milk
6 £80
7 7 people

8 80 Francs = £47.62. 55 euros = £49.11.
   It is cheaper to pay in francs.

Review exercise

  1 2 : 3
  2 21 : 4
  3 5 : 2
  4 45 apples
  5 10 British cars
  6 **a** 50 cm   **b** 16 m
  7 110 green counters
  8 900 g of sugar, 18 g of butter, 720 g of condensed milk, 135 m*l* of milk
  9 **a** €280   **b** £8
  10 2:33 pm
  11 Paris
  12 45 litres

Chapter 25 Answers

25.1 Get ready

1 **a** (0, 1) and (4, 4)
  b

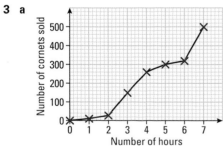

Exercise 25A

1 **a** £12.80
  b 21 minutes
2 **a** **i** 68°F   **ii** 212°F   **iii** 97°F
  b **i** 60°C   **ii** 16°C   **iii** 31°C
3 **a**

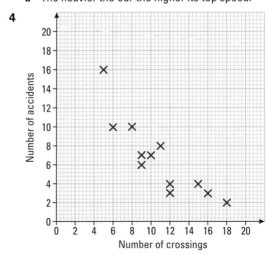

  b 7th hour

25.2 Get ready

1 A = 3.8
2 B = 60.25

3 C = 44.75
   D = 45.5

Exercise 25B

1 The higher the number of breaths per minute the higher the pulse rate.

2 **a**

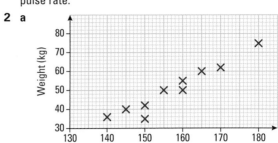

  b The greater the height the larger the weight.

3 **a**

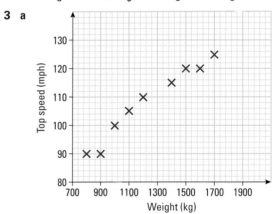

  b The heavier the car the higher its top speed.

4

The more crossings there are the fewer the number of accidents.

25.3 Get ready

1 August, September, October, November, December

Answers

Exercise 25C

1 No correlation

2 a Negative correlation
   b As the age of the car increases the cost decreases.

3

Variables	Positive correlation	Negative correlation	No correlation
Height and Weight of people	✓		
Intelligence and Weight of people			✓
Size of garden and Number of birds	✓		
Age and Running speed of adults		✓	
Height and Shoe size of people	✓		
Age of cars and Engine size			✓
Arm length and Leg length of people	✓		

4 a

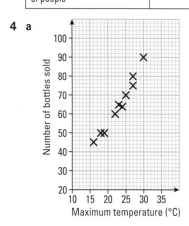

b There is positive correlation. Jacob's claim is correct: the higher the temperature the more bottles of water Jacob sells.

25.4 Get ready

1 About £6000

Exercise 25D

1 a Negative correlation
   b, c

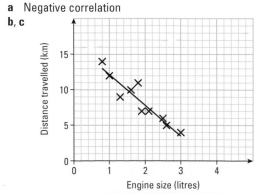

d The larger the engine size the lower the distance travelled on a litre of petrol.

2 a, c

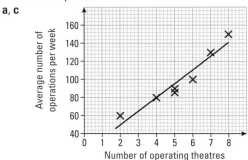

b Positive correlation
d The greater the town size the bigger the profits. He should expand to a town with a size of more than 50 000.

3 a, c

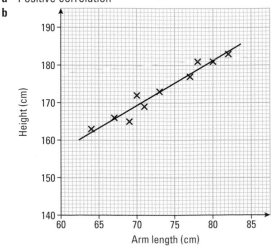

b Positive correlation
d The higher the number of operating theatres the higher the average number of operations per week.

25.5 Get ready

1 (5, 4.7)

Exercise 25E

1 a Positive correlation
   b i £70 000   **ii** 25 000 models

2 a Negative correlation
   b i 31 units   **ii** 4.8°C

3 a 147
   b 52 years
   c Positive correlation: the older a man is the higher his blood pressure.

Review exercise

1 a Positive correlation
   b

c 175 cm

2 a Negative correlation

b

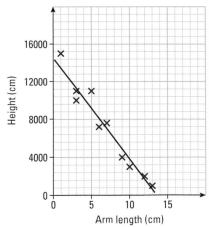
Height (cm) vs Arm length (cm)

3 a, c

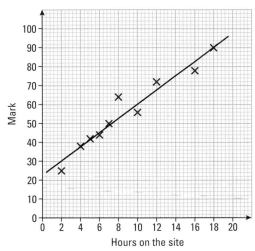
Mark vs Hours on the site

b Positive correlation

4 a About −1
b 9.6 minutes
c Negative correlation
d The older the apprentices the quicker they learn skills

5 a, b

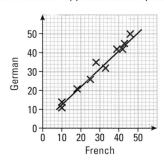
German vs French

c About 1
d Students A, B, C and D got less than 26 in at least one exam.
$\frac{4}{10} = \frac{2}{5}$ students

6 a, c

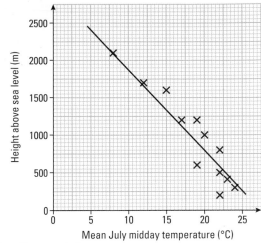
Height above sea level (m) vs Mean July midday temperature (°C)

b Negative correlation
d 11°C   **e** 1000 m

7 a Negative correlation: the older the car the lower its value.

b

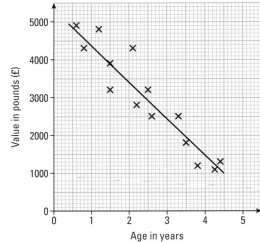
Value in pounds (£) vs Age in years

c £2450   **d** 1.9 years

8 a Positive correlation: the greater the weight the taller the height.

b

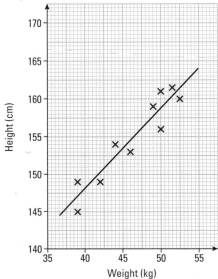
Height (cm) vs Weight (kg)

c 155.5 cm

Answers

9 a Negative correlation

b

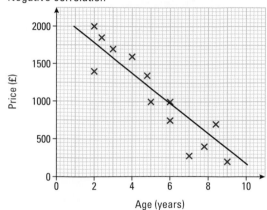

c 3.4 years

10 a Approximately £9000

b The one-year-old Mini is more expensive than expected. It may have extra features or a very low milage.

Chapter 26 Answers

26.1 Get ready

1 Certain
Impossible

Exercise 26A

1 Probability scales showing:
  a certain   **b** impossible   **c** even chance
  d likely   **e** unlikely
2 Students' examples
3 Probability scale showing 'unlikely'
4 Students' opinions

26.2 Get ready

1

Fraction	$\frac{3}{10}$	$\frac{6}{10}$	$\frac{3}{8}$	$\frac{13}{20}$
Decimal	0.3	0.6	0.375	0.65
Percentage	30%	60%	37.5%	65%

Exercise 26B

1 a $\frac{3}{8}$ **b** $\frac{3}{8}$ **c** $\frac{6}{8} = \frac{3}{4}$

2 a $\frac{3}{8}$ **b** $\frac{2}{8} = \frac{1}{4}$ **c** $\frac{2}{8} = \frac{1}{4}$ **d** $\frac{1}{8}$

3 a $\frac{3}{10}$ **b** $\frac{7}{10}$ **c** 0

4 a $\frac{1}{10}$ **b** $\frac{3}{10}$ **c** $\frac{2}{10} = \frac{1}{5}$
  d $\frac{6}{10} = \frac{3}{5}$ **e** 0

5 a $\frac{1}{6}$ **b** $\frac{2}{6} = \frac{1}{3}$ **c** $\frac{3}{6} = \frac{1}{2}$ **d** $\frac{4}{6} = \frac{2}{3}$
  e 0   **f** $\frac{3}{6} = \frac{1}{2}$

6 a $\frac{67}{120}$ **b** $\frac{53}{120}$

7 a $\frac{3}{15} = \frac{1}{5}$ **b** $\frac{7}{15}$ **c** $\frac{5}{15} = \frac{1}{3}$ **d** $\frac{10}{15} = \frac{2}{3}$
  e $\frac{8}{15}$ **f** $\frac{1}{5}$

26.3 Get ready

1 0.3   **2** 0.4   **3** $\frac{1}{2}$ **4** $\frac{1}{4}$

Exercise 26C

1 0.7
2 0.85
3 $\frac{1}{3}$
4 $\frac{50}{53}$
5 0.3 or 30%
6 0.36
7 0.675
8 P(train is not late) $= 1 - 0.32 = 0.68$

Exercise 26D

1 0.3
2 0.1
3 a 0.17   **b** 0.81   **c** 0.64
4 0.25
5 $\frac{1}{3}$
6 $\frac{1}{10}$

26.4 Get ready

1 a heads or tails   **b** 1, 2, 3, 4, 5 or 6   **c** win, lose or draw

Exercise 26E

1

Dice		
6	(H, 6)	(T, 6)
5	(H, 5)	(T, 5)
4	(H, 4)	(T, 4)
3	(H, 3)	(T, 3)
2	(H, 2)	(T, 2)
1	(H, 1)	(T, 1)
	H	T

Coin

2 a $\frac{1}{8}$ **b** $\frac{1}{6}$ **c** $\frac{5}{12}$

3 a $\frac{1}{16}$ **b** $\frac{4}{16} = \frac{1}{4}$ **c** $\frac{6}{16} = \frac{3}{8}$ **d** $\frac{7}{16}$
  e 1   **f** $\frac{9}{16}$ **g** 0

4

Spinner 1			
3	(1, 3)	(2, 3)	(3, 3)
2	(1, 2)	(2, 2)	(3, 2)
1	(1, 1)	(2, 1)	(3, 1)
	1	2	3

Spinner 2

5 a

Sheet			
W	(W, W)	(G, W)	(O, W)
Y	(W, Y)	(G, Y)	(O, Y)
B	(W, B)	(G, B)	(O, B)
G	(W, G)	(G, G)	(O, G)
	W	G	O

Pillow case

b i $\frac{2}{12} = \frac{1}{6}$ ii $\frac{5}{6}$

6 P(2) $= \frac{1}{2}$, P(3) $= \frac{2}{12}$, P(4) $= \frac{3}{12}$, P(5) $= \frac{3}{12}$, P(6) $= \frac{2}{12}$
Scores of 4 and 5 are both most likely.

26.5 Get ready

1 a $\frac{1}{2}$　　**b** $\frac{2}{3}$　　**c** $\frac{2}{3}$　　**d** $\frac{1}{2}$

Exercise 26F

1 a Students' results
b i $\frac{1}{6}$　　ii $\frac{3}{6} = \frac{1}{2}$　　iii $\frac{2}{6} = \frac{1}{3}$
c Students' conclusions
2 a $\frac{1}{2}$
b, c Students' results and conclusions
3 a Students' results
b Throw the drawing pin more times.
4 Students' results
5 a Students' results
b Yes, because different languages use letters differently, affecting how often each letter appears.

26.6 Get ready

1 a 3　　**b** 5　　**c** 10

Exercise 26G

1

	Butterflies	Moths	Total
May	9	4	13
June	9	3	12
Total	18	7	25

2 a

	Walk	Bus	Cycle	Total
Boys	4	5	3	12
Girls	7	4	2	13
Total	11	9	5	25

b i $\frac{13}{25}$　　ii $\frac{7}{25}$　　iii $\frac{3}{25}$　　iv $\frac{9}{25}$

3 a

	Cinema	Club	Bowling	Total
Boys	5	6	12	23
Girls	9	9	4	22
Total	14	15	16	45

b i $\frac{23}{45}$　　ii $\frac{14}{45}$　　iii $\frac{4}{45}$

4 a

	Orange	Lemonade	Milk	Total
Sandwiches	5	7	1	13
Biscuits	4	4	5	13
Crisps	8	7	1	16
Total	17	18	7	42

b i $\frac{18}{42} = \frac{3}{7}$　　ii $\frac{16}{42} = \frac{8}{21}$　　iii $\frac{4}{42} = \frac{2}{21}$　　iv $\frac{4}{42} = \frac{2}{21}$

c John is right.
P(milk and biscuits) $= \frac{5}{42}$
P(orange and sandwiches) $= \frac{5}{42}$

26.7 Get ready

1 Students' tally

Exercise 26H

1 50 times
2 30 times
3 a 15 times
b 45 times
c 30 times
4 a 16 games
b 14 games
5 a $\frac{3}{8} \times 60 = 22.5$; 22 or 23 red balls
b $\frac{5}{8} \times 60 = 37.5$; 37 or 38 blue balls
6 95 claims
7 Students' numbers on star
(average of numbers on star $= x - 5$)
For example, if it costs 15p to play the game, $x = 15$.
Average win should be $x - 5 = 10$p.
Therefore numbers add up to $10 \times 6 = 60$

Review exercise

1 a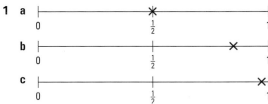

b

c

2 $\frac{2}{8} = \frac{1}{4}$

3

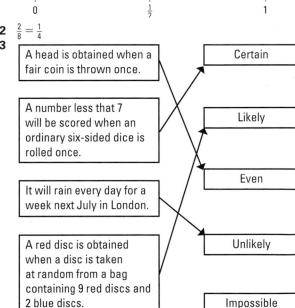

| A head is obtained when a fair coin is thrown once. |
| A number less that 7 will be scored when an ordinary six-sided dice is rolled once. |
| It will rain every day for a week next July in London. |
| A red disc is obtained when a disc is taken at random from a bag containing 9 red discs and 2 blue discs. |

Certain
Likely
Even
Unlikely
Impossible

4 (fish, apple)　　(fish, banana)
(fish, pear)　　(lamb, apple)
(lamb, banana)　　(lamb, pear)
(salad, apple)　　(salad, banana)
(salad, pear)

5 **a**

	Black	White	Total
3-sided shape	1	4	5
4-sided shape	2	2	4
5-sided shape	2	0	2
Total	5	6	11

b $\frac{4}{11}$

6 **a** (1, H), (2, H), (3, H), (4, H), (5, H), (1, T), (2, T), (3, T), (4, T), (5, T)

b $\frac{1}{10}$

7 **a**

	London	York	Total
Boys	23	14	37
Girls	19	24	43
Total	42	38	80

b $\frac{42}{80} = \frac{21}{40}$

8 **a**

b (green, heads), (green, tails), (blue, heads), (blue, tails), (yellow, heads), (yellow, tails)

9 $\frac{5}{12}$

10 $\frac{3}{8}$

11 **a** $\frac{7}{20}$ **b** $\frac{9}{20}$ **c** 0

12 **a** 0.2 **b** 15 green counters

13 $x = 0.35$

14 0.4

15 **a** 0.25 **b** 35 times

16 $\frac{8}{15}$

17 Robbie

18 The yellow horse has a 50% chance of moving. The other two each have only a 25% chance of moving.

19 Horse 1 cannot win as a score of 1 is not possible. Horse 7 is most likely to win.

20 Students' own experiments.

Chapter 27 Answers

27.1 Get ready

1 **a** 16 **b** 5.29 **c** 625
2 **a** 5 **b** 13 **c** 0.5
3

a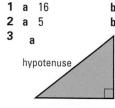
hypotenuse

b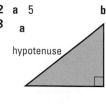
hypotenuse

c
hypotenuse

d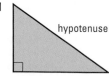
hypotenuse

Exercise 27A

1 **a** $h = 13$ cm **b** $p = 5.83$ cm **c** $s = 25$ cm
d $v = 8.38$ cm **e** $q = 6.19$ cm **f** $t = 7.4$ cm
g $m = 11.7$ cm **h** $n = 5.66$ cm **i** $w = 6.08$ cm
2 **a** 15.7 cm **b** 6.32 cm **c** 26.6 cm
d 11.0 cm **e** 9.79 cm **f** 10.0 cm
g 13.6 cm **h** 8.44 cm **i** 8.32 cm
3 **a** 33.2 cm **b** 27.0 cm
4 £1467.40

27.2 Get ready

1 **a** 9 **b** 16 **c** 26.07
d 144 **e** 576
2 **a** 3 **b** 4 **c** 13
d 2.5 **e** 24
3 **a** **b**

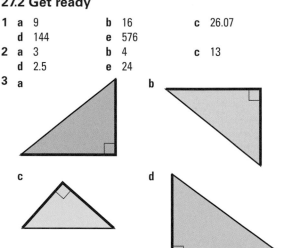

c **d**

Exercise 27B

1 **a** $x = 12$ cm **b** $y = 3$ cm
c $s = 7$ cm
d $r = 11.5$ cm **e** $v = 4.50$ cm **f** $t = 9.30$ cm
g $m = 6$ cm **h** $n = 3.75$ cm **i** $w = 4.85$ cm
2 **a** 5 cm **b** 4 cm **c** 0.7 cm
d 9.60 cm **e** 4.8 cm **f** 6.58 cm
g 10.4 cm **h** 7.49 cm **i** 4.62 cm
3 7.35 m
4 £54 127

27.3 Get ready

1 and **3**

Exercise 27C

1 **a** Yes **b** No **c** No
d No **e** Yes **f** No
2 Jenny has squared the sum of the shorter sides. She should have squared the shorter sides *then* added.
$12^2 = 144$; $6^2 + 6^2 = 36 + 36 = 72$
Since $12^2 \neq 6^2 + 6^2$, PQR is not a right-angled triangle.
3 Jason has squared the sum of the shorter sides. He should have squared the shorter sides *then* added.
$10^2 = 100$; $6^2 + 8^2 = 36 + 64 = 100$
Since $10^2 = 6^2 + 8^2$, PQR is a right-angled triangle.
4 **a** Acute-angled **b** Obtuse-angled
c Acute-angled **d** Obtuse-angled
e Right-angled **f** Acute-angled

Exercise 27D

1 a 4.47 b 5.83 c 5 d 4.24
  e 6.40 f 3.61 g 9.90 h 9.06
  i 7.21 j 5.83
2 a 7.07 b 8.49 c 10.6 d 11.4
3 a 14.1 b 15.3 c 11.4 d 9.43
  e 13.6 f 12.7

Review exercise

1 10.63 cm
2 a 30 cm² b 169 cm²
3 11.9 cm
4 56 cm
5 230 km
6 19.4 cm
7 6.71 cm
8 $2.5^2 = 6.25$; $1.5^2 + 2^2 = 2.25 + 4 = 6.25$
  Since $2.5^2 = 1.5^2 + 2^2$, ABC is a right-angled triangle.
9 5.10 cm
10 7.6 cm
11 8.8 km
12 84 cm²
13 28.6 cm²
14 £54 000
15 AB is 10

Chapter 28 Answers

28.1 Get ready

1 £3.75 2 £212.40 3 37.68

Exercise 28A

1 35 cm
2 £185
3 40 cm²
4 22 cm
5 £19.20
6 a Total cost = number of stamps × cost of 1 stamp
  b £5.04
7 a Number left = starting number − number sold
  b 29
8 £184
9 a Number per person = number of sweets ÷ number of people = 21 sweets
  b Number of people = number of slices ÷ number of slices received = 12
10 43 mph
11 900°
12 45°
13 50 cm² (to the nearest whole number)

28.2 Get ready

a 12 cm b 16 cm c 24 cm

Exercise 28B

1 a 8 b 0 c 5 d 30
  e 9 f 66 g 40 h 18
  i 9 j 26 k 9 l 54
  m 64 n 1
2 a 1 b $3\frac{1}{2}$ c $1\frac{1}{2}$ d $\frac{1}{2}$
  e $\frac{5}{16}$ f $1\frac{1}{16}$ g 0 h 0
  i $\frac{9}{16}$ j 22 k 4 l $\frac{3}{32}$
  m 1
3 a 1 b 11 c 13.5 d 9.5
  e 9.75 f −8.5 g 12.5 h 7
  i 1.25 j 27 k 27 l 9.25
  m −7.875

Exercise 28C

1 1 2 −11 3 11
4 −3 5 8 6 −1
7 −8 8 −2 9 2
10 0 11 28 12 −30
13 5 14 −36 15 2
16 −18 17 −30 18 54
19 30 20 −6 21 18
22 −2 23 −15 24 −20
25 25 26 6 27 97
28 14 29 65 30 16
31 1 32 1 33 49
34 9 35 432 36 −375
37 −48 38 32 39 −11
40 −27 41 0.25 42 20.25

28.3 Get ready

1 −2 2 −15 3 2

Exercise 28D

1 a 11 b 15 c 23 d 18
2 a 21 b 63 c 22.2 d 37.8
3 a 6 b 16 c 12 d 29
4 a 12 b 33
5 a 50 b 212 c −22 d 32
6 a 40 b 140 c 240 d 199.206
7 a 20 b 20 c 20 d 59
8 a 90 b 135 c 195 d 285
9 a 20 b 480.2 c 26.875

28.4 Get ready

1 9.42 2 25.1 3 31.4

Exercise 28E

1 $P = 6l$
  a 18 b 42 c 174 d 51.6
2 $P = 4a + b$
  a 28 b 55 c 24.6 d 27.3
3 a $P = 70n$ b i £2.80 ii £4.20 iii £8.40
4 $v = s^3$
  a 8 cm³ b 91.125 cm³
5 $A = 6s^2$
  a 24 cm² b 121.5 cm²
6 $s = \sqrt{A}$
  a 2 cm b 1.2 cm

Answers

5 $A = 6s^2$
   a 24 cm²   **b** 121.5 cm²
6 $s = \sqrt{A}$
   a 2 cm   **b** 1.2 cm

28.5 Get ready

1 $x = 3$ **2** $a = 16$ **3** $x = 5$

Exercise 28F

1 **a** 4   **b** 9   **c** 23   **d** 5.9
2 **a i** 9   **ii** 18   **b i** 5   **ii** 15
3 **a i** 6   **ii** 14   **b i** 7   **ii** 20
4 **a** 6   **b** 12   **c** $-6\frac{1}{2}$ **d** -3
5 **a i** 3   **ii** 14   **b i** 3   **ii** $5\frac{1}{2}$
6 **a** 2   **b** $-6\frac{1}{2}$ **c** 20   **d** $-\frac{1}{4}$
7 **a** 4   **b** 5   **c** 4
8 **a i** 5   **ii** 7   **b** 6   **c** 4
9 **a** 20   **b** 85   **c** 37   **d** 0
10 **a** 15   **b** -72 **c** 45   **d** -58.24

28.6 Get ready

1 $x = -\frac{1}{3}$ **2** $x = 5$ **3** $x = 9$

Exercise 28G

1 $d = \dfrac{P}{5}$ **2** $I = \dfrac{P}{V}$

3 $w = \dfrac{A}{l}$ **4** $d = \dfrac{C}{\pi}$

5 $h = \dfrac{V}{lw}$ **6** $r = \dfrac{A}{\pi l}$

7 $x = \dfrac{y + 3}{4}$ **8** $n = \dfrac{t - 5}{3}$

9 $y = P - 2x$ **10** $m = \dfrac{y - c}{x}$

11 $u = v + gt$ **12** $t = \dfrac{u - v}{g}$

13 $b = 2\dfrac{A}{h}$ **14** $T = \dfrac{100I}{PR}$

15 $V = \dfrac{D}{T}$ **16** $V = \dfrac{kT}{P}$

17 $T = \dfrac{PV}{k}$ **18** $v = \dfrac{I}{m} + u$

19 $b = \dfrac{2A}{h} - a$ **20** $x = 3(y + 2)$

21 $x = \frac{1}{2}y + 1$ **22** $y = \frac{1}{3}x - 2$

23 $A = 2(17 - H)$ **24** $x = \dfrac{2y + 6}{3}$

25 $y = \dfrac{3x - 6}{2}$ **26** $q = P + 12$

27 $x = \dfrac{y^2 + 12y}{2}$

Review Exercise

1 **a** $n = b - 3$ **b** $d = 2b$
2 **a** $b = 2a$ **b** $c = a - 7$
3 **a** $b = 3a$ **b** $c = a + 5$
4 **a** £22   **b** 8 days
5 **a** £45   **b** £5.50
6 **a** £2000   **b** £700   **c** 7200 miles
7 **a** £50
   b 40 represents a fixed cost of £40
     0.05 is a cost of 5p per leaflet
8 **a** 5   **b** -23
9 **a** 87   **b** 22   **c** 7.5   **d** 8
10 **a** 15   **b** 28
11 **a** $A = l^2$ **b** 81
12 **a** $P = 2(l + w)$
   b **i** 26
      ii 20.2
13 9 trees
14 **a** 57.6   **b** -133.95 **c** -594
15 **a** 300 cm²   **b** 195 cm²   **c** 10.585 cm²   **d** 1.495 cm²
16 **a** 56   **b** -52 **c** -124
17 20
18 **a** 80   **b** 29.5
19 **a** -5 **b** 73
20 **a** 6 hours   **b** $P = 35h + 50$
21 **a** £130   **b** $C = 90 + \frac{1}{2}m$
22 **a** 25°C   **b** 32°F   **c** Fahrenheit, 100°F = 37.8°C
23 **a** Student's own answers
   b 3.5 hours
   c 4.5 hours
   d 4 hours 4 minutes
   e 4 hours 21 minutes
24 **a** 31   **b** 28
25 **a** **i** 68°F
      ii 113°F
     iii 158°F
   b **i** 100°C
      ii 50°C
     iii 25°C
26 **a** $P = 4x + 8$ **b** 15.5 cm
27 $P = 5a + 2b$
28 **a** 261.8 cm³   **b** 769.7 cm³   **c** 203.9 cm³
29 **a** 7   **b** 5
30 $b = \dfrac{P - 2a}{2}$
31 $x = \frac{1}{2}(P - 2y)$
32 $a = 2s - b - c$
33 $D = TV$
34 $h = \dfrac{2A}{a + b}$
35 $x = 5 - 2y$